AF323430

A Course in
Real Analysis

A Course in
Real Analysis

Mainak Mukherjee

Alpha Science International Ltd.
Oxford, U.K.

A Course in
Real Analysis

385 pgs. | 105 figs.

Mainak Mukherjee
Associate Professor
Department of Mathematics
Sri Venkateswara College
University of Delhi
New Delhi

Copyright © 2011

ALPHA SCIENCE INTERNATIONAL LTD.
7200 The Quorum, Oxford Business Park North
Garsington Road, Oxford OX4 2JZ, U.K.

www.alphasci.com

ISBN 978-1-84265-389-0

Printed in India

Preface

The thought of this present volume first came to my mind while I started teaching Real Analysis to the under graduate students of the Delhi University some ten years ago. While teaching this course I interacted with various cross sections of students taking Analysis Course with different subjects as their major and found their demands were different. While students with Mathematics as major wanted motivations for the concepts and more rigors in the proofs, students with Physics as major were interested in solving different varieties of problems rather than the proofs. I have tried to blend these two ideas in this book, while care has been taken to provide motivations for concepts introduced at the same time enough examples have been included to clarify the concepts. At the end of each chapter good numbers of exercises are given and serious students are expected to solve them as one can learn mathematics only by doing mathematics.

The material in this book covers topics of Real Analysis appearing in the undergraduate syllabus of any university. The chapters are more or less arranged in order of dependence but depending on the need and priority one can easily postpone some of the chapters for example after going through Chapter 2 (Limits, continuity and differentiability) one can skip subsequent chapters and go to Chapter 7 (Sequence and Series of functions) without facing much problem.

As such no prerequisites are necessary to follow this book and basic concepts of Natural Numbers, Real Numbers, Sets, Mapping and other important basic topics are briefly discussed in the Introduction chapter.

Chapter 1 deals with the most fundamental concept of Analysis 'Sequence and Series' large number of examples have been included to make reader familiar with sequences & series and basic concepts concerning them. As we proceed to later chapters one observes the power of the use of sequences in simplifying proofs of some theorems and also while doing some exercises. Limits, continuity and differentiability of functions of one variable have been discussed with great detail in Chapter 2 before Riemann Integral is introduced in Chapter 3. To develop the concept Riemann Integral in a natural way we have given a brief motivation for the definition of Riemann Integral at the beginning. As a generalization of Riemann Integral we have dealt with Riemann Stieltjes Integral in Chapter 4. Riemann Integral finally has been extended as Improper Integral in Chapter 5. In Chapter 6 properties of very important functions Beta and Gamma functions were discussed with examples. Chapter 7 deals with 'Sequence and Series of real functions' where the most important concept is uniform convergence. Sufficient examples have been included to make students understand the need and power of uniform convergence. The use of uniform convergence can be seen in the theory of Power series introduced in Chapter 8, sufficient number of examples have been included to make reader familiar with the theory. Concepts introduced in Chapter 3 have been extended to functions of two variables in Chapter 9. In Chapter 10 Multiple integrals are developed just like Riemann Integrals and theorems are similar to that given in Chapter 3. Large numbers of examples have been included to exhibit the techniques involved in evaluating Multiple Integrals. Chapter 11 deals with Line, Surface Integrals important theorems connecting them. In this chapter care has been taken so that the student feels how naturally ordinary integrals with whom he is familiar with so far, has been generalized to surface integrals. Sufficient number worked out examples are included so that students will find it easy to handle surface and line integrals.

This project would not have seen the day of light without the help of numerous students whom I have taught over the years at Sri Venkateswara College. I am indebted to them for more than one reasons. In this regard I wish to mention the names of Miss R. Charanya and Miss Nisha Bhora of 2006-2009 batch who took pains to read some of the initial chapters of my book.

I sincerely thank our principal Dr.(Mrs) P. Hemalatha Reddy for providing me the permission to use the infrastructure of the college needed to complete the book.

Special thanks is due to Dr. B Biswal, Associate professor, Department of Physics Sri Venkateswara College who has spent several hours of his precious time to provide me technical assistance.

Finally I thank Mr. N.K. Mehra, managing director of Narosa Publishing House for bringing out a nice edition of the present volume.

Mainak Mukherjee

Contents

Preface   *v*

0. Prerequisites   **0.1-0.7**
   0.1   Introduction   0.1
   0.2   Basic Definitions and Symbols of Set Theory   0.1
   0.3   Field Axioms, Order   0.2
   0.4   Supremum, Infimum and Order Completeness Property   0.4
   0.5   Basic Definitions of Point Set Topology   0.5
   0.6   Functions   0.6
   0.7   The Distance Function on Set of Reals   0.7

1. Sequence and Series of Real Numbers   **1.1-1.23**
   1.1   Introduction   1.1
   1.2   Definitions and Examples   1.1
   1.3   Types of Sequences   1.2
   1.4   Convergent and Non-Convergent Sequences   1.3
   1.5   Subsequences & Bolzano-Weierstrass Theorem   1.9
   1.6   Cluster Points, Limit Superior, Limit Inferior   1.10
   1.7   Cauchy Criterion   1.13
   1.8   Convergence of Series of Real Numbers   1.15
   1.9   Cauchy's Criterion for Convergence of an Infinite Series   1.15
   1.10 Test for Convergence of Series of Non-negative Terms   1.16
   1.11 Series of Arbitrary Terms: Conditional and Absolute Convergence   1.19
       Problems   *1.22*

2. Functions of One Variable   **2.1-2.46**
   2.1   Introduction   2.1
   2.2   Limit of Functions   2.1
   2.3   Continuity   2.11
   2.4   Uniform Continuity   2.20
   2.5   Derivability and Differentiability   2.22
   2.6   Sufficient Conditions for Extremum of Functions   2.34
   2.7   Higher Order Derivatives and Taylor's Theorem   2.36
   2.8   Taylor's Series and Maclaurin's Series   2.38
   2.9   L'Hospital's Rule   2.39
       Problems   *2.42*

3. The Riemann Integral **3.1-3.45**

3.1 Introduction 3.1

3.2 Motivations for Analytic Definition of Integral 3.1

3.3 Notations, Definitions and Basic Lemmas 3.3

3.4 Definition of Riemann Integrability 3.7

3.5 Riemann Approximating Sum 3.7

3.6 A Necessary and Sufficient Condition of Riemann Integrability in Terms of Riemann Approximating Sum 3.11

3.7 Conditions for Integrability 3.13

3.8 Classes of Riemann Integrable Functions 3.15

3.9 Properties of Riemann Integrable Functions and Riemann Integrals 3.19

3.10 Fundamental Theorem and Mean Value Theorems of Integral Calculus 3.28

3.11 Method of Substitution in an Integral 3.37

3.12 Integration by Parts 3.38

3.13 Characterization of Riemann Integrability 3.39

Problems *3.43*

4. Riemann-Stieltjes Integral **4.1-4.21**

4.1 Introduction 4.1

4.2 Notations and Definitions 4.1

4.3 Some Basic Lemmas and Theorems 4.2

4.4 Functions of Bounded Variations and Riemann-Stieltjes Integral 4.12

4.5 Some Miscellaneous Problems 4.16

Problems *4.19*

5. Improper Integrals **5.1-5.29**

5.1 Introduction 5.1

5.2 Types of Improper Integrals 5.2

5.3 Type I Integrals 5.3

5.4 Absolute Convergence 5.5

5.5 Test for Type I integrals 5.7

5.6 Type II Integrals 5.9

5.7 Absolute Convergence 5.13

5.8 Test for Type II integrals 5.14

5.9 Absolute and Ordinary Convergence of Integrant as Product of Functions 5.14

5.10 Some Miscellaneous Problems 5.17

Problems *5.27*

6. Beta and Gamma Functions **6.1-6.10**

6.1 Introduction 6.1

6.2 Beta Function 6.1

6.3 Properties of Beta Function 6.1

6.4 Gamma Function — 6.3
6.5 Properties of Gamma Function — 6.3
6.6 Relation Between Beta and Gamma Function — 6.4
6.7 Some Miscellaneous Problems — 6.7
Problems — *6.9*

7. Sequence and Series of Functions — **7.1-7.39**
7.1 Introduction — 7.1
7.2 Definitions and Examples — 7.1
7.3 Pointwise Convergence and Uniform Convergence — 7.2
7.4 The Cauchy Criterion for Uniform Convergence of a Sequence of Functions — 7.9
7.5 Series of Functions and Uniform Convergence — 7.10
7.6 Uniform Convergence and Continuity — 7.14
7.7 Uniform Convergence and Integration — 7.17
7.8 Uniform Convergence and Differentiation — 7.21
7.9 Uniform Convergence of Series of Product of Two Functions — 7.30
7.10 Weierstrass Approximation Theorem — 7.34
Problems — *7.37*

8. Power Series — **8.1-8.12**
8.1 Introduction — 8.1
8.2 Definition — 8.1
8.3 Radius of Convergence — 8.2
8.4 Formulae for finding Radius of Convergence — 8.4
8.5 Significance of Radius of Convergence: Term by Term Integration and Differentiation of a Power Series — 8.6
8.6 Real Analytic Functions and Power Series — 8.8
8.7 Abel's Theorem — 8.9
Problems — *8.11*

9. Function of Two Variables — **9.1-9.38**
9.1 Introduction — 9.1
9.2 Basic Facts of $\mathbb{R}^2$ — 9.1
9.3 Limit of a Function — 9.3
9.4 Continuity of Functions — 9.9
9.5 Directional Derivatives and Partial Derivatives — 9.10
9.6 Homogeneous Functions and Euler's Theorem — 9.15
9.7 Differentiability — 9.17
9.8 Higher Order Partial Derivatives and Taylor's Theorem — 9.23
9.9 Some Applications of Partial Derivatives: Implicit Function Theorem, Maxima Minima — 9.30
Problems — *9.34*

10.   Double and Triple Integrals   **10.1-10.39**

    10.1   Introduction   10.1

    10.2   Double (Triple) Integral over a Closed Rectangle (Box)   10.1

    10.3   Classes of Integrable Functions   10.5

    10.4   Properties of Double Integral   10.8

    10.5   Evaluation of Double Integrals: Cavalieri's Principle   10.8

    10.6   Double Integrals of Functions Defined on any Arbitrary Bounded Domain   10.15

    10.7   Iterated Integrals over Arbitrary Bounded Domains Change
           in the Order of Integration   10.15

    10.8   Change of Variable Formula in Double Integrals   10.25

    10.9   Some Miscellaneous Problems   10.31

          Problems   *10.36*

11.   Line and Surface Integrals   **11.1-11.55**

    11.1   Introduction   11.1

    11.2   Scalar and Vector Field   11.1

    11.3   Divergence and Curl of a Vector Field   11.2

    11.4   Curves in Plane and Space   11.3

    11.5   Rectifiable Curves (Arc Length)   11.7

    11.6   Simple and Multiply Connected Regions of Plane   11.8

    11.7   Line Integrals   11.10

    11.8   Some Properties of Line Integrals   11.13

    11.9   Green's Theorem in Plane   11.17

    11.10 Parametric Representation of a Surface   11.27

    11.11 Notion of Surface Area   11.29

    11.12 Surface Integral of First Kind   11.34

    11.13 Orientable Surfaces   11.38

    11.14 Surface Integral of Second Kind   11.39

    11.15 Stoke's Theorem   11.43

    11.16 Gauss' Divergence Theorem   11.48

          Problems   *11.53*

Index   *I.1-I.3*

Chapter 0

Prerequisites

0.1 INTRODUCTION

Present chapter consists of certain basic definitions and results from miscellaneous topics like basic set theory, system of reals, basic point set topology, which forms a necessary prerequisite for this book. Readers may be familiar with many of these notions but still a sweeping glance of this chapter is recommended, which will not only help them to refresh their memories but also they will get chance to be acquainted with the notations and symbols used in this book. In this chapter no attempt have been made to give proofs of any results stated but are explained with examples where ever needed.

0.2 BASIC DEFINITIONS AND SYMBOLS OF SET THEORY

We begin with the definition of a set. A *set* is defined to a well defined collection of objects. We note that above is the most commonly used definition but not the most rigorous one but it suffices the need of this book. Interested readers are referred to Dugundji [7] for the axiomatic definition of a set and the need for such definition.

We will denote usually a set by letters A, B, C, ... or some time by symbols Ω, $\mathfrak{S}$, ... etc. The elements of sets will be denoted by a, b, c, ... If is a set comprising of elements a, b & c we will write $A = \{a, b, c\}$. If a is an element of A we will write $a \in A$.

For two sets A & B we call A is a *subset* of B if every element of A is an element of B and we write $A \subset B$, equivalently we will call the set B a *superset* of A and write $B \supset A$. In any situation involving sets there will be a fixed set called the *universal set* which is the superset of every other set arising in the given situation. *Null set or empty set* is denoted by ϕ and is assumed to be subset of every set. We say two sets A & B are **equal** if $A \subset B$ & $B \subset A$ and write $A = B$. That is two sets are equal if they have same elements. If $A \subset B$ and $A \neq B$ then we say A is a proper subset of B.

For the rest of the section Ω will denote the universal set. We may denote a set by writing all the elements of the set specifically or we may write it as collection of some elements of Ω satisfying certain property. We give an example, suppose Ω comprises of two red balls and two black balls. Let us call them as R_1, R_2 and B_1, B_2 respectively. Now suppose A be the subset of Ω comprising of two black balls only then we can write $A = \{B_1, B_2\}$ or we may say A consists of those elements of Ω such that they satisfies the property that they are black, in symbols we will write

$$A = \{x \in \Omega : x \text{ is black}\}$$

For a set $A \subset \Omega$ we define ***complement of*** as the set of all elements of Ω which does not belong to A and is denoted by A^c (or $\Omega - A$) that is,

$$A^c = \{x \in \Omega : x \text{ is not in } A\}$$

For two subsets A & B of Ω we define two operations ***union*** and ***intersection*** denoted by $A \cup B$ & $A \cap B$ as,

$$A \cup B = \{x \in \Omega : \text{either } x \text{ is in } A \text{ or in } B\}$$

$$A \cap B = \{x \in \Omega : \text{either } x \text{ is in } A \text{ and } B\}$$

Suppose a, b be two distinct element then the sets $\{a, b\}$ and $\{b, a\}$ are equal as there is no order involved. But there are situations where we need to specify an order in which the elements are to be written. For example, we are all familiar with two dimensional cartesian co-ordinate system where any point on the plane has a co-ordinate of the form (a, b). a is called the X-co-ordinate and b is called the Y-co-ordinate of the point. Hence in this case $(a, b) \neq (b, a)$. In general whenever we want to specify order among the elements a, b we will write (a, b) meaning a is the first element and b is the second element and call it an ***ordered pair***.

If A & B are two non empty sets then the set of all ordered pair (a, b) such that $a \in A$ & $b \in B$ is called the ***cartesian product*** of the sets A & B and is denoted by $A \times B$.

Next we intend to discuss some properties of real numbers. Our goal is not to give the construction of real numbers rather we will assume the existence of such a system which will be denoted by $\mathbb{R}$ with usual operations addition and multiplication defined on it. We will also assume that reader is familiar with special subsets of R namely the set of ***integers, natural numbers and rational numbers*** denoted by

$$\mathbb{Z} = \{\ldots -3, -2, -1, 0, 1, 2, 3 \ldots\}$$

$$\mathbb{N} = \{1, 2, 3 \ldots\}$$

$\mathbb{Q} = \left\{\dfrac{m}{n} : m, n \in \mathbb{Z} \text{ & } n \neq 0\right\}$ respectively. From definition it is clear that $\mathbb{Z} \subset \mathbb{Q}$. The set $\mathbb{R} - \mathbb{Q}$ is known as ***irrational numbers***. Note that the set of positive real numbers is denoted by $\mathbb{R}^+$.

The first property of $\mathbb{R}$ that we want to highlight is that it is an ordered field hence we begin with field axioms and define order.

0.3 FIELD AXIOMS, ORDER

Definition   A non empty F set with two operations '+' & '·' called addition and multiplication is said to be a ***field*** if it satisfies the following

Set of axioms:

(A) Field Axioms for Addition

A1   If $x, y \in F$ then $x + y \in F$

A2   If $x, y, z \in F$ then $(x + y) + z = x + (y + z)$

A3   There exists an element $0 \in F$ such that $0 + x = x + 0$ for all $x \in F$

A4   For every element $x \in F$ there exists a corresponding element $-x \in F$ such that

$$x + (-x) = 0$$

A5 $x + y = y + x \,\, \forall \text{ (for all) } x, y \in F$

(M) Field Axioms for Multiplication

M1   If $x, y \in F$ then $x \cdot y \in F$

M2   If $x, y, z \in F$ then $(x \cdot y) \cdot z = x \cdot (y \cdot z)$

M3   There exists an element $1 \neq 0 \in F$ such that $1 \cdot x = x \cdot 1$ for all $x \in F$

M4   Suppose $x \in F$ and $x \neq 0$ then there exists a corresponding element $x^{-1} \in F$ such that
$$x \cdot (x^{-1}) = 1$$

M5 $x \cdot y = y \cdot x \ \forall \ x, y \in F$

(D) Field Axiom for Distribution of Two Operations

For all $x, y, z, \in F$ following equation holds,
$$x \cdot (y + z) = x \cdot y + x \cdot z$$

One can easily check that the set of real numbers $\mathbb{R}$ and rational numbers $\mathbb{Q}$ forms a field with usual operation of addition and multiplication. Hence we say that the **set of real numbers is a field containing the rationals as subfield**.

Definition   Let A be a non empty set, a relation '$<$' on the set is said to be an *order* if it satisfies two properties

    (i)   For $x, y, \in A$ only one of the statement
$$x < y, \quad x = y, \quad y < x$$

    (ii)   For $x, y, z \in A$ then $x < y$ and $y < z$ implies $x < z$

Any set where some order is defined is called an *ordered set*.

Note that $x < z$ is also some times written as $y < z$.

Definition   Suppose '$<$' be an order defined on a field F satisfying following conditions

    (a)   If $x < y$ then for every $z \in F$ we have $x + z < y + z$

    (b) $x > 0, y > 0$ then $xy > 0$

Then $\mathbb{F}$ is called an *ordered field*.

We define a relation '$<$' on $\mathbb{R}$ as $x < y$ (we pronounce x is *less than* y) if $y - x$ is a positive real number. It follows that with this relation $\mathbb{R}$ **is an ordered field**.

The order '$<$' on $\mathbb{R}$ is called *usual order* of $\mathbb{R}$. Note that we write $x \geq y$ (we pronounce x is *more than or equal to* y) equivalently $y \leq x$ (we pronounce y is *less than or equal to* x) if either $y < x$ or $y = x$ without specifying which one of this will hold.

We now intend to define certain important subsets of $\mathbb{R}$ called **intervals.**

Definition   With usual order on $\mathbb{R}$ and $a < b$ we define the sets (a, b) and $[a, b]$ (called *open* and *closed intervals*) as follows,
$$(a, b) = \{x \in \mathbb{R} : a < x < b\}$$
$$[a, b] = \{x \in \mathbb{R} : a \leq x \leq b\}$$

Similarly *half open or half closed intervals* are defined as
$$[a, b] = \{x \in \mathbb{R} : a < x \leq b\}$$
$$[a, b) = \{x \in \mathbb{R} : a \leq x < b\}$$

Infinite intervals (a, ∞) and (a, ∞) are defined as
$$[a, \infty) = \{x \in \mathbb{R} : x \geq a\}$$
$$(a, \infty) = \{x \in \mathbb{R} : x > a\}$$

Similarly,

$$(-\infty, b) = \{x \in \mathbb{R} : x < b\}$$
$$(-\infty, b] = \{x \in \mathbb{R} : x \le b\}$$

Finally we write

$$(-\infty, \infty) = \mathbb{R}$$

We will now state what is called order completeness property of real numbers but for that we have to define important concepts of supremum and infimum.

0.4 SUPREMUM, INFIMUM & ORDER COMPLETENESS PROPERTY

Suppose $M \subset \mathbb{R}$, an element $\xi \in M$ is called the ***minimum element*** of the set if $\xi \le x$ for all $x \in M$ and we write $\xi = \min M$. Similarly an element $\eta \in M$ is called the ***maximum element*** of the set if $x \le \eta$ for all $x \in M$ and denote $\eta = \max M$.

Note that,

    (i) a and b are minimum and maximum element of the set $[a, b]$.

    (ii) a is the minimum element of the sets $[a, b)$ & $[a, \infty)$ both of which has no maximum element

    (iii) b is the maximum element of the sets $(a, b]$ & $(-\infty, b]$ both which has no minimum element

    (iv) There are no maximum and minimum elements of the sets (a, b), $(-\infty, b)$ & (a, ∞)

We now define what are called bounded sets of real numbers.

Definition   A set $M \subset \mathbb{R}$ is said to be bounded ***below*** if there exists a real number l such that $l \le x$ for all $x \in M$. In that case l is called a ***lower bound*** of M. If M has no lower bound the M is said to be unbounded below.

Note that one should not confuse lower bound with minimum element for example $(a, b]$ has no minimum element but a is a lower bound of the set. However if minimum element of a set exists the definitely that serves as a lower bound.

Definition $M \subset \mathbb{R}$ is said to be ***bounded above*** if there exists a real number L such that $x \le L$ for all $x \in M$ and in that case L is called an upper bound of M. If M has no upper bound we say M is ***unbounded above***.

Again we notice that the set $[a, b)$ has no maximum element but b is an upper bound of the set so that maximum element and upper bound of a set are different.

Definition $M \subset \mathbb{R}$ is said to be a ***bounded*** set if it is both bounded above and below.

Next we define two very important terms of analysis called ***supremum*** and ***infimum*** of subsets of real numbers.

Definition   Let M be a subset of real numbers bounded above. A real number λ is called the ***supremum*** (or ***least upper bound***) of M if it satisfies following properties

    (i) λ is an upper bound of the set M

    (ii) Any number less than λ is not an upper bound of M

In that case we write $\lambda = \sup M$

Definition   Let M be a subset of real numbers bounded below. A real number μ is called the ***infimum*** (or ***greatest lower bound***) of M if it satisfies following properties

    (i) μ is an upper bound of the set M

    (ii) Any number more than μ is not a lower bound of M

In that case we write $\mu = \inf M$

Next we state what is called **order completeness axiom** of real numbers:

Every non-empty set of real numbers bounded above must have a supremum.

We express the above statement by saying that the system of real numbers is **order complete.** As a consequence of above axiom we get that

Every non-empty set of real numbers bounded below must have an infimum.

We now wish to give few basic definitions associated with basic point set topology such as open sets, closed sets, dense sets etc. for ready reference for future.

0.5 BASIC DEFINITIONS OF POINT SET TOPOLOGY

For a point $x \in \mathbb{R}$ any open interval I containing the point is called an ***open neighbourhood*** of the point. A point $x \in S \subset \mathbb{R}$ is said to an ***interior point*** of S if there exists an open neighbourhood I of x such that $I \subset S$. A set $U \subset \mathbb{R}$ is said to be ***open*** if every point of U is its interior point. Trivially an open interval is an open set. A set $F \subset \mathbb{R}$ is said to be ***closed*** if F^c is open. A closed interval is an example of a closed set.

Definition A real number ξ is said to be a ***limit point*** of $E \subset R$ if every open neighbourhood of ξ contains at least one element of E other than ξ.

The set of all limit points of E is called the ***derived set*** of E and is denoted by E'.

Note that the derived set of (a, b) is $[a, b]$.

Definition A set $D \subset \mathbb{R}$ is said to be ***dense*** in $\mathbb{R}$ if every real number is either is a member of D or a limit point of D.

The sets Q and $\mathbb{R} - \mathbb{Q}$ are dense in $\mathbb{R}$. This is known as the ***density property*** of rationals and irrational numbers. One of the direct consequence of this property is that between any two real number there exists a rational number and similarly between any two real number there is a irrational number. This in turn implies that we can find a rational number arbitrary close to a irrational number that is irrationals can be approximated by rational numbers. Another consequence is that every open neighbourhood of a rational number contains infinitely many irrationals and similarly every open neighbourhood of a irrational number contains infinitely many rationals.

We will end this section by stating a very useful result known as Bolzano-Weierstrass' Theorem which is in fact a consequence of order completeness property of real numbers.

Theorem 0.5.1 (Bolzano-Weierstrass' Theorem)

Every infinite bounded set of real numbers must have a limit point.

We want to state another important theorem which provides a characterization of closed and bounded subsets sets of real numbers called Hiene-Borel Theorem. For that we have to define certain terms necessary to state the theorem.

Definition The family of subsets of $\mathbb{R}$, $\{A_\alpha : \alpha \in I\}$ is called a ***covering*** of $H \subset \mathbb{R}$ if, $H \subset \bigcup_{\alpha \in I} A_\alpha$. The family $\{A_\alpha : \in I\}$ is then said to cover of H.

Further, if every member of the cover is open we say that the family is an ***open cover***.

Definition A subset E of $\mathbb{R}$ is called ***compact*** if any open cover of E has a finite ***subcover,*** that is a finite sub collection which also covers E.

Following theorem characterizes all compact subsets of $\mathbb{R}$.

Theorem 0.5.2 (Hiene-Borel)

A subset of $\mathbb{R}$ is compact if and only if it is closed and bounded.

0.6 FUNCTIONS

Suppose A and B be two sets, if by some *rule* denoted by f we can associate every element x of A an unique element $f(x)$ of B (called the **image** of x under the rule) then f is called a **function or mapping** from A to B and symbolically write,

$$f : A \to B$$

A is called the **domain** denoted by $D(f)$ and B is known as the **co-domain**. The set defined by, $R(f) = \{f(x) : x \in A\}$ is called the **range** of f and is certainly a subset of B.

For example if to every positive real number we associate another real number by taking its positive square root we write the association as,

$$f(x) = \sqrt{x}$$

Notice that the domain and range of this function is set of positive real numbers. The co-domain of the function being set of real numbers. Note that the functions whose co-domain is the set of real numbers we call such functions as **real valued**. If further, the domain of such functions are also real numbers we call such functions as **real valued functions of real variable**.

Let $f : A \to B$ be a function. If distinct elements of A has distinct images that is, $f(x) = f(y) \Rightarrow x = y$ we say in that case f is a **one to one** (**injective**) function.

If every element of B is image of some element of A, that is for every $y \in B$ there exists some $x \in A$ such that $y = f(x)$ then f is called **onto** (**surjective**) function.

If $f : \mathbb{N} \to \mathbb{N}$ be defined as,

$$f(n) = n + 1$$

then it is easy to see that f is one to one but certainly it is not onto as the natural number 1 is not image of any natural number.

Again if $f : \mathbb{R} \to \mathbb{R}^+$ be defined as,

$$f(x) = x^2$$

then f is not one to one as 4 is the image of distinct elements 2 & –2 But the function is definitely onto as every positive number is the image of its square root under f.

See that the function $f : A \to A$ defined by,

$$f(x) = x$$

for any non empty sets A is both one to one and onto. Such a function which is both one to one and onto is called a **bijective** function.

If $f : A \to B$ is a bijective map then observe that every element of B can be written in the form $f(x)$ for some $x \in A$. Then the function $f^{-1} : B \to A$ defined by,

$f^{-1}(y) = x$ where $y = f(x)$ is called the **inverse** of f.

Suppose $f : A \to B$ and $g : C \to D$ be two functions such that $C \in R(f)$ then the function $g \, o \, f : A \to D$ defined as,

$$(g \, o \, f)(x) = g(f(x))$$

is called the **composition function** of f & g which takes every element of A to D.

Finally we define what is meant by equality of two functions. Suppose f & g be two functions with same domain and co-domain A & B respectively. We say that two functions f & g are **equal** and write $f = g$ if $f(x) = g(x) \ \forall \ x \in A$.

0.7 THE DISTANCE FUNCTION ON SET OF REALS

We first define a function $|\,|: \mathbb{R} \to \mathbb{R}^+$ as follows,

$$|x| = \begin{cases} x & \text{if } x \geq 0 \\ -x & \text{if } x < 0 \end{cases}$$

This is known as **modulus function** on $\mathbb{R}$. If we look closely to the above definition we find that modulus of a real number measures the distance of the real number from the origin. For any $x, y \in \mathbb{R}$ the modulus function satisfies,

$$|x \pm y| \leq |x| + |y|$$

The function $d(x, y) = |x - y|$ is called the **distance function** on the set of reals possessing the following properties:

   (i) $d(x, y) \geq 0 \; \forall \; x, y \in \mathbb{R}$

  (ii) $d(x, y) = 0$ if and only if $x = y$

 (iii) $d(x, y) = d(y, x) \; \forall \; x, y \in \mathbb{R}$

 (iv) $d(x, y) \leq d(x, z) + d(y, z)$ for any $x, y, z \in \mathbb{R}$

Chapter 1

Sequence and Series of Real Numbers

1.1 INTRODUCTION

By a sequence of real numbers we mean a succession of real numbers for which we can determine the first element, the second element and so on. Generally, a sequence of real numbers is denoted by $a_1, a_2, \ldots a_n, \ldots$. In this chapter we will study various types of sequences and their properties. Main focus will be on what are called convergent sequences. We will further see using the concepts of finite sum and convergent sequences, how naturally one can talk about the sum of an infinite sequence of real numbers called infinite series. We will also study various properties of infinite series.

1.2 DEFINITIONS AND EXAMPLES

In this section we will give few basic definitions and give examples to illustrate them. We begin with the precise definition a sequence of real number.

Definition A sequence of real numbers is a function from the set of natural numbers to set of real numbers.

The usual convention is to denote a sequence which is a function by x. That is $x : \mathbb{N} \to \mathbb{R}$ will denote a sequence. Then $x(n)$ denotes the image of the natural number n under x and will be called the nth term of the sequence, by convention we will denote it by x_n. A sequence will always be denoted by $\{x_n\}_{n=1}^{\infty}$ or just by $\{x_n\}$. We give some examples of sequences:

Example 1.1 Most trivial example of a sequence is $\mathbb{N}$ the set of natural numbers. We denote it by $\{n\}$ observe that the function here is the identity map from $\mathbb{N}$ to $\mathbb{N}$.

Example 1.2 The function defined by $x(n) = a \ \forall n$ where a is any real number is called a constant sequence and is denoted by $\{a\}$.

Example 1.3 The sequence denoted by $\{2n + 1\}$ is nothing but the odd natural numbers the function here is clear from the definition $x(n) = 2n + 1$. Similarly the sequence $\{2n\}$ is the sequence of positive even natural numbers.

We may denote a sequence by writing the terms of sequence explicitly or sometimes a recurrence relation may also determine a sequence.

Example 1.4 The sequence $\{n^2\}$ may be denoted by $\{1, 4, 9, \dots n^2, \dots\}$

Example 1.5 The sequence $\{1, 1, 2, 3, 5, 8, 13, 21, \dots\}$ is known as ***Fibonacci Sequence*** and the terms are generated by the recurrence relation $x_1 = x_2 = 1$ and $x_n = x_{n-1} + x_{n-2}\, n \geq 3$.

Definition A sequence is a ***finite sequence*** if the range of the sequence is finite otherwise it is termed as an ***infinite sequence***.

In this chapter by a sequence we will always mean an infinite sequence unless otherwise stated.

Example 1.6 The sequence $\{(-1)^n\}$ is a finite sequence as the range4 of the sequence is the set $\{1, -1\}$.

Example 1.7 As the range of the sequence $\{2^n\}$ is not finite it is an infinite sequence.

1.3 TYPES OF SEQUENCES

In this section we will be looking at various kinds of sequences, their definitions and examples.

Definition A sequence $\{x_n\}$ is said to be ***bounded above*** if there exists a real number M such that $x_n \leq M\ \forall n$ M is called an ***upper bound*** of the sequence. A sequence $\{x_n\}$ is said to be ***bounded below*** if there exists a real number k such that $x_n \geq k\ \forall n$ k is called a ***lower bound*** of the sequence. A sequence $\{x_n\}$ is said to be ***bounded*** if it is both bounded above and below.

Remark A sequence may be bounded above but not bounded below and similarly it may be bounded below but not bounded above. There exists sequences which is neither bounded above nor below.

Definition A sequence $\{x_n\}$ is not bounded above (unbounded above) if for any real number G no matter how large there exists at least one natural number n such that $x_n > G$.

Similarly a sequence $\{x_n\}$ is not bounded below (unbounded below) if for any real number h no matter how small there exists at least one natural number n such that $x_n < h$.

Example 1.8 The sequence $\{-n^2\}$ is bounded above but not bounded below as 0 is an upper bound of the sequence whereas $\{n^2\}$ is bounded below, with 0 as a lower bound but the sequence is not bounded above.

Example 1.9 The sequence $\{\sin n\}$ is a bounded sequence as $-1 \leq \sin n \leq 1$

Example 1.10 The sequence $\{1, -1, 2, -2, 3, -3, \dots n, -n, \dots\}$ is neither bounded above nor bounded below.

We note two very short yet important results:

Theorem 1.2.1

If $\{x_n\}$ is bounded above if and only if $\{-x_n\}$ is bounded below.

Proof: Easy exercise.

Theorem 1.2.2

A sequence $\{x_n\}$ is bounded if and only if there exists a real number $K > 0$ such that $|x_n| \leq K\ \forall n$.

Proof: One way the proof is trivial. If such $K > 0$ exists then we see that $-K \leq x_n \leq K$ hence the sequence is both bounded above and below that is the $\{x_n\}$ is bounded. Suppose that the sequence is bounded then by definition there exist real numbers k & M such that $k \leq x_n \leq M\ \forall\, n$. Let $K = \max\ \{|k|, |M|\}$ then $K > 0$ and for any n we have, $-K \leq -|k| \leq k \leq x_n \leq M \leq |M| \leq K$ that is, $|x_n| \leq K$.

Definition A sequence $\{x_n\}$ is said to be ***monotone increasing (monotone decreasing)*** if $x_{n+1} \geq x_n$ $(x_{n+1} \leq x_n)$ for all n.

If the inequality $\geq$ ($\leq$) above is replaced by strict inequality $>$ ($<$) we call the sequence ***strictly increasing (strictly decreasing)***.

Example 1.11 Any constant sequence $\{a\}$ is both monotone increasing and monotone decreasing.

Example 1.12 The sequence $\{1, 1, 2, 2, 3, 3.... n, n...\}$ is a monotone increasing sequence but not strictly monotone increasing.

Example 1.13 The sequence $\{n^2 + 1\}$ is strictly monotone increasing.

Example 1.14 The sequence $\left\{1, 1, \dfrac{1}{2}, \dfrac{1}{2}, \dfrac{1}{3}, \dfrac{1}{3}, \dots \dfrac{1}{n}, \dfrac{1}{n}, \dots\right\}$ is monotone decreasing but not strictly.

Example 1.15 The sequence $\{e^{-n}\}$ is strictly monotone decreasing.

Example 1.16 The sequence $\{(-1)^n\, 3^n\}$ is neither monotone increasing nor monotone decreasing.

We again note a small but important result whose proof is trivial and is left as an exercise.

Remark A sequence $\{x_n\}$ is monotone increasing (strictly) if and only if $\{-x_n\}$ is monotone decreasing (strictly)

1.4 CONVERGENT AND NON-CONVERGENT SEQUENCES

Before actually coming to the definition of convergent sequences let us first look at the special sequence $\left\{\dfrac{1}{n}\right\}$. Let us plot first five terms of the sequence (bold) on the real line (Fig. 2.1)

Fig. 1.1

We observe that for large values of n the members of the sequence is getting closer and closer to the value zero. It is also intuitively clear from the picture that we can get a member of the sequence as close to zero as we desire. Further, if we look more closely it will be clear that for any positive real number say ε however small, all members of the sequence except finitely many will lie in the interval $(0, \varepsilon)$. The above observations are in fact characteristics of a *convergent sequence* which we define below.

Definition A sequence $\{x_n\}$ is said to converge to a real number l written in symbol as $\lim\limits_{n \to \infty} x_n = l$ or $x_n \to l$ as $n \to \infty$ if given any $\varepsilon > 0$ there exists a positive integer n_0 (ε) depending on ε only such that $|x_n - l| < \varepsilon$ whenever $n > n_0$ (ε). In this case we say that the sequence $\{x_n\}$ has a finite limit and the real number l is said to be the limit of the sequence.

From the definition it is clear that, if the sequence $\{x_n\}$ is convergent then leaving only finitely many members $(x_1, x_2, \dots x_{n_0})$ all members of the sequence lie in the interval $(l - \varepsilon, l + \varepsilon)$. Note that in the above discussion of the sequence $\left\{\dfrac{1}{n}\right\}$, $l = 0$. Pictorially we can represent the concept of convergence as follows: We plot the natural numbers along X-axis and members of sequence $\{x_n\}$ along Y-axis. If the sequence is convergent, then for any $\varepsilon > 0$ there exists a positive integer n_0 such that all x_n for $n > n_0$ lies between the lines $l - \varepsilon$ and $l + \varepsilon$.

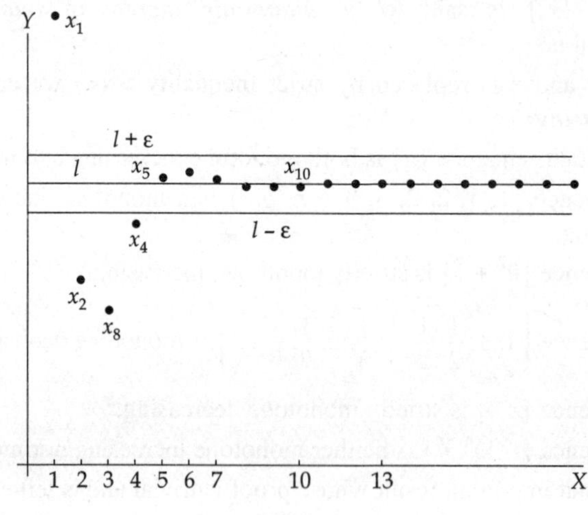

Fig. 1.2

See that in the above figure from x_5 onwards all members of the sequence lies between the lines $l + \varepsilon$ and $l - \varepsilon$.

Example 1.17   A constant sequence $\{x_n = a\}$ is a convergent sequence with limit as a.

Sol: Note that $|x_n - a| = 0 \; \forall \; n$.

Example 1.18   We show that $\left\{\dfrac{1}{n}\right\}$ converges to 0. For $\varepsilon > 0$ we have, $\left|\dfrac{1}{n} - 0\right| < \varepsilon$ whenever $n > \dfrac{1}{\varepsilon}$. If we choose $n_0 = \left[\dfrac{1}{\varepsilon}\right] + 1$ ([ ] denoting the greatest integer function) then it is clear that for all $n > n_0$, $\left|\dfrac{1}{n} - 0\right| < \varepsilon$ that is $\lim\limits_{n \to \infty} \dfrac{1}{n} = 0$

Definition   A convergent sequence whose limit is zero is called null a sequence

Note that Example 18 shows that $\left\{\dfrac{1}{n}\right\}$ is a null sequence.

Example 1.19   Show that (x^n) is a null sequence for $|x| < 1$.

Sol: We see that if $|x| < 1$ then $|x| = \dfrac{1}{1 + t}$ for some $t > 0$. Now,

$$|x^n - 0| = |x|^n = \frac{1}{(1 + t)^n}$$

$$= \frac{1}{1 + nt + \dfrac{n(n - 1)}{2} t^2 + \dots t^n} \qquad \text{(expanding Binomially as } n \text{ is a positive integer)}$$

$$< \frac{1}{1 + nt}$$

Now for any $0 < \varepsilon < 1$ we have $|x^n - 0| < \varepsilon$ if $\dfrac{1}{1 + nt} < \varepsilon$ that is if $nt > \dfrac{1}{\varepsilon} - 1$. Now if we choose $n_0 = \left[\dfrac{1 - \varepsilon}{\varepsilon}\right]$ then for $n > n_0$ we have $|x^n - 0| < \varepsilon$ in other words $\lim\limits_{n \to \infty} x^n = 0$

To understand convergence we must understand the contra positive statement of the definition of convergence of a sequence to it's limit. We will say that x_n does not converge to the real number l if there exists a real number $\varepsilon > 0$ such that for every given natural number n_0 there exists another natural number n satisfying $n > n_0$ such that, $|x_n - l| > \varepsilon$.

To illustrate we consider the following example:

Example 1.20   Show that $\dfrac{1}{n^2}$ does not tend to any positive real number say a.

Sol: Let $\varepsilon = \dfrac{a}{2}$. It is again easy to see that $\dfrac{1}{n^2}$ is monotone decreasing sequence (in fact it tends to zero as $n \to \infty$) hence for any given n_0 we can find another positive integer satisfying $n_1 > n_0$ such that $\dfrac{1}{n_1^2} < \dfrac{a}{2}$

in other words,

$$\left| \frac{1}{n_1^2} - a \right| > \frac{\varepsilon}{2} \qquad \text{(See Fig 2.3)}$$

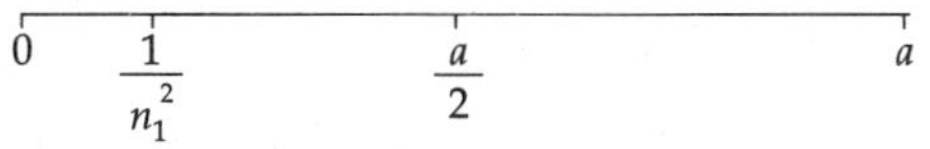

$$\begin{array}{cccc} 0 & \dfrac{1}{n_1^2} & \dfrac{a}{2} & a \end{array}$$

Fig. 1.3

That is $\dfrac{1}{n^2}$ does not tend to a.

At this point we will talk about sequences which are not convergent. There are two types of non-convergent sequences namely ***Divergent Sequences*** and ***Oscillating Sequences*** which we define below.

Definition   A sequence $\{x_n\}$ is said to diverge to $+\infty$ if given any real number K no matter how large there exists a natural number n_0 such that, $x_n > K$ whenever $n > n_0$. In this case we write, $\lim\limits_{n \to \infty} x_n = \infty$

Analogously, a sequence $\{x_n\}$ is said to diverge to $-\infty$ if given real number L no matter how small there exists a natural number n_0 such that, $x_n < L$ whenever $n > n_0$. In this case we write, $\lim\limits_{n \to \infty} x_n = -\infty$

Remarks

1. Note that although there is a significant similarity between sequences which are unbounded above and sequences which diverge to $+\infty$ but they are not same. Precisely, we can say every sequence which diverge to $+\infty$ are unbounded above but converse is not true.

2. Note that in the both the concept of convergence and divergence of a sequence $\{x_n\}$ we look for the behavior of the terms for large values of n. In other words, a sequence does not chance it's nature if we add or delete finitely many terms from the sequence that is, for example if a sequence is convergent then even after deletion or addition of finitely many terms the sequence continues to convergent sequence with same limit as before same is true for divergent or oscillating sequence.

Example 1.21   It is easy to see that the sequence $\{n\}$ is a divergent sequence.

Example 1.22   Show that the sequence defined by $\{(-1)^n \, n\}$ is unbounded above but certainly it does not diverge to $+\infty$.

Sol: We know that above sequence is unbounded above. Now suppose above sequence diverge to $+\infty$ then by definition, for any large positive number K there exists a natural number such that, $x_n > K$ whenever $n > n_0$. Now choose any natural number $n_1 > n_0$ then $x_{n_1} > k$ and $x_{n_1 + 1} > K$ but from definition of the sequence it is clear that either x_{n_1} or x_{n+1} is negative. This is a contradiction as $K > 0$. So above sequence does not diverge to $+\infty$.

Definition A sequence which neither converges nor diverges is called an oscillating sequence.

Example 1.23 The sequence $\{(-1)^n\}$ is an oscillating sequence.

Definition An oscillating sequence with finite range is said to oscillate finitely and if it has infinite range it is said to oscillate infinitely.

It is easy to see that $\{(-1)^n\}$ oscillate finitely whereas $\{(-1)^n n\}$ oscillates infinitely. As we will not be dealing much with divergent or oscillating sequence let us turn our attention again to convergent sequences.

We now ask a natural question if a sequence converges can it have two distinct limits? Let us try to see this intuitively, if a sequence $\{x_n\}$ have two limits say l & l' then by definition for sufficiently large n, x_n is arbitrarily close to both l & l' this is possible if l & l' are arbitrarily close to each other which must imply that l and l' must be equal. Hence we get the next theorem.

Theorem 1.4.1

A convergent sequence can have at most one limit.

Proof: Suppose $x_n \to l$ and $x_n \to l'$. From definition for $\varepsilon > 0$ there exists natural numbers n_1 & n_2 such that $|x_n - l| < \dfrac{\varepsilon}{2}$ and $|x_n - l'| < \dfrac{\varepsilon}{2}$ for all n greater than n_1 & n_2 respectively. If we choose $n_0 = \max(n_1, n_2)$ then for $n > n_0$ both above inequalities are true. For $n > n_0$ we choose x_n and keep it fixed then we get,

$$|l - l'| < |l - x_n| + |l' - x_n| < \frac{\varepsilon}{2} + \frac{\varepsilon}{2} = \varepsilon \tag{1.4.1.2}$$

Hence we conclude that $l = l'$.

Since from the definition of convergence of sequence it is clear that leaving finitely many elements of the sequence are very close to its limit hence it is expected that a convergent sequence should be bounded which is exactly the next theorem.

Theorem 1.4.2

Every convergent sequence bounded.

Proof: Let $x_n \to l$ be a sequence convergent to l then for $\varepsilon = 1$ there exists a positive integer n_0 such that, $|x_n - l| < 1$ whenever $n \geq n_0$. Hence,

$$|x_n| \geq |x_n - l| + |l| < 1 + |l| \tag{1.4.2.1}$$

whenever $n \geq n_0$. Now let us choose $M = \max\{|x_1|, |x_2|, \ldots |x_{n_0-2}| \, |x_{n_0-1}|, |l| + 1\}$ then it is clear from (1.4.2.1) that, $|x_n| < M$ for all n. That is $\{x_n\}$ is bounded.

Note that converse of Theorem 1.4.2 may not be true that is every bounded sequence is not convergent for example $\{(-1)^n\}$ is a bounded sequence but not convergent.

The following theorem gives some important properties of convergent sequences which are also called *Algebra of Limits*.

Theorem 1.4.3

Suppose $\{x_n\}$ & $\{y_n\}$ are convergent sequences with limits say l & m respectively and suppose c be any real number. Then,

(i) $\lim_{n \to \infty} cx_n = c \lim_{n \to \infty} x_n = cl$

(ii) $\lim_{n \to \infty} (x_n + y_n) = \lim_{n \to \infty} x_n + \lim_{n \to \infty} y_n = l + m$

(iii) $\lim_{n \to \infty} (x_n\, y_n) = (\lim_{n \to \infty} x_n)(\lim_{n \to \infty} y_n) = lm$

(iv) *Further if we assume that $l \neq 0$ then we have,*

$$\lim_{n \to \infty} \frac{1}{x_n} = \frac{1}{\lim\limits_{n \to \infty} x_n} = \frac{1}{l}$$

Proof: (i) and (ii) are easy to prove and are left as an exercise. To prove (iii) we note that as $\{x_n\}$ is convergent by Theorem 1.3.2 it is bounded hence there exists $M > 0$ such that,

$$|y_n| < M \text{ for all } n. \tag{1.4.3.1}$$

Next choose an arbitrary number $\varepsilon > 0$. Corresponding to this ε there exist natural numbers n_1 & n_2 such that for all $n > n_0 = \max(n_1, n_2)$ we have,

$$|x_n - l| < \frac{\varepsilon}{2} K \text{ and } |y_n - m| < \frac{\varepsilon}{2K} \quad \text{where } K \text{ is a real number such that } K > M + |l| \tag{1.4.3.2}$$

Now we see that,

$$|x_n\, y_n - l_m| = |(x_n - l)\, y_n + (y_n - m)l|$$
$$\leq |x_n - l|\,|y_n| + |y_n - m|\,|l|$$
$$< \frac{\varepsilon}{K}(M + |l|) < \varepsilon \quad \text{(using (1.4.3.1) and (1.4.3.2))}$$

Which proves that,

$$\lim_{n \to \infty} (x_n\, y_n) = l_m$$

To prove (iv) we first choose $\varepsilon = \dfrac{|l|}{2}$, then as $\{x_n\}$ is convergent, there exists a natural number n' such that $|x_n - l| < \dfrac{|l|}{2}$ whenever $n > n'$ (1.4.3.3)

Now we observe that,

$$|x_n| \geq |l| - |x_n - l| \geq |l| - \frac{|l|}{2} = \frac{|l|}{2} \text{ for } n > n \tag{1.4.3.4}$$

As $l \neq 0$ (1.4.3.4) shows that all but finitely many elements of the sequence are not equal to zero and hence without loss of generality we can assume that all members sequence are not equal to zero as we may always delete those members which are equal to zero keeping the same nature of the sequence (follows from Remark 2 after definition of divergent sequence).

Next choose any arbitrary positive number ε. corresponding to this ε there exists another natural number n'' such that,

$$|x_n - l| < \frac{\varepsilon\,|l|^2}{2} \quad \text{for all } n > n'' \tag{1.4.3.5}$$

If now we choose $n_0 = \max(n', n'')$ then using (1.4.3.3) and (1.4.3.4) we get,

$$\left| \frac{1}{x_n} - \frac{1}{l} \right| = \frac{|x_n - l|}{|x_n|\,|l|} < \frac{\dfrac{\varepsilon|l|^2}{2}}{\dfrac{|l|}{2}\,|l|} = \varepsilon \quad \text{for all } n > n_0.$$

Hence we conclude that,

$$\lim_{n \to \infty} \frac{1}{x_n} = \frac{1}{l}$$

Remark

Let c_0 denotes the space of all convergent sequences. We introduce two operations '+' and '·' on c_0 as $\{x_n\} + \{y_n\} = \{x_n + y_n\}$ and $\{x_n\} \cdot \{y_n\} = \{x_n \, y_n\}$ which is nothing but coordinate wise addition and multiplication. See that above theorem tells us that c_0 is closed under above two operations and further, it is a real vector space with the operation '+'.

If we give a close look to the equation (1.3.6.3) we can easily see that all but finitely many terms of the sequence keep the same sign of its limit. Following theorem is some sort of converse of this observation.

Theorem 1.4.4

If $\lim_{n \to \infty} x_n = l$ *and* $x_n > 0$ *for all n then* $l \geq 0$.

Proof: If possible suppose let $l < 0$. Then from (1.3.6.3) we get for all $n > n'$

$$\frac{3}{2} l < x_n < \frac{1}{2} l \tag{1.4.4.1}$$

This a contradiction as (1.4.3.1) shows that x_n lies between two negative numbers for all $n > n'$.

Corollary 1.4.5

If $\lim_{n \to \infty} x_n = l$ *and* $x_n > a$ *for all n then* $l \geq a$.

Proof: Easy exercise.

Theorem 1.4.6 (Sandwich Theorem)

Let $\{x_n\}$, $\{y_n\}$ *&* $\{z_n\}$ *are sequences such that* $x_n \leq z_n \leq y_n$ *for all n and suppose that, both* $\{x_n\}$ *and* $\{y_n\}$ *converges to the same limit l (say). Then* $z_n \to l$ *as* $n \to \infty$.

Proof: As $\{x_n\}$ and $\{y_n\}$ are convergent sequences, for an arbitrary positive real number $\varepsilon > 0$ we can always choose a natural number n_0 such that, $l - \varepsilon < x_n < z_n < y_n < l + \varepsilon$ whenever $n > n_0$. That is,

$$l - \varepsilon < z_n < l + \varepsilon$$

whenever $n > n_0$. In other words $\lim_{n \to \infty} z_n = l$.

We know that every bounded sequence may not be convergent but does there exists any class of sequence for which boundedness implies convergence? The answer is provided by following theorems.

Theorem 1.4.7

A monotone increasing sequence bounded above is convergent.

Proof: Suppose $\{x_n\}$ is a monotone increasing sequence and bounded above. We will show that $\{x_n\}$ converges to its supremum m say. Let $\varepsilon > 0$ be any real number then by definition of supremum there exists at least one member of the sequence say x_{n_1}, such that $m - \varepsilon < x_{n_1} < m$. As the

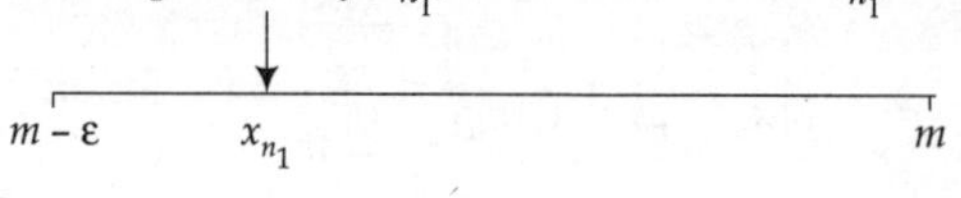

$$m - \varepsilon \qquad x_{n_1} \qquad\qquad\qquad m$$

Fig. 2.4

sequence is monotone increasing we will have $x_n \geq x_{n_1}$ for all $n \geq n_1$ also as m is the supremum we can write, $m - \varepsilon < x_n < m < m + \varepsilon$ for $n \geq n_1$.

That is, $|x_n - m| < \varepsilon$ for $n \geq n_1$ or in other words, $\lim_{n \to \infty} x_n \to m$.

Similarly we can prove the following theorem:

Theorem 1.4.8

A monotone decreasing sequence bounded below converges.

Proof: Proceed as the proof of above theorem showing that such a sequence converges to its infimum.

As an immediate consequence of the above to theorems we can conclude the following:

Corollary 1.4.9

Bounded monotone sequences are convergent.

Proof: Combine Theorem 1.4.7 & Theorem 1.4.8.

Next theorem shows that a monotone sequence has a definite character that, is either it converges or it diverges properly.

Theorem 1.4.10

A monotone sequence can never oscillate.

Proof: Let $\{x_n\}$ is a monotone increasing sequence, in case it is bounded above then by Theorem 1.3.2 it converges. Otherwise, suppose it is unbounded above then for any real number K no matter how large there exists there exists a natural number n_1 such that $x_{n_1} > K$. As the sequence is monotone for all $n > n_1$, $x_n \geq x_{n_1} > K$ in other words $\lim_{n \to \infty} x_n = \infty$.

Similarly, if the sequence is monotone decreasing and bounded below it converges otherwise we can prove that it diverges to $-\infty$. Hence we have proved that a monotone sequence can never oscillate.

1.5 SUBSEQUENCES AND BOLZANO-WEIERSTRASS THEOREM

We have seen that every bounded monotone sequence is convergent and but bounded sequence may not be monotone. We will show that from every sequence we can always extract a monotone sequence so, if the sequence is bounded this extracted sequence will be convergent by Corollary 1.3.4 which is exactly Bolzano-Weierstrass Theorem for sequences. Before going to prove the theorem we need the concept of subsequences.

Definition Let $\{x_n\}$ be a sequence and let $n_1 < n_2 < n_3 \ldots < n_k < \ldots$ be a collection of natural numbers then the sequence $x_{n_1}, x_{n_2}, x_{n_3}, \ldots, x_{n_k}, \ldots$ denoted by $\{x_{n_k}\}$ is called a subsequence of $\{x_n\}$.

Example 1.25 The sequence $\{n^2\}$ is a subsequence of the sequence $\{n\}$. Similarly $\{\cos 3n\}$ is subsequence of $\{\cos n\}$.

Next we prove a very important theorem which essentially tells us that "inside" every sequence there exists a monotone sequence.

Theorem 1.5.1

Every sequence has a monotone subsequence.

Proof: For a sequence $\{x_n\}$ the kth term x_k is called a peak if $x_k > x_m$ for $m > k$. We denote by P the set of all peaks of $\{x_n\}$. That is,

$$P = \{x_n : x_n > x_m \quad \text{whenever } m > n\}.$$

If P is infinite we choose a subsequence $\{x_{n_k}\}$ consisting entirely of elements from P then we claim that this subsequence is monotone decreasing indeed, if $n_{k_1} > n_{k_2}$ then $x_{k_{n1}} < x_{k_{n2}}$ as $x_{k_{n1}}$ and $x_{k_{n2}}$ are elements of P.

If P is finite we define $n_0 = \max \{n : x_n \in P\}$. The required subsequence will be defined inductively. We choose the first element x_{n_1} such that $n_1 > n_0$. Certainly $x_{n_1} \notin P$ hence there exists some $n_2 > n_1$ such that $x_{n_1} < x_{n_2}$. After choosing $x_{n_1}, x_{n_2}, \dots x_{n_k}$, where $x_{n_1} < x_{n_2} < \dots < x_{n_k}$ and $n_1 < n_2 \dots < n_k$ we can always choose $n_{k+1} > n_k$ such that $x_{n_1} < x_{n_{k+1}}$ as $x_{n_k} \notin P$. Hence inductively we have defined the subsequence $\{x_{n_k}\}$ which is monotone increasing.

So in any case $\{x_n\}$ contains a monotone subsequence.

As a consequence of the above theorem we get the following:

Theorem 1.5.2 (Bolzano-Weierstrass Theorem)

Every bounded sequence has a convergent subsequence.

Proof: Suppose $\{x_n\}$ is a bounded sequence. By previous theorem it has a monotone subsequence $\{x_{n_k}\}$ say and this subsequence is bounded as the whole sequence is bounded. Consequently $\{x_{n_k}\}$ is convergent by Corollary 1.4.9.

1.6 CLUSTER POINTS, LIMIT SUPERIOR, LIMIT INFERIOR

Even if a sequence as a whole may not converge but it may contain many subsequences which are convergent. Naturally, members of the sequence 'cluster' around any neighborhood of the limits of these convergent subsequences. These are what are limit points or cluster points of the sequence which we define formally below.

Definition ξ is called a limit point (or cluster point or accumulation point) of $\{x_n\}$ if there exists a subsequence $\{x_{n_k}\}$ such that, $x_{n_k} \to \xi$ as $n \to \infty$.

A close look at the above definition reveals the following: ξ is a *limit* point of the sequence $\{x_n\}$ if for any $\varepsilon > 0$ and for any natural number n_0 there exists a natural n, $n > n_0$, such that, $|x_n - \xi| < \varepsilon$

It is also clear from the definition that ξ is a *limit* point of the sequence $\{x_n\}$ if it is a limit point of the set $\{x_n : n \in \mathbb{N}\}$

Note that a *limit* is a *limit point* as the whole sequence treated as a subsequence converges to the limit. Converse is not true is evident from the following example.

Example 1.26 Define the sequence $\{x_n\}$ as,

$$x_n = \begin{cases} 1 + \dfrac{1}{n} & \text{if } n \text{ is odd} \\[2mm] n & \text{if } n \text{ is even} \end{cases}$$

We see that the subsequence $\left\{1 + \dfrac{1}{2n - 1}\right\}$ converges to 1 hence 1 is a limit point of the sequence.

However, notice that for any given any ε-neighborhood of 1 and any positive integer n_0 we can choose an even natural n, $n > n_0$ such that x_n lies outside the given ε-neighbourhood (as the subsequence $\{2_n\}$ is divergent). Hence 1 is not the limit of the sequence.

Next example shows that a sequence can have more than one limit point.

Example 1.27   Show that -1 and $+1$ are the limit points of the sequence $\{x_n = (-1)^n\}$.

Sol: It is clear that the subsequence $\{x_{2n-1} = -1\}$ converges to -1 and the subsequence $\{x_{2n} = 1\}$ converges to -1. Hence -1 and $+1$ are limit points of the above sequence.

Next we introduce the concept of limit superior and limit inferior of a sequence which are nothing but the supremum and infimum of the set of limit points of the sequence.

Definition   Let $\{x_n\}$ be a sequence we define *limit superior* and *limit inferior* of the sequence denoted by $\limsup x_n$ $(\overline{\lim}\, x_n)$ and $\liminf x_n$ $(\underline{\lim}\, x_n)$ respectively as,

$$\limsup x_n = \begin{cases} \lim\limits_{n \to \infty} [\sup \{x_m \; m > n\}] & \text{if } \{x_n\} \text{ is bounded above and } \{x_n\} \text{ does not tend to } -\infty \\ \infty \text{ if } \{x_n\} \text{ is unbounded above} \\ -\infty \text{ if } \{x_n\} \text{ tends to } -\infty \end{cases}$$

$$\liminf x_n = \begin{cases} \lim\limits_{n \to \infty} [\inf \{x_m \; m > n\}] & \text{if } \{x_n\} \text{ is bounded below and } \{x_n\} \text{ does not tend to } \infty \\ -\infty \text{ if } \{x_n\} \text{ is unbounded below} \\ \infty \text{ if } \{x_n\} \text{ tends to } \infty \end{cases}$$

Let us analyze the definition of *limit superior* closely. Suppose $\{x_n\}$ is bounded and let $S_n = \sup \{x_m \, m > n\}$. If $\limsup x_n = S$ then $S = \lim\limits_{n \to \infty} S_n$. Hence for any $\varepsilon > 0$ there exists a natural number n_0, such that, $S - \varepsilon < S_n < S + \varepsilon$ for all $n \geq n_0$.

But this implies that $\sup \{x_m \; m > n\} < S + \varepsilon$ for all $n \geq n_0$. Hence in particular this implies that there are all but finitely many elements of the sequence which are greater than $S + \varepsilon$.

Again as $s_n = \sup \{x_m \; m > n\}$ and $S - \varepsilon < s_n$ it implies that there exists infinitely many elements of the sequence which are greater than $S - \varepsilon$.

Hence limit superior S of a bounded sequence $\{x_n\}$ must satisfy following conditions:

(i)   For every $\varepsilon > 0$ there exists a natural number n_0 such that for all $n \geq n_0$, $x_n \leq S + \varepsilon$, that is for every $\varepsilon > 0$ there exists finitely many elements of the sequence are greater than $S + \varepsilon$.

(ii)   Given any $\varepsilon > 0$ and a natural number m there exists another positive integer $n > m$ such that $x_n > S - \varepsilon$, that is there exists infinite number of elements of the sequence greater than $S - \varepsilon$.

Now suppose a real number S satisfies condition (i) and (ii) as above we will show that S is the limit superior of the sequence. For that let $S_n = \sup \{x_m : m > n\}$ we will show $\lim\limits_{n \to \infty} S_n = S$.

We see that for any arbitrary $\varepsilon > 0$ by condition (i) there exists a natural number n_0 such that that for all $n \geq n_0$, $x_n \leq S + \dfrac{\varepsilon}{2}$ which implies that,

$$\sup \{x_m : m > n\} = S_n \leq S + \frac{\varepsilon}{2} \qquad \text{for } n \geq n_0$$

or,
$$S_n < S + \varepsilon \qquad \text{for } n \geq n_0$$

Again using condition (ii) we see that for any $n \geq n_0$ there exists another $m > n$ such that $x_m > S - \varepsilon$ so that we get $\sup \{x_m : m > n\} = S_n > S - \varepsilon$ whenever $n \geq n_0$. So we conclude that for any $\varepsilon > 0$ there exists a positive integer n_0 such that,

$$S - \varepsilon < S_n < S + \varepsilon \quad \text{for all } n \geq n_0$$

Summarizing the above observation above we get the following:

Theorem 1.6.1

*A real number is **limit superior** of a bounded sequence if and only if it satisfies following conditions:*

(i) *For every $\varepsilon > 0$ there exists a natural number n_0 such that for all $n \geq n_0$, $x_n \leq S + \varepsilon$, that is for every $\varepsilon > 0$ there exists finitely many elements of the sequence are greater than $S + \varepsilon$.*

(ii) *Given any $\varepsilon > 0$ and a natural number m there exists another positive integer $n > m$ such that $x_n > S - \varepsilon$, that is there exists infinite number of elements of the sequence greater than $S - \varepsilon$.*

Similarly we get the following:

Theorem 1.6.2

*A real number is **limit inferior** of a bounded sequence $\{x_n\}$ if and only if it satisfies following conditions:*

(i) *For every $\varepsilon > 0$ there exists a natural number n_0 such that for all $n \geq n_0$, $x_n \geq s - \varepsilon$, that is for every $\varepsilon > 0$ there exists only finitely many elements of the sequence which are less than $S - \varepsilon$.*

(ii) *Given any $\varepsilon > 0$ and a natural number m there exists another positive integer $n > m$ such that, $x_n < s + \varepsilon$, that is there exists infinite number of elements of the sequence less than $s + \varepsilon$.*

Next we prove some important theorems which are easy consequences of the above definitions,

Theorem 1.6.3

Suppose E be set of all cluster points of a bounded sequence $\{x_n\}$ then,

(i) $\overline{\lim} \, x_n = \max E$

and (ii) $\underline{\lim} \, x_n = \min E$

Proof: To prove (i) we must show that maximum of the set E exists. As $\{x_n\}$ is bounded E is non empty by Bolzano-Weierstrass Theorem for sequences (Theorem 1.4.2). Let $\mu = \sup E$ then for any $\varepsilon > 0$ we observe two things,

(a) There exist only finitely many elements of $\{x_n\}$ greater than, otherwise by Bolzano-Weierstrass Theorem for sequences we will get a cluster point of the sequence lying beyond $\mu + \varepsilon$ contradicting the fact that μ is the supremum of the set of cluster points of the sequence.

(b) From the definition of supremum we get there exists one member of E say $\eta > \mu - \varepsilon$. Now as η is a limit point of the sequence there exists infinite number of n such that $\eta > x_n \, \mu - \varepsilon$.

From (a) and (b) we conclude that $\mu = \sup E = \overline{\lim} \, x_n$. Again from (b) we see that for any $\varepsilon > 0$ there exists infinite number of elements x_n of the sequence satisfying $\mu - \varepsilon < x_n < \mu + \varepsilon$ hence we can easily choose a subsequence $\{x_{n_k}\}$ such that $x_{n_k} \to \mu$ proving that μ itself is a cluster point of $\{x_n\}$ hence $\mu \in E$ and $\mu = \max E$.

We can prove (ii) with similar arguments as above and is kept as an exercise.

Theorem 1.6.4

Suppose $\{x_n\}$ be a sequence then,

(i) $\underline{\lim} \, x_n \leq \overline{\lim} \, x_n$

(ii) *$\{x_n\}$ is convergent if and only if $\overline{\lim} \, x_n = \underline{\lim} \, x_n = a$ real number and in that case,*

$$\overline{\lim} \, x_n = \underline{\lim} \, x_n = \lim_{n \to \infty} x_n$$

Proof: (i) If the sequence is unbounded the result is clear. Suppose the sequence is bounded and if

$S_n = \sup\, [x_m : m > n]$ and $s_n = \inf\, [x_m : m > n]$ then it follows that,

$$\lim_{n \to \infty} S_n \le \lim_{n \to \infty} S_n \text{ and consequently } \underline{\lim}\, x_n \le \overline{\lim}\, x_n.$$

(ii) Suppose $\{x_n\}$ is convergent having limit l hence bounded. Then it is easy to see that l satisfies both conditions of Theorem 1.6.1 as well as that of Theorem 1.6.2 hence l is both upper a well as lower limit that is

$$\overline{\lim}\, x_n = \underline{\lim}\, x_n = l$$

Conversely, let $\overline{\lim}\, x_n = \underline{\lim}\, x_n = l$ (say). For $\varepsilon > 0$ using condition (ii) of Theorem 1.6.1 and Theorem 1.6.2 there exists natural numbers n_1 and n_2 such that,

$$x_n \le l + \frac{\varepsilon}{2} < l + \varepsilon \;\forall\; n \ge n_1 \tag{1.6.3.1}$$

and

$$x_n \ge l - \frac{\varepsilon}{2} > l - \varepsilon \;\forall\; n \ge n_2 \tag{1.6.3.2}$$

Now if we choose $n_0 = \max\,(n_1, n_2)$ using (1.6.3.1) & (1.6.3.2) we get,

$$l - \varepsilon < x_n < l + \varepsilon \;\forall\; n \ge n_0$$

Hence $\lim_{n \to \infty} x_n = l$

Theorem 1.6.5

If $\{x_n\}$ & $\{y_n\}$ are two bounded sequences then,

(i) $\overline{\lim}\, (x_n + y_n) \le \overline{\lim}\, x_n + \overline{\lim}\, y_n$

(ii) $\underline{\lim}\, (x_n + y_n) \ge \underline{\lim}\, x_n + \overline{\lim}\, y_n$

Proof: We will prove only (i) and (ii) will be left as an left as an exercise.

First note that as we are dealing with bounded sequences all limit superior and limit inferior in the above statement are real numbers. Let n be arbitrary but fixed natural number. Then for any natural number $m > n$, it follows from the definition of limit superior that,

$$x_m \le \sup\, \{x_k : k > n\} \tag{1.6.5.1}$$

and

$$y_m \le \sup\, \{y_k : k > n\} \tag{1.6.5.2}$$

Adding equations (1.5.5.1) & (1.5.5.2) we get for $m > n$,

$$x_m + y_m \le \sup\, \{x_k : k > n\} + \sup\, \{y_k : k > n\} \tag{1.6.5.3}$$

Now first taking supremum of the left side of equation (1.5.5.3) for all $m > n$ and then taking limit yields,

$$\overline{\lim}\, (x_n + y_n) \le \overline{\lim}\, x_n + \overline{\lim}\, y_n$$

1.7 CAUCHY CRITERION

We will now formulate a necessary and sufficient condition for convergence of a sequence. For that we will begin with the following definition:

Definition   A sequence is said to be a **Cauchy sequence** or is said to satisfy **Cauchy Criterion,** if for any $\varepsilon > 0$ there exists a positive integer $n_0\,(\varepsilon)$ such that whenever $m > n > n_0$ we have $|x_m - x_n| < \varepsilon$.

The definition implies that if a sequence is Cauchy then for sufficiently large m & n the distance between x_n & x_m is arbitrarily small. Next theorem tells us that such sequences are convergent and the converse is also true.

Theorem 1.7.1

A necessary and sufficient condition that a sequence $\{x_n\}$ is convergent if and only if it satisfies Cauchy criterion or in other words if and only if it is a Cauchy sequence.

Proof: We first prove the necessary part. For that suppose $x_n \to x$ as $n \to \infty$. Then by definition for any $\varepsilon > 0$ there exists a positive integer $n_0 (\varepsilon)$ such that whenever $n > n_0$

$$|x_n - x| < \frac{\varepsilon}{2} \tag{1.7.1.1}$$

Now if $m > n > n_0$ both x_n & x_m satisfy equation (1.1). Hence we have,

$$|x_m - x_n| = |x_m - x + x - x_n| < |x_m - x| + |x_m - x|$$

$$< \frac{\varepsilon}{2} + \frac{\varepsilon}{2} = \varepsilon \tag{1.7.1.2}$$

(1.6.1.2) shows that $\{x_n\}$ is a cauchy sequence.

Next for the sufficient part we assume the sequence $\{x_n\}$ is a cauchy sequence. For that we take two cases:

Case 1 Suppose the range R of the sequence is finite say $R = \{x_1, x_2, \dots x_n\}$ we show in this case that after finite number of terms all members of the sequence are equal. For that let us choose $\varepsilon > 0$ such that $\varepsilon < \min_{i, j \in 1,2 \dots n} \{|x_i - x_j|\}$. By cauchy criterion there must exist some n_0 such that whenever $m > n > n_0$ we have $|x_m - x_n| < \varepsilon$ but this is possible only when $x_n = x_m = x_t$ for some t such that $1 \le t \le n$ and for all $m > n > n_0$. This shows that $x_n \to x_t$ as $n \to \infty$.

Case 2 Suppose now the range R of the sequence is not finite. In this case again using the definition of cauchy sequence we have for any $\varepsilon = 1$, a positive integer $n_0 (\varepsilon)$ such that whenever $m > n > n_0$ we have $|x_m - x_n| < 1$. We first prove that this sequence is bounded for that we choose an $N > n_0$ and keep it fixed. Then we have for any $n > n_0$

$$|x_n| = |x_n - x_N + x_N| \le |x_n - x_N| + |x_N| < 1 + |x_n| \tag{1.7.1.3}$$

If we now choose $K = \max \{|x_1|, |x_2|, \dots |x_{n_0}|, 1 + |x_N|\}$ then $K > 0$

and
$$|x_n| \le K \ \forall \ n \tag{1.7.1.4}$$

This shows that $\{x_n\}$ is bounded. Now it follows that the range of the sequence is an infinite set of bounded real numbers hence by Bolzano-Weierstrass Theorem R has a limit point say x. We show that x is the limit of the sequence $\{x_n\}$. For any $\varepsilon > 0$ we choose a positive integer $n_0 (\varepsilon)$ such whenever $m > n > n_0$ we have $|x_m - x_n| < \frac{\varepsilon}{2}$. By definition of limit point there exists a positive integer $m > n_0$ such that,

$$|x_m - x| < \frac{\varepsilon}{2} \tag{1.7.1.5}$$

Now let $n > n_0$ be any positive integer then we have,

$$|x_n - x| = |x_n - x_m + x_m - x| \le |x_n - x_m| + |x_m - x| < \frac{\varepsilon}{2} + \frac{\varepsilon}{2} = \varepsilon \tag{1.7.1.6}$$

This shows $x_n \to x$ as $n \to \infty$ which concludes the theorem.

1.8 CONVERGENCE OF SERIES OF REAL NUMBERS

Suppose we are to add a finite number of real numbers say $u_1, u_2, u_3 \ldots u_n$ then we know the answer is $u_1 + u_2 + \ldots + u_n$ which is called the finite sum. In present section we will be interested to build the notion of infinite sum based on the concept of finite sum and using the knowledge of sequences. If $\{u_n\}$ is a sequence of real numbers then the symbol $\sum\limits_{n-1}^{\infty} u_n = u_1 + u_2 + \ldots + u_n + \ldots$ is called an infinite series with n^{th} term as u_n. In fact, in this section we will try to give a meaning to the above symbol. It is natural to deal first with the series of positive terms which is much easy to handle.

Definition If $\sum\limits_{n=1}^{\infty} u_n$ be an infinite series of real numbers then the sequence $\{s_n\}$ defined as $s_n = u_1 + u_2 + \ldots + u_n$ is defined as the sequence of partial sums for the above series and s_n will be called the nth partial sum.

An infinite series $\sum\limits_{n=1}^{\infty} u_n$ of real numbers is said to converge or diverge if the corresponding sequence of partial sums $\{s_n\}$ as defined above converge or diverge. In case $s_n \to s$ we say the series $\sum\limits_{n-1}^{\infty} u_n$ converges and has sum s in symbols we write $\sum\limits_{n=1}^{\infty} u_n = s$

Remark

It is clear from the definition that the nature of the series does not change even if we delete or add finitely many terms to the given series. That is if $\sum\limits_{n=1}^{\infty} u_n$ converges (diverges) then the series $\sum\limits_{n=k}^{\infty} u_n$ and $\sum\limits_{i=1}^{k} v_i + \sum\limits_{n=1}^{\infty} u_n$ converges (diverges) and conversely.

Also we note that as an easy consequence of above definition we conclude that series $\sum\limits_{n=1}^{\infty} u_n$ and $\sum\limits_{n=k}^{\infty} au_n$ converges or diverges simultaneously.

1.9 CAUCHY'S CRITERION FOR CONVERGENCE OF AN INFINITE SERIES

Let $\sum\limits_{n=1}^{\infty} u_n$ be an infinite series with $\{s_n\}$ as the corresponding sequence of partial sums. Let $\varepsilon > 0$ then by Theorem 1.6.1 (Cauchy Criterion) $\{s_n\}$ is convergent if and only if there exists a natural number $n_0 (\varepsilon)$ such that whenever $m > n > n_0 (\varepsilon)$ we must have,

$$|S_m - S_n| = |U_{m+1} + u_{m+2} + \ldots u_n| < \varepsilon$$

Now by definition $\sum\limits_{n=1}^{\infty} u_n$ converges if $\{S_n\}$ converges hence we conclude the following:

Theorem 1.9.1 (Cauchy's Criterion for Convergence of an Infinite Series)

The infinite series $\sum\limits_{n=1}^{\infty} u_n$ converges if and only if given any $\varepsilon > 0$ there exists a natural number $n_0 (\varepsilon)$ such that whenever $m > n > n_0 (\varepsilon)$ we must have,

$$|u_{m+1} + u_{m+2} + \dots u_n| < \varepsilon$$

As an immediate corollary to above theorem we get,

Theorem 1.9.2 (n^{th}-Term Test)

If the series $\sum\limits_{n=1}^{\infty} u_n$ converges then the sequence $\{u_n\}$ is a null sequence.

Proof: If $\{S_n\}$ denotes the sequence of partial sums of the above series then by given condition it is cauchy. Hence for $\varepsilon > 0$ there exists a natural number n_0 such that whenever $n, m > n_0$ we have,

$$|S_n - S_m| < \varepsilon$$

In particular, if we choose $n - 1 > n_0$ that is if $n > n_0 + 1 = n_1$ (say) we get, $|S_n - S_{n-1}| = |u_n| < \varepsilon$ which shows that, $\lim\limits_{n \to \infty} x_n = 0$

Remark

By negation of above theorem we conclude that if n^{th} term of a series does not tend to zero as n tends to infinity then the series cannot converge.

1.10   TEST FOR CONVERGENCE OF SERIES OF NON-NEGATIVE TERMS

An infinite series $\sum\limits_{n=1}^{\infty} u_n$ where $u_n \geq 0$ for all n, is known as series of non-negative terms. In this section we will study various conditions under which a series of non-negative terms converges or diverges. First we observe that if $\sum\limits_{n=1}^{\infty} u_n$ be a series of positive terms with corresponding sequence of partial sums as $\{S_n\}$, then $S_{n+1} - S_n = u_{n+1} \geq 0$ that is, $\{S_n\}$ is monotone increasing. Hence either the sequence of partial sums converges or diverges to $+\infty$. As an immediate consequence of Theorem 1.3.2 we have the following:

Theorem 1.10.1

A series of non-negative terms converges if and only if the sequence of partial sums of the series is bounded above.

Theorem 1.10.2 (Comparison Test)

Suppose $\sum\limits_{n=1}^{\infty} u_n$ and $\sum\limits_{n=1}^{\infty} v_n$ be two series of non-negative terms satisfying, $u_n \leq v_n \; \forall \, n$ then,

(i) $\sum\limits_{n=1}^{\infty} u_n$ *converges if* $\sum\limits_{n=1}^{\infty} v_n$ *converges.*

(ii) $\sum\limits_{n=1}^{\infty} v_n$ *diverges if* $\sum\limits_{n=1}^{\infty} u_n$ *diverges.*

Proof: If $\{S_n\}$ and $\{S_n'\}$ denote the sequence of partial sums of the series $\sum\limits_{n=1}^{\infty} u_n$ and $\sum\limits_{n=1}^{\infty} v_n$ respectively then from the given condition it is clear that, $s_n \leq s_n' \; \forall \, n$ (1.10.2.1)

Now if $\sum\limits_{n=1}^{\infty} v_n$ converges then by necessity part of Theorem 1.10.1 $\{s_n'\}$ is bounded above and using (1.10.2.1) we get that $\{s_n\}$ is bounded above. By sufficiency part of Theorem 1.10.1 we conclude that $\sum\limits_{n=1}^{\infty} u_n$ converges. This proves (i).

Part (ii) can be proved similarly.

Remark

Note that the conclusion of above theorem remains true even if we assume that,

$$u_n \le v_n \ \forall \ n \ge k.$$

where, k is any positive integer.

Theorem 1.10.3 (Limit Comparison Test)

If $\sum\limits_{n=1}^{\infty} u_n$ and $\sum\limits_{n=1}^{\infty} v_n$ be two series of positive terms if,

(i) $\lim\limits_{n \to \infty} \dfrac{u_n}{v_n} = l$, *where l is positive real number*

then, above two series converges and diverges simultaneously.

(ii) $\lim\limits_{n \to \infty} = \dfrac{u_n}{v_n} = 0$ *then $\sum\limits_{n=1}^{\infty} v_n$ convergent implies $\sum\limits_{n=1}^{\infty} u_n$ is also convergent.*

(iii) $\lim\limits_{n \to \infty} \dfrac{u_n}{v_n} = \infty$ *then $\sum\limits_{n=1}^{\infty} v_n$ divergent implies $\sum\limits_{n=1}^{\infty} u_n$ is also divergent.*

Proof: (i) Let us choose ε such that $0 < \varepsilon < l$. As $\dfrac{u_n}{v_n} \to l$ corresponding to this ε we will get a positive integer n_0 such that,

$$\left| \dfrac{u_n}{v_n} - l \right| < \varepsilon \ \forall \ n \ge n_0 \text{ that is,}$$

$$(l - \varepsilon) v_n < u_n < (l + \varepsilon) v_n \ \forall \ n \ge n_0 \tag{1.10.3.1}$$

Now if $\sum\limits_{n=1}^{\infty} u_n$ converges then by using left inequality of (1.10.3.1) and Theorem 1.10.2 we conclude $\sum\limits_{n=1}^{\infty} (l - \varepsilon) v_n$ converges but that implies $\sum\limits_{n=1}^{\infty} v_n$ is convergent. Similarly we can prove if $\sum\limits_{n=1}^{\infty} v_n$ is convergent then $\sum\limits_{n=1}^{\infty} u_n$ is convergent by using right inequality of (1.10.3.1) and Theorem 1.10.2. Using similar argument we can prove that $\sum\limits_{n=1}^{\infty} u_n$ is divergent if and only if $\sum\limits_{n=1}^{\infty} v_n$ is divergent.

(ii) If $\lim\limits_{n \to \infty} \dfrac{u_n}{v_n} = 0$ then for any $\varepsilon > 0$ there exists a positive integer n_0 such that

$$u_n \le \varepsilon \, v_n \quad \forall \, n \ge n_0 \tag{1.10.3.2}$$

Hence if $\sum\limits_{n=1}^{\infty} v_n$ is convergent then $\sum\limits_{n=1}^{\infty} u_n$ converges by Theorem 1.10.2.

(iii) If $\lim\limits_{n \to \infty} \dfrac{u_n}{v_n} = \infty$ then for any arbitrary large number M there exists a positive integer n_0 such that,

$$u_n \geq M v_n \quad \forall \, n \geq n_0 \tag{1.10.3.3}$$

Hence if $\sum\limits_{n=1}^{\infty} v_n$ is divergent then $\sum\limits_{n=1}^{\infty} u_n$ diverges by Theorem 1.9.2.

Theorem 1.10.4 (Ratio Test)

Suppose for a series of non-negative terms $\sum\limits_{n=1}^{\infty} u_n$

$$\lim_{n \to \infty} \frac{u_n + 1}{u_n} = l$$

Then,

$$\sum_{n=1}^{\infty} u_n \text{ converges if } l < 1$$

$$\sum_{n=1}^{\infty} u_n \text{ diverges if } l > 1$$

Proof: (i) We choose $\varepsilon > 0$ so small such that $0 < \lambda = l + \varepsilon < 1$. Corresponding to this ε there exists a positive integer n_0 such that,

$$(l - \varepsilon)\, u_n < u_{n+1} < (l + \varepsilon)\, u_n \; \forall \, n \geq n_0 \tag{1.10.4.1}$$

Now choose any $n > n_0$, then repeated use of the right hand equality of (1.10.4.1) yields,

$$u_{n+1} < \lambda^2 u_n < \lambda^2 u_{n-1} \ldots < \lambda^{n-n_0+1} u_{n_0} = M\lambda^n \tag{1.10.4.2}$$

where $M = \lambda^{1-n_0} u_{n_0}$ fixed a positive number. Hence we get, $u_{n+1} < M\lambda^n \; \forall \, n \geq n_0$ $\qquad$ (1.10.4.3)

Now we know that $\sum\limits_{n=1}^{\infty} \lambda^n$ converges as $\lambda < 1$ (Example) hence by comparison test we conclude that $\sum\limits_{n=1}^{\infty} u_n$ converges.

To prove (ii) we choose $\varepsilon > 0$ so that $\mu = l - \varepsilon > 1$. By similar argument a given in (i) we will obtain a positive integer n_0 such that, $u_{n+1} > K\mu^n \; \forall \, n \geq n_0$ $\qquad$ (1.10.4.4)

where $K > 0$ is a constant. Now $\sum\limits_{n=1}^{\infty} \mu^n$ diverges as $\mu > 1$ it follows from comparison test that $\sum\limits_{n=1}^{\infty} u_n$ diverges.

Theorem 1.10.5 (Cauchy's Root Test)

For a series of non-negative terms $\sum\limits_{n=1}^{\infty} u_n$ let $\rho = \overline{\lim\limits_{n \to \infty}} \, (u_n)^{\frac{1}{n}}$ then,

(i) $\sum\limits_{n=1}^{\infty} u_n$ *converges if* $\rho < 1$

(ii) $\sum\limits_{n=1}^{\infty} u_n$ *diverges if* $\rho > 1$

Proof: (i) We choose $\varepsilon > 0$ such that $\mu = \rho + \varepsilon < 1$ then by definition of ρ there exists a positive integer n_0 such that,

$$(u_n)^{\frac{1}{n}} < \mu \; \forall \, n \geq n_0$$

that is,
$$u_n < \mu^n \ \forall \ n \geq n_0 \qquad\qquad (1.10.5.1)$$

As $\mu < 1$ we get $\sum_{n=1}^{\infty} \mu^n$ converges hence by comparison test $\sum_{n=1}^{\infty} u_n$ converges.

(ii) In this case it follows that there exists infinite number of the members of the sequence $\left\{ (u_n)^{\frac{1}{n}} \right\}$ are greater than 1. This implies $u_n > 1$ for infinitely many n. This shows that the sequence $\{u_n\}$ does not tend to 0 as which is a necessary condition for $\sum_{n=1}^{\infty} u_n$ to converge. Hence $\sum_{n=1}^{\infty} u_n$ cannot converge but as it is a series of positive terms it must diverge.

1.11 SERIES OF ARBITRARY TERMS: CONDITIONAL AND ABSOLUTE CONVERGENCE

So far we have studied convergence and divergence of infinite series of non-negative terms only. In this section we will be interested to study infinite series of arbitrary terms (i.e. terms of series can be both positive and negative). First we introduce two important definitions.

Definition   The series $\sum_{n=1}^{\infty} u_n$ is said to converge *absolutely* (or $\sum_{n=1}^{\infty} u_n$ is *absolutely convergent*) if the series $\sum_{n=1}^{\infty} |u_n|$ converges.

Definition   The series $\sum_{n=1}^{\infty} u_n$ is said to converge *conditionally* (or $\sum_{n=1}^{\infty} u_n$ is *conditionally* convergent) if the series converges but does not converge absolutely.

Following theorem shows that absolute convergence is stronger than ordinary convergence. In other words absolute convergence implies ordinary convergence.

Theorem 1.11.1

If the series $\sum_{n=1}^{\infty} u_n$ is absolutely is absolutely convergent then it is convergent.

Proof: If $\sum_{n=1}^{\infty} |u_n|$ is convergent then for any $\varepsilon > 0$ using necessity part of Theorem 1.9.1 we will get some positive integer n_0 such that whenever $m > n > n_0$ we have,

$$|u_n| + |u_{n+1}| + |u_{n+2}| \dots + |u_m| < \varepsilon \qquad\qquad (1.11.1.1)$$

This implies that,

$$|u_n + u_{n+1} + u_{n+2} \dots + u_m| \leq |u_n| + |u_{n+1}| + |u_{n+2}| \dots + |u_m| < \varepsilon, \text{ whenever } m > n > n_0 \qquad (1.11.1.2)$$

Now using the sufficiency part of Theorem 1.8.1 we conclude that $\sum_{n=1}^{\infty} u_n$ is convergent.

Converse of above Theorem 1.11.1 is not true. (See Example 28)

We will now be interested in the study of convergence of the series of the form $\sum_{n=1}^{\infty} u_n v_n$ where $\{u_n\}$ and $\{v_n\}$ are sequence of real numbers. In fact we will be proving two theorems in this regard known as Dirichlet's Test and Abel's Test.

Theorem 1.11.2 (Dirichlet's Test)

Suppose sequence of partial sums $\{s_n\}$ of the series $\sum\limits_{n=1}^{\infty} u_n$ is bounded and if $\{v_n\}$ be a monotone decreasing sequence of real numbers converging to 0 then $\sum\limits_{n=1}^{\infty} u_n v_n$ converges.

Proof: First we observe that, $u_r = s_r - s_{r-1}$. Again if $\{s_n'\}$ denotes the partial sums of the series $\sum\limits_{n=1}^{\infty} u_n v_n$ then,

$$s_n' = \sum_{r=1}^{\infty} u_r v_r = (v_1 - v_2)\, s_1 + (v_2 - v_3)\, s_2 + \ldots + (v_{n-1}\, v_n)\, s_{n-1} + v_n s_n$$

$$= \sum_{r=1}^{n-1} (v_r - v_{r+1})\, s_r + v_n s_n \tag{1.11.2.1}$$

Now as $\{s_n\}$ is bounded there exists some $M > 0$ such that $|s_n| < M \;\forall\; n$. As $\{v_n\}$ is monotone decreasing we get,

$$\sum_{r=1}^{n-1} |(v_r - v_{r+1})\, s_r| \le M \left| \sum_{r=1}^{n-1} (v_r - v_{r+1}) \right| = M\,|v_1 - v_n| \le 2M\,|v_1| \tag{1.11.2.2}$$

From (1.11.2.2) it clear that the partial sums of the series $\sum\limits_{n=1}^{\infty} |(v_r - v_{r+1})|\, s_r$ is bounded. This implies that the series $\sum\limits_{r=1}^{\infty} (v_r - v_{r+1})\, s_r$ is absolutely convergent and hence by Theorem 1.11.1 the series is convergent. Suppose,

$$S = \sum_{r=1}^{\infty} (v_r - v_{r+1})\, s_r \tag{1.11.2.3}$$

Again we note that,

$$|v_n s_n| \le M\,|v_n| \tag{1.11.2.4}$$

Now we know $v_n \to 0$ as $n \to \infty$ as hence from (1.11.2.3) we conclude

$$v_n s_n \to 0 \text{ as } n \to \infty \tag{1.11.2.5}$$

Taking limit $n \to \infty$ on both sides of (1.11.2.1) and using (1.11.2.3) & (1.11.2.5) we get,

$$\lim_{n \to \infty} s_n' = S + 0$$

In other words $\sum\limits_{n=1}^{\infty} u_n v_n$ converges.

Example 1.28   Show that $\sum\limits_{n=1}^{\infty} \dfrac{(-2)^{n+1}}{n}$ is convergent but not absolutely.

Sol: Let $u_n = (-2)^{n+1}$ then s_n the nth partial sum of $\sum\limits_{n=1}^{\infty} u_n$ is given by,

$$s_n = \begin{cases} -2 & \text{if } n \text{ is odd} \\ 0 & \text{if } n \text{ is even} \end{cases}$$

Clearly s_n is bounded. If we choose $v_n = \dfrac{1}{n}$ then v_n is monotone decreasing and $v_n \to 0$ as $n \to \infty$. Hence by above theorem $\sum\limits_{n=1}^{\infty} \dfrac{(-2)^{n+1}}{n}$ is convergent. It is easy to see that the series $\sum\limits_{n=1}^{\infty} \left| \dfrac{(-2)^{n+1}}{n} \right|$

$= \sum\limits_{n=1}^{\infty} \dfrac{2^{n+1}}{n}$ is not convergent by ratio test.

Theorem 1.11.3 (Abel's Test)

The series $\sum\limits_{n=1}^{\infty} u_n v_n$ converges if

    (i) $\sum\limits_{n=1}^{\infty} u_n$ *is convergent and*

    (ii) $\{v_n\}$ *is monotone and bounded*

Proof: Let $\{s_n\}$ and $\{s'_n\}$ be the sequence of partial sums of the series $\sum\limits_{n=1}^{\infty} u_n$ and $\sum\limits_{n=1}^{\infty} u_n v_n$ and then proceeding as in previous theorem we obtain equation (1.11.2.1). First let us suppose that $\{v_n\}$ is monotone decreasing sequence. Now as $\sum\limits_{n=1}^{\infty} u_n$ is convergent it follows that $\{s_n\}$ is bounded hence there exists some $M > 0$ such that $|s_n| \leq M \ \forall \ n$, consequently we again arrive at equation (1.11.2.2) by which we conclude that the series $\sum\limits_{r=1}^{\infty} (v_r - v_{r+1})\, s_r$ is convergent and again as before we assume that,

$$S = \sum\limits_{r=1}^{\infty} (v_r - v_{r+1})\, s_r$$

As $\{v_n\}$ is bounded there exists some $M_1 > 0$ such that, $|s_n| \leq M_1$ hence we obtain

$$|v_n\, s_n| \leq M_1\, |s_n| \tag{1.11.3.1}$$

Since $\{s_n\}$ is convergent it is cauchy, consequently by using (1.11.3.1) we conclude that $\{v_n\, s_n\}$ is also cauchy and hence convergent. Let $v_n\, s_n \to S'$ as $n \to \infty$. Now by taking limit $n \to \infty$ of both sides of equation (1.11.2.1) we obtain,

$$\lim_{n \to \infty} s'_n = S + S'$$

In other words $\sum\limits_{n=1}^{\infty} u_n v_n$ is convergent.

We now study convergence of a special type of series called alternating series which is defined below,

Definition   If $\{u_k\}$ be a sequence of positive real number then the series of the form $\sum\limits_{n=1}^{\infty} (-1)^{n-1} u_k$ is called an *alternating series*.

Theorem 1.11.4 (Leibniz Test for Alternating Series)

If $\{u_k\}$ is monotone decreasing sequence converging to 0, then the alternating series $\sum\limits_{n=1}^{\infty} (-1)^{n-1} u_k$ is convergent.

Proof: Let $s_n = \sum\limits_{i=1}^{n} (-1)^{i-1}$. Then,

$$s_n = \begin{cases} 0 & \text{if } n \text{ is even} \\ -1 & \text{if } n \text{ is odd} \end{cases}$$

So that $|s_n| \leq 1 \ \forall n$. Now using Dirichlet's theorem we conclude that $\sum\limits_{n=1}^{\infty} (-1)^{n-1} u_k$ is convergent.

Example 1.29   By above theorem $\sum\limits_{n=1}^{\infty} \dfrac{(-1)^{n+1}}{n}$ is convergent.

PROBLEMS

1. Using definition of convergence show that,

    (i) $\lim\limits_{n \to \infty} \dfrac{\sqrt{n}}{n+2} = 0$

    (ii) $\lim\limits_{n \to \infty} \dfrac{n^2}{n^3 + 1} = 0$

    (iii) $\lim\limits_{n \to \infty} \dfrac{n+1}{n} = 1$

    (iv) $\lim\limits_{n \to \infty} \dfrac{\cos n}{n} = 0$

2. Show that if $x_n \to 0$ as $n \to \infty$ then $|x_n| \to 0$ as $n \to \infty$. Is the converse true?

3. If $\{x_n\}$ is a convergent sequence then show that $\{|x_n|\}$ is also convergent. Also give an example to justify that the converse may not be true.

4. Suppose $x_n \to 0$ as $n \to \infty$ and $\{y_n\}$ be a bounded sequence then show that $x_n y_n \to 0$ as $n \to \infty$. Give an example to show that above result is not necessarily true if we do not assume $\{y_n\}$ a bounded sequence.

5. Let $\{x_n\}$ and $\{y_n\}$ be two sequences such that $z_n \to 0$ as $n \to \infty$ where $z_n = x_n - y_n$. If $\{x_n\}$ is convergent show that $\{y_n\}$ is also convergent.

6. If $x_n \to x$ and $y_n \to y$ then show that $z_n \to \max(x, y)$ where $z_n = \{x_n, y_n\}$.

7. Show that,

    (i) $\lim\limits_{n \to \infty} (a^n + b^n + c^n)^{\frac{1}{n}} = c$ where a, b, c are positive integers satisfying $c > b > a$

    (ii) $\lim\limits_{n \to \infty} \left(\dfrac{1}{\sqrt{n^2 + 1}} + \dfrac{1}{\sqrt{n^2 + 2}} + \dots + \dfrac{1}{\sqrt{n^2 + n}} \right) = 1$ 　　　[Hint: Use Sandwich Theorem]

8. Show that the sequence $\{x_n\}$ defined by,

$$x_{n+1} = \frac{6(1 + x_n)}{7 + x_n}$$

    is monotone decreasing or monotone increasing according as $x_1 \geq 2$ or $x_1 < 2$ and in either case $\lim\limits_{n \to \infty} x_n = 2$

9. Find cluster points of the following sequences

    (i) $\left\{ -1, 1, -\dfrac{1}{2}, \dfrac{1}{2}, -\dfrac{1}{3}, \dfrac{1}{3}, -\dfrac{1}{4}, \dfrac{1}{4} \dots \right\}$

    (ii) $\left\{ x_n = (-1) \cos\left(\dfrac{n\pi}{3} \right) \right\}$

10. Give an example of a sequence having infinite number of limit points

11. If ξ is a limit point of a sequence $\{x_n\}$ then show that there is a subsequence of the given sequence converging to ξ.

12. Show that a sequence $\{x_n\}$ converges if and only if $\overline{\lim}\, x_n = \underline{\lim}\, x_n$

13. If $\sum\limits_{n=1}^{\infty} u_n$ and $\sum\limits_{n=1}^{\infty} v_n$ be two convergent series then show that $\sum\limits_{n=1}^{\infty} (u_n + v_n)$ is convergent. Is the converse true?

14. Let $\sum\limits_{n=1}^{\infty} u_n$ be a convergent series of non-negative terms then show that the seris $\sum\limits_{n=1}^{\infty} v_n$ converges where $v_n = u_{n+k}$ for some fixed positive integer k.

15. If $\sum\limits_{n=1}^{\infty} u_n$ is a absolutely convergent and $\{v_n\}$ be a bounded sequence then show that $\sum\limits_{n=1}^{\infty} u_n v_n$ is absolutely convergent.

16. If $\sum\limits_{n=1}^{\infty} u_n$ is a convergent series of non-negative terms then show that both series $\sum\limits_{n=1}^{\infty} u_n^2$ and $\sum\limits_{n=1}^{\infty} \dfrac{u_n}{1+u_n}$ converges.

17. Test for convergence of the following series

   (i) $\sum\limits_{n=1}^{\infty} \dfrac{1}{(\log n)^n}$

   (ii) $\sum\limits_{n=1}^{\infty} \dfrac{\sqrt{n+1} - \sqrt{n-1}}{n}$

   (iii) $\sum\limits_{n=1}^{\infty} \dfrac{4n^2 - n + 3}{n^3 + 2n}$

   (iv) $\sum\limits_{n=1}^{\infty} \dfrac{n^4}{e^{n^2}}$

18. Test for convergence (absolute or conditional) of the following series

   (i) $\sum\limits_{n=1}^{\infty} (-1)^n \tan\left(\dfrac{1}{n}\right)$

   (ii) $\sum\limits_{n=1}^{\infty} (-1)^{n+1} \dfrac{\log(n+1)}{(n+1)^2}$

   (iii) $\sum\limits_{n=1}^{\infty} \dfrac{(-1)^{n-1}}{x^2 + n}$

   (iv) $\sum\limits_{n=1}^{\infty} (-1)^{n+1} \dfrac{n^2}{(n+1)!}$

Functions of One Variable

2.1 INTRODUCTION

In the last chapter we have studied sequences which are functions defined on a particular subset of real numbers called *natural numbers*. In this chapter we will study functions with domain as arbitrary subset of real numbers. The central point in study of sequences was the concept of convergence that is the behaviour of the function x_n when the value of n is large when we consider the sequence as a function of independent variable n. For functions defined on an arbitrary subset of real numbers we will begin our of study by introducing similar kind of concept called limit of functions which examines the behavior of the function as the independent variable approaches any real number or $\pm\infty$. With the concept of limit of function being the focal point, we study as well other important properties of such functions like continuity, uniform continuity and differentiability in this chapter.

2.2 LIMIT OF FUNCTIONS

Let us begin with two examples.

Example 2.1 Consider the function $f : [-3, 3] \to \mathbb{R}$ defined as,

$$f(x) = x^2$$

Now as we approach $x = 2$ from any side (that is if the independent variable x takes successive values which increases or decreases as the case may be to the real number 2) we see that the values of $f(x)$ increases or decreases along the curve $y = x^2$ and ultimately '***assumes***' the value 4 (Fig. 2)

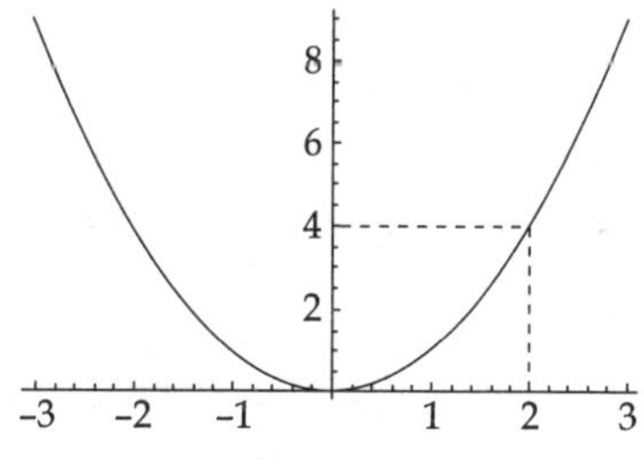

Fig. 2.1

and observe that $f(2) = 4$.

Example 2.2   Next consider the function $f : \mathbb{R} \to \mathbb{R}$ defined as follows,

$$f(x) = \frac{x^2 - 4}{x - 2}$$

Clearly the function is defined every where except at the point $x = 2$. In fact,

$$f(x) = x + 2, \, x \neq 2$$

Again as we approach $x = 2$ in this case we see that the values of $f(x)$ increases or decreases

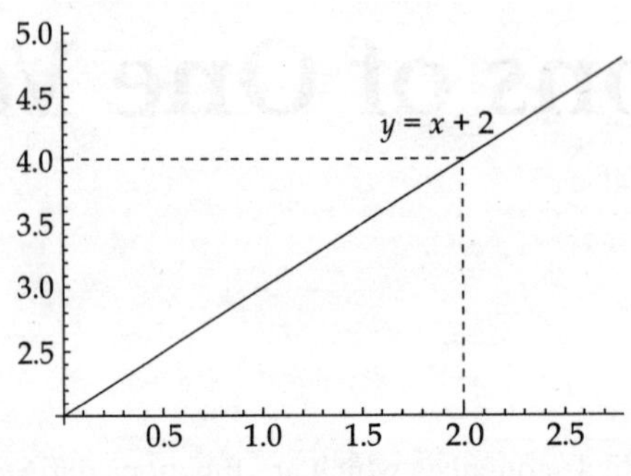

Fig. 2.2

(As the case may be) along the line $y = x + 2$ and is very *'close'* to value 4 (Fig. 1). But observe that in this case $f(2) \neq 4$ (as $f(2)$ does not exist).

In both examples we see that as the independent variable x is very near to the value 2, $f(x)$ is close to 4 which is the value of the function in the first case but not in the second case. In both the cases we say that $f(x)$ approaches 4 as x approaches 2 or limit of $f(x)$ as x tends to 2 is 4. We formally give the following definition:

Definition   Let D be a subset of $\mathbb{R}$ and f be a real valued function defined on D. Suppose $a \in D$ such that there is an open interval $I \subset D$ containing a, then we write,

$$f(x) \to l \quad \text{as} \quad x \to a \quad \text{or} \quad \lim_{x \to a} f(x) = l$$

if there exists a real number l having the following property: For any arbitrary positive number ε no matter how small there exists a $\delta > 0$ such that,

$$|f(x) - l| < \varepsilon$$

for all $x \in D$ satisfying $0 < |x - a| < \delta$.

The above definition which is some time referred as $\varepsilon - \delta$ form tells us that if $f(x) \to l$ as $x \to a$ then we can find a deleted or punctured neighbourhood of a such that for all x in that neighbourhood the absolute difference between values of $f(x)$ and l is arbitrarily small.

In the following examples we exhibit how to use $(\varepsilon - \delta)$ definition

Example 2.3   Show that $\lim_{x \to 2} f(x) = 4$ where f the function as defined in Example 2.1.

Sol: For $x \in [-3, 3]$ we have,

$$|f(x) - 4| = |x^2 - 4| = |x + 2| \, |x - 2|$$
$$\leq |x - 2| \, (|x| + |2|)$$
$$\leq 5|x - 2| \quad (\text{as } |x| \leq 3)$$

Now for any arbitrary $\varepsilon > 0$, $|f(x) - 4| < \varepsilon$ if $5|x - 2| < \varepsilon$ that is if $|x - 2| < \dfrac{\varepsilon}{5}$. Thus if we choose $\delta = \dfrac{\varepsilon}{5}$ we obtain, $|f(x) - 4| < \varepsilon$ whenever $|x - 2| < \delta$. Hence by above definition we write $\lim\limits_{x \to 2} f(x) = 4$

Example 2.4 Show that $\lim\limits_{x \to 2} \dfrac{x^2 - 4}{x - 2} = 4$

Sol: Let $f(x) = \dfrac{x^2 - 4}{x - 2}$, as $x \neq 2$ we get,

$$|f(x) - 4| = \left| \frac{x^2 - 4}{x - 2} - 4 \right|$$
$$= |x + 2 - 4|$$
$$= |x - 2|$$

Hence for any $\varepsilon > 0$ $|f(x) - 4| < \varepsilon$ if $|x - 2| < \varepsilon$

Thus in this case we see that $\delta = \varepsilon$ satisfies the condition of the above definition so that,

$$\lim_{x \to 2} f(x) = 4$$

Next theorem will provide us an equivalent definition of limit of a function which is sometimes referred as *sequential form* (of definition).

Theorem 2.2.1

Let D be a subset of $\mathbb{R}$ and f be a real valued function defined on D. Suppose $a \in D$ be a point such that there exists is a deleted neighbourhood I of with $I \subset D$ then $\lim\limits_{x \to a} f(x) = l$ if and only if for any sequence $\{a_n\}$ in $D - \{a\}$ with $a_n \to a$ implies $f(x_n) \to l$.

Proof: Suppose $\lim\limits_{x \to a} f(x) = l$ then for $\varepsilon > 0$ there exists a $\delta > 0$ such that,

$$|f(x) - l| < \varepsilon \ \forall \ x \text{ satisfying } 0 < |x - a| < \delta \tag{2.2.1.1}$$

If $\{a_n\}$ is a sequence in $D - \{a\}$ with $a_n \to a$ then corresponding to $\delta > 0$ as obtained above there exists a positive integer $n_0 (\varepsilon)$ such that,

$$0 < |a_n - a| < \delta \ \forall \ n \geq n_0 \tag{2.2.1.2}$$

From (2.1.1.1) & (2.1.1.2) we conclude that,

$$|f(a_n) - l| < \varepsilon, \ \forall \ n \geq n_0$$

In other words $\lim\limits_{n \to \infty} f(a_n) \to l$. This proves 'if' part.

To prove the only if part we use the method of contradiction. Suppose that for any sequence $\{a_n\}$ in $D - \{a\}$ with $a_n \to a$ implies $f(x_n) \to l$ assume if possible l is not the limit of $f(x)$ as $x \to a$.

This means that there exists $\varepsilon > 0$ such that for every $\delta > 0$ we can find some x satisfying $0 < |x - a| < \delta$ but $|f(x) - l| \geq \varepsilon$. Now for each natural number n we chose a sequence $\{a_n\} \in D - \{a\}$ such that,

$$a_n = a + \frac{1}{n} \quad n = 1, 2, 3 \ldots$$

Then see that, $0 < |a_n - a| < \dfrac{1}{n}$ that is $\lim\limits_{n \to \infty} a_n = a$ whereas $|f(a_n) - l| \geq \varepsilon$ that is $f(a_n)$ does not tend to l as $n \to \infty$ which a contradiction to the given condition. Hence we conclude that

$$\lim_{x \to a} f(x) \to l$$

Next we want to introduce the concept of one sided limits for that consider the function $f : [-2, 2] \to \mathbb{R}$ defined as,

$$f(x) = \quad -x, \ -2 \le x \le 0$$
$$1 - x, \ \ 0 < x \le -2$$

We observe that if x approaches 0 from the left then the values of $f(x)$ decreases along the

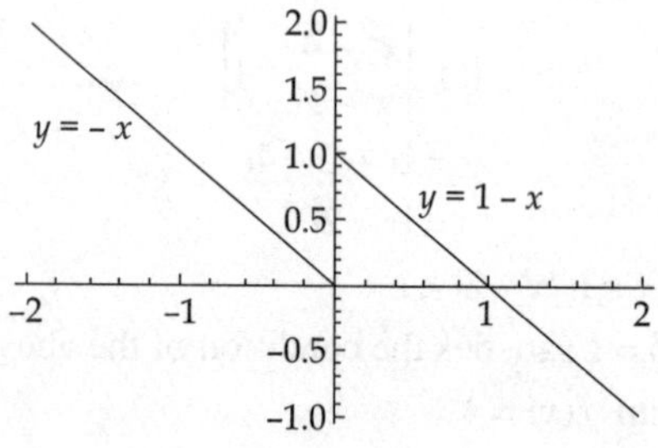

Fig. 2.3

the line $y = -x$ and ultimately **assumes** the value 0 whereas if x approaches 0 from the right then the values of $f(x)$ increases along the line $y = 1 - x$ and come very close to 1. Above example shows that the values of the function f may get 'close' to two different values say l_1 & l_2 as the independent variable x approaches the real number 'a' from two different directions, left and right that is through values which are less or more than 'a' respectively. This gives rise to the concept of what are called right and left hand limits we define formally below.

Definition   Let D be a subset of $\mathbb{R}$ and f be a real valued function defined on D. For $a \in D$ we say that $l_1 \in R$ is the left hand limit of f as x approaches a from left and write,

$$\lim_{x \to a^-} f(x) = l_1$$

if for any $\varepsilon > 0$ there exists some positive number $\delta > 0$ such that whenever x satisfies $a - \delta < x < a$ we have $|f(x) - l_1| < \varepsilon$.

We say that $l_2 \in \mathbb{R}$ is the right hand limit of f as x approaches a from right and write

$$\lim_{x \to a^+} f(x) = l_2$$

if for any $\varepsilon > 0$ there exists some positive number $\delta > 0$ such that whenever x satisfies $a < x < a + \delta$ we have $|f(x) - l_1| < \varepsilon$.

Remark

If $l_1 \ne l_2$ we say that $\lim_{x \to a} f(x)$ does not exist and in case $l_1 = l_2 = l$ it is easy to see that $\lim_{x \to a} f(x) = l$.

Example 2.5   Show that $\lim_{x \to a^+} f(x) = l$ if and only if $\lim_{h \to 0^+} f(a + h) = l$.

Sol: Suppose $\lim_{h \to 0^+} f(a + h) = l$ then for $\varepsilon > 0$ there exist $\delta > 0$ such that whenever $0 < h < \delta$

we have $|f(a + h) - l| < \varepsilon$. Now suppose $x \to a^+$ then $u \to 0^+$ where $u = x - a$. Hence whenever $0 < u < \delta$ we have $|f(a + u) - l| < \varepsilon$ that is whenever $a < x < a + \delta$ we have $|f(x) - l| < \varepsilon$. Thus, $\lim_{x \to a^+} f(x) = l$. Reversing the argument we can show that if $\lim_{x \to a^+} f(x) = l$ then $\lim_{h \to 0^+} f(a + h) = l$.

Similarly one can show $\lim_{x \to a^-} f(x) = l$. In fact $\lim_{x \to a} f(x)$ exists if and only if $\lim_{h \to 0} f(a + h)$

Example 2.6 Show that for the function $f: [-2, 2] \to \mathbb{R}$ defined as,

$$f(x) = \quad -x, -2 \le x \le 0$$
$$1 - x, \quad 0 < x \le -2$$

$$\lim_{x \to 0^-} f(x) = 0 \ \& \ \lim_{x \to 0^+} f(x) = 1$$

Sol: Observe that if we approach 0 from the left then the values of x is always negative and close to 0 so that in this case,

$$|f(x) - 0| = |-x| = |x|$$

Hence for an arbitrary real number $\varepsilon > 0$ if we choose $\delta = \varepsilon$ then whenever x satisfies $0 - \delta < x < 0$ we have $|x| < \varepsilon$ and hence from above we conclude,

$$|f(x)| < \varepsilon$$

that is $\lim_{x \to 0^-} f(x) = 0$.

Now if we approach 0 from the right then the values of x is always positive and close to 0 hence in this case we have,

$$|f(x) - 1| = |1 - x - 1| = |-x| = |x|$$

Again as before for $\varepsilon > 0$ we choose $\delta = \varepsilon$ then whenever x satisfies $0 < x < \delta$ we have $|x| < \varepsilon$ and hence,

$$|f(x) - 1| < \varepsilon$$

In other words, $\lim_{x \to 0^+} f(x) = 1$

Next we want to know the behavior of the function when the independent variable approach infinity (positive or negative) or also what do we mean by a function possessing infinite limit. These queries lead us to the concept of what are known as ***limit at infinity*** and ***infinite limits*** respectively as defined and illustrated below.

Definition (Limits at infinity) Let D be a subset of $\mathbb{R}$ and f be a real valued function defined on D. For a real number l we write,

$$f(x) \to l \text{ as } x \to \infty \text{ or } \lim_{x \to \infty} f(x) = l$$

if given any $\varepsilon > 0$, there exists some $M > 0$ such that for all x satisfying $x > M$ we must have,

$$|f(x) - l| < \varepsilon$$

Similarly, for a real number m we write $f(x) \to m$ as $x \to -\infty$ or

$$\lim_{x \to -\infty} f(x) = m$$

if given any $\varepsilon > 0$, there exists some $m < 0$ such that for all x satisfying $x < m$ we must have,

$$|f(x) - m| < \varepsilon$$

Example 2.7 Show that $\lim_{x \to \infty} \left(\dfrac{1}{x^2} \right) = 0 = \lim_{x \to \infty} \left(\dfrac{1}{x^2} \right)$

Sol: Let be any arbitrary number then,

$$\left(\frac{1}{x^2} \right) - 0 < \varepsilon \quad \text{if } x^2 > \frac{1}{\varepsilon}$$

that is either $x > \dfrac{1}{\sqrt{\varepsilon}}$ or $x - \dfrac{1}{\sqrt{\varepsilon}}$

Hence if we choose $M = \dfrac{1}{\sqrt{\varepsilon}}$ and $m = -\dfrac{1}{\sqrt{\varepsilon}}$ it is clear from the definition that

$$\lim_{x \to -\infty} \left(\frac{1}{x^2} \right) = 0 \; \& \; \lim_{x \to -\infty} \left(\frac{1}{x^2} \right) = 0$$

Definition   (Infinite Limits) Let D be a subset of $\mathbb{R}$ and f be a real valued function defined on D. For $a \in D$ we write,

$$f(x) \to \infty \text{ as } x \to a^- \text{ or } \lim_{x \to a^-} f(x) = \infty$$

if given any $M \in \mathbb{R}$ there exists $\delta > 0$ such that for all x satisfying $a - \delta < x < a$ we have,

$$f(x) > M$$

Similarly we write $f(x) \to \infty$ as $x \to a^+$ or

$$\lim_{x \to a^+} f(x) = \infty$$

if given any arbitrarily large $M > 0$ there exists $\delta > 0$ such that for all x satisfying $a < x < a + \delta$ we have,

$$f(x) > M$$

Again we write $\lim\limits_{x \to a^-} f(x) = -\infty$ if for any arbitrary large positive number M we can find a $\delta > 0$ such that for all x satisfying $a - \delta < x < a$ we have,

$$f(x) < -M$$

Analogously we can define $\lim\limits_{x \to a^+} f(x) = -\infty$.

Example 2.8   Show that $\lim\limits_{x \to 0^-} \dfrac{1}{x} = -\infty$ and $\lim\limits_{x \to 0^+} \dfrac{1}{x} = \infty$.

Sol: Let $M > 0$ be any arbitrary large number. Then for $x < 0$ we have,

$$\frac{1}{x} < -M \text{ if } x > -\frac{1}{M}$$

Hence if we choose $\delta = \dfrac{1}{M}$ we obtain for x satisfying $-\delta < x < 0$ we get $\dfrac{1}{x} < -M$ or in other words $\lim\limits_{x \to 0^-} \dfrac{1}{x} = -\infty$

Similarly we can show that $\lim\limits_{x \to 0^+} \dfrac{1}{x} = \infty$

Note that this example show $\lim\limits_{x \to 0} \dfrac{1}{x}$ does not exist.

Example 2.9   Let f be a real valued function defined on $D \subseteq \mathbb{R}$ and suppose $a \in D$ be a point such that there exists is a deleted neighbourhood I of a with $I \subset D$. Show that $\lim\limits_{x \to a^+} f(x) = \infty$ if and only if for every sequence $\{a_n\} \in D - \{a\}$ satisfying $a_n > a \; \forall \, n \; \& \; a_n \to a$ we have $f(a_n) \to \infty$ as $n \to \infty$.

Sol: Suppose $\lim\limits_{x \to a} f(x) = \infty$ then for any arbitrary large number $G > 0$ there exists $\delta > 0$ such that $f(x) > G$ whenever $a < x < a + \delta$. Let $\{a_n\} \in D - \{a\}$ satisfying $a_n > a \; \forall \, n \; \& \; a_n \to a$. Then corresponding to above $\delta > 0$ we will obtain a positive integer n_0 such that,

$$a < a_n < a + \delta \; \forall \, n \geq n_0$$

hence $f(a_n) > G \; \forall \, n \geq n_0$. In other words $f(a_n) \to \infty$ as $n \to \infty$.

For the converse part let us assume if possible $f(x)$ does not tend to ∞ as $x \to a^+$. This means that there exists some $G > 0$ such that for every $\delta > 0$ $f(x) \leq G$ for at least one x satisfying $a < x < a + \delta$. Now if we choose $\delta_1 = 1$ there exists $a_1 \in D - \{a\}$ satisfying $a < a_1 < a + 1$ for which $f(a_1) \leq G$. Next we choose $\delta_2 = \min \left\{ x_1 - a, \dfrac{1}{2} \right\}$ then there exists $a_2 \in D - \{a\}$ satisfying $a < a_2 < a + \delta_2$ for which $f(a_2) \leq G$. Observe that $a_1 \neq a_2$ and $a < a_2 < a + \dfrac{1}{2}$

Continuing this way we will obtain a sequence $\{a_n\} \in D - \{a\}$ satisfying $a_n > a \ \forall \ n, a < x < a + \dfrac{1}{n}$ and $f(a_n) \leq G$. In other words we have got a sequence $\{a_n\} \in D - \{a\}$ satisfying $a_n > a \ \forall \ n$ & $a_n \to a$ but $\{f(a_n)\}$ does not tend to ∞ as $n \to \infty$. This contradicts the given condition so we conclude that $\lim\limits_{x \to a^+} f(x) = \infty$ in this case.

Next theorem is what are called algebra of limits and are very similar to what we have seen in case of sequences.

Theorem 2.2.2

Let f, g be two functions defined on $D \subseteq \mathbb{R}$ and suppose $a \in D$ be a point such that there exists is a deleted neighbourhood I of with $I \subset D$. If $\lim\limits_{x \to a} f(x) = l$ & $\lim\limits_{x \to a} g(x) = m$ then,

(a) $\lim\limits_{x \to a} [f(x) \pm g(x)] = l \pm m$

(b) $\lim\limits_{x \to a} [f(x) \, g(x)] = lm$

(c) $\lim\limits_{x \to a} \left[\dfrac{1}{f(x)} \right] = \dfrac{1}{l}$ when, $l \neq 0$

Proof: Let $\{a_n\}$ be any sequence in $D - \{a\}$ such that $a_n \to a$ as $n \to \infty$ then using necessary part of Theorem 1.2.1 we get,

$$\lim_{n \to \infty} f(a_n) = l \ \& \ \lim_{n \to \infty} g(a_n) = m$$

Now by Theorem 1.3.6 (ii) & (iii) we get

$$\lim_{n \to \infty} [(f \pm g) \, (a_n)] = l \pm m \ \& \ \lim_{x \to \infty} [(fg) \, (a_n)] = lm \qquad (2.2.2.1)$$

Again by using (1.2.2.1) and sufficient condition of Theorem 1.2.1 we get

$$\lim_{x \to a} [(f \pm g) \, (x)] = \lim_{x \to a} [f(x) \pm g(x)] = l \pm m \quad \text{and}$$

$$\lim_{x \to a} [(fg) \, (x)] = \lim_{x \to a} [f(x) \, g(x)] = lm$$

This proves (a) and (b)

(c) As $l \neq 0$ for $\varepsilon = \dfrac{|l|}{2}$ there exists a $\delta_1 > 0$ such that for x satisfying $0 < |x - a| < \delta$ we have,

$$|f(x) - l| < \frac{|l|}{2}$$

Now,

$$|f(x)| = |f(x) - l + l| \geq |l| - |f(x) - l|$$

$$> |l| - \frac{|l|}{2} = \frac{|l|}{2} \qquad (2.2.2.2)$$

Again given any $\varepsilon > 0$ there exists $\delta_2 > 0$ such that for x satisfying $0 < |x - a| < \delta_2$ we have,

$$|f(x) - l| < \frac{|l|\,\varepsilon}{2} \qquad\qquad (2.2.2.3)$$

Now see that for x satisfying $0 < |x - a| < \delta$ where $\delta = \min(\delta_1, \delta_2)$ we have,

$$\left|\frac{1}{f(x)} - \frac{1}{l}\right| = \frac{|f(x) - l|}{|f(x)|}$$

$$< \frac{2|l|\,\varepsilon}{|l|\,2} = \varepsilon \quad \text{Using (1.2.2.2) \& (1.2.2.3)}$$

That is $\lim\limits_{x \to a} \dfrac{1}{f(x)} = \dfrac{l}{1}$. This proves (c)

We work out some examples illustrating the use of algebra of limits

Example 2.10   Let $p(x)$ be a polynomial then show that for any real number a

$$\lim_{x \to a} p(x) = p(a)$$

Further if f be any function such that $\lim\limits_{x \to a} f(x) = b$ then show that

$$\lim_{x \to a} p(f(x)) = p\left(\lim_{x \to a} f(x)\right) = p(b)$$

Sol: First that if we write $\phi(x) = x$ then $\lim\limits_{x \to a} \phi(x) = \lim\limits_{x \to a} x = a$ Now for any positive integer m we have,

$$\lim_{x \to a} x^m = \lim_{x \to a} (\phi(x))^m = \lim_{x \to a} \underbrace{[\phi(x) \cdot \phi(x) \ldots = \phi(x)]}_{m}$$

$$= \underbrace{\lim_{x \to a} \phi(x) \cdot \lim_{x \to a} \phi(x) \ldots \lim_{x \to a} \phi(x)}_{m} \quad \text{(by algebra of limits)}$$

$$= a^m$$

Now if $p(x) = a_0 + a_1 x + a_2 x^2 + \ldots + a_n x^n$ then by using algebra of limits we get,

$$\lim_{x \to a} p(x) = \lim_{x \to a} (a_0 + a_1 x + a_2 x^2 + \ldots + a_n x^n)$$

$$= a_0 + a_1 \lim_{x \to a} x + a_2 \lim_{x \to a} x^2 + \ldots + a_n \lim_{x \to a} x^n$$

$$= a_0 + a_1 a + a_2 a^2 + \ldots a_n a^n$$

$$= p(a)$$

Next see that for any positive integer m

$$\lim_{x \to a} (f(x))^m = \underbrace{\lim_{x \to a} f(x) \cdot \lim_{x \to a} (f(x)) \ldots \lim_{x \to a} (f(x))}_{m}$$

$$= l^m$$

Now $\qquad p(f(x)) = a_0 + a_1 (f(x)) + a_2 (f(x))^2 + \ldots + a_n (f(x))^n$ hence,

$$\lim_{x \to a} p(f(x)) = \lim_{x \to a} (a_0 + a_1(f(x)) + a_2(f(x))^2 + \ldots + a_n(f(x))^n)$$

$$= a_0 + a_1 \lim_{x \to a} (f(x)) + a_2 \lim_{x \to a} (f(x))^2 + \ldots + a_n \lim_{x \to a} (f(x))^n$$

$$= a_0 + a_1 b + a_2 b^2 + \dots + a_n b^n$$

$$= p(b) = p\left(\lim_{x \to a} f(x)\right)$$

Example 2.11 Evaluate $\lim\limits_{x \to 1} \dfrac{x^{10} + x^3 - 3x}{4x^2 - 3x + 1}$

Sol: Observe we are to find limit of a function which of the form $\dfrac{p(x)}{q(x)}$ where $p(x)$ & $q(x)$ are polynomials

we know from above example that $\lim\limits_{x \to 1} p(x) = p(1) = -1$ and $\lim\limits_{x \to 1} q(x) = q(1) = 2$. Now as $\lim\limits_{x \to 1} q(x) = 2 \neq 0$

we have by using algebra of limits,

$$\lim_{x \to 1} \frac{x^{10} + x^3 - 3x}{4x^2 - 3x + 1} = \frac{\lim\limits_{x \to 1} (x^{10} + x^3 - 3)}{\lim\limits_{x \to 1} (4x^2 - 3x + 1)} = \frac{-1}{2} = -\frac{1}{2}$$

Example 2.12 Evaluate $\lim\limits_{x \to \infty} \dfrac{4x^8 - 2x^6 + 3}{5x^9 - 2}$

Sol: As in previous example,

$$\lim_{x \to \infty} \frac{4x^8 - 2x^6 + 3}{5x^9 - 2} = \lim_{x \to \infty} \frac{\dfrac{4}{x} - \dfrac{2}{x^3} + \dfrac{3}{x^9}}{5 - \dfrac{2}{x^9}} \quad \text{(by dividing numerator and denominator by } x^9)$$

$$= \frac{\lim\limits_{x \to \infty} \left(\dfrac{4}{x} - \dfrac{2}{x^3} + \dfrac{3}{x^9}\right)}{\lim\limits_{x \to \infty} \left(5 - \dfrac{2}{x^9}\right)}$$

$$= \frac{0 + 0 + 0}{5 - 0} = 0$$

Note that it is easy to prove that $\lim\limits_{x \to \infty} \dfrac{1}{x^n} = 0$ for any positive integer n.

Next theorem tells us that the sign of values of $f(x)$ is not different from that of l in a sufficiently small neighbourhood of a when $f(x) \to l (\neq 0)$ as $x \to a$.

Theorem 2.2.3

Let f be a real valued function defined on $D \subseteq \mathbb{R}$ and suppose $a \in D$ be a point such that there exists is a deleted neighbourhood I of a with $I \subset D$. If $\lim\limits_{x \to a} f(x) = l \, (\neq 0)$ then exists $\delta > 0$ such that for all x satisfying $0 < |x - a| < \delta$, $f(x)$, $f(x)$ retains same sign of l.

Proof: Suppose $l > 0$ then corresponding to $\varepsilon = \dfrac{l}{2}$ there exists a $\delta > 0$ such that for all x satisfying $0 < |x - a| < \delta$ we have,

$$|f(x) - l| < \frac{l}{2}$$

or,
$$\frac{l}{2} < f(x) < \frac{3l}{2}$$

As $f(x)$ lies between two positive numbers we get $f(x) > 0$ when $0 < |x - a| < \delta$.

Similarly if $l < 0$ one can argue that $f(x) < 0$ when x satisfies $0 < |x - a| < \delta$ for some $\delta > 0$.

Following theorem is an analogue of Theorem 1.3.9

Theorem 2.2.4 (Sandwich Theorem)

Let f, g & h be real valued function defined on $D \subseteq \mathbb{R}$ satisfying,

$$f(x) \le g(x) \le h(x)\, x \in D$$

and suppose $a \in D$ be a point such that there exists is a deleted neighbourhood I of a with $I \subset D$. If $\lim\limits_{x \to a} f(x) = \lim\limits_{x \to a} h(x) = l$ then $\lim\limits_{x \to a} g(x) = l$.

Proof: Let $\{a_n\}$ be a sequence in $D - \{a\}$ such that $a_n \to a$ as $n \to \infty$ then by Theorem 2.2.1 we get, $\lim\limits_{n \to \infty} f(a_n) = \lim\limits_{n \to \infty} h(a_n) = l$. Now by give condition,

$$f(a_n) \le g(a_n) \le h(a_n), \ \forall n$$

Hence using Sandwich Theorem for sequences we get

$$\lim_{n \to \infty} h(a_n) = l$$

Again using Theorem 2.2.1 we conclude $\lim\limits_{x \to a} h(x) = l$.

This proves the theorem.

Next theorem gives a necessary and sufficient condition for the existence for limit of a function called Cauchy Criteria.

Theorem 2.2.5

Let f be a real valued function defined on $D \subseteq \mathbb{R}$ and suppose $a \in D$ be a point such that there exists is a deleted neighbourhood I of with $I \subset D$. Then $\lim\limits_{x \to a} f(x)$ exists if and only if given $\varepsilon > 0$ there exists $\delta(\varepsilon)$ such that,

$$|f(x) - f(y)| < \varepsilon$$

for all x, y satisfying $0 < |x - a| < \delta$ & $0 < |y - a| < \delta$

Proof: First we prove that the condition is necessary. For that suppose $\lim\limits_{x \to a} f(x) = l$. Then by $\varepsilon - \delta$ definition of limit given any $\varepsilon > 0$ we will get some $\delta(\varepsilon) > 0$ such that,

$$|f(x) - l| < \frac{\varepsilon}{2} \ \& \ |f(y) - l| < \frac{\varepsilon}{2} \quad \text{whenever } x \ \& \ y \text{ satisfies } 0 < |x - a| < \delta \ \& \ 0 < |y - a| < \delta \qquad (2.2.5.1)$$

Hence for x & y satisfying $0 < |x - a| < \delta$ & $0 < |y - a| < \delta$ we have,

$$|f(x) - l + l - f(y)| < |f(x) - l| + |f(y) - l|$$

$$< \frac{\varepsilon}{2} + \frac{\varepsilon}{2} = \varepsilon$$

This proves that the condition is necessary

To prove the sufficiency part we assume the condition that for any $\varepsilon > 0$ there exists $\delta(\varepsilon)$ such that,

$$|f(x) - f(y)| < \frac{\varepsilon}{3} \qquad (2.2.5.2)$$

for all x, y satisfying $0 < |x - a| < \delta$ & $0 < |y - a| < \delta$.

Now let $\{a_n\}$ a sequence in $D - \{a\}$ such that $a_n \to a$. Then corresponding to above $\delta > 0$ there exists a positive integer n_0 such that,

$$0 < |a_n - a| < \delta \;\forall\; n \geq n_0 \tag{2.2.5.3}$$

Hence using (2.2.5.2) & (2.2.5.3) we can write,

$$|f(a_n) - f(a_m)| < \frac{\varepsilon}{3} \;\forall\; n, m \geq n_0 \tag{2.2.5.4}$$

That is $\{f(a_n)\}$ is a Cauchy sequence hence by Theorem 1.6.1 there exists $l_1 \in \mathbb{R}$ such that,

$$\lim_{n \to \infty} f(a_n) = l_1$$

Again if $\{b_n\}$ a sequence in $D - \{a\}$ such that $b_n \to a$ then by similar argument as given above we can show $\{f(b_n)\}$ is a Cauchy sequence hence so that there exists $l_2 \in \mathbb{R}$ such that

$$\lim_{n \to \infty} f(b_n) = l_2$$

We will prove $l_1 = l_2$. We observe that for $\varepsilon > 0$ we can choose a positive integer N large enough so that for $m, n > N$ we have,

$$0 < |a_n - a| < \delta, \; 0 < |b_m - a| < \delta, \; |f(a_n) - l_1| < \frac{\varepsilon}{3} \;\&\; |f(b_n) - l_2| < \frac{\varepsilon}{3}. \tag{2.2.5.5}$$

Now see that if we choose $m, n > N$,

$$|l_1 - l_2| = |l_1 - f(a_n) + f(b_m) - l_2 + f(a_n) - f(b_m)|$$

$$< |f(a_n) - l_1| + |f(b_n) - l_2| + |f(a_n) - f(b_m)|$$

$$< \frac{\varepsilon}{3} + \frac{\varepsilon}{3} + \frac{\varepsilon}{3} - \varepsilon \quad \text{(Using (2.2.5.4) & (2.2.5.5))}$$

That is $l_1 = l_2 = l$ (say)

Hence we conclude from above that if $\{a_n\}$ is any arbitrary sequence in $D - \{a\}$ such that $a_n \to a$ then $f\{(a_n)\}$ converges to l which proves the existence of $\lim_{x \to a} f(x)$.

Example 2.13 Using Cauchy Criteria show that $\lim_{x \to 0} \frac{1}{x}$ does not exist.

Sol: Let $f(x) = \frac{1}{x}$ and $\varepsilon = 1$. Now given any $\delta > 0$ we can choose a positive integer $n > 1$ such that

$$-\delta < x_2 = -\frac{1}{n} < 0 < x_1 = \frac{1}{n} < \delta$$

but see that

$$|f(x_1) - f(x_2)| = |n + n| = 2n > 1 = \varepsilon$$

Hence by Cauchy Criteria we conclude that $\lim_{x \to 0} \frac{1}{x}$ does not exist.

2.3 CONTINUITY

We refer to the function $f : [-2, 2] \to \mathbb{R}$ defined as,

$$f(x) = \begin{cases} -x, & 2 \leq x \leq 0 \\ 1 - x, & 0 < x \leq -2 \end{cases}$$

If we look at the graph of the above function (Fig. 3) we find that at the point 0 there is a 'break' on the curve or we say that the 'continuity' of the curve is lost at the point 0. In this case this has happened due to the basic fact that the right hand and left hand limit at the point 0 are unequal. So what do we need so that the continuity a function at a point is not lost? The example above suggests that at least the limit of the function for the given point must exist. We see that it is also quite natural to demand that the value of this limit must be equal to the functional value at that point. We finally give the $\varepsilon - \delta$ definition of continuity of a function at a point.

Definition   Let D be a subset of $\mathbb{R}$ and suppose f be a real valued function defined on D. We say that f is **continuous at a point** $a \in D$ if given any $\varepsilon > 0$ there exists a $\delta > 0$ (in general δ depends both on ε and a) such that,

$$|f(x) - f(a)| < \varepsilon \quad \text{for all } x \text{ in } D \text{ satisfying } |x - a| < \delta.$$

We say a function is **discontinuous** at $a \in D$ if it is not continuous at that point.

Definition   A real valued function f is said to be continuous on the set $D \subseteq \mathbb{R}$ if it is continuous at every point of D.

By comparing the definition of limit of a function at a point with the above definition we conclude that if a function is continuous at a point $a \in D$ if

$$\lim_{x \to a} f(x) = f(a)$$

Which we can write as,

$$\lim_{x \to a} f(x) = f(\lim_{x \to a} x)$$

This tells us that if a function is continuous at a point then we can interchange the symbols lim and f or in other words we have limit of the function is equal to function of limit at that point.

Example 2.14   Show that any constant function is continuous

Sol: Let c be any real number and define a function f by,

$$f(x) = c \; x \in \mathbb{R}$$

As for any $\varepsilon > 0$ $|f(x) - f(y)| = |c - c| = 0 < \varepsilon \; \forall \; x, y \in \mathbb{R}$ trivially f is continuous at every real point.

Example 2.14   Show the function defined by,

$$f(x) = |x| \; x \in \mathbb{R}$$

is continuous everywhere.

Sol: Let a_0 be any real number then,

$$|f(x) - f(a_0)| = ||x| - a_0||$$
$$< |x - a|$$

Hence for any $\varepsilon > 0$ if we choose $\varepsilon = \delta$ we have

$$|f(x) - f(a_0)| < \varepsilon \quad \text{whenever } |x - a| < \delta.$$

Example 2.15   Show that the function defined by,

$$f(x) = \begin{cases} x \sin\left(\dfrac{1}{x}\right) & \text{if } x \neq 0 \\ 0 & \text{if } x = 0 \end{cases}$$

is continuous at $x = 0$.

Sol: Let $\varepsilon > 0$ be any real number. We see that,

$$|f(x) - f(0)| = \left| x \sin\left(\frac{1}{x}\right) \right|$$

$$\leq |x|$$

Hence $|f(x) - f(0)| < \varepsilon$ whenever $|x| < \varepsilon$. That is the definition of continuity is satisfied with $\delta = \varepsilon$. So we conclude that f is continuous at $x = 0$.

Example 2.16 Show that $f(x) = \dfrac{1}{x}$ is continuous in $\mathbb{R} - \{0\}$

Sol: Let $x_0 > 0$ be any positive real number we show that f is continuous at x_0. For that choose $\varepsilon > 0$. Now,

$$|f(x) - f(x_0) < \varepsilon$$

$\Rightarrow$
$$\frac{1}{x_0} - \varepsilon < \frac{1}{x} < \varepsilon + \frac{1}{x_0}$$

or
$$\frac{x_0}{1 + \varepsilon x_0} < x < \frac{x_0}{1 - \varepsilon x_0}$$

or,
$$\frac{-\varepsilon x_0^2}{1 + \varepsilon x_0} < x - x_0 < \frac{\varepsilon x_0^2}{1 - \varepsilon x_0}$$

or
$$\frac{-\varepsilon x_0^2}{1 - \varepsilon x_0} < x - x_0 < \frac{\varepsilon x_0^2}{1 - \varepsilon x_0} \qquad \left(\text{since } \frac{-\varepsilon x_0^2}{1 + \varepsilon x_0} > \frac{-\varepsilon x_0^2}{1 - \varepsilon x_0} \text{ as } \varepsilon, x_0 > 0 \right)$$

Now hence from above we get,

$$|f(x) - f(x_0)| < \varepsilon \text{ whenever } |x - x_0| < \delta \text{ where } \delta = \frac{\varepsilon x_0^2}{1 - \varepsilon x_0}$$

This shows that f is continuous at x_0. As $x_0 > 0$ is arbitrary we conclude f is continuous at all positive real numbers. By similar arguments we can show f is continuous at all negative real numbers.

We recall that if $\lim\limits_{x \to a} f(x) = l \neq 0$ then f retains the same sign of l in a certain deleted neighbourhood of a. Now if a function is continuous at $x = a$ we know $\lim\limits_{x \to a} f(x) = f(a)$ hence if we assume $f(a) \neq 0$ we expect that f should retain the same sign of $f(a)$ in certain neighbourhood of a. In fact we have the following theorem which is called the ***sign preserving property*** of continuous function.

Theorem 2.3.1

Let $f : D \subseteq \mathbb{R} \to \mathbb{R}$ be a function continuous at a point $a \in D$. If $f(a) \neq 0$ we can find some $\delta > 0$ such that $f(x)$ has same sign of $f(a)$ $\forall \, x \in D \cap |x - a| < \delta$.

Proof: First suppose that $f(a) > 0$ then corresponding to $\varepsilon = \dfrac{1}{2} f(a)$ there exists $\delta > 0$ such that,

$$|f(x) - f(a)| < \frac{1}{2} f(a) \qquad\qquad \forall \, x \in D \cap |x - a| < \delta$$

or
$$f(a) - \frac{1}{2} f(a) < f(x) < f(a) + \frac{1}{2} f(a) \qquad\qquad \forall \, x \in D \cap |x - a| < \delta$$

or
$$\frac{1}{2} f(a) < f(x) < \frac{3}{2} f(a) \qquad\qquad \forall \, x \in D \cap |x - a| < \delta$$

We see that $\forall\, x \in D \cap |x - a| < \delta$ value of $f(x)$ lie between two positive numbers hence is positive.

If $f(a) < 0$ we proceed in the same way as above choosing by $\varepsilon = -\dfrac{1}{2} f(a)$ and obtain $\delta > 0$ such that $\dfrac{3}{2} f(a) < f(x) < \dfrac{1}{2} f(a)\, \forall\, x \in D \cap |x - a| < \delta$. Since of $f(x)$ lie between two negative numbers in this hence is negative.

Next theorem will provide a very useful necessary and sufficient condition for a function to be continuous at a point in terms of sequences.

Theorem 2.3.2

Let f be a real valued function defined on $D \subseteq \mathbb{R}$ and $a \in D$ be any point. Then f is continuous at a if and only if for any sequence $\{a_n\}$ in D such that $a_n \to a$ as $n \to \infty$ implies $\lim\limits_{n \to \infty} f(a_n) = f(a)$.

Proof: Suppose f is continuous at a. Then for any $\varepsilon > 0$ there exists some $\delta > 0$ such that,

$$|f(x) - f(a)| < \varepsilon \quad \text{for } x \text{ satisfying } |x - a| < \delta \tag{2.3.1.1}$$

Let $\{a_n\}$ be any sequence in D such that $a_n \to a$ as $n \to \infty$ then corresponding to above δ there exists a positive integer n_0 such that, $|a_n - a| < \delta\ \forall\ n \geq n_0$ (2.3.1.2)

Using (2.3.1.1) & (2.3.1.2) we get,

$$|f(a_n) - f(a)| < \varepsilon\ \forall\ n \geq n_0.$$

Hence $\lim\limits_{n \to \infty} f(a_n) = f(a)$ and this proves 'if' part.

For the converse part let us assume if possible f is not continuous at a. Then there exists some $\varepsilon > 0$ such that for every $\delta > 0$ $|f(x) - f(a)| \geq \varepsilon$ for at least one x which satisfies $|x - a| < \delta$.

Now choose $\delta_1 - 1$ then there exists $a_1 \in D$ satisfying $|a_1 - a| < 1$ such that $|f(x_1) - f(a)| \geq \varepsilon$.

Again for $\delta_2 = \min\left\{|a_1 - a|, \dfrac{1}{2}\right\}$ there exists $a_2 \in D$ satisfying $|a_2 - a| < \delta_2$ such that $|f(a_2) - f(a) \geq \varepsilon$. Observe that $a_1 \neq a_2$ and $|a_2 - a| < \dfrac{1}{2}$.

Continuing this way we will obtain a sequence $\{a_n\}$ in D such that $|a_n - a| < \dfrac{1}{n} < \dfrac{1}{n}\ \forall\ n$ and $|f(a_n) - f(a)| \geq \varepsilon$. In other words $a_n \to a$ as $n \to \infty$ but $f(a_n)$ does not tend to $f(a)$. This contradicts the given condition and hence the assumption that f is not continuous at a is not correct and so we conclude that f is continuous at a.

Note that the above theorem gives an useful criterion for a function ***not*** continuous at a given point. Let f is a function defined on $D \subseteq \mathbb{R}$ and suppose that for a point $a \in D$ there exists a sequence $\{a_n\} \in D$ such that $a_n \to a$ but $f(a_n)$ does not tend to $f(a)$ the we will conclude that f is not continuous at a

Example 2.18   Let $f : \mathbb{R} \to \mathbb{R}$ be defined as,

$$f(x) = \begin{cases} 0 \text{ if } x \text{ is rational} \\ 1 \text{ if } x \text{ is irrational} \end{cases}$$

then show that f is not continuous at any point of $\mathbb{R}$.

Sol: Let q be any rational number. As irrational numbers are dense in $\mathbb{R}$ there exists a sequence of irrational numbers $\{i_n\}$ such that $i_n \to q$ as $n \to \infty$. Now by definition $f(i_n) = 1\ \forall\ n$ and $f(q) = 0$ that is $\lim\limits_{n \to \infty} f(i_n) = 1 \neq f(q) = 0$ as $n \to \infty$. Hence we conclude that f is not continuous at any rational point of

$\mathbb{R}$. Again if p be any irrational number then as rational numbers are dense in $\mathbb{R}$ we can choose a sequence $\{q_n\}$ of rational number such that $q_n \to p$ as $n \to \infty$. Now by definition $f(q_n) = 0 \; \forall \; n$ and $f(p) = 0$ that is $\lim_{n \to \infty} f(q_n) = 0 \neq f(p) = 1$ as $n \to \infty$. So that f is not continuous at any irrational point of $\mathbb{R}$. Consequently the function is not continuous at any real point.

Example 2.19 Let $f : [0, 1] \to [0, 1]$ be defined as,

$$f(x) = \begin{cases} x & \text{if } x \text{ is rational} \\ 0 & \text{if } x \text{ is irrational} \end{cases}$$

Show that f is continuous only at $x = 0$.

Sol: Let q be any non zero rational number then as in previous example we can choose a sequence of irrational numbers $\{i_n\} \subseteq [0, 1]$ such that $i_n \to q$ then we have,

$$\lim_{n \to \infty} f(i_n) = 0 \neq q = f(q)$$

and hence f is not continuous at any non zero rational point of $\mathbb{R}$. Next suppose p be any non zero irrational number then we can choose a sequence $\{q_n\}$ of non-zero rational number in $[0, 1]$ such that $q_n \to p$ as $n \to \infty$. Now by definition $f(q_n) = q_n \; \forall \; n$ and $f(p) = 0$ that is $\lim_{n \to \infty} f(q_n) \to p \neq f(p) = 0$ as $n \to \infty$. So that f is not continuous at any non zero irrational point of R.

We now show that f is continuous at $x = 0$. For that let $\varepsilon > 0$ be any real number and choose $\delta = \varepsilon$ then for any x satisfying $|x| < \delta$ we have,

$$|f(x) - f(0)| = \begin{cases} |0 - 0| = 0 < \varepsilon & \text{if } x \text{ is irrational} \\ |x - 0| = |x| < \varepsilon & \text{if } x \text{ is rational } |x - 0| = |x| < \varepsilon \end{cases}$$

Hence in any case we find $|f(x) - f(0)| < \varepsilon$ whenever $|x| < \delta$ for $\delta = \varepsilon$, this shows that f is continuous at $x = 0$.

Example 2.20 Show that the function f defined on $\mathbb{R}$ as,

$$f(x) = \begin{cases} 1 & \text{if } x \text{ is rational} \\ -1 & \text{if } x \text{ is irrational} \end{cases}$$

is not continuous at any point of $\mathbb{R}$ (no where continuous).

Sol: We will show that f is discontinuous at every rational and irrational points. Let p be any rational number. As irrational numbers are dense in $\mathbb{R}$ we can always find a sequence $\{q_n\}$ of irrational numbers such that, $q_n \to p$ as $n \to \infty$. Now by definition $f(q_n) = -1 \; \forall n$ hence as $n \to \infty \, f(q_n) \to -1 \neq f(p) = 1$. From Theorem 2.3.1 we see that f not continuous at $x = p$ and as p is arbitrary we conclude f not continuous at any rational point.

Next let q be any irrational number and as rationals are dense in $\mathbb{R}$ we choose a sequence $\{p_n\}$ of rational numbers such that $p_n \to q$ as $n \to \infty$. Again we see that as $n \to \infty \, f(p_n) \to 1 \neq f(q) = -1$ so that again by Theorem 1.3.1 we see that f not continuous at $x = q$ and as q is arbitrary we conclude f not continuous at any irrational point.

Example 2.21 Show that the function f defined on $\mathbb{R}$ as,

$$f(x) = \begin{cases} x & \text{if } x \text{ is rational} \\ 1 - x & \text{if } x \text{ is irrational} \end{cases}$$

is continuous only at $x = \dfrac{1}{2}$.

Sol: We first prove that f is continuous at $x = \frac{1}{2}$. For any real number $\varepsilon > 0$ we choose $\delta = \varepsilon$. Now for a rational number x satisfying $\left| x - \frac{1}{2} \right| < \delta = \varepsilon$ we have,

$$\left| f(x) - f\left(\frac{1}{2}\right) \right| = \left| x - \frac{1}{2} \right| < \varepsilon$$

Again for a irrational number x satisfying $\left| x - \frac{1}{2} \right| < \delta = \varepsilon$ we have,

$$\left| f(x) - f\left(\frac{1}{2}\right) \right| = \left| 1 - x - \frac{1}{2} \right| = \left| x - \frac{1}{2} \right| < \varepsilon$$

So that we find that,

$$\left| f(x) - f\left(\frac{1}{2}\right) \right| < \varepsilon \text{ for any } x \text{ satisfying } \left| x - \frac{1}{2} \right| < \delta$$

Thus f is continuous at $x = \frac{1}{2}$.

Now suppose $p \neq \frac{1}{2}$ be any rational number and suppose $\{q_n\}$ be a sequence of irrational numbers such that, $q_n \to p$ as $n \to \infty$. Then we see that as $n \to \infty$

$$f(q_n) = 1 - q_n \to 1 - p \neq p = f(p)$$

In other words f is not continuous at $x = p$ and hence at any rational point.

Let now q be any irrational number and $\{p_n\}$ be a sequence of rational numbers such that, $p_n \to q$ as $n \to \infty$. Then as $n \to \infty$

$$f(p_n) = p_n \to q \neq 1 - q = f(q)$$

Hence f is not continuous at $x = q$ and hence at any irrational point.

Example 2.22   Show that the function $f : [0, 1] \to \mathbb{R}$ defined by,

$$f(x) = \begin{cases} \dfrac{1}{q^2} & \text{if } x \neq 0 \text{ is rational of the form } \dfrac{p}{q} \; p, q \text{ are integers and are in lowest terms} \\[3mm] 0 & \text{if } x \text{ is a irrational number or } x = 0 \end{cases}$$

is continuous at every point of $[0, 1]$ except non-zero rational points.

Sol: We show that f is continuous at any irrational number i. We see that $\frac{1}{q^2} \to 0$ when $q \to \infty$ hence for

$\varepsilon > 0$ there can exist only finitely many rational numbers say $\dfrac{p_1}{q_1}, \dfrac{p_2}{q_2}, \dots \dfrac{p_m}{q_m}$ such that $\dfrac{1}{q_k^2} \geq \varepsilon \;\; k = 1, 2, \dots, m$.

Now we define $\delta_k = \left| i - \dfrac{p_k}{q_k} \right| \; k = 1, 2, \dots, m$ then and let $\delta = \min \{\delta_1, \delta_2, \dots \delta_m\}$ so that $\delta > 0$. Now if x be any irrational number satisfying $|i - x| < \delta$ then we have,

$$|f(i) - f(x)| = |0 - 0| < \varepsilon$$

Again if $x = \dfrac{p}{q}$ be any rational number satisfying $|i - x| < \delta$ then by construction $x \neq \dfrac{p_k}{q_k} \; k = 1, 2, \dots, m$
we have,

$$|f(x) - f(i)| = \left| \frac{1}{q^2} - 0 \right| = \frac{1}{q^2} < \varepsilon$$

Hence for any $\varepsilon > 0$ we have obtained $\delta > 0$ such that,

$$|f(x) - f(i)| < \varepsilon \quad \text{whenever } x \text{ satisfies } |i - x| < \delta$$

In other words f is continuous at $x = i$. Similar argument will prove that f is continuous at $x = 0$

Suppose $r = \dfrac{p}{q}$ be any non zero rational number. Let $\{i_n\}$ be a sequence of irrationals in $[0, 1]$ converging to r. Now we see that $f(i_n) = 0 \neq \dfrac{1}{q} = f(r)$ hence f is not continuous at $x = r$. As r is arbitrary we conclude that f is not continuous at any non zero rational.

For $X \subseteq \mathbb{R}$, $C(X)$ will denote the set of all continuous functions on X. Observe that $C(X)$ is non empty as constant functions defined on X are in $C(X)$. We can define natural operations on $C(X)$ called pointwise addition, pointwise multiplication and multiplication by real number. Next theorem tells us that under above operations the space $C(X)$ is closed.

Theorem 2.3.2

Let $f, g \in C(X)$ for $X \subseteq \mathbb{R}$ and let r be any real number. Then we have the following:

    (i) $f + g \in C(X)$

    (ii) $rf \in C(X)$

    (iii) $fg \in C(X)$

    (iv) $\dfrac{1}{f} \in C(X)$ *provided f does not vanish at any point of X^+.*

Proof: Let $a_0 \in X$ be arbitrary. Suppose $\{a_n\}$ be a sequence in X such that $a_n \to a_0$ then as $f, g \in C(X)$ we have $f(a_n) \to f(a_0)$ and $g(a_n) \to g(a_0)$ as $n \to \infty$ then by Theorem 1.3.6 part (i), (ii) & (iii) we get,

$$\lim_{n \to \infty} (f + g)(a_n) = \lim_{n \to \infty} (f(a_n) + g(a_n)) = \lim_{n \to \infty} f(a_n) + g(a_n)$$

$$= f(a_0) + g(a_0) = (f + g)(a_0)$$

$$\lim_{n \to \infty} (rf)(a_n) = r \lim_{n \to \infty} f(a_n) = rf(a_0) = (rf)(a_0)$$

$$\lim_{n \to \infty} (fg)(a_n) = \lim_{n \to \infty} f(a_n) g(a_n))$$

$$= \lim_{n \to \infty} f(a_n) \lim_{n \to \infty} g(a_n) = f(a_0) g(a_0) = (fg)(a_0)$$

This proves, $f + g$, rf, & fg are continuous at a_0. As a_0 is arbitrary we conclude $f + g$, rf, & fg are continuous in whole of X and hence belong to $C(X)$. This proves (i), (ii) & (iii).

To prove (iv) we see first that $\dfrac{1}{f}$ is defined for every $x \in X$ as f does not vanish at any point of X also in particular $f(a) \neq 0$. Hence by Theorem 2.3.6 $\dfrac{1}{f(a_n)} \to \dfrac{1}{f(a_0)} \Rightarrow \dfrac{1}{f}$ is continuous at $x = a_0$. Since a_0 is arbitrary we conclude $\dfrac{1}{f}$ is continuous is continuous for every point of X that is $\dfrac{1}{f} \in C(X)$.

Part (i) & (ii) of above theorem shows that $C(X)$ is a real vector space with operations as pointwise addition and multiplication with a real number. By (iii) we get that $C(X)$ has a multiplicative structure we want to enquire whether it is a group with respect to 'multiplication'. It is easy to see that the constant function $e(x) = 1$ acts as an identity, question is whether every element of $C(X)$ has an inverse (multiplicative). In other words if $f \in C(X)$ does $\dfrac{1}{f} \in C(X)$? Certainly answer is no as we see that $\dfrac{1}{f}$ is not defined for those values for which $f(x) = 0$ (iv) provides a sufficient condition for $f \in C(X)$ to have a multiplicative inverse.

When $X = [a, b]$, the space $C(X)$ is denoted by $C[a, b]$. Following theorems give some interesting properties of $C[a, b]$. We first define what are called bounded functions.

Definition   A function $f : D \to \mathbb{R}$ is said to be bounded if the set $\{f(x) : x \in D\}$ is bounded.

Theorem 2.3.3

If $f \in C[a, b]$ then it is bounded. Moreover the bounds of f are attained.

Proof: Let $f \in C[a, b]$ we show that f is bounded by method of contradiction. Suppose if possible f is not bounded. Then for every natural number n there exists some $x_n \in [a, b]$ such that,

$$|f(x_n)| > n$$

In other words for the sequence $\{x_n\}$ in $[a, b]$ the sequence $\{f(x_n)\}$ is unbounded. Now as $x_n \in [a, b]$ $\forall\, n$ the sequence $\{x_n\}$ is bounded so that by Bolzano-Weierstrass' Theorem there exists a convergent subsequence say $\{x_{n_k}\}$ of $\{x_n\}$ such that $x_{n_k} \to x_0$ as $n \to \infty$. Again as $[a, b]$ is closed we have $a \in [a, b]$ and since f is continuous at $x = a$ we have $f(x_{n_k}) \to f(a)$ as $n \to \infty$. This contradicts the fact that $\{f(x_{n_k})\}$ is unbounded which proves that f must be bounded.

Next we show that the bounds of f are attained. Let m & M are defined as,

$$m = \inf_{x \in [a, b]} f(x) \ \& \ M = \sup_{x \in [a, b]} f(x)$$

Then we have $m \le f(x) \le M$. Suppose if possible $f(x) \neq M$ for any $x \in [a, b]$ then the function $g(x) = M - f(x)$ is in $C[a, b]$ and strictly positive. Hence by Theorem 1.3.2 (iv) we have $\dfrac{1}{g} \in C[a, b]$ and consequently by what we have proved we conclude $\dfrac{1}{g}$ is bounded. Thus there exists $K > 0$ such that,

$$\frac{1}{g(x)} < K \ \forall\, x \in [a, b]$$

$$\Rightarrow \qquad f(x) < M - \frac{1}{K} \ \forall\, x \in [a, b]$$

This contradicts the fact that M is supremum of the set $\{f(x) : x \in [a, b]\}$. Hence we conclude that there must exist one point x_1 such that $f(x_1) = M$ that is the supremum of the function is attained.

To prove that infimum of the function is attained we note that if $f \in C[a, b]$ then $-f \in C[a, b]$ and $\inf\limits_{x \in [a, b]} f(x) = \sup\limits_{x \in [a, b]} (-f(x))$. By what we have just now proved $\sup\limits_{x \in [a, b]} (-f(x))$ is attained hence m is attained that is there exists one point x_2 such that $f(x_2) = m$.

Given a function f defined on a subset D of reals an important aspect of study of the function is to find the so called *zeroes* of the function. That is the set of points of D where f vanishes. For example, if f be a polynomial of order n defined on $\mathbb{R}$ it can have at most n zeroes called the roots of the polynomial.

It is always interesting to enquire whether a given function vanishes on a closed interval or how many zeroes the function has in the given interval. Next theorem gives us a sufficient condition under which an element of $C[a, b]$ has a zero in the interval $[a, b]$.

Theorem 2.3.4

Let $f \in C[a, b]$ and suppose $f(a) . f(b) < 0$ then $f(\xi) = 0$ for at least one $\xi \in (a, b)$.

Proof: The given condition tells us that $f(a)$ and $f(b)$ must be of opposite signs. The geometrical implication of the theorem is very clear it says that if $f(a)$ and $f(b)$ are of opposite sign and if f is continuous it has to cross X–axis at least once as shown in the figure (Fig.). We now give the formal proof. Without loss of generality we can assume $f(a) > 0$ & $f(b) < 0$ as $-f$ and f vanish simultaneously and both belong to $C[a, b]$. We define the set G as follows,

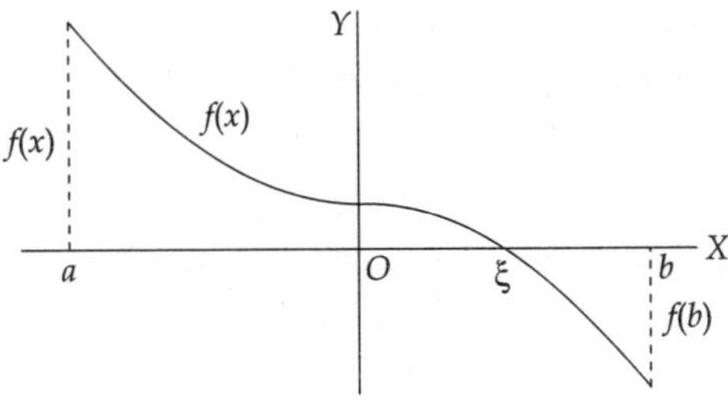

Fig. 2.4

$$G = \{x \in [a, b] : f(x) > 0\}$$

G is non-empty as $a \in G$. Let $\xi = \sup G$. As f is continuous at $x = a$ there exists a right neighbourhood of a such that for every x in that neighbourhood we have $f(x) > 0$ hence $\xi \neq a$ again as f is continuous at $x = b$ there exists a left neighbourhood of b such that for every x in that neighbourhood $f(x) < 0$ so that $\xi \neq b$ in other words we get $a < \xi < b$. Suppose if possible $f(\xi) > 0$. Now again by continuity of f at $x = \xi$ we will get a right neighbourhood of ξ such that for every x there we have $f(x) > 0$, that is that is there exists $\xi_1 > \xi$ such that $f(\xi_1) > 0$ which contradicts the fact that $\xi = \sup G$ hence $f(\xi)$ cannot be positive. Again if possible we assume that $f(\xi) < 0$. Then as before there exists $\delta > 0$ such that $f(x) < 0$ for $\xi - \delta < x < \xi$ but as $\xi = \sup G$ there exists some $\xi_2 \in G$ satisfying $\xi - \delta < \xi_2 < \xi$ that is $f(\xi_2) > 0$. This is clearly a contradiction which shows that $f(\xi)$ cannot be negative. Hence we conclude that $f(\xi) = 0$.

As an easy fall out of the above theorem we get a very important result which states that if f is a non constant element of $C[a, b]$ then f assumes every intermediate value between $f(a)$ and $f(b)$. In other words we say that every element of $C[a, b]$ possesses intermediate value property.

Example 2.23 Using above theorem show that there exists one zero of the function $f(x) = xe^x - \cos x$ in the interval $(0, 1)$.

Sol: Clearly $f(x)$ is a continuous function. Also $f(0) = -1$ and $f(1) = e - \cos 1 > 0$. Hence by above theorem there exists one $\xi \in (0, 1)$ such that $f(\xi) = 0$.

Theorem 2.3.5

Suppose f is a non constant element of $C[a, b]$ and let λ be any value between $f(a)$ and $f(b)$ then there exists at least one real number say $\xi \in (a, b)$ such that $f(\xi) = \lambda$.

Proof: We define $F : [a, b] \to \mathbb{R}$ as, $F(x) = f(x) - \lambda$

Then clearly $F \in C[a, b]$ and $F(a) . F(b) < 0$

Hence by above theorem there exist $\xi \in (a, b)$ such that,

$$F(\xi) = 0$$

So that $f(\xi) = \lambda$

Corollary 2.3.6

Suppose is a non constant element of $C[a,b]$ then f attains every value between its supremum and infimum.

Proof: Follows easily from Theorem 2.3.3 & Theorem 2.3.5.

Before proceeding to the next theorem we introduce the notion of monotonic functions on a subset D of real numbers.

Definition   Suppose $f : D \subseteq \mathbb{R} \to \mathbb{R}$ be a function and let $x_1, x_2 \in D$ then,

   (i) f is ***monotone increasing*** if $x_1 > x_2 \Rightarrow f(x_1) \geq f(x_2)$

  (ii) f is ***strictly monotone increasing*** if $x_1 > x_2 \Rightarrow f(x_1) > f(x_2)$

 (iii) f is ***monotone decreasing*** if $x_1 > x_2 \Rightarrow f(x_1) \leq f(x_2)$

 (iv) f is ***strictly monotone decreasing*** if $x_1 > x_2 \Rightarrow f(x_1) < f(x_2)$

It is easy to see that every strictly monotone function on D is one to one. Next theorem shows that the inverse of a strictly monotone element of $C[a, b]$ on is not only strictly monotone but also continuous.

Theorem 2.3.7

Let $f \in C[a, b]$ and is strictly monotone increasing. If g denotes the inverse of f then $g \in C[m, M]$ and is strictly monotone increasing where m & M denotes the infimum and supremum of f respectively.

Proof: Note that as f is monotone increasing it is one to one and as it is continuous Corollary 1.3.6 tells us that it is onto with range $[m, M]$. Hence domain of g is $[m, M]$. We now prove that g is strictly monotone increasing. For that let y_1 & $y_2 \in [m, M]$ such that $y_1 > y_2$. Now there exists x_1 & $x_2 \in [a, b]$ such that $y_1 = f(x_1)$ & $y_2 = f(x_2)$ but as f is monotone increasing the relation $y_1 > y_2$ implies $x_1 > x_2$ or in other words $g(y_1) > g(y_2)$. Thus g is strictly monotone increasing. Let $y_0 \in [m, M]$ be any interior point and $\varepsilon > 0$ be any positive number. If $y_0 = f(x_0)$ then consider $\delta = \min\{f(x_0) - f(x_0 - \varepsilon), f(x_0 + \varepsilon) - f(x_0)$. Then it is easy to see that whenever y satisfies we have $g(y_0) - \varepsilon < g(y) < g(y_0) + \varepsilon$ that is, g is continuous at y_0 and hence at any interior point of $[m, M]$. By slightly modifying the above arguments we can prove the continuity of g at m & M. This proves $g \in C[m, M]$ and is strictly monotone increasing.

2.4   UNIFORM CONTINUITY

Recall the $\varepsilon - \delta$ definition of continuity of a function f at a point x_0 in its domain. We have seen that for a given in general the δ which is obtained depends on both ε and x_0. In fact Example 14 shows the dependence of δ on x_0. In case it is possible to find δ which does not depend on x_0 or in other words if the δ works uniformly for all x in the domain of definition of the function we say that the function is ***uniformly continuous*** which we define below formally.

Definition   A function $f : D \to \mathbb{R}$ is said to uniformly continuous on D if given any $\varepsilon > 0$ there exists a $\delta > 0$ such that $|f(x) - f(y)| < \varepsilon$ for all x, y in D satisfying $|x - y| < \delta$.

Clearly uniform continuity implies ordinary continuity but the converse is not true as is shown by the following example

Example 2.24   The function $f : (0, \infty) \to R$ defined by

$$f(x) = \frac{1}{x}$$

is continuous but not uniformly continuous.

Sol: We have already shown that above function is continuous (Example 14) so here we show that the function is not uniformly continuous. Let us choose $\varepsilon = \frac{1}{2}$ and suppose $\delta > 0$ be any real. If $\delta > 1$ then choose $x_1 = 1$ & $x_2 = 3$ & then $|x_1 - x_2| = 1 < \delta$ but

$$|f(x_1) - f(x_2)| = \left| \frac{1}{x_1} - \frac{1}{x_2} \right|$$

$$= 1 - \frac{1}{3} = \frac{2}{3} > \frac{1}{2}$$

Again if $\delta \leq 1$ then choose $x_1 = \delta$ & $x_2 = \dfrac{\delta}{1 + \dfrac{1}{2}}$ then,

$$|x_1 - x_2| = \left| \delta - \frac{\delta}{1 + \dfrac{1}{2}} \right| = \frac{\dfrac{1}{2}}{1 + \dfrac{1}{2}} \, \delta < \delta \text{ but,}$$

$$|f(x_1) - f(x_2)| = \left| \frac{1}{x_1} - \frac{1}{x_2} \right|$$

$$= \left| \frac{1}{\delta} - \frac{1 + \dfrac{1}{2}}{\delta} \right| = \frac{\dfrac{1}{2}}{\delta} > \frac{1}{2} \quad (\text{as } \delta \leq 1)$$

Hence we see that for $\varepsilon = \frac{1}{2}$ no δ works so we conclude that f is not uniformly continuous.

We now give a necessary sufficient condition for a function defined on a subset of real number to be uniformly continuous.

Theorem 2.4.1

A function $f : D \subseteq \mathbb{R} \to \mathbb{R}$ is uniformly continuous if and only for every sequences $\{x_n\}$ & $\{y_n\}$ in D satisfying $\lim\limits_{n \to \infty} x_n = \lim\limits_{n \to \infty} y_n$ implies $\lim\limits_{n \to \infty} f(x_n) = \lim\limits_{n \to \infty} f(y_n)$.

Proof: Suppose f is uniformly continuous then for $\varepsilon > 0$ there exists a $\delta > 0$ such that $|f(x) - f(y)| < \varepsilon$ for all x, y in D satisfying $|x - y| < \delta$. Now let $\{x_n\}$ & $\{y_n\}$ be two sequences in D satisfying $\lim\limits_{n \to \infty} x_n = \lim\limits_{n \to \infty} y_n$ then there exists a positive integer n_0 such that whenever $n \geq n_0$ we have $|x_n - y_n| < \delta$ and consequently $|f(x_n) - f(y_n)| < \varepsilon$. That is $\lim\limits_{n \to \infty} f(x_n) - f(y_n) = 0$ or equivalently $\lim\limits_{n \to \infty} f(x_n) = \lim\limits_{n \to \infty} f(y_n)$.

For the converse, let f satisfies the condition that whenever we have sequences $\{x_n\}$ & $\{y_n\}$ in D with $\lim\limits_{n \to \infty} x_n = \lim\limits_{n \to \infty} y_n$ implies $\lim\limits_{n \to \infty} f(x_n) = \lim\limits_{n \to \infty} f(y_n)$. If possible suppose f is not uniformly continuous. Hence there exists an $\varepsilon > 0$ such that for every n there exists x_n & y_n in D such that $|x_n - y_n| < \frac{1}{n}$ but

$|f(x_n) - f(y_n)| \geq \varepsilon$. Thus $\lim\limits_{n \to \infty} x_n = \lim\limits_{n \to \infty} y_n$ but $\lim\limits_{n \to \infty} (f(x_n) - f(y_n)) \neq 0$ which is a contradiction to the given condition. This contradiction shows that our assumption that f is not uniformly continuous is not correct hence we conclude that f is uniformly continuous.

We have observed already that uniform continuity implies ordinary continuity and converse is not true. Next theorem shows that the converse is true for a special class of continuous functions.

Theorem 2.4.2

Every element of C[a, b] is uniformly continuous on [a, b].

Proof: Let $f \in C[a, b]$ and if possible suppose f is not uniformly continuous. Then there exists an $\varepsilon > 0$ such that for every n there exists x_n & y_n in $[a, b]$ such that,

$$|x_n - y_n| < \frac{1}{n} \quad \text{and} \quad |f(x_n) - f(y_n)| \geq \varepsilon \tag{2.4.2.1}$$

It is clear that both $\{x_n\}$ & $\{y_n\}$ are bounded sequences hence by Bolzano-Weierstrass Theorem (Theorem 1.4.2) there exists subsequences $\{x_{n_k}\}$ & $\{y_{n_k}\}$ such that $x_{n_k} \to x$ & $y_{n_k} \to y$ for some x & y in D. From (2.4.2.1) we conclude $x = \lim\limits_{k \to \infty} x_{n_k} = \lim\limits_{k \to \infty} y_{n_k} = y$. Again as f is continuous at x & y we have $f(x) = \lim\limits_{k \to \infty} f(x_{n_k})$ & $f(y) = \lim\limits_{k \to \infty} y_{n_k}$ that is

$$\lim_{k \to \infty} f(x_{n_k}) = \lim_{k \to \infty} f(y_{n_k})$$

This contradicts (2.4.2.1). Hence we conclude that f is uniformly continuous in $[a, b]$.

2.5 DERIVABILITY AND DIFFERENTIABILITY

Suppose we write $y = f(x)$ then x is independent variable and is dependent variable. We want to study the rate of change in the value of dependent variable when there is a small change in the value of independent variable. Let us describe a situation which will motivate us in to do this kind of study. Suppose a particle is moving on a circle in the plane. Then the position of this particle is a function of time. See that at every instant of time the direction of motion of the particle is changing. If we want to know the speed of the particle at instant of time we have to consider the rate of change of the distance by the particle with respect to time. In fact, we are looking for the value of the quotient $\dfrac{\Delta S}{\Delta t}$ where ΔS denotes the distance covered by the particle for a small increment of time Δt or in other words we want to know the value of $\lim\limits_{\Delta t \to 0} \dfrac{\Delta S}{\Delta t}$. This leads to the concept of derivability of functions we define formally below.

Definition Let f be a real valued function defined on an open interval I. We say that the function is **derivable** at a point $c \in I$ if

$$\lim_{h \to 0} \frac{f(c + h) - f(c)}{h}$$

exists (in the sense of Definition 2.2.1) and in that case the above limit is called the **derivative** of the function at c (or the differential coefficient of the function at c) and is denoted by $f'(c)$ or $\left(\dfrac{df}{dx}\right)_{x = c}$.

If the derivative at every point of I exists the function is said to be derivable on the entire interval.

Remarks

1. Note that existence of $\lim\limits_{h \to 0} \dfrac{f(c + h) - f(c)}{h}$ implies the existence of the limits

   $\lim\limits_{h \to 0^+} \dfrac{f(c + h) - f(c)}{h}$ & $\lim\limits_{h \to 0^-} \dfrac{f(c + h) - f(c)}{h}$ which are called the ***right-hand derivative*** and ***left-hand derivative*** of the function and are denoted by $f'_+(c)$ and $f'_-(c)$ respectively.

2. We can represent $f'(c)$ equivalently as,

   $$f'(c) = \lim\limits_{x \to c} \frac{f(x) - f(c)}{x - c}$$

   where $x \neq c$ belongs to the domain of f.

3. If $y = f(x)$ denotes a function defined in some open interval then the derivative of the function at any point x of the domain if exists is sometimes denoted by $\dfrac{dy}{dx}$ that is $\dfrac{dy}{dx} = f'(x)$

Example 2.25 Let $f : (a, b) \to \mathbb{R}$ be defined by

$$f(x) = k, \quad k \text{ being a constant}$$

then $f'(x) = 0 \ \forall \ x \in (a, b)$. That is constant functions defined on any open interval is differentiable with differential coefficient vanishing at every point of the interval.

Sol: See that for any $x \in (a, b)$ such that $x + h \in (a, b)$ we have,

$$\lim\limits_{h \to 0} \frac{f(x + h) - f(x)}{h} = \lim\limits_{h \to 0} \frac{k - k}{h} = 0$$

Hence $f'(x) = 0 \ \forall \ x \in (a, b)$.

Example 2.26 Let $f : \mathbb{R} \to \mathbb{R}$ be defined by,

$$f(x) = x^n \ n \in \mathbb{N}$$

then $f'(x) = nx^{n-1} \ \forall \ x \in \mathbb{R}$.

Sol: For any real x we have,

$$\lim\limits_{h \to 0} \frac{f(x + h) - f(x)}{h} = \lim\limits_{h \to 0} \frac{(x + h)^n - x^n}{h}$$

$$= \lim\limits_{h \to 0} \frac{x^n + {}^nC_1 x^{n-1} h + {}^nC_2 x^{n-2} h^2 + \ldots + {}^nC_n h^n - x^n}{h}$$

$$= nx^{n-1} \quad (\text{using } \lim\limits_{h \to 0} h^k = 0 \text{ for any positive integer } k \text{ and limit theorems})$$

Let us try to see the geometrical interpretation of $f'(c)$. Suppose $f : (a, b) \to \mathbb{R}$ be a function and assume that the slope of the tangent at a point P with coordinate $(c, f(c))$ be ψ where $a < c < b$ (see Fig.). Let $Q(c + h, f(c + h))$ be a point near P with slope of PQ being ϕ. We see that $\tan \phi = \dfrac{QT}{PT}$

$= \dfrac{f(c + h) - f(c)}{h}$. Now as $h \to 0$ we have $Q \to P$ and hence the limiting position of PQ is the

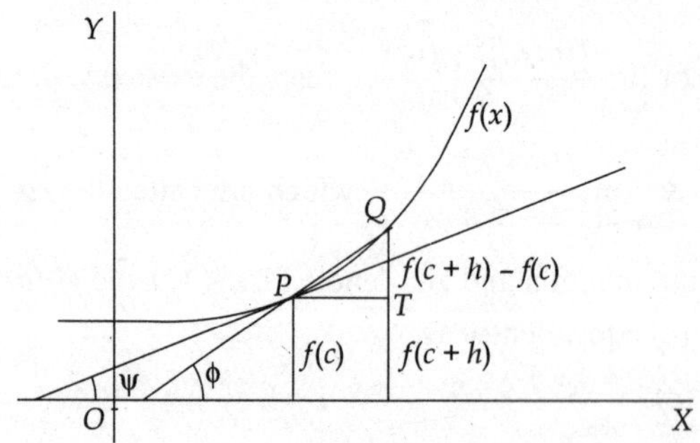

Fig. 2.5

$\tan \phi = \dfrac{QT}{PT} = \dfrac{f(c + h) - f(c)}{h}$. Now as $h \to 0$ we have $Q \to P$ and hence the limiting position of PQ is the

tangent at P so that $f'(c) = \lim\limits_{h \to 0} \dfrac{f(c + h) - f(c)}{h} = \tan \psi$. Thus the derivative of f at any interior point c of

(a, b) gives the gradient of the tangent line of the curve at the point $(c, f(c))$.

Next theorem shows that if a function is derivable at a point then the function must be continuous there other words derivability implies continuity.

Theorem 2.5.1

Let $f : (a, b) \to \mathbb{R}$ be a function derivable at $c \in (a, b)$ then the function is also continuous at that point.

Proof: We have,

$$f(c + h) - f(c) = \frac{f(c + h) - f(c)}{h}\, h$$

Hence,

$$\lim_{h \to 0} \left[f(c + h) - f(c) \right] = \lim_{h \to 0} \left[\frac{f(c + h) - f(c)}{h}\, h \right]$$

As f is derivable at $c \in (a, b)$ $\lim\limits_{h \to 0} \dfrac{f(c + h) - f(c)}{h}$ exists and is equal to $f'(c)$ so by using algebra of limit we get,

$$\lim_{h \to 0} \left[f(c + h) - f(c) \right] = \lim_{h \to 0} \left[\frac{f(c + h) - f(c)}{h} \right] \lim_{h \to 0} h$$

$$f(c) \times 0 = 0$$

That is of is continuous at c.

The converse of the above theorem is not true as evident form the following example.

Example 2.27 Let $f : (0, 2) \to \mathbb{R}$ defined as,

$$f(x) = |x - 1|$$

then f is continuous at $x = 1$ but is not derivable there.

Sol: To show f is continuous at $x = 1$ we consider $\varepsilon > 0$ to be any real number and choose $\delta = \varepsilon$. Then for $|x - 1|\ \delta$ we trivially obtain,

$$|f(x) - f(1)| = |x - 1| < \varepsilon$$

This shows f is continuous at $x = 1$.

Now for h such that $1 + h \in (0, 2)$ we have,

$$f_+'(1) = \lim_{h \to 0^+} \frac{f(1+h) - f(1)}{h} = \lim_{h \to 0^+} \frac{|h|}{h} = \lim_{h \to 0^+} \frac{h}{h} = 1$$

Similarly,

$$f_-'(1) = \lim_{h \to 0^-} \frac{f(1+h) - f(1)}{h} = \lim_{h \to 0^-} \frac{|h|}{h} = \lim_{h \to 0^+} \frac{-h}{h} = -1$$

Thus as the right hand and left hand derivatives are not equal at $x = 1$ we conclude that f is not derivable at $x = 1$

Observe that at $x = 1$ the function takes a sharp turn. In general, the function is not differentiable at those points on the curve where there are sharp edges.

We now extend the definition of derivatives for functions defined in a closed interval and also definition of infinite derivatives.

Definition The function $f : [a, b] \to \mathbb{R}$ is said to be derivable in the interval $[a, b]$ if $f_+'(a)$ & $f_-'(b)$ exists and it is derivable at every interior point of the interval.

We write $f'(c) = \infty$ for an interior point c of $[a, b]$ if $f_+'(c) = f_-'(c)$ and f is said to posses infinite derivatives at a & b if $f_+'(a) = \infty$ and $f_-'(b) = \infty$ respectively.

Next we introduce the important concept of differentiability,

Definition A function $f : I \to \mathbb{R}$ is said to be differentiable at point of c of the open interval I if for every real number h such that $c + h \in I$ we can write,

$$f(c + h) - f(c) = Ah + \varepsilon h$$

where A is a constant independent of h and ε is a function of h which tends to zero as $h \to 0$.

Remark Above definition has a useful equivalent form:

A function $f : I \to \mathbb{R}$ is said to be differentiable at point of c of the open interval I if for every $x \in I$ we have,

$$f(x) - f(c) = A(x - c) + \varepsilon(x - c)$$

where A is a constant independent of x and ε is a function of x which tends to zero as $x \to c$.

Form the above definition it is clear that if a function is differentiable at any point c then the 'increment' function $f(c + h) - f(c)$ is approximately equal to the function Ah for small values of h. Note that functions of the form $\phi(h) = Ah$ for a constant A are called 'linear' functions of $\mathbb{R}$ and εh can be taken as 'error' committed in approximating the increment function by Ah. Finally we can say that f is differentiable at the point c if $f(c + h) - f(c)$ can be written as a sum of a linear function and an 'error' term near the point.

Next theorem shows that the concept of derivability and differentiability coincide in the case of functions of single variable. But we will see that this is not true in case of functions of two variables (Chapter 9).

Theorem 2.5.2

A function $f : (a, b) \to \mathbb{R}$ is differentiable at $c \in (a, b)$ if and only if the derivative of the function exists at that point.

Proof: Suppose f is differentiable at c. Then if $c + h \in (a, b)$ we can write,

$$f(c + h) - f(c) = Ah + \varepsilon h \qquad (2.5.2.1)$$

where A is a constant independent of h and $\varepsilon \to 0$ as $h \to 0$.

Using (2.5.2.1) we get,

$$\lim_{h \to 0} \frac{f(c + h) - f(c)}{h} = A$$

or,
$$f'(c) = A$$

Thus derivative of f at c exists.

For the converse let us assume that $f'(c)$ exists. Then for $c + h \in (a, b)$ we have,

$$f(c + h) - f(c) = f'(c)\, h + \left[\frac{f(c + h) - f(c)}{h} - f'(c) \right] h$$

$$= Ah + \varepsilon h$$

where $A = f'(c)$ independent of h and $\varepsilon = \left[\dfrac{f(c + h) - f(c)}{h} - f'(c) \right] \to 0$ as $h \to 0$. This shows that f is differentiable at c.

Suppose $f : (a, b) \to \mathbb{R}$ is differentiable for every point of the interval then we can define a function $f' : (a, b) \to \mathbb{R}$ by the relation,

$$f'(x) = \lim_{h \to 0} \frac{f(x + h) - f(x)}{h} \quad x \in (a, b)$$

then f' is called the ***derived function*** of f. Note that f' may or may not be continuous. In case f' is continuous we say f is ***continuously differentiable***. $C^1(a, b)$ & $D^1(a, b)$ will denote the space of all functions which are continuously differentiable and differentiable in the interval (a, b) respectively. It is clear that $C^1(a, b) \subset D^1(a, b)$. Next theorem will show that $D^1(a, b)$ is closed under point-wise addition, point-wise multiplication and multiplication by a real number.

Theorem 2.5.3

Let f & g be two functions defined in the interval (a, b) and are differentiable at $c \in (a, b)$. If k is any real number then,

    (i)   *$f + g$ is also differentiable at c and $(f + g)'(c) = f'(c) + g'(c)$*

    (ii)   *fg is also differentiable at c and $(fg)'(c) = f'(c)\, g(c) + f(c)\, g'(c)$*

    (iii)   *kf is also differentiable at c and $(kf)'(c) = kf'(c)$*

    (iv)   *If $f(c) \neq 0$, $\dfrac{1}{f}$ is differentiable at c and $\left(\dfrac{1}{f}\right)'(c) = \dfrac{f'(c)}{(f(c))^2}$*

Proof: As f & g are differentiable at c we choose h such that $c + h \, \varepsilon \, (a, b)$ then using Theorem 2.5.2 we get,

$$\left.\begin{array}{l} f(c + h) - f(c) = f'(c)\,h + \varepsilon_1 h \\ g(c + h) - g(c) = g'(c)\,h + \varepsilon_2 h \end{array}\right\} \tag{2.5.3.1}$$

where ε_1 & ε_2 tends to zero as $h \to 0$. Now,

$$\begin{aligned} (f + g)\,(c + h) - (f + g)\,(c) &= (f(c + h) + g(c + h) - f(c) - g(c)) \\ &= (f(c + h) - f(c)) + (g(c + h) - g(c)) \\ &= f'(c)\,h + \varepsilon_1 h + g'(c)\,h + \varepsilon_2\,h \qquad \text{(Using (2.5.3.1))} \\ &= (f'(c) + g'(c))\,h + \varepsilon h \end{aligned}$$

where $\varepsilon = \varepsilon_1 + \varepsilon_2$ which tends to zero as $h \to 0$ also we note that $f'(c) + g'(c)$ is independent of h. Thus it follows from definition that $f + g$ is differentiable at c moreover from the first part of Theorem 2.5.2 we get,

$$(f + g)'\,(c) = f'(c) + g'(c)$$

which proves (i)

To prove (ii) consider,

$$\begin{aligned} (fg)\,(c + h) - (fg)\,(c) &= (f(c + h)\,g(c + h) - f(c)\,g(c)) \\ &= g(c + h)\,(f(c + g) - f(c)) + f(c)\,(g(c + h) - g(c)) \\ &= g(c + h)\,(f'(c)\,h + \varepsilon_1\,h) + f(c)\,(g'(c)\,h + \varepsilon_2 h) \qquad \text{(Using (2.5.3.1))} \end{aligned}$$

Therefore,

$$\begin{aligned} \lim_{h \to 0} \frac{(fg)\,(c + h) - (fg)(c)}{h} &= \lim_{h \to 0} [g(c + h)(f'(c) + \varepsilon_1)] + \lim_{h \to 0} f(c)\,(g'(c) + \varepsilon_2)] \\ &= f'(c)\,g(c) + f(c)\,g'(c) \qquad (\text{As } g \text{ is continuous at } c \lim_{h \to 0} g(c + h) - g(c)) \end{aligned}$$

This proves (ii)

Part (iii) is easy and is kept as an exercise.

To prove (iv) we note that as f is continuous at c and $f(c) \neq 0$ there exists an open neighbourhood U of c such that $U \subset (a, b)$ and $f(x) \neq 0 \; \forall x \in U$. Also note that as $f(c) \neq 0 \; \lim_{h \to 0} \dfrac{1}{f(c + h)} = \dfrac{1}{f(c)}$. We choose h such that $c + h \in U$ then,

$$\begin{aligned} \frac{1}{f(c + h)} - \frac{1}{f(c)} &= -\frac{(f(c + h) - f(c))}{f(c + h)\,f(c)} \\[2mm] &= \frac{-f'(c)h + \varepsilon_1\,h}{f(c + h)\,f(c)} \qquad \text{(using 2.5.3.1)} \end{aligned}$$

Therefore,

$$\begin{aligned} \lim_{h \to 0} \frac{1}{h}\left(\frac{1}{f(c + h)} - \frac{1}{f(c)}\right) &= -\lim_{h \to 0} \frac{f'(c) + \varepsilon_1}{f(c + h)\,f(c)} \\[2mm] &= -\frac{f'(c)}{f(c)\,f(c)} = -\frac{f'(c)}{(f(c))^2} \end{aligned}$$

So that we get $\dfrac{1}{f}$ is differentiable at c and $\left(\dfrac{1}{f}\right)'(c) = \dfrac{f'(c)}{(f(c))^2}$. This proves (iv)

Next theorem tells us that composition of two differentiable functions is also differentiable moreover provides a rule how to differentiate such functions.

Theorem 2.5.4

Let $f : (a, b) \to \mathbb{R}$ and $g : D \to \mathbb{R}$ be two functions such that $R(f) \subset D$. Suppose $c \in (a, b)$ be a point such that $f(c)$ is an interior point of D if we assume f is differentiable at c and g differentiable at $f(c)$ then $g \circ f$ is differentiable at c and moreover we have,

$$(gof)'(c) = g'(f(c)).f'(c)$$

Proof: As f is differentiable at c then for $x \in (a, b)$ we have,

$$f(x) - f(c) = (x - c)f'(c) + \varepsilon_1(x - c) \tag{2.5.4.1}$$

where $\varepsilon_1 \to 0$ as $x \to c$.

Let $p = f(c)$ then as g is differentiable at p for $q \in D$ we have,

$$g(q) - g(p) = (q - p)g'(p) + \varepsilon_2(q - p) \tag{2.5.4.2}$$

where $\varepsilon_2 \to 0$ as $q \to p$.

In particular putting $q = f(x)$ in (2.5.4.2) we get,

$$g(f(x)) - g(f(c)) = (f(x) - f(c))g'(f(c)) + \varepsilon_2(f(x) - f(c))$$

$$= [(x - c)f'(c) + \varepsilon_1(x - c))g'(f(c)] + \varepsilon_2[(x - c)f'(c) + \varepsilon_1(x - c)] \quad \text{(Using (2.5.4.2))}$$

Therefore,

$$\lim_{x \to c} \frac{g(f(x)) - g(f(c))}{x - c} = \lim_{x \to c}[f'(c) + \varepsilon_1]\,g'(f(c)) + \lim_{x \to c}\varepsilon_2[f'(c) + \varepsilon_1] \tag{2.5.4.3}$$

Now as $x \to c\ \varepsilon_1 \to 0$ also as f is continuous at c, $x \to c \Rightarrow q \to p$, so that $\varepsilon_2 \to 0$. Hence from (2.5.4.3) we get,

$$\lim_{x \to c} \frac{g(f(x)) - g(f(c))}{x - c} = g'(f(c)).f'(c)$$

In other words gof is differentiable at c and

$$(gof)'(c) = g'(f(c)).f'(c)$$

Note that if $z = g(y)$ and $y = f(x)$ where f & g are differentiable functions then the above theorem gives the rule

$$\frac{dz}{dx} = \frac{dz}{dy}\frac{dy}{dx}$$

which is some times referred as **Chain Rule**.

We will now interpret the sign of derivative of a function sun at a given point. Before that we need to have following definition.

Definition Let $f : I = [a, b] \to \mathbb{R}$ be a function and let c be an interior point of I. The function is said to be

(a) ***increasing*** at c if there exists $\delta > 0$ such that,

$$f(x) < f(c) \text{ for } c - \delta < x < c \text{ and}$$
$$f(x) > f(c) \text{ for } c < x < c + \delta$$

(b) ***decreasing*** at c if there exists $\delta > 0$ such that,

$$f(x) > f(c) \text{ for } c - \delta < x < c \text{ and}$$
$$f(x) < f(c) \text{ for } c < x < c + \delta$$

(c) ***increasing*** at a if there exists $\delta > 0$ such that,

$$f(x) > f(a) \text{ for } a < x < a + \delta$$

(d) ***decreasing*** at a if there exists $\delta > 0$ such that,

$$f(x) < f(a) \text{ for } a < x < a + \delta$$

(e) ***increasing*** at b if there exists $\delta > 0$ such that,

$$f(x) < f(b) \text{ for } b - \delta < x < b$$

(f) ***decreasing*** at b if there exists $\delta > 0$ such that,

$$f(x) > f(b) \text{ for } b - \delta < x < b$$

Theorem 2.5.5

Suppose $f : [a, b] \to \mathbb{R}$ be a differentiable function and let for $a \leq c \leq b$ $f'(c)$ be a positive real number or ∞ then the function is increasing at c.

Proof: First suppose $a < c < b$ and $f'(c) = \lambda > 0$. That is we have,

$$\lim_{x \to c} \frac{f(x) - f(c)}{x - c} = \lambda$$

If we choose $\varepsilon = \dfrac{\lambda}{2}$ then from definition of limit we will obtain some $\delta > 0$ such that,

$$\left| \frac{f(x) - f(c)}{x - c} - \lambda \right| < \frac{\lambda}{2} \text{ whenever } 0 < |x - c| < \delta$$

Thus,

$$\frac{\lambda}{2} < \frac{f(x) - f(c)}{x - c} < \frac{3\lambda}{2} \text{ whenever } 0 < |x - c| < \delta$$

As $\lambda > 0$ we see that $\dfrac{f(x) - f(c)}{x - c}$ always lies between two positive numbers hence is positive for all values of x satisfying $0 < |x - c| < \delta$. Now this means

$$f(c) > f(x) \text{ if } c - \delta < x < c \text{ as } x - c < 0$$

Similarly,

$$f(c) > f(x) \text{ if } c < x < c + \delta \text{ as } x - c > 0$$

This shows that f is increasing at c.

Next suppose that $a < c < b$ and $f'(c) = \infty$. Again by definition for a large positive number K we will obtain some $\delta > 0$ such that,

$$\frac{f(x) - f(c)}{x - c} > K > 0 \quad \text{whenever } 0 < |x - c| < \delta$$

Again we see that $\dfrac{f(x) - f(c)}{x - c}$ is positive for all values of x satisfying $0 < |x - c| < \delta$. So as before we get f is increasing at c.

By suitably modifying above argument we can prove the result for the end points a & b.

By arguing as above we can easily prove the following theorem:

Theorem 2.5.6

Suppose $f : [a, b] \to \mathbb{R}$ be a differentiable function and let for $a \le c \le b$ $f'(c)$ be a negative real number or $-\infty$ then the function is decreasing at c.

As an immediate corollary of the above theorem we get an analogue of Theorem 2.3.4

Theorem 2.5.7

Suppose $f : [a, b] \to \mathbb{R}$ be a differentiable function then if $f'(a)$ & $f'(b)$ are of opposite sign there exists at least one $\xi \in (a, b)$ such that $f'(\xi) = 0$.

Proof: We note that as f & $-f$ vanish simultaneously without loss of generality we can assume $f'(a) > 0$ and $f'(b) < 0$ otherwise we may proceed with $-f$. Hence using Theorem 2.5.5 & Theorem 2.5.6 we will obtain positive numbers δ_1 & δ_2 such that,

$$f(x) > f(a) \quad \text{for } a < x < a + \delta_1 \tag{2.5.7.1}$$

$$f(x) > f(b) \quad \text{for } b - \delta_2 < x < b \tag{2.5.7.2}$$

Now as f is differentiable in $[a, b]$ it is continuous there hence there exists $\xi \in [a, b]$ such that $f(\xi) = M$ where $M = \sup\limits_{x \in [a, b]} f(x)$. From (2.5.7.1) & (2.5.7.2) we conclude $\xi \ne a$ & $\xi \ne b$ hence $a < \xi < b$. If possible suppose $f'(\xi) > 0$ then by Theorem 2.5.6 there exists $\delta_3 > 0$ such that $(x) > f(\xi)$ for $\xi < x < \xi + \delta_3$ which contradicts the fact that f attains supremum at ξ hence $f'(\xi)$ cannot be positive. If possible suppose $f'(\xi) < 0$. Then by Theorem 2.5.7 there exists $\delta_4 > 0$ such that

$$f(x) > f(\xi) \quad \text{for } \xi - \delta_3 < x < \xi$$

which again contradicts the fact that f attains supremum at ξ hence $f'(x)$ cannot be negative. So that we conclude $f'(\xi) = 0$.

The above theorem leads to the important theorem called the intermediate value property of derivatives as given below

Theorem 2.5.8

Suppose $f : [a, b] \to \mathbb{R}$ be a differentiable function and let λ be any value between $f'(a)$ and $f'(b)$ then there exists at least one $\xi \in (a, b)$ such that $f'(\xi) = \lambda$.

Proof: We consider the function,

$$g(x) = f(x) - \lambda x$$

Then g is differentiable $[a, b]$ as f is so and

$$g'(x) = f'(x) - \lambda \qquad (2.5.8.1)$$

Also $g'(a)$ & $g'(b)$ are of opposite sign hence using Theorem 2.5.7 we must have at least one $\xi \in (a, b)$ such that $g'(\xi) = 0$ that is,

$$f'(\xi) = \lambda \qquad \text{(by (2.5.8.1))}$$

We observe that if a function is differentiable and the derivative does not vanish at a point in an interval then either the function is increasing or decreasing at that point. Hence it is not possible for the function to attain maximum or minimum at points where the derivative does not vanish. This is precisely the next theorem but before formally stating the theorem we need to know the following definition.

Definition   A function $f : D \subseteq \mathbb{R} \to \mathbb{R}$ is said to have a **local maximum** at $c \in D$ if there exists a $\delta > 0$ such that $f(x) < f(c) \ \forall \ x \in (c - \delta, c + \delta) \cap D$.

Similarly f is said to have a **local minimum** at $c \in D$ if exists a $\delta > 0$ such that $f(x) > f(c) \ \forall \ x \in (c - \delta, c + \delta) \cap D$.

f is said to have a **local extremum** at $c \in D$ if it either has a local maximum or local minimum at that point.

Theorem 2.5.9

Let $f : (a, b) \to \mathbb{R}$ has a local extremum at an interior point of the interval c. If f is differentiable at c then $f'(c) = 0$.

Proof: Suppose $f'(c) > 0$ then we can find some $\delta > 0$ such that,

$$f(x) \geq f(c) \ \forall \ x \in (c, c + \delta) \in (a, b)$$

and

$$f(x) \leq f(c) \ \forall \ x \in (c - \delta, c) \in (a, b)$$

which shows that f cannot attain local maximum or minimum at c.

By similar argument as above we can show that if $f'(c) < 0$ f cannot attain local extremum at c. Hence we conclude that if f attains local extremum at c then $f'(c) = 0$.

Note that $f'(c) = 0$ is only a necessary condition for the function to have a local extremum but it is not at all sufficient as evident from the following example.

Example 2.28   Let $f : (-1, 1) \to \mathbb{R}$ be a function defined as,

$$f(x) = x^3$$

then it is easy to see that $f'(0) = 0$. We find that $f(x) < f(0) = 0$ if $x < 0$ and

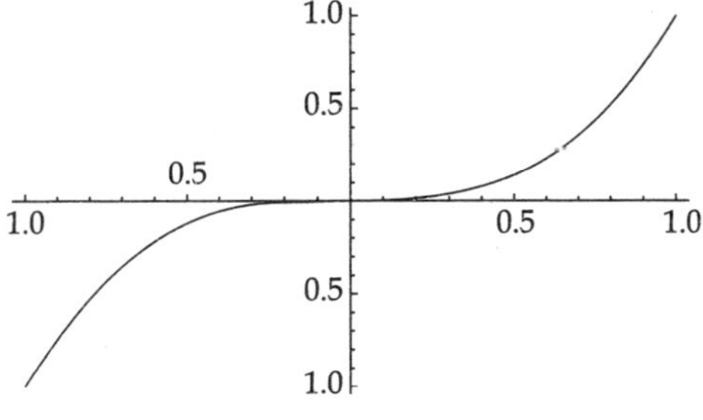

Fig. 2.6

$f(x) > f(0) = 0$ if $x > 0$ hence $x = 0$ is not a point of extremum.

We recall that the derivative of a function at any point denotes the slope of the tangent at that point. So by above theorem the tangent at the point where the function attains local extremum is parallel to X–axis.

Next theorem is known as **Rolle's Theorem** which is immediate corollary of above theorem .

Theorem 2.5.10 (Rolle's Theorem)

Let $f : [a, b] \to \mathbb{R}$ be a continuous function and suppose $f(a) = f(b)$. If further, the derivative of the function exists at every point of (a, b) then there exists at least one $c \in (a, b)$ such that $f'(c) = 0$.

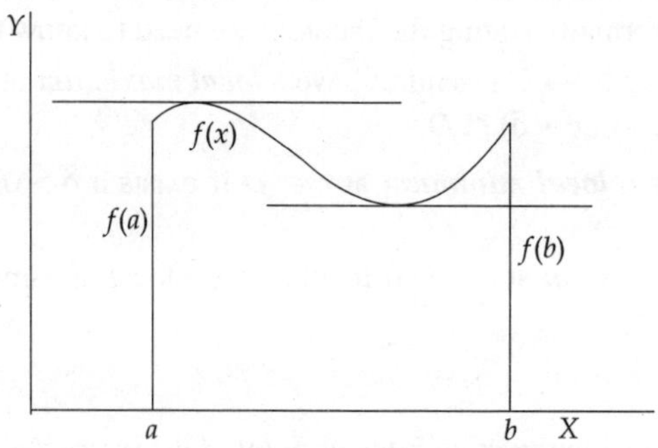

Fig. 2.7

Proof: We define $M = \sup\limits_{x \in [a, b]} f(x)$ & $m = \inf\limits_{x \in [a, b]} f(x)$ as f is continuous there exists points c_1 & $c_2 \in [a, b]$ such that $M = f(c_1)$ and $m = f(c_2)$. Now if $M = m$ the function is a constant and $f'(c) = 0 \ \forall \ x \in (a, b)$ and hence conclusion of the theorem is true in this case. Now if $M \neq m$ then at least one among $f(c_1), f(c_2)$ is distinct from $f(a), f(b)$. If $f(c_1) \neq f(a)$ then we have $a < c_1 < b$ and as f attains maximum at c_1, by previous theorem $f'(c_1) = 0$. Similarly if $f(c_2) \neq f(a)$ then we have $a < c_2 < b$ and as f attains minimum at c_1, by previous theorem $f'(c_2) = 0$. Hence in any case we see that there exists $c \in (a, b)$ such that $f'(c) = 0$.

Rolle's Theorem basically asserts the existence of a point on the curve $y = f(x)$ where the tangent is parallel to X-axis, provided f is sufficiently well behaved function. As an application of above Rolle's Theorem we get what is called **Mean Value Theorem** of differential Calculus.

Theorem 2.5.11 (Mean Value Theorem)

Let $f : [a, b] \to \mathbb{R}$ be a continuous function and suppose the derivative of the function exists at every point of (a, b) then there exists at least one $c \to (a, b)$ such that,

$$f'(c) = \frac{f(b) - f(a)}{b - a}.$$

Proof: If case $f(a) = f(b)$ then above theorem is nothing but Rolle's Theorem. If $f(a) \neq f(b)$ then we consider the $g : [a, b] \to \mathbb{R}$ as follows,

$$g(x) = f(x) \, (b - a) - x \, [f(b) - f(a)] \qquad (2.5.11.1)$$

We see that g is continuous in the interval $[a, b]$ and derivable in (a, b) as f is so. Also it is easy to see that $g(a) = g(b)$, then g satisfies all conditions of Rolle's Theorem and hence there exists some $c \in (a, b)$ such that $g'(c) = 0$ that is

$$f'(c) = \frac{f(b) - f(a)}{b - a} \quad \text{(using (2.5.11.1))}$$

Geometrically Mean Value theorem tells us that for a function f continuous in $[a, b]$ and derivable in (a, b) there exists at least one point P on the curve $y = f(x)$ such that the tangent at P is parallel to the line joining the points $(a, f(a))$ and $(b, f(b))$ (Fig.).

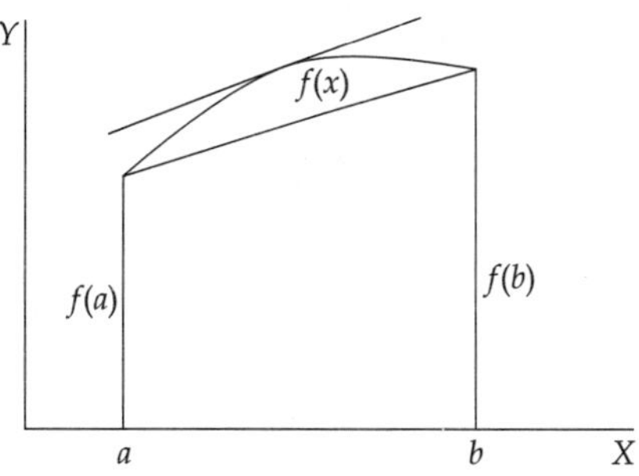

Fig. 2.8

Also note that as any point in (a, b) can be expressed as $a + \theta(b - a)$ for some θ satisfying $0 < \theta < 1$ we can state above theorem as,

Let $f : [a, b] \to \mathbb{R}$ be a continuous function and suppose the derivative of the function exists at every point of (a, b) then there exists at least one $\theta \in (0, 1)$ such that,

$$f'(a + \theta(b - a)) = \frac{f(b) - f(a)}{b - a}.$$

Example 2.29 Let $f : (a, b) \to \mathbb{R}$ be a differentiable function and suppose that $f'(x) = 0 \; \forall \, x \in (a, b)$ then show that $f(x)$ is a constant.

Sol: Let $x_1 > x_2$ be any two arbitrary points in (a, b) then by Mean Value Theorem we have,

$$f(x_1) - f(x_2) = (x_1 - x_2) f'(c) \text{ for some } c \in (x_1, x_2)$$

But as by given condition we have $f'(c) = 0$ which implies

$$f(x_1) = f(x_2)$$

Since x_1, x_2 arbitrary we conclude f is a constant.

As an immediate application of Mean Value theorem we get the following theorem which gives a sufficient condition under which a function is monotone increasing or decreasing.

Theorem 2.5.12

Let $f : (a, b) \to \mathbb{R}$ be a differentiable function then if,

(i) $f'(x) \geq 0$ $(f'(x) > 0)$ $\forall \, x \in (a, b)$ *then f is monotone (strictly monotone) increasing in (a, b)*

(ii) $f'(x) \leq 0$ $(f'(x) < 0)$ $\forall \, x \in (a, b)$ *then f is monotone (strictly monotone) decreasing in (a, b).*

Proof: Let $x_1, x_2 \in (a, b)$ such that $x_2 > x_1$ then as f is differentiable at x_1, x_2 it is also continuous there hence using Mean Value Theorem in $[x_1, x_2]$ we get,

$$f(x_2) - f(x_1) = f'(c)\,(x_2 - x_1) \text{ for some } c \in (x_1, x_2) \qquad (2.5.12.1)$$

If we assume $f'(x) \geq 0 \ \forall\, x \in (a, b)$ then $f'(c) \geq 0$ hence using (2.5.12.1) we conclude

$$f(x_2) - f(x_1) \geq 0 \text{ (as } x_2 - x_1 > 0)$$

That is f is monotone increasing in (a, b). If we assume $f'(x) > 0$ it follows from (2.5.12.1) that,

$$f(x_2) - f(x_1) > 0$$

that is, f is strictly monotone increasing in (a, b). This proves (i)

Similarly we can prove (ii).

Example 2.30   Show that,

$$\frac{\tan x}{x} > \frac{x}{\sin x} \text{ for } x \in \left(0, \frac{\pi}{2}\right)$$

Sol: We have to show $\dfrac{\tan x}{x} > \dfrac{x}{\sin x}$ for $x \in \left(0, \dfrac{\pi}{2}\right)$

Equivalently to show $\dfrac{\tan x \sin x - x^2}{x \sin x} > 0$ for $x \in \left(0, \dfrac{\pi}{2}\right)$

Now we observe that as $x \sin x > 0$ for $x \in \left(0, \dfrac{\pi}{2}\right)$ it is sufficient if we show,

$$\tan x \sin x - x^2 > 0 \text{ for } x \in \left(0, \frac{\pi}{2}\right)$$

Let
$$\phi(x) = \tan x \sin x - x^2 \text{ then,}$$
$$\phi'(x) = \tan x \cos x + \sec^2 x,\ \sin x - 2x$$
$$= \sin x + \sec^2 x \cdot \sin x - 2x$$

or,
$$\phi''(x) = \cos x + \sec x + 2 \sin x \sec^3 x \tan x - 2$$
$$= (\sqrt{\sec x} - \sqrt{\cos x})^2 + 2 \sin^2 x \sec^3 x$$

Now as $\sin^2 x \sec^3 x > 0$ for $x \in \left(0, \dfrac{\pi}{2}\right)$ we get,

$$\phi''(x) > 0 \text{ for } x \in \left(0, \frac{\pi}{2}\right)$$

This implies that $\phi'(x)$ is monotone increasing but as $\phi'(0) = 0$ we obtain $\phi'(x) > 0$. This again implies $\phi'(x)$ is monotone increasing but as $\phi(0) = 0$ we have $\phi(x) > 0$.

2.6   SUFFICIENT CONDITIONS FOR EXTREMUM OF FUNCTIONS

We will now look for sufficient conditions for a function having extreme values. First theorem uses the sign of first derivative near a point of the extreme value whereas the second theorem uses the sign of the second derivative at the point of extrme value.

Theorem 2.6.1 (Using first derivative)

Let $f : I \to \mathbb{R}$ where I is an interval and $a \in I$ be such that $f'(x)$ exists in $\{x : 0 < |x - a| < \delta\} \cap I$ for some $\delta > 0$ ($f'(a)$ (may or may not exist) also suppose that $f'(x)$ has a fixed sign for $x < a$ and $x > a$. Then as x increases through a,

    (i) *$f(x)$ has a minimum at if sign of $f'(x)$ changes from positive to negative.*

    (ii) *$f(x)$ has a maximum at $x = a$ if sign of $f'(x)$ changes from negative to positive.*

    (iii) *$f(x)$ has no extreme value at as $f'(x) > 0$ to the left of a, $f(x)$ is increasing $(a - \delta, a)$ if sign of $f'(x)$ does not change.*

Proof: (i) In this case as $f'(x) > 0$ to the left of a, $f(x)$ is increasing in $(a - \delta, a)$ similarly as $f'(x) < 0$ to the right of a, $f(x)$ is decreasing $(a, a + \theta)$. Hence at $x = a$ $f(x)$ is minimum.

    (ii) In this case as $f'(x) < 0$ to the left of a, $f(x)$ is decreasing in $(a - \delta, a)$ similarly as $f'(x) > 0$ to the right of a, $f(x)$ is increasing in $(a, a + \delta)$. Hence at $x = a$ is $f(x)$ maximum.

    (iii) In this case as $f(x) - f(a)$ has a constant sign in $0 < |x - a| < \delta$ we conclude

$f(x)$ has neither minimum or maximum at $x = a$.

Example 2.31   Show that $f(x) = |x|$ has a minimum at $x = 0$.

Sol: First note that $|x|$ is not differentiable at $x = 0$. Also we know that $f'(x) = -1$ if $x < 0$ and $f'(x) = 1$ if $x > 0$. So we see that as x passes through 0 the sign of $f'(x)$ changes from negative to positive hence from above theorem we conclude that $f(x)$ has a minimum at $x = 0$.

Example 2.32   Show that $x = 0$ is not an extreme value of $f(x) = x^3$.

Sol: As $f'(x) = 2x^2$ we see that $f'(x) > 0$ for $0 < |x| < \delta$ where $\delta > 0$ any number. So that as x passes through 0 $f'(x)$ does not change sign hence by above theorem we conclude that $f(x)$ has no extreme value at $x = 0$.

Theorem 2.6.2 (Second derivative test)

Let $f : I \to \mathbb{R}$ where I is an interval and a be a interior point of I. Suppose that

    (i) *$f'(a) = 0$*

    (ii) *$f'(a)$ exists and not equal to 0*

Then if,

    (a) *$f''(a) > 0$, f has a minimum at $x = a$*

    (b) *$f''(a) < 0$, f has a maximum at $x = a$*

Proof: As $f''(a)$ exists f, f' exists in suitably small neighbourhood of a say $N(a)$. If we now assume $f''(a) > 0$ then by, Theorem 2.5.5 there exists $\delta > 0$ such that $(a, a + \delta) \subset N(a)$ and we have,

$$f'(x) > f'(a) = 0 \; \forall \, x \in (a, a + \delta) \tag{2.6.2.1}$$

$$f'(x) < f'(a) = 0 \; \forall \, x \in (a, a + \delta) \tag{2.6.2.2}$$

Now by Mean Value Theorem we have,

$$f(a + \delta) = f(a) + hf'(a + \theta_1 \delta) \text{ for some } 0 < \theta_1 < 1 \tag{2.6.2.3}$$

$$f(a - \delta) = f(a) - hf'(a - \theta_2 h) \text{ for some } 0 < \theta_2 < 1 \tag{2.6.2.4}$$

By (2.6.2.1) & (2.6.2.2) we see that $f'(a + \theta_1\delta) > 0$ and $f'(a - \theta_2\delta) < 0$ hence from (2.6.2.3) & (2.6.2.4) we conclude $f(a + \delta) > f(a)$ and $f(a - \delta) > f(a)$ which proves that f attains minimum at $x = a$.

Similarly if $f''(a) < 0$, one can show f has a maximum at $x = a$

(**Note:** Above test has a generalized version see Problem 35)

Example 2.33   Find the extreme values of the function defined by,

$$f(x) = x \log x, \; x > 0$$

Sol: We have,

$$f'(x) = \log x + x \times \frac{1}{x} = 1 + \log x$$

$$f''(x) = \frac{1}{x}$$

Now $f'(x) = 0 \Rightarrow x = \frac{1}{e}$ so that, $f''\left(\frac{1}{e}\right) = e > 0$

Hence f as minimum at $x = \frac{1}{e}$.

2.7   HIGHER ORDER DERIVATIVES AND TAYLOR'S THEOREM

If f' denotes the derived function of $f : (a, b) \to \mathbb{R}$ and if at every point of the open interval f' is differentiable then we denote the derived function of f' as f'' and so on. In general for $n \in \mathbb{N} \, f^{(n)}$ denotes the derived function of $f^{(n-1)}$ whenever it exists. If $c \in (a, b), f'(c), f''(c), ..., ..., ..., f^{(n)}(c)$ are known as the first order derivative, second order derivative, ... n^{th} order derivatives of the function f at c respectively. Note that from definition it is clear that for $n \in \mathbb{N} \, f''(c)$ is the first order derivative of $f^{(n-1)}$ at c. We have seen that if $f'(c)$ exists then $f(c + h) - f(c)$ is approximately equal to the linear function $f'(c)h$ for small values of h. **Taylor's Theorem** is some what generalization of the above fact, it says that if the higher order derivatives of the function at c exists then $f(c + h) - f(c)$ is approximately equal to a polynomial in h provided h is small.

Theorem 2.7.1 (Taylor's Theorem)

Let $f : [a, a + h] \to \mathbb{R}$ be a function such that f^{n-1} continuous in the $[a, a + h]$ and f^n exists in $(a, a + h)$ then there exists $\xi \in (a, b)$ such that,

$$f(a + h) = f(a) + f'(a)h + \frac{h^2}{2!}f'(a) + \frac{h^3}{3!}f''(a) + ... + \frac{h^{n-1}}{(n-1)!}f^{n-1}(a) + \frac{h^n}{n!}f^n(\xi)$$

Proof: The continuity of f^{n-1} implies the existence and continuity of $f, f', f'' ... f^{n-2}$. We consider the function,

$$g(x) = f(x) + (a + h - x)f'(x) + \frac{(a + h - x)^2}{2!}f''(x) + ... + \frac{(a + h - x)^{n-1}}{(n-1)!}f^{(n-1)}(x) + (a + h - x)^n C$$

$$(2.7.1.1)$$

where C is a function chosen in such that,

$$g(a) = g(a + h).$$

That is,

$$f(a + h) = f(a) + hf'(a) + \frac{h^2}{2!}f''(a) + ... + \frac{h^{n-1}}{(n-1)!}f^{(n-1)}(a) + h^n C \qquad (2.7.1.2)$$

We see that,

$$g'(x) = \frac{(a+h-x)^{n-1}}{(n-1)!} f^n(x) - n(a+h-x)^{n-1} C. \tag{2.7.1.3}$$

Hence $g'(x)$ exists in the open interval $(a, a+h)$ as $f^n(x)$ is so and $(a+h-x)^{n-1}$ is everywhere differentiable also it is clear from (2.6.1.1) that g is continuous. By choice of C

$$g(a) = g(a+h)$$

As g satisfies all conditions of Rolle's Theorem in $[a, a+h]$ we get some $\xi \in (a, b)$ such that,

$$g'(c) = 0$$

which implies $C = \dfrac{f^n(c)}{n!}$ (using 2.7.1.3)

Again putting the value C in (2.7.1.2) we get.

$$f(a+h) = f(a) + hf'(a) + \frac{h^2}{2!}f^n(a) + \ldots + \frac{h^{n-1}}{(n-1)!}f^{(n-1)}(a) + \frac{h^n}{n!}f^n(c).$$

Note: The term $R_n = \dfrac{h^n}{n!}f^n(c)$ is called the ***Lagrange's form of remainder*** after terms and can also be written as $\dfrac{h^n}{n!}f^n(a+\theta h)$ where $0 < \theta < 1$.

There are other versions of Taylor's Theorem available where we have different different forms of remainder after n terms.

Theorem 2.7.2 (Taylor's Theorem with Schlomich-Roche's form of remainder)

Let $f : [a, a+h] \to \mathbb{R}$ be a function such that f^{n-1} continuous in the $[a, a+h]$ and f^n exists in $(a, a+h)$ then for any natural number p there exists $\theta \in (0, 1)$ such that,

$$f(a+h) = f(a) + f'(a)\,h + \frac{h^2}{2!}f'(a) + \frac{h^3}{3!}f''(a) + \ldots + \frac{h^{n-1}}{(n-1)!}f^{n-1}(a) + \frac{(1-\theta)^{n-p}}{(n-1)!\,p}f^n(a+\theta h).$$

Proof: We define $g(x)$ a follows,

$$g(x) = f(x) + (a+h-x)f'(x) + \frac{(a+h-x)^2}{2!}f''(x) + \ldots + \frac{(a+h-x)^{n-1}}{(n-1)!}f^{(n-1)}(x) + (a+h-x)^p C \tag{2.7.2.1}$$

Then as in Theorem 2.7.2 we see that all conditions of Rolle's Theorem in $[a, a+h]$ we get some $\theta \in (0, 1)$ such that,

$$g'(a+\theta h) = 0 \tag{2.7.2.2}$$

$$g'(x) = \frac{(a+h-x)^{n-1}f^n(x)}{(n-1)!} - Cp(a+h-x)^{p-1} \tag{2.7.2.3}$$

Using (2.7.2.1) & (2.7.2.2) we get,

$$C = \frac{h^{n-p}(1-\theta)^{n-p}f^n(a+\theta h)}{(n-1)!\,p} \tag{2.7.2.4}$$

From (2.7.2.1) & (2.7.2.2) we obtain,

$$f(a + h) = f(a)\,f'(a)h + \frac{h^2}{2!}f'(a) + \frac{h^3}{3!}f''(a) + \ldots + \frac{h^{n-1}}{(n-1)!}f^{n-1}(a) + \frac{(1-\theta)^{n-p}}{(n-1)!\,p}f^n(a + \theta h)$$

Note: The term $R_n = \dfrac{(1-\theta)^{n-p}}{(n-1)!\,p}f^n(a + \theta h)$ is the Schlomich-Roche's form of remainder after n terms.

Remarks

1. If we put $p = n$, $R_n = \dfrac{1}{(n-1)!\,n}f^n(a + \theta h) = \dfrac{f^n(a + \theta h)}{n!}$ which is **Lagranges** form of remainder

2. If we put $p = 1$ we get $R_n = \dfrac{(1-\theta)^{n-1}}{(n-1)!}f^n(a + \theta h)$ which is called **Cauchy** form of remainder.

2.8 TAYLOR'S SERIES AND MACLAURIN'S SERIES

Let I be an open interval and $f : I \to \mathbb{R}$ is infinitely differentiable function at a point $a \in I$ then the series $\displaystyle\sum_{n=0}^{\infty} \frac{f^n(a)}{n!}(x - a)^n$, $x \in I$ is called Taylor's Infinite series of f around a. In particular if $a = 0$ the above seires becomes $\displaystyle\sum_{n=0}^{\infty} \frac{f^n(0)}{n!}x^n$ and is called Maclaurin's series of f.

Now we may ask the following question: If $\displaystyle\sum_{n=0}^{\infty} \frac{f^n(0)}{n!}x^n$ is the Maclaurin's series of f defined in an open interval I containing 0 then is it true that $f(x) = \displaystyle\sum_{n=0}^{\infty} \frac{f^n(0)}{n!}x^n$, $\forall\, x \in I$?

First note that for $x = 0$ the Maclaurin's series of f converges to $f(0)$. But it may happen that for other values of $x \in I$ the Maclaurin's series may not converge to $f(x)$ as is evident from Problem 34.

Now observe that using Taylor's Theorem we have,

$$f(x) = f(0) + f'(0)x + \frac{x^2}{2!}f'(0) + \frac{h^3}{3!}f''(0) + \ldots + \frac{x^{n-1}}{(n-1)!}f^{n-1}(0) + R_n, \; x \in I$$

where R_n is the remainder after n terms.

That is,

$$f(x) - P_n(x) = R_n$$

where $P_n(x) = f(0) + f'(0)x + \dfrac{x^2}{2!}f'(0) + \dfrac{x^3}{3!}f''(0) + \ldots + \dfrac{x^{n-1}}{(n-1)}f^{n-1}(0)$ is the nth partial sum of Maclaurin's Series of the function. Hence if $R_n \to 0$ as $n \to \infty$ then the Maclaurin's Series of the function converges to the function.

Example 2.34 Find the Maclaurin's Series of the function $f(x) = \sin x$ and show that it converges to the function.

Sol: We know that,

$$f^n(x) = \sin\left(x + \frac{n\pi}{2}\right) \forall\, x \in \mathbb{R} \text{ and } n \in \mathbb{N}$$

Now the remainder after n terms in the Taylor's expansion of $f(x)$ is given by,

$$R_n = \frac{x^n}{n!} \sin\left(\theta x + \frac{n\pi}{2}\right)$$

Hence,

$$|R_n| \leq \left|\frac{x^n}{n!}\right| = \frac{|x^n|}{n!} \to 0 \text{ as } n \to \infty.$$

So that Maclaurin's Series of $f(x)$ converges to $f(x)$. Now

$$f^n(0) = \sin\left(\frac{n\pi}{2}\right) = \begin{cases} 0 & \text{if } n \text{ is even} \\ (-1)^{k+1} & \text{if } n = 2k-1 \end{cases}$$

Hence the Maclaurin's Series of $f(x)$ is given by,
and we can write

$$x - \frac{x^3}{3!} + \frac{x^5}{5!} - \cdots \frac{(-1)^{2k-1}}{(2k-1)!} x^{2k-1} + \ldots \ \forall \ x \in \mathbb{R}$$

2.9 L'HOSPITAL'S RULE

In this section we are going to discuss the existence of the limits of the form $\lim\limits_{x \to a^+} \dfrac{f(x)}{g(x)}$ when $f(x)$, $g(x)$ both tends to zero or ∞ as $x \to a^+$ and also we will give a technique to evaluate such limits whenever they exist known as **L'Hospital's Rule**. We will need the following theorem know as Cauchy's Mean Value Theorem.

Theorem 2.9.1

Let f, g be two functions both continuous in the clued interval $[a, b]$ and derivable in the open interval (a, b). Further suppose that $g'(x) \neq 0$ for $x \in (a, b)$ then there exists $\xi \in (a, b)$ such that $\dfrac{f(b) - f(a)}{g(b) - g(a)} = \dfrac{f'(\xi)}{g'(\xi)}$

Proof: We consider the function

$$F(x) = f(x)\,[g(b) - g(a)] + g(x)\,[f(b) - f(a)]$$

It is easy to see that,

   (i) F is continuous in $[a, b]$

   (ii) F is derivable in (a, b)

   (iii) $F(a) = F(b)$

Hence by Rolle's Theorem there exists $\xi \in (a, b)$ such that $F'(\xi) = 0$.
Thus we have,

$$\frac{f(b) - f(a)}{g(b) - g(a)} = \frac{f'(\xi)}{g'(\xi)}$$

Theorem 2.9.2

Let f, g be two functions both continuous in $[a, a + h]$ and derivable in $(a, a + h)$ and suppose $g'(x) \neq 0$

$\forall\, x \in (a, a + h)$. *If $f(a) = g(a) = 0$ and $\displaystyle\lim_{x \to a^+} \frac{f'(a)}{g'(a)} = l$ then,*

$$\lim_{x \to a^+} \frac{f(x)}{g(x)} = l. \text{ Here } l \text{ is a real number or } +\infty \text{ or } -\infty.$$

Proof: We see that f & g satisfies all conditions of Cauchy's Mean Value Theorem hence, there exists $\xi \in (a, a + h)$ such that,

$$\frac{f'(\xi)}{g'(\xi)} = \frac{f(a + h) - f(a)}{g(a + h) - g(a)} = \frac{f(a + h)}{g(a + h)} \tag{2.9.2.1}$$

As $h \to 0$ we have $a + h \to a^+$ & $\xi \to a^+$ hence from (2.9.2.1) we get,

$$\lim_{x \to a^+} \frac{f(x)}{g(x)} = \lim_{x \to a^+} \frac{f'(x)}{g'(x)} = l$$

Note that above form of limit is called $\dfrac{0}{0}$ form.

Theorem 2.9.3

Let f, g be two functions both continuous in $[a, a + h]$ and derivable in $(a, a + h)$. Further suppose $g'(x) \neq 0 \; \forall\, x \in (a, a + h)$ and $\displaystyle\lim_{x \to a^+} f(x) = \lim_{x \to a^+} g(x) = \infty$ then if,

$$\lim_{x \to a^+} \frac{f'(x)}{g'(x)} = l \text{ or } \pm\infty \text{ then, } \lim_{x \to a^+} \frac{f(x)}{g(x)} = l \text{ or } \pm\infty.$$

Proof: First let us assume l is finite and not equal to zero. As $\displaystyle\lim_{x \to a^+} \frac{f'(x)}{g'(x)} = l$. For $\varepsilon > 0$ there exists some $h_1 > 0$ such that,

$$\left| \frac{f'(x)}{g'(x)} - 1 \right| < \frac{\varepsilon}{2} \quad \text{whenever } x \in (a, a + h_1) \tag{2.9.3.1}$$

also there exists some $h_2 > 0$ such that,

$$\frac{f'(x)}{g'(x)} < \frac{l}{4} \quad \text{whenever } x \in (a, a + h_2) \tag{2.9.3.2}$$

Choose $\delta = \min\,(h, h_1, h_2) > 0$ and let $a < x < b < a + \delta$. We observe that f, g satisfies all conditions of Cauchy's Mean Value Theorem in the interval $[x, b]$ hence there exists $\xi \in (x, b)$ such that,

$$\frac{f(x) - f(b)}{g(x) - g(b)} = \frac{f(x)}{g(x)} \cdot \frac{1 - \dfrac{f(b)}{f(x)}}{1 - \dfrac{g(b)}{g(x)}} = \frac{f'(\xi)}{g'(\xi)}$$

Hence,

$$\frac{f(x)}{g(x)} = \frac{f'(\xi)}{g'(\xi)} \frac{1 - \dfrac{g(b)}{g(x)}}{1 - \dfrac{f(b)}{f(x)}} = \frac{f'(\xi)}{g'(\xi)} \phi(x) \tag{2.9.3.3}$$

where,

$$\phi(x) = \frac{1 - \dfrac{g(b)}{g(x)}}{1 - \dfrac{f(b)}{f(x)}}$$

Now from given condition we have,

$$\lim_{x \to a^+} \frac{f(b)}{f(x)} = 0 \ \& \ \lim_{x \to a^+} \frac{g(b)}{g(x)} = 0 \tag{2.9.3.4}$$

So that,

$$\phi(x) \to 1 \text{ as } x \to a^+$$

Then for $\varepsilon > 0$ there exists $\delta_1 > 0$ such that ,

$$|\phi(x) - 1| < \frac{2\varepsilon}{l} \quad \text{for } a < x < a + \delta_1 \tag{2.9.3.5}$$

Hence there exists a sufficiently small neighbourhood of a say $(a, a + \gamma)$ such that, for $x \in (a, a + \gamma)$ we have,

$$\left| \frac{f(x)}{g(x)} - l \right| = \left| \frac{f'(x)}{g'(x)} \phi(x) - \frac{f'(x)}{g'(x)} + \frac{f'(x)}{g'(x)} - l \right| \quad \text{(using (2.9.3.3))}$$

$$\leq \left| \frac{f'(x)}{g'(x)} \right| |\phi(x) - 1| + \left| \frac{f'(x)}{g'(x)} - l \right|$$

$$< \frac{l}{4} \frac{2\varepsilon}{l} + \frac{\varepsilon}{2} = \varepsilon$$

That is,

$$\lim_{x \to a^+} \frac{f(x)}{g(x)} = l$$

The case $l = 0$ or $\pm\infty$ can be disposed by slightly modifying above argument.

Note that the above limit is popularly known is $\frac{\infty}{\infty}$ form.

Example 2.34 Evaluate $\lim\limits_{x \to 0} \dfrac{x - \sin x}{x^3}$

Sol: Clearly above is $\dfrac{0}{0}$ form. Let $g(x) = \dfrac{x - \sin x}{x^3}$ then g satisfies all conditions of L'Hospital's Rule (Theorem 2.7.2) hence by the differentiating numerator and denominator of g we get,

$$\lim_{x \to 0} \frac{x - \sin x}{x^3} = \lim_{x \to 0} \frac{1 - \cos x}{3x^2} \left(\frac{0}{0} \right)$$

$$= \lim_{x \to 0} \frac{\sin x}{6x} \quad \left(\frac{0}{0}\right)$$

$$= \lim_{x \to 0} \frac{\cos x}{6} = \frac{1}{6} \quad (\text{as } \lim_{x \to 0} \cos x = 1)$$

Example 2.35   Evaluate $\lim\limits_{x \to 0} \dfrac{e^{ax} - e^{bx}x}{\log(1 + bx)}$

Sol: Again as before the above limit is in $\left(\dfrac{0}{0}\right)$ form and using L'Hospital's Rule, (as all conditions of Theorem 2.7.2 are satisfied) we get,

$$\lim_{x \to 0} \frac{e^{ax} - e^{bx}}{\log(1 + bx)} = \lim_{x \to 0} \frac{ae^{ax} - be^{bx}}{\dfrac{b}{1 + bx}}$$

$$= \frac{a - b}{b}$$

Example 2.36   Evaluate $\lim\limits_{x \to 0} \dfrac{\log x}{\log (e^x - 1)}$

Sol: The above limit is clearly in $\left(\dfrac{\infty}{\infty}\right)$ form. Using L'Hospital's Rule as all conditions of Theorem 2.9.3 are satisfied we get,

$$\lim_{x \to 0} \frac{\log x}{\log (e^x - 1)} = \lim_{x \to 0} \frac{\dfrac{1}{x}}{\dfrac{e^x}{e^x - 1}}$$

$$= \lim_{x \to 0} \frac{e^x - 1}{xe^x} \quad \left(\frac{0}{0}\right) \text{ form and hence using L'Hospital's Rule}$$

we get,

$$= \lim_{x \to 0} \frac{e^x}{xe^x + e^x} = 1.$$

PROBLEMS

1. From the definition of limit show the following:

   (i) $\lim\limits_{x \to 2} (x^2 - 2) = 2$

   (ii) $\lim\limits_{x \to 0} x^p \cos \dfrac{1}{x} = 0, \quad p > 0$

   (iii) $\lim\limits_{x \to \infty} \dfrac{1}{x^2 - 1} = 0$

   (iv) $\lim\limits_{x \to 2^+} \dfrac{1}{x^2 - 4} = \infty$

2. Using algebra of limits show the following:

   (i) $\lim\limits_{x \to 0} \dfrac{\sqrt{1 + x} - \sqrt{1 - x}}{x} = 1$

   (ii) $\lim\limits_{x \to \infty} \dfrac{x^3 + 3x - 2}{2x^4 - 3x^3 + 4x^2 - 1} = 0$

3. Show that $\lim\limits_{x \to 0} \dfrac{1}{1 + e^{\frac{1}{x}}}$ does not exist

4. Suppose f, g are two real valued functions and if

$$h(x) = \frac{f(x) + \{2g(x) - f(x)\} \cos^{2n} \pi x}{1 + \cos^{2n} \pi x}$$

then show that,

$$h(x) = \begin{cases} g(x) & \text{if } x \text{ is an integer} \\ f(x) & \text{otherwise} \end{cases}$$

5. Let $f : [0, 1] \to [0, 1]$ be defined as follows:

$$f(x) = \begin{cases} 0 & \text{if } x \text{ is a irrational number or zero} \\ \dfrac{1}{p} & \text{if } x = \dfrac{p}{q}\, p, q \text{ in lowest terms}, p \neq 0 \end{cases}$$

6. Show that f is discontinuous at every rational point of $[0, 1]$ and continuous at every irrational number and zero.

7. If $f : [0, 1] \to [0, 1]$ be defined as follows:

$$f(x) = \begin{cases} 0 & \text{if } x \text{ is a irrational number or zero} \\ 1 & \text{if } x \text{ is a rational number} \end{cases}$$

   Show that f is discontinuous at every point of $[0, 1]$.

8. Show that the function $f : [0, 1] \to [0, 1]$ defined by

$$f(x) = \begin{cases} -x & \text{if } x \text{ is a irrational number or zero} \\ x & \text{if } x \text{ is a rational number} \end{cases}$$

   is continuous only at $x = 0$

9. Find the set of points of continuity of the function $f : \mathbb{R} \to \mathbb{R}$ defined by

$$f(x) = \begin{cases} x & \text{if } x \text{ is a irrational number or zero} \\ 0 & \text{if } x \text{ is a rational number} \end{cases}$$

10. Let $f : \mathbb{R} \to \mathbb{R}$ be any function and E be any subset of $\mathbb{R}$. We define the set $f^{-1}(E)$ as $f^{-1}(E) = \{x \in \mathbb{R} : f(x) \in E\}$. Show that f is continuous at $x = a$ if and only if for every open neighbourhood E of $f(a)$, $f^{-1}(E)$ is an open neighbourhood of a.

11. Let f, g be two functions defined on the set of reals. We define two functions h & g as, $h(x) = \max\,(f(x), g(x))$ & $g(x) = \min\,(f(x), g(x))$. Show that if f, g is continuous at a then h & g are also continuous at a.

12. Suppose f, g be two functions defined on the set of reals satisfies $f(x) = g(x)$ whenever x is rational. Show that $f(x) = g(x)$ for all real number x.

13. If a continuous function $f : \mathbb{R} \to \mathbb{R}$ satisfies the functional equation,

$$f(x + y) = f(x) + f(y)\ \forall\ x, y \in \mathbb{R}$$

    when show that there exists a real number k such that,

$$f(x) = kx\ \forall\ x \in \mathbb{R}$$

14. Let $f : D \to \mathbb{R}$ and $g : E \to \mathbb{R}$ be two real valued functions such that the range of f is contained in E. If f is continuous at $x = a$ and g is continuous at $f(a)$ then show that $f \circ g$ is continuous at $x = a$.

15. Show that the above result is not true if we do not assume f is continuous at $x = a$ or is continuous at $f(a)$.

16. For any non empty set $E \subset \mathbb{R}$. Let $f : \mathbb{R} \to \mathbb{R}$ be defined as follows,

    $$f(x) = \inf_{y \in A} |x - y|$$

    Show that f is continuous.

17. Let $f : \mathbb{R} \to \mathbb{R}$ be a continuous function. Define $E = \{x \in \mathbb{R} : f(x) = 0\}$ then if $\{y_n\}$ is a sequence in E such that $\lim_{n \to \infty} = y$ then show that $y \in E$.

18. For a function $f : \mathbb{R} \to \mathbb{R}$ we define 'Graph' of f denoted by G_f as,

    $$G_f = \{(x, f(x)) : x \in \mathbb{R}\}$$

    If f is continuous and if $(x_n, y_n) \in G_f$ such that $x_n \to x$ & $y_n \to y$ then show that $y = f(x)$.

19. A function $f : \mathbb{R} \to \mathbb{R}$ satisfies the following condition

    $$|f(x) - f(y)| < K |x - y| \ \forall \ x, y \in \mathbb{R} \text{ for some } K > 0$$

    Show that f is continuous at every point of $\mathbb{R}$.

20. A function $f : [a, b] \to \mathbb{R}$ satisfies the following condition

    $$|f(x) \to f(y)| < K |x - y| \ \forall \ x, y \in \mathbb{R} \text{ for some } K > 0$$

    Show that f is uniformly continuous.

21. Show that $f(x) = x^2$ is uniformly continuous on any closed interval $[a, b]$ but not in the whole real line.

22. Let $I \subset \mathbb{R}$ and $f : I \to \mathbb{R}$ be a uniformly continuous function. Then if $\{x_n\}$ is a cauchy sequence in I show that $\{f(x_n)\}$ is also a cauchy sequence.

23. Let $f, g : \mathbb{R} \to \mathbb{R}$ be uniformly continuous. Show that kf & $f + g$ are uniformly continuous for $k \in \mathbb{R}$. What can you say about the product fg?

24. Give an example of a uniformly continuous function $f \neq 0$ on a set but $\dfrac{1}{f}$ is not uniformly continuous.

25. If $f : \mathbb{R} \to \mathbb{R}$ is defined as,

    $$f(x) = |x - 1| + |x - 3|$$

    Find the points where f is not derivable. Can you construct a function which is derivable at every point of $\mathbb{R}$ except finitely many points $a_1, a_2, \dots a_n$.

26. Let $f : (a, b) \to \mathbb{R}$ be a differentiable function. If $c \in (a, b)$ such that $f(c) = 0$ then show that $g'(c)$ exists if and only if $f'(c) = 0$ where $g(x) = |f(x)|$.

27. A function $f : (a, b) \to \mathbb{R}$ satisfies the following condition:

    $$|f(x) - f(y)| < K |x - y| \ \forall \ x, y \in (a, b) \text{ for some } K > 0$$

    Show that f is derivable at every point of I.

28. Suppose $f : (a, b) \to \mathbb{R}$ is differentiable such that $|f'(x)| < K \ \forall \ x \in (a, b)$. Then show that $|f(x) - f(y)| < K |x - y| \ \forall \ x, y \in (a, b)$.

29. Suppose $f : (a, b) \to \mathbb{R}$ with finite derivative at each point of the interval. If for $c \in (a, b) \lim_{x \to c} f'(x) = A$ then show that $A = f'(c)$.

30. If $f : \mathbb{R} \to \mathbb{R}$ is a differentiable function with $f(0) = 0$ & $f'(x) \le 0 \;\forall x$. Show that $f(x) \le 0$ for all $x \in \mathbb{R}$.

31. Let $f : \mathbb{R} \to \mathbb{R}$ be a differentiable function then by applying Mean Value Theorem on $f(x)$ in $[0, h]$ one obtains,

    $f(h) = f(0) + hf'(\theta h)$ for some $\theta(h)$ such that $0 < \theta(h) < 1$

    Show that $\lim\limits_{x \to 0} \theta(h) = \dfrac{1}{2}$ if $f(x) = \cos x$

32. Let the functions $f,\, g,\, f',\, g'$ be continuous in $[a, b]$ such that,

    $$f(x)\, g'(x) - f'(x)\, g(x) \ne 0$$

    Show that between any two zeros of the function $f(x)$ in $[a, b]$ there lies one zero of the function $g(x)$ and conversely.

33. Use Mean Value Theorem to show,

    (i) $\dfrac{x}{1+x} < \log(1+x) < x, \quad x > 0$

    (ii) $0 < \dfrac{1}{x} \log\left(\dfrac{e^x - 1}{x}\right) < 1, \quad x > 0$

34. Show that the function $f : \mathbb{R} \to \mathbb{R}$ defined by,

    $$f(x) = \begin{cases} e^{\frac{-1}{x^2}} & \text{if } x \ne 0 \\ 0 & \text{if } x = 0 \end{cases}$$

    is infinitely differentiable function but the Maclaurin's series of this function does not converge to $f(x)$ for any value of x except 0.

    [Hint: For any $n \in \mathbb{N}$ and $x \ne 0$ we can show by induction that,

    $$f^n(x) = \dfrac{f(x)}{p\left(\dfrac{1}{x}\right)} \text{ where } p\left(\dfrac{1}{x}\right) \text{ is some polynomial in } \left(\dfrac{1}{x}\right).$$

    Again, it is not difficult to show that $\lim\limits_{x \to 0} \dfrac{f(x)}{p\left(\dfrac{1}{x}\right)} = 0$

    Again use induction to show that $f^n(0) = 0 \;\forall\, n$

    Hence Maclaurin's series of the function is identically 0 but $f(x)$ is not (Identically zero function)

35. (Test for Extremum of a function)

    Let f be a function defined on an open interval I if for any point $a \, Î \, I$ we have,

    (i) $f'(a) = f''(a) = f''(a) \ldots = f^{n-1}(a) = 0$

    (ii) $f^n(a) \ne 0$

then show that at $x = a$ the function f

  (a) no extreme value if n an odd integer

  (b) has extreme value if n an even integer and in that case $f(a)$ is a maximum if $f''(a) < 0$ and a minimum if $f''(a) > 0$.

    (Hint: Use Taylor's Theorem)

36. Evaluate the following limits using L'Hospitals Rules:

(i) $\displaystyle\lim_{x \to 0} \frac{\tan x - x}{x - \sin x}$

(ii) $\displaystyle\lim_{x \to 0^+} \left(\frac{1}{x} - \frac{1}{\sin x} \right)$

(iii) $\displaystyle\lim_{x \to 0} \left(\frac{\sin x}{x} \right)^{\frac{1}{x}}$

(iv) $\displaystyle\lim_{x \to \infty} x^{\frac{1}{x}}$

(v) $\displaystyle\lim_{x \to \infty} \frac{\log x}{\sqrt{x}}$

(vi) $\displaystyle\lim_{x \to \infty} \frac{x^n + x^{n-1} + x^{n-2} + \ldots + 1}{e^{ax}}$,

$a > 0$ & n is a positive integer.

Chapter 3

The Riemann Integral

3.1 INTRODUCTION

The origin of process of integration can be traced back to the days of Archimedes, who derived formulae for calculating areas of several kinds of figures. He used an argument based on "method of exhaustion" to derive these formulae, which is a technique by which a figure is approximated by finite union of simple figures whose areas are known.

It was *B. Riemann* who in 1854 proposed a rigorous mathematical theory of process of integration. Later *G. Darboux* presented a more elegant definition of the integral and showed it is equivalent to that of Riemann's definition. It is his approach which is used in most contemporary presentation.

3.2 MOTIVATIONS FOR ANALYTIC DEFINITION OF INTEGRAL

We here present ideas which leads to the definition of Integral of a function. Suppose A is the area bounded by $x = a$, $y = b$ and graph of a positive function $f(x)$

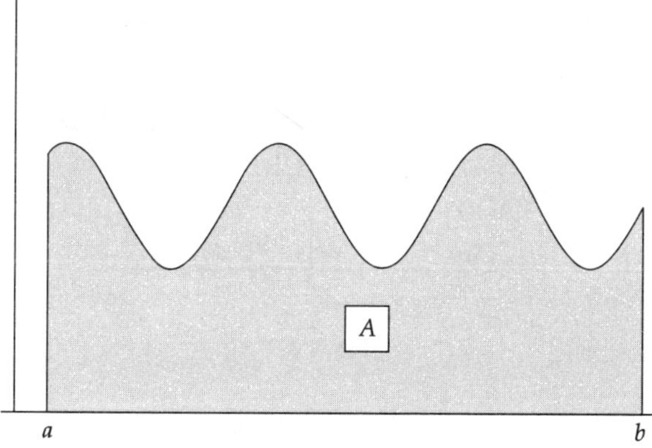

Fig. 3.1

(Fig. 3.1). We call this as area "under the curve $f(x)$". Assuming that area of A is a real number our aim is to find this real number. We here use argument based on "exhaustion principle". We first divide the interval $[a, b]$ into n small intervals (not necessarily equal) by introducing n points in between a and b (Fig. 3.2).

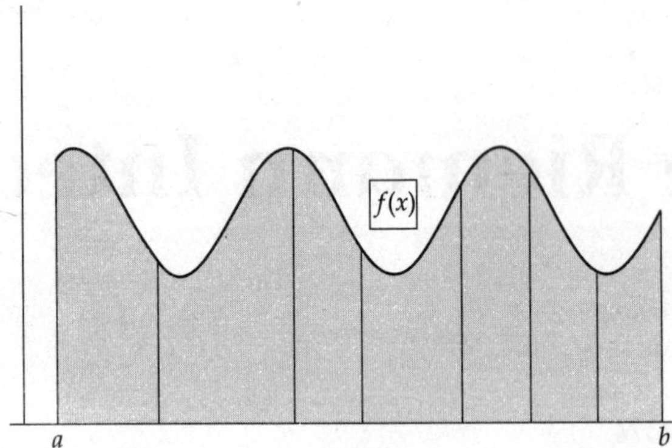

Fig. 3.2

We now draw vertical lines through these points which divide the whole area A into n stripes. Now we see that it is as difficult to calculate the areas of these smaller stripes because of the presence of curve side. So we approximate areas of these strips both from above and from below by areas of the circumscribed and inscribed rectangle with same base, where the curved boundary of the strip is replaced by a horizontal line at a distance from X-axis which is either the greatest or the smallest value of $f(x)$ in the strip. Let A_n^1 denotes sum of areas n circumscribed rectangles and A_n^2 denotes sum of areas n inscribed rectangles then definitely we have $A_n^1 \leq A \leq A_n^2$ (Fig. 3.3, Fig. 3.4)

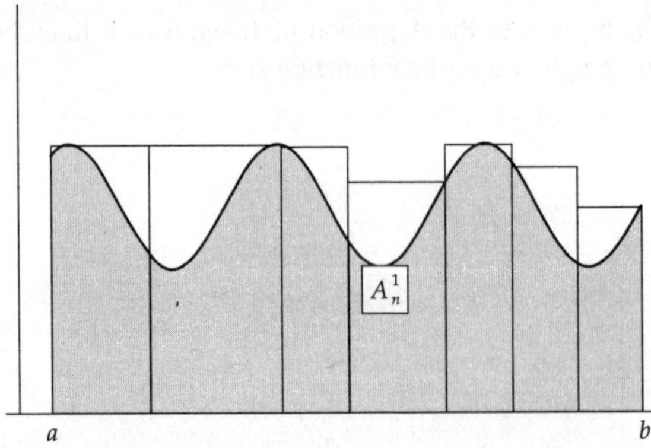

Fig. 3.3

Intuitively we can see that as we increase the numbers of subintervals i.e. if strips get finer and finer then both A_n^1 and A_n^2 will tend to A from above and below respectively.

Above ideas when put in proper mathematical form gives rise to the analytical definition of Riemann Integral of a function.

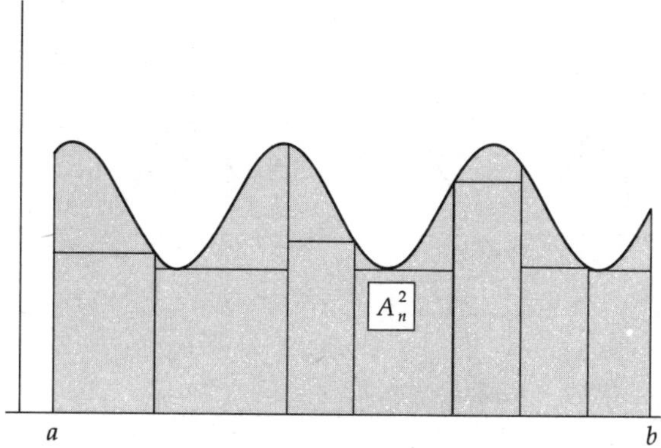

Fig .3.4

3.3 NOTATIONS, DEFINITIONS AND BASIC LEMMAS

In this section we will give some basic definitions and prove some Lemmas which are necessary to define Riemann Integral of a function. We begin with some notations which will be fixed through out this chapter.

Let f be a bounded function whose domain is the closed interval $[a, b]$. By partition P of $[a, b]$ [we mean collection of points $\{x_0, x_2, ..., x_r, ..., x_n\}$ in $[a, b]$ such that $a = x_0 < x_2 < ... < x_r < ... < x_n = b$. The rth sub-interval of $P[x_{r-1}, x_r]$ will be denoted by Δ_r and δ_r will denote the length of the rth sub-interval that is $\delta_r = x_r - x_{r-1}$. By norm of the partition P we mean maximum value of δ_r, $r = 1, 2, ..., n$ and denoted by $\|P\|$. As f is a bounded function in $[a, b]$ we define $M, m, M_r,$ and m_r as follows,

$$M = \sup_{x \in [a, b]}, m = \inf_{x \in [a, b]} f(x), M_r = \sup_{x \in \delta_r} f(x) \text{ and } m_r = \inf_{x \in \delta_r} f(x).$$

For the partition P the quantities $\sum_{r=1}^{n} M_r \delta_r$ and $\sum_{r=1}^{n} m_r \delta_r$ will be called the *Upper Sum* and *Lower Sum* for P and will be denoted by $U(P, f)$ and $L(P, f)$ respectively.

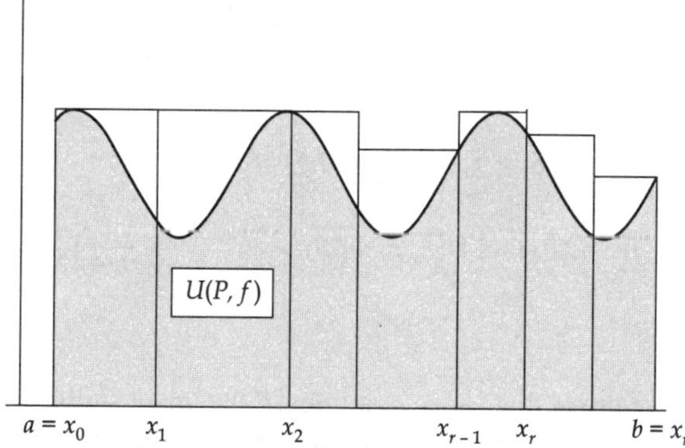

Fig. 3.5

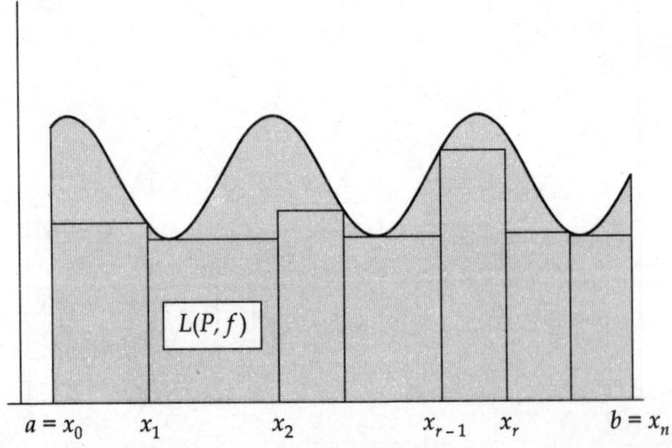

Fig. 3.6

Definition A partition P' is said to be a *refinement* of the partition P of the interval $[a, b]$ if P' contains some more points of $[a, b]$ than P, that is $P' \supset P$.

Remark: It is easy to see that $\|P'\| \le \|P\|$.

Definition For two partitions P and P' of $[a, b]$ their *common refinement* is defined to be a partition of $[a, b]$ consisting of all points of P and P', that is if P'' is common refinement of P and P' then $P'' = P \cup P'$.

Remark: Observe that in this case $\|P''\| \le \|P\|$ and also $\|P''\| \le \|P'\|$.

Now we will prove some lemmas which will be useful in the development of the theory. For the rest of this section we will take f to be a bounded function on $[a, b]$.

Lemma 3.3.1

For a partition P of $[a, b]$ have, $m(b - a) \le L(P, f) \le U(P, f) \le M(b - a)$.

Proof: We see that for the bounded function f we have,

$$m \le m_r \le M_r \le M$$

or,

$$m\,\delta_r \le m_r\,\delta_r \le M_r\,\delta_r \le M\,\delta_r$$

or,

$$\sum_{r=1}^{n} m\,\delta_r \le \sum_{r=1}^{n} m_r\,\delta_r \le \sum_{r=1}^{n} M_r\,\delta_r \le \sum_{r=1}^{n} M\,\delta_r$$

or,

$$m \sum_{r=1}^{n} \delta_r \le \sum_{r=1}^{n} m_r\,\delta_r \le \sum_{r=1}^{n} M_r\,\delta_r \le M \sum_{r=1}^{n} \delta_r$$

or,

$$m(b - a) \le L(P, f) \le U(P, f) \le M(b - a)$$

observing that $\sum_{r=1}^{n} \delta_r = b - a$.

Remark: In this lemma we prove that for any partition P the upper sum and lower sum of a bounded function are always bounded and moreover, we see that upper sum can never be less than the lower sum which we have already observed while giving motivations for the integral.

In the next lemma we will see that refinement of a partition increases lower sum and decreases upper sum.

Lemma   3.3.2

If P_1 be a refinement of P having one more additional point other than the points of P then we have the following,

$$U(P_1, f) \leq U(P, f)$$

$$L(P, f) \leq L(P_1, f)$$

Also we have,

$$0 \leq U(P, f) - U(P_1, f) \leq (M - m)\delta$$

$$0 \leq L(P_1, f) - L(P, f) \leq (M - m)\delta$$

where, δ is the norm of the partition P.

Proof: Let the additional point ξ_r of P_1 be in r^{th} subinterval of P i.e. in Δ_r. Then we have,

$$x_{r-1} \qquad \xi_r \qquad\qquad\qquad\qquad x_r$$

Fig. 3.7

$$U(P_1, f) - U(P, f) = M_r(x_r - x_{r-1}) - M_r'(\xi_r' - x_{r-1}) - M_r''(x_r - \xi_r)$$

$$= M_r(x_r - \xi_r + \xi_r - x_{r-1}) - M_r'(\xi_r - x_{r-1}) - M_r'' (x_r - \xi_r)$$

$$= (M_r - M_r')\,(\xi_r - x_{r-1}) + (M_r - M_r'')\,(x_r - \xi_r) \qquad\qquad ...(3.3.2.1)$$

where M_r' and M_r'' are supremum of the function on $[x_{r-1}, \xi_r]$ and $[\xi_r, x_r]$. Now we have $M_r \geq M_r'$ and $M_r \geq M_r''$ also we note that $\xi_r - x_{r-1} \geq 0$ and $x_r - \xi_r \geq 0$ hence from (3.3.2.1) we get $U(P_1, f) - U(P, f) \geq 0$. Let m_r and m_r' be the infimum of f in the interval $[x_{r-1}, \xi_r]$ and $[\xi_r, x_r]$ respectively. Then we note that $M \geq M_r \geq M_r' \geq m_r' \geq m$ and $M \geq M_r \geq M_r'' \geq m_r'' \geq m$ hence we have $M_r - M_r' \leq M - m$. and $M_r - M_r'' \leq M - m$. Now using [3.2.1] we get,

$$U(P, f) - L(P, f) \leq (M - m)\,(\xi_r - x_{r-1}) + (M - m)\,(x_r - \xi_r)$$

$$\leq (M - m)\,(x_r - x_{r-1})$$

$$\leq (M - m)\delta$$

Proceeding similarly we can prove the other inequality.

In the next lemma we will generalize above result when the refined partition contains at most k additional points.

Lemma   3.3.3

Let P_k be a refinement of a partition P of $[a, b]$ containing k more additional points other than the points of P then we have,

$$0 \leq U(P, f) - U(P_k, f) \leq (M - m)k\delta$$

$$0 \leq L(P_k, f) - L(P, f) \leq (M - m)k\delta$$

where, δ denotes the norm of the partition P.

Proof: We introduce the points one by one and denote by P_r the refinement of P_{r-1} which contains one more additional point other than those of P_{r-1} for $r = 1, 2, 3, ...k$. Where P_0 denotes the partition P. We now repeatedly use

Lemma 3.3.2 and get,

$$0 \le U(P,f) - U(P_1,f) \le (M-m)\delta$$

$$0 \le U(P_1,f) - U(P_2,f) \le (M-m)\delta_1$$

where δ_1 denotes the norm of partition P_1, but as P_1 is refinement of P we have $\delta_1 \le \delta$ hence we have

$$0 \le U(P_1,f) - U(P_2,f) \quad \le (M-m)\delta$$

similarly we get,

$$0 \le U(P_2,f) - U(P_3,f) \quad \le (M-m)\delta$$

$$\cdots\cdots\cdots\cdots\cdots\cdots\cdots\cdots\cdots\cdots\cdots\cdots$$

$$0 \le U(P_{k-1},f) - U(P_k,f) \le (M-m)\delta.$$

Now by adding the above k inequalities we get,

$$0 \le U(P,f) - U(P_k,f) \le (M-m)k\delta.$$

Similarly we can prove the other inequality.

Next lemma gives a comparison of upper sum and lower sum for any two partition of $[a, b]$.

Lemma 3.3.4

For any two partitions P and P' of $[a, b]$ we have,

$$U(P,f) \ge L(P',f)$$

Proof: Let P'' be the common refinement of P and P' then,

$$U(P,f) \ge U(P'',f) \ge L(P'',f) \ge L(P,f)$$

We observe that Lemma 3.3.1 says that sets of Upper and Lower sums corresponding to different partitions are bounded and Lemma 3.3.4 tells us that any Upper sum is an upper bound for set of all Lower sums similarly any Lower sum is an lower bound of the set of all Upper sums. This motivate us to define Upper and Lower integrals.

Definition The infimum of the set of all Upper sums of f over all modes of partition P of $[a, b]$ is called the *Upper Riemann Integral* of and is denoted by J or $\int_b^a f(x)\,dx$ and the supremum of the set of all Lower sums of f over all modes of partitions of $[a, b]$ is called the *Lower Riemann Integral* of f and is denoted by I or $\int_a^b f(x)\,dx$.

That is $J = \int_b^a f(x)\,dx = \inf \{U(P,f): P$ a partition of $[a, b]\}$ and

$$I = \int_a^b f(x)\,dx = \sup \{L(P,f): P \text{ a partition of } [a, b]\}$$

Lemma   3.3.5

The Upper Riemann Integral is not less than the Lower Riemann Integral that is $I \leq J$.

Proof: For any two partitions P and P' we have,

$$U(P, f) \geq L(P', f).$$

We keep P' fixed and take infimum of $U(P', f)$ for all possible modes of partition P of $[a, b]$ then, we get from definition,

$$J \geq L(P', f)$$

Now we take supremum of $L(P', f)$ over all possible modes of partitions P' of $[a, b]$ and we obtain $J \geq I$.

3.4   DEFINITION OF RIEMANN INTEGRABILITY

The theorem of the above section motivates us to give the natural definition of Riemann Integrability of a bounded function f over the interval $[a, b]$.

Definition   A bounded function f defined over $[a, b]$ is said to be *Riemann Integrable* or just *R-Integrable* if the Upper Riemann Integral is equal to the Lower Riemann Integral, that is if $J = I$ In that case, the Riemann Integral of the function f is denoted by $\int_a^b f(x)\,dx$ and is equal to the common value of J and I.

The class of Riemann Integrable functions on the interval $[a, b]$ is denoted by $R[a, b]$.

Does every bounded function defined on $[a, b]$ is R-Integrable? The answer is certainly no which is shown by next example.

Example 1.1   We consider the Dirichlet's Function defined on the interval $[0, 1]$

$$f(x) = \begin{cases} 1 & \text{if } x \text{ is rational} \\ 0 & \text{if } x \text{ is irrational} \end{cases}$$

We will show that f is not R-Integrable.

For any partition $P : 0 = a_0 < a_2 < \ldots < a_r < \ldots < a_n = 1$ of $[0, 1]$ it is easy to see that $m_r = 0$, $M_r = 1$, hence $\sum_{r=1}^{n} M_r \delta_r = 1 \sum_{r=1}^{n} \delta_r$ and $\sum_{r=1}^{n} m_r \delta_r = 0 \sum_{r=1}^{n} \delta_r = 0$. So by definition $\int_a^b f(x)\,dx = 0$ and $\int_b^a f(x)\,dx = 1$. That is $I \neq J$. Hence f is not R-Integrable.

3.5   RIEMANN APPROXIMATING SUM

Consider f be a function defined on $[a, b]$ and P be a partition of $[a, b]$. In each subinterval Δ_r of P we choose a point x_r then the quantity

$$\sum_{r=1}^{n} f(\xi_r)\, \delta r$$

is called a *Riemann Approximating Sum* or *Riemann Sum*.

Definition   Suppose f be a boundebfunction on $[a, b]$. Then the Riemann approximating sums of f is said to be tend to a limit as norm of the partition tends to zero if there exists a real number A with the following property: For every $\varepsilon > 0$ there is a $\delta > 0$ such that for every partition P of $[a, b]$ with $\|P\| < \delta$, the inequality

$$\left\| \sum_{r=1}^{n} f(\xi_r)\delta_r - A \right\| < \varepsilon$$

holds for every possible choice of ξ_r in Δ_r.

In symbols we write,

$$\lim_{\delta \to 0} \sum_{r=1}^{n} f(\xi_r)\delta_r = A$$

where $\|P\|$ is the norm of the partition.

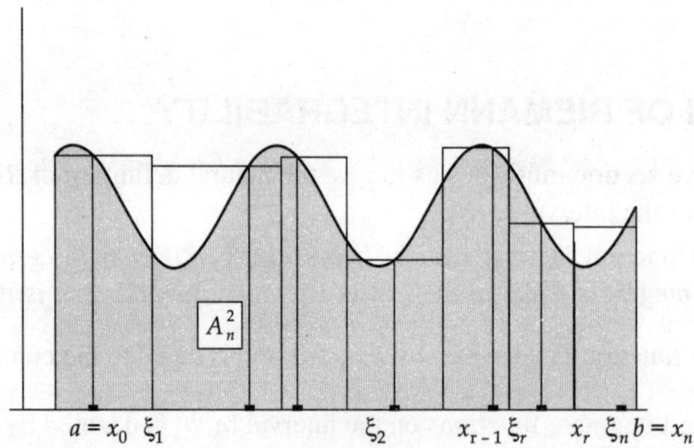

Riemann Approximating Sum

Fig. 3.8

We will show that if the Reimann approximating sums tend to a limit as norm of patition tends to zero then the function is Riemann integrable and conversely. To prove it we first need to prove a important theorem known as Darboux's Theorem.

Theorem 3.5.1(Darboux's Theorem)

Let f be a bounded function on the closed interval [a, b] then given any $\varepsilon > 0$, there exists a $\delta > 0$, such that, for all partition P with $\|P\| < \delta$,

$$U(P,f) < \int_{b}^{\bar{a}} f(x)\,dx + \varepsilon \text{ and } L(P,f) > \int_{a_-}^{b} f(x)\,dx - \varepsilon.$$

Proof: By definition of inf. there exists a partition P_1 such that for $\varepsilon > 0$, $U(P_1,f) < \int_{b}^{a} f(x)\,dx + \dfrac{\varepsilon}{2}$.

Let P_1 has K points other than the end points a & b. We may assume that $K \geq 1$ if possible by allowing refinement of P_1. Let $\delta = \dfrac{\varepsilon}{2(M-m)K}$. Let P be a partition with $\|P\| = \delta_1 < \delta$. We will show that the

conclusion of the theorem holds for this partition P. Let P_2 be the common refinement of P_1 and P. Let P_2 has r more points than P. We see that these points are points of P_1 and as P_1 has K points other than end points we have $0 \leq r \leq K$. Now we have,

$$0 \leq U(P,f) - U(P_2,f) \leq (M-m)r\delta_1 \qquad \qquad \text{...(3.5.1.1)}$$

and also we have

$$U(P_2, f) \le U(P_1, f) < \int_b^{\overline{a}} f(x)\,dx + \frac{\varepsilon}{2} \qquad\qquad ...(3.5.1.2)$$

Combining (3.5.1.1) and (3.5.1.2) we get,

$$U(P, f) < \int_b^{\overline{a}} f(x)\,dx + \frac{\varepsilon}{2} + (M - m)\,r\,\delta_1$$

$$\le \int_b^{\overline{a}} f(x)\,dx + \frac{\varepsilon}{2} + (M - m)\,\delta K \quad \text{as } (0 \le r \le K \,\&\, \delta_1 < \delta)$$

$$\le \int_b^{\overline{a}} f(x)\,dx + \frac{\varepsilon}{2} + \frac{\varepsilon}{2}$$

$$= \int_b^{\overline{a}} f(x)\,dx + \varepsilon.$$

Similarly we can prove for $L(P, f)$.

Remark: We see that Darboux's Theorem actually tells us that,

$$\lim_{\delta \to 0} U(P, f) = J \text{ and } \lim_{\delta \to 0} L(P, f) = I$$

where δ is the norm of the partition P.

Following example shows use of the above theorem:

Example 1.2 Consider the function $f(x) = x$ in the interval $[0, 1]$. We show that $f \in R[0, 1]$ and $\int_0^1 f(x)\,dx = \frac{1}{2}$.

For any positive integer n we consider the partition P_n of $[0, 1]$ as $\left\{ 0, \frac{1}{n}, \frac{2}{n}, \frac{3}{n}, \ldots \frac{n-1}{n}, 1 \right\}$. As the function is monotone increasing in $\left[\frac{r-1}{n}, \frac{r}{n} \right]$ we have, $M_r = \frac{r}{n},\ m_r = \frac{r-1}{n}$

Also we have $\delta_r = \frac{1}{n}$. Hence,

$$U(P_n, f) = \sum_{r=1}^{n} \frac{r}{n} \cdot \frac{1}{n}$$

$$= \sum_{r=1}^{n} \frac{1}{n^2} (1 + 2 + \ldots + n)$$

$$= \frac{n(n+1)}{2n^2} = \frac{1}{2} + \frac{1}{2n}$$

and

$$L(P_n, f) = \sum_{r=1}^{n} \frac{r-1}{n} \cdot \frac{1}{n}$$

$$= \frac{1}{n^2} (1 + 2 + \ldots + (n-1))$$

$$= \frac{n(n-1)}{2n^2} = \frac{1}{2} - \frac{1}{2n}.$$

So we see that,

$$U(P_n, f) - L(P_n, f) = \frac{1}{n}.$$

Hence for any $\varepsilon > 0$ we can find a suitable n such that $\frac{1}{n} < \varepsilon$ and the corresponding partition P_n for this

n satisfies $U(P_n, f) - L(P_n, f) = \frac{1}{n} < \varepsilon$. Using Theorem 7.1 we conclude, $f \in R\,[0, 1]$.

Now to show that $\int_0^1 f(x)\,dx = 1$ we note that $\|P_n\| = \frac{1}{n} \to 0$ as $n \to \infty$ hence it follows from the Remark of Darboux's Theorem that,

$$J = \lim_{n \to \infty} U(P_n, f) = \frac{1}{2} \text{ and } I = \lim_{n \to \infty} L(P_n, f) = \frac{1}{2} \text{ that is, } I = J = \int_0^1 f(x)\,dx = \frac{1}{2}$$

Example 1.3   For the integral $\int_{-1}^{1} |x|\,dx$,

find the upper and lower sums corresponding to the partition P_{2n} of $[-1, 1]$ into $2n$ equal intervals and hence compute the value of the integral.

Sol: Let $f(x) = |x|$. We first write $[-1, 1]$ as $[-1, 0] \cup [0, 1]$. Now we subdivide $[-1, 0]$, into n equal subintervals by,

$$P_1 = \left\{ -1, -1 + \frac{1}{n}, -1 + \frac{2}{n}, \dots, -1 + \frac{r-1}{n}, -1 + \frac{r}{n}, \dots -1 + \frac{n}{n} = 0 \right\}. \text{ With usual notations we have}$$

$$\Delta_r = \left[-1 + \frac{r-1}{n}, -1 + \frac{r}{n} \right] \text{ and } \delta_r = \frac{1}{n}. \text{ Now we note that } |x| \text{ is decreasing in } [-1, 0], \text{ so we have}$$

$$M_r = \left| -1 + \frac{r-1}{n} \right| = 1 - \frac{r-1}{n} \text{ and } m_r = \left| -1 + \frac{r}{n} \right| = 1 - \frac{r}{n}.$$

In the same way we partition $[0, 1]$ by $P_2 = \left\{ 0, \frac{1}{n}, \frac{2}{n}, \dots, \frac{i-1}{n}, \frac{i}{n}, \dots \frac{n}{n} = 1 \right\}$ and we have

$$\Delta'_n = \left[\frac{i-1}{n}, \frac{i}{n} \right], \delta'_r = \frac{1}{n}, M'_r = \frac{i}{n} \text{ and } m'_r = \frac{i-1}{n}. \text{ Now } P_{2n} = P_1 \cup P_2.$$

$$U(P_{2n}, f) = \sum_{r=1}^{n} M_r \delta_r + \sum_{r=1}^{n} M'_r \delta'_r$$

$$= \frac{1}{n} \left[\left\{ 1 + \left(1 - \frac{1}{n} \right) + \left(1 - \frac{2}{n} \right) + \dots + \left(1 - \frac{n-1}{n} \right) \right\} + \left\{ \frac{1}{n} + \frac{2}{n} + \dots + \frac{n}{n} \right\} \right]$$

$$= \frac{1}{n} \left(n + \frac{n}{n} \right) = 1 + \frac{1}{n}$$

Similarly

$$L(P_{2n}, f) = \sum_{r=1}^{n} m_r \delta_r + \sum_{r=1}^{n} m'_r \delta'_r$$

$$= \frac{1}{n}\left[\left\{\left(1 - \frac{1}{n}\right) + \left(1 - \frac{2}{n}\right) + \ldots + \left(1 - \frac{n}{n}\right)\right\} + \left\{0 + \frac{1}{n} + \frac{2}{n} + \ldots \frac{n-1}{n}\right\}\right]$$

$$\frac{1}{n}\left(n - \frac{n}{n}\right) = 1 - \frac{1}{n}$$

Now we see that $\|P_{2n}\| = \frac{1}{n} \to 0$ as $n \to \infty$. Also from above $U(P_{2n}, f) \to 1$ and $L(P_{2n}, f) \to 1$ as $n \to \infty$.

So by Darboux Theorem $I = J = 1$. This proves that f is R-Integrable as well as integral of f is 1.

Example 1.4   A function $f(x)$ is defined on [0, 1] as,

$$f(x) = \begin{cases} x & \text{if is rational} \\ 0 & \text{if is irrational} \end{cases}$$

Find the upper and lower sums corresponding to the partition $P_n = \left\{0, \frac{1}{n}, \frac{2}{n}, \frac{3}{n}, \ldots, \frac{n}{n}\right\}$ of [0, 1]. Hence find I & J and show that $f(x)$ is not integrable.

Sol: With usual notation we have here,

$$\Delta_r = \left[\frac{r-1}{n}, \frac{r}{n}\right], \ \delta_r = \frac{1}{n}.$$

Now as Δ_r contains infinite rationals and irrationals we have,

$M_r - \frac{r}{n}$ and $m_r = 0$, for all r. So we have,

$$U(P_n, f) = \sum_{r=1}^{n} M_r \delta_r = \frac{1}{n}\left(\frac{1}{n} + \frac{2}{n} + \ldots \frac{n}{n}\right) = \frac{n(n+1)}{2n^2} = \frac{1}{2} + \frac{1}{2n}$$

$$L(P_n, f) = \sum_{r=1}^{n} m_r \delta_r = 0.$$

We see that as $n \to \infty$ $\|P\| \to 0$ and $U(P_n, f) \to \frac{1}{2}$ & $L(P_n, f) \to 0$ hence by Darboux Theorem, $I = 0$

and $J = \frac{1}{2}$. Hence we conclude that $f(x)$ is not integrable.

3.6   A NECESSARY AND SUFFICIENT CONDITION OF RIEMANN INTEGRABILITY IN TERMS OF RIEMANN APPROXIMATING SUM

Theorem   3.6.1

A bounded real valued function f defined on the closed interval [a, b] is integrable if and only if

$$\lim_{\delta \to 0} \sum_{r=1}^{n} f(\xi_r) \delta_r \to A.$$

Proof: The condition is necessary:

Suppose the function is integrable then $I = A = J$ (say). By Darboux theorem for $\varepsilon > 0$, there exists $\delta_1 > 0$ such for any partition P with norm less than d_1 we have $U(P, f) < J + \varepsilon$ and $L(P, f) > I - \varepsilon - \varepsilon$ i.e. $U(P, f) < A + \varepsilon$ and $L(P, f) > A - \varepsilon$. Again as ξ_r in Δ_r we have $m_r \leq f(\xi_r) \leq M_r$ hence

$$\sum_{r=1}^{n} m_r \delta_r \le \sum_{r=1}^{n} f(\xi_r)\delta_r \le \sum_{r=v1}^{n} M_r \delta_r$$

or,
$$U(P,f) \le \sum_{r=1}^{n} f(\xi_r)\delta_r \le L(P,f)$$

Hence from above we get,

$$A - \varepsilon < L(P,f) \le \sum_{r=1}^{n} f(\xi_r)\delta_r \le U(P,f) < A + \varepsilon$$

That is we have,

$$A - \varepsilon < \sum_{r=1}^{n} f(\xi_r)\delta_r < A + \varepsilon$$

or,
$$\left| \sum_{r=1}^{n} f(\xi_r)\delta_r - A \right| < \varepsilon$$

Hence the condition is necessary.

The condition is sufficient

Suppose for $\varepsilon > 0$ there exists a $\delta_1 > 0$ such that for any partition P with $\|P\| < \delta < \delta_1$ we have,

$$A - \frac{\varepsilon}{2} < \sum_{r=1}^{n} f(\xi_r)\delta_r < A + \frac{\varepsilon}{2} \tag{3.6.1.1}$$

for every choice of ξ_r in Δ_r. Now by the definition of supremum and infimum we have α_r and β_r in Δ_r such that, $f(\alpha_r) > M_r - \dfrac{\varepsilon}{2(b-a)}$ and $f(\beta_r) < m_r + \dfrac{\varepsilon}{2(b-a)}$.

Hence
$$\sum_{r=1}^{n} f(\alpha_r)\delta_r > \sum_{r=1}^{n} M_r \delta_r - \frac{\varepsilon}{2}$$

or,
$$\sum_{r=1}^{n} f(\alpha_r)\delta_r > U(P,f) - \frac{\varepsilon}{2} \tag{3.6.1.2}$$

Using (3.6.1.1) with $\xi_r = \alpha_r$ we have

$$A - \frac{\varepsilon}{2} < \sum_{r=1}^{n} f(\alpha_r)\delta_r < A + \frac{\varepsilon}{2} \tag{3.6.1.3}$$

From (3.6.1.2) and (3.6.1.3) we have

$$U(P,f) < \sum_{r=1}^{n} f(\alpha_r)\delta_r + \frac{\varepsilon}{2}$$

or,
$$U(P,f) < A + \frac{\varepsilon}{2} + \frac{\varepsilon}{2}$$

Hence,
$$U(P,f) < A + \varepsilon$$

Similarly as we have $A - \varepsilon < L(P,f)$. We obtain,

$$A - \varepsilon < L(P,f) \le I \le J \le U(P,f) < A + \varepsilon \quad \text{therefore}$$
$$|I - A| < \varepsilon \ \& \ |J - A| < \varepsilon$$
$$I = J = A.$$

Remark:

1. In view of above theorem we can formulate an alternate equivalent definition of Riemann Integrability as follows:

   A bounded function f is Riemann Integrable on $[a, b]$ if there exists a real number A with the following property: For every $\varepsilon > 0$ there is a $\delta > 0$ such that for every partition P of $[a, b]$ with $\|P\| < \delta$, the inequality

   $$\left| \sum_{r=1}^{n} f(\xi_r)\delta_r - A \right| < \varepsilon$$

   holds for every possible choice of ξ_r in Δ_r. In that case we write $A = \int\limits_{a}^{b} f(x)\, dx$

2. If a function f is Riemann Integrable on $[a, b]$ then we can use above Theorem to find the integral of f as follows:

   Let $P\{a = x_0 < x_2 < ... < x_{r-1} < x_r ... < x_n = b\}$ be a partition of $\{a, b\}$ such that

   $\delta_r = x_r - x_{r-1} = \dfrac{b-a}{n}\; r = 1, 2 \,...\, n$. We choose ξ_r to be the right hand end point of Δ_r. Then the Riemann approximating sum for this partition is given by,

   $$\sum_{r=1}^{n} f(\xi_r)\,\delta_r = \sum_{r=1}^{n} f\left(a + \frac{r}{n}(b-a)\right)\frac{b-a}{n}.$$

   Now as $n \to \infty$ we see that $\|P\| \to 0$ hence by definition,

   $$\int\limits_{a}^{b} f(x)\, dx = \lim_{n \to \infty}\left(\frac{b-a}{n}\right) \sum_{r=1}^{n} f\left(a + \frac{r}{n}(b-a)\right).$$

   In particular, if we take the interval to be $[0, 1]$ we have,

   $$\int\limits_{0}^{1} f(x)\, dx = \lim_{n \to \infty} \frac{1}{n} \sum_{r=1}^{n} f\left(\frac{r}{n}\right).$$

   In the next section we present two more necessary and sufficient conditions for Riemann integrability. These conditions will prove very handy to determine the class of functions which are integrable.

3.7 CONDITIONS FOR INTEGRABILITY

Theorem 3.7.1

A bounded function f defined on the closed interval $[a, b]$ is Riemann Integrable if and only if for each $\varepsilon > 0$ there must exist a partition P of $[a, b]$ such that,

$$0 \le U(P, f) - L(P, f) < \varepsilon.$$

Proof: First let us assume that the function f is R-Integrable, then by definition we have,

$$I = J = \int\limits_{a}^{b} f(x)\, dx.$$

Now by definition of I and J we can find partitions P_1 and P_2 of $[a, b]$ such that,

$$U(P_1, f) < J + \frac{\varepsilon}{2}$$

$$L(P_2, f) > I - \frac{\varepsilon}{2}$$

Let P be the common refinement of P_1 and P_2 then we have for this P we have,

$$U(P, f) < J + \frac{\varepsilon}{2}$$

$$L(P, f) > I - \frac{\varepsilon}{2}$$

Now we have from above,

$$0 \le U(P, f) - L(P, f) < (J - I) + \frac{\varepsilon}{2} + \frac{\varepsilon}{2}$$

but $I = J$ by our assumption that f is integrable. Hence for this partition P of $[a, b]$ we get as desired, $0 \le U(P, f) - L(P, f) < \varepsilon$.

This proves that the condition is necessary.

For the sufficiency part, let $\varepsilon > 0$ be any arbitrary positive real number then from given condition we can get a partition P of $[a, b]$ such that $0 \le U(P, f) - L(P, f) < \varepsilon$. Now since we know, $0 \le L(P, f) \le I \le J \le U(P, f)$ hence we have,

$$I - J \le U(P, f) - L(P, f) < \varepsilon$$

or,
$$I - J < \varepsilon$$

now as $\varepsilon > 0$ is arbitrary we conclude that $I = J$ that is f is R-Integrable.

We now give a second necessary and sufficient condition.

Theorem   3.7.2

A necessary and sufficient condition that a bounded function f defined on the closed interval $[a, b]$ is Riemann Integrable is, for each $\varepsilon > 0$ there must exist a $\delta > 0$ such that for every partition P of $[a, b]$ with $\|P\| < \delta$ we must have,

$$0 \le U(P, f) - L(P, f) < \varepsilon.$$

Proof: Note that the proof for sufficiency part of the condition is exactly the same as Theorem.7.1. For necessary part we assume that f Riemann Integrable that is $I = J$. Let $\varepsilon > 0$ be any given positive number, then by Darboux's Theorem (Theorem 3.5.1) we can find a $\delta > 0$, such that for every partition P of $[a, b]$ with $\|P\| < \delta$ we have,

$$U(P, f) < J + \frac{\varepsilon}{2}$$

$$L(P, f) > I - \frac{\varepsilon}{2}$$

Hence we have,

$$0 \le U(P, f) - L(P, f) < (J - I) + \frac{\varepsilon}{2} + \frac{\varepsilon}{2}$$

or,
$$0 \le U(P, f) - L(P, f) < \varepsilon.$$

Remarks:

1. Obseve that in the second necessary and sufficient condition we have used Darbox's Theorem where as in first necessary and sufficient condition we have not used Darbox's Theorem.

2. We note that in both necessary and sufficient conditions require the sum

$$U(P,f) - L(P,f)^+ = \sum_{r=1}^{n} (M_r - m_r)\delta_r \text{ must be arbitrarily small. This sum is called the } \textbf{\textit{Oscillatory}}$$

Sum of the function f and the difference $M - m$ is known as **Oscillation** of the function f.

Example 1.5   We again consider the Dirichlet's Function (Example 4.1) defined on the interval [0, 1]:

$$f(x) = \begin{cases} 0 & \text{if } x \text{ is rational} \\ 1 & \text{if } x \text{ is irrational} \end{cases}$$

We use above theorem to show that f is not R-Integrable

We choose $\varepsilon = \dfrac{1}{2}$. Now we have seen that for any partition P of interval [0, 1] we have

$M_r = 1, m_r = 0$ for $r = 1, 2, \dots n$. Hence,

$$U(P,f) - L(P,f) = \sum_{r=1}^{n} (M_r - m_r)\delta_r$$

$$= \sum_{r=1}^{n} (1 - 0)\delta_r$$

$$= \sum_{r=1}^{n} \delta_r = 1 > \frac{1}{2} = \varepsilon$$

So by the necessary part of Theorem 7.1 f is not R-Integrable.

3.8   CLASSES OF RIEMANN INTEGRABLE FUNCTIONS

We will now look for class of functions which are R-Integrable. We will see that every continuous function is R-Integrable and also functions which are some what "close" to continuous function are R-Integrable.

Theorem   3.8.1

Every continuous function defined on a closed interval [a, b] is R-Integrable.

Proof: Let f be a continuous function defined on [a, b] then f is uniformly continuous. For any arbitray $\varepsilon > 0$ there exists a $\delta > 0$ such that, for any x and y in [a, b] satisfying $|x - y| < \delta$ we must have, $|f(x) - f(y)| < \varepsilon/b - u$. We now consider a partition P of [a, b] such that $\|P\| < \delta$. We also note that as the function is continuous in [a, b] it is continuous in all the sub intervals $\Delta_r, r = 1, 2, \dots, n$. Since continuous functions attains its bounds so we must have p_r and q_r in Δ_r such that $M_r = f(p_r)$ and $m_r = f(q_r)$; $r = 1, 2, \dots, n$. Let us now consider the oscillatory sum of f,

$$U(P,f) - L(P,f) = \sum_{r=1}^{n} (M_r - m_r)\delta_r$$

$$= \sum_{r=1}^{n} (f(p_r) - f(q_r))\delta_r$$

As $\|P\| < \delta$ we have $|P_r - q_r| < \delta$ hence $|f(p_r) - f(q_r)| < \varepsilon/b - a$ using this in the above equation we get,

$$U(P,f) - L(P,f) < \sum_{r=1}^{n} \left(\frac{\varepsilon}{b-a}\right) \delta_r = \frac{\varepsilon}{b-a} \sum_{r=1}^{n} \delta r = \varepsilon$$

Hence using the sufficiency part of Theorem 3.7.2 we get is R-Integrable.

We will now see that there are functions which are not continuous but still are R-Integrable.

Theorem 3.8.2

Every monotone function defined on a closed bounded interval is R-Integrable.

Proof: We first note that every monotone function on $[a, b]$ is bounded. Without loss of generality we may assume that is not a constant function, as all constant functions are continuous and by previous theorem continuous functions are R-Integrable hence we may take $f(b) - f(a) \neq 0$. Next we suppose that the function is monotone increasing. Let $\varepsilon > 0$ be any positive real number. We consider a partition of such that $\|P\| < \dfrac{\varepsilon}{f(b) - f(a)}$. As the function is monotone increasing we have $M_r = f(x_r)$ and $m_r = f(x_{r-1})$ $r = 1, 2..., n$. Now let us consider the oscillatory sum of f,

$$U(P,f) - L(P,f) = \sum_{r=1}^{n} (M_r - m_r)\delta_r$$

$$= \sum_{r=1}^{n} (f(x_r) - f(x_{r-1}))\delta_r < \frac{\varepsilon}{f(b) - f(a)} \sum_{r=1}^{n} (f(x_r) - f(x_{r-1}))$$

We see that the above summation is a telescopic sum so all terms get cancelled except the first and the last term. But we know that $x_0 = a$ and $x_n = b$ so using these in the above equation we get

$$U(P,f) - L(P,f) < \frac{\varepsilon}{f(b) - f(a)} \cdot (f(b) - f(a)) = \varepsilon$$

Hence using the sufficiency part of Theorem 3.7.2 we get f is R-Integrable.

By modifying above argument slightly one can prove that monotone decreasing functions are also R-Integrable.

Next we consider the following example:

Example 1.6 Let be a function defined on by

$$f(x) = \begin{cases} \sin\left(\dfrac{1}{x}\right) & \text{if } x \neq 0 \\ 0 & \text{if } x \neq 0 \end{cases}$$

Does $f \in R\,[-1, 1]$?

We will show that the answer is yes that is, $f \in R[-1, 1]$ but observe that as f is neither continuous at nor it is monotone on $[-1, 1]$ so that we cannot use previous theorems. We employ a special technique to prove Integrability of f and the reader is advised to go through it very carefully as it uses a technique which is very important and will be used in the proof of some subsequent theorems and also will be employed in solving some problems.

Let $\varepsilon < 0$ be any arbitrary real number. As the function has only discontinuity at $x = 0$ the function is continuous in $\left[-1, -\dfrac{\varepsilon}{8}\right]$ and $\left[\dfrac{\varepsilon}{8}, 1\right]$. Hence there exists partitions P_1 and P_2 of $\left[-1, -\dfrac{\varepsilon}{8}\right]$ and $\left[\dfrac{\varepsilon}{8}, 1\right]$ respectively such that, and

$$U(P_1, f) - L(P_1, f) < \frac{\varepsilon}{4} \quad \text{and}$$

$$U(P_2, f) - L(P_2, f) < \frac{\varepsilon}{4}.$$

We now consider the partition P of $[-1, 1]$ given by $P_1 \cup P_2 \cup \left[-\dfrac{\varepsilon}{8}, \dfrac{\varepsilon}{8}\right]$. Then if M' and m' are sup and inf of f in $\left[-\dfrac{\varepsilon}{8}, \dfrac{\varepsilon}{8}\right]$ we have,

$$U(P, f) - L(P, f) = [U(P_1, f) - L(P_1, f)] + U(P_2, f) - L(P_2, f)] + (M' - m')\frac{\varepsilon}{4}$$

$$< \frac{\varepsilon}{4} + \frac{\varepsilon}{4} + 2 \times \frac{\varepsilon}{4} = \varepsilon$$

We have as $(M' - m') \le 2$ as $\left|\sin\left(\dfrac{1}{x}\right)\right| \le 1$. Hence $f \in R[-1,1]$ by Theorem 3.7.1.

From the above problem it is clear that if a bounded function has one point of discontinuity in a closed bounded interval it is R-Integrable clearly this can be extended to finite number of points of discontinuities using similar argument. So we get the following theorem:

Theorem 3.8.4

A bounded function f defined on [a, b] having finite number of discontinuities is R-Integrable.

Proof: Let function f be bounded and has discontinuities at $a \le x_1 < x_2 < x_3 < ... < x_1 \le b$. As the number of point where there are discontinuities is finite we can enclose those points with closed non-overlapping intervals whose total length is less than $\dfrac{\varepsilon}{2(M - m)}$ where M and m are bounds of the function and $\varepsilon > 0$ being arbitrary. We now see that in the complementary portion of these l intervals which is at most $l + 1$ intervals, there are no discontinuities of f hence by Theorem 3.8.2 for each of these $l + 1$ intervals there exists partitions such that the oscillatory sum of each of them is less than $\dfrac{\varepsilon}{2(l + 1)}$. Consider a partition P of $[a, b]$ which consists of l intervals enclosing all discontinuities together with the partitions of complementary $l + 1$ intervals. We now consider the oscillatory sum of P. The contribution of oscillatory sum coming from discontinuous part is,

$$< \frac{\varepsilon}{2(M - m)} \times (M - m) = \frac{\varepsilon}{2}$$

and contribution coming from the continuous part is,

$$< \frac{\varepsilon}{2(l + 1)} \times (l + 1) = \frac{\varepsilon}{2}$$

Hence the oscillatory sum of P is,

$$< \frac{\varepsilon}{2} + \frac{\varepsilon}{2} = \varepsilon,$$

so by necessary and sufficient condition f is R-Integrable.

Next we will see that not only functions only finitely many discontinuities but also there are some special functions with infinite many points of discontinuities are R-Integrable.

Theorem 3.8.5

If set of points of discontinuities of a bounded function f defined on [a, b] has finite number of limit points then the function is R-Integrable.

Proof: We use similar technique as before. Let the number of limit points be l. We now enclose these l points non overlapping intervals say, $[k_1'', k_1'] [k_2', k_2'']$................ $[k_l', k_l'']$, whose total length is less than $\frac{\varepsilon}{2(M - m)}$ where M and m are bounds of the function and $\varepsilon > 0$ being arbitrary. Now outside these l intervals the function can have only finitely many discontinuities in $[a, b]$ otherwise it will have more than l limit points contradicting our hypothesis. Hence by Theorem 3.8.4 we can partition the complement of these l intervals such that the oscillatory sum of this partition is less than $\frac{\varepsilon}{2}$. Now consider the partition P of $[a, b]$ consisting of the above partition together with the above l intervals. The contribution of the oscillatory sum of P coming from l intervals is,

$$< \frac{\varepsilon}{2(M - m)} \times (M - m) = \frac{\varepsilon}{2}$$

and we know the contribution of the oscillatory sum of P coming from the complementary part of l intervals is $\frac{\varepsilon}{2}$. Hence total oscillatory sum P of is $< \varepsilon$. This proves that f is R-Integrable.

We now illustrate by some examples how to use above theorems:

Example 1.7   Show that the function defined by

$$f(x) = \frac{1}{2^n} \ \text{ for } \frac{1}{2^{n+1}} < x \le \frac{1}{2^n}, \quad n = 1, 2, 3...$$

is R-Integrable over $[0, 1]$, although it has infinite number of discontinuities.

Sol: Here we see that f has infinite number of discontinuities. In fact the set of discontinuities of f is given by $S = \left\{ \frac{1}{2}, \frac{1}{2^2}, ..., \frac{1}{2^n}, ... \right\}$. Obviously has zero as only limit point. Hence by Theorem 3.8.5 the function is R-Integrable.

Note that the above function is monotone decreasing hence one can use Theorem 3.8.2 to show f is R-Integrable.

Example 1.8 A function f is defined on $[0, 1]$ as follows:

$$f(x) = \begin{cases} 0 & \text{if } x \text{ is irrational} \\ \dfrac{1}{q} & \text{if } x = \dfrac{p}{q} \text{ where } p, q \text{ are in lowest terms} \end{cases}$$

Show f that f is R-Integrable and $\displaystyle\int_0^1 f(x)\, dx = 0$.

Sol: We note that for any $\varepsilon > 0$, there can exist only finitely many rational numbers $\dfrac{p}{q}$ in $[0, 1]$ such that $\dfrac{1}{q} > \varepsilon$. We call these points as exceptional points. As there are only finitely many exceptional points we can cover these points by disjoint non-overlapping intervals whose total length is less than $\dfrac{\varepsilon}{2}$. Let P be the partition of $[0, 1]$ induced by these non-overlapping intervals. Now denote by $\sum_1 (M_r - m_r)\delta_r$ the contribution of the oscillatory sum coming from the intervals containing exceptional points then we have,

$$\sum_1 (M_r - m_r)\delta_r \leq \sum_1 \delta_r < \frac{\varepsilon}{2} \quad \text{as } M_r \leq 1 \text{ and } m_r = 0.$$

Now we note that for the intervals which does not have any exceptional points $M_r \leq \dfrac{\varepsilon}{2}$ and of course, $m_r = 0$. Hence contribution of the oscillatory sum coming from intervals which does not have any exceptional points, denoted by $\sum_2 (M_r - m_r)\delta_r$ is given by,

$$\sum_2 (M_r - m_r)\delta_r \leq \frac{\varepsilon}{2} \sum_2 \delta_r < \frac{\varepsilon}{2}, \quad \text{as } \sum_2 \delta_r < 1$$

So the total oscillatory sum for this partition P given by,

$$\sum_{r=1}^{n} (M_r - m_r)\delta_r = \sum_1 (M_r - m_r)\delta_r + \sum_2 (M_r - m_r)\delta_r < \frac{\varepsilon}{2} + \frac{\varepsilon}{2} = \varepsilon.$$

Hence by Theorem 3.7.1 is R-Integrable.

Now as $m_r = 0$ for $\forall r$ hence $L(P, f) = 0$ as this true for all P we conclude that $I = 0$. Since f is R-Integrable, $I = J = \displaystyle\int_a^b f(x)\, dx = 0$.

We discuss now some well known and obvious properties of R-Integrable functions.

3.9 PROPERTIES OF RIEMANN INTEGRABLE FUNCTIONS & RIEMANN INTEGRALS

Theorem 3.9.1

If $f \in R[a, b]$ and λ be a real number then $\lambda f \in R[a, b]$. Also

$$\int_a^b \lambda f(x)\, dx = \lambda \int_a^b f(x)\, dx$$

Proof: First we show $\lambda f \in R[a, b]$ that for any real λ. We consider three cases,

Case1: $\lambda = 0$, then $\lambda f = 0$ and hence it is R-Integrable as every constant function is R-Integrable.

Case2: $\lambda > 0$. $f \in R[a, b]$, As $f \in R[a, b]$, for $\varepsilon > 0$, there exists a partition P of $[a, b]$ such that,

$$U(P, f) - L(P, f) < \frac{\varepsilon}{\lambda}.$$

Now,

$$U(P, \lambda f) - L(P, \lambda f) = \lambda [U(P, f)] < \lambda \cdot \frac{\varepsilon}{\lambda} = \varepsilon.$$

Hence $\qquad\qquad\qquad\qquad\qquad \lambda f \in R[a, b].$

Case3: $\lambda < 0$. This case can be disposed by similar argument as above using $-\lambda > 0$.

Now we show, $\int_{a}^{b} \lambda f(x)\, dx = \lambda \int_{a}^{b} f(x)\, dx$ for this we again consider three cases,

Case1: $\lambda = 0$. There is nothing to prove here as both sides equals 0.

Case2: $\lambda > 0$. We know $\overline{\int_{a}^{b}} \lambda f(x)\, dx = \inf \{U(P, \lambda f)\}$, by definition

$$= \inf\{1\, U(P, f)\} \text{ as } \lambda > 0$$

$$= \lambda \inf\{U(P, f)\} \text{ as } \lambda > 0$$

$$= \lambda \overline{\int_{a}^{b}} f(x)\, dx$$

Case3: $\lambda < 0$. Again $\overline{\int_{a}^{b}} \lambda f(x)\, dx = \inf\{U(P, \lambda f)\}$, by definition

$$= \inf\{\lambda L(P, f)\} \text{ as } \lambda < 0$$

$$= \lambda \sup\{L(P, f)\} \text{ as } \lambda < 0$$

$$= \lambda \underline{\int_{a}^{b}} f(x)\, dx.$$

Before proceeding further we will prove a very useful lemma,

Lemma   3.9.2

If f is a bounded function on $[a, b]$, the oscillation of f is given by,

$$M - m = \sup \{|f(x_1) - f(x_2)|: x_1, x_2 \in [a, b]\}$$

Proof: We have, $m \leq f(x_1), f(x_2) \leq M$ hence,

$$M - m \geq |f(x_1) - f(x_2)|, \ \forall x_1, x_2 \in [a, b].$$

Now as $M = \sup f(x)$ and $m = \inf f(x)$ there exists x_1 and x_2 in $[a, b]$ such that for any $\varepsilon > 0$,

$$f(x_1) > M - \frac{\varepsilon}{2} \quad \text{and,} \quad f(x_2) < m + \frac{\varepsilon}{2}, \text{ so that}$$

$$f(x_1) - f(x_2) > \left(M - \frac{\varepsilon}{2}\right) - \left(m + \frac{\varepsilon}{2}\right) = (M - m) - \varepsilon \text{ that is,}$$

$$|f(x_1) - f(x_2)| > (M - m) - \varepsilon.$$

This proves that $M - m = \sup\{|f(x_1) - f(x_2)|: x_1, x_2 \in [a, b]\}$.

Theorem   3.9.3

If f & $g \in R\{a, b\}$, then $f \pm g \in R\{a, b\}$

and
$$\int_a^b [f(x) \pm g(x)] = \int_a^b f(x)dx \pm \int_a^b g(x)dx.$$

Proof: First we prove that for f & $g \in R[a, b]$ $f + g \in R[a, b]$. Let P be a partition of $[a, b]$ which is applied to the functions $f + g$, f and g. Let (M_r, m_r), (M_r^1, m_r^1) and (M_r^2, m_r^2) be (sup, inf) of the functions $f + g, f,$ and g respectively on the interval Δ of P, $r = 1, 2, ...n$. Now for $x_1, x_2 \in \Delta_r$ we have,

$$|(f + g)(x_1) - (f + g)(x_2)| \leq |f(x_1) - f(x_2)| + |g(x_1) - g(x_2)| \text{ hence,}$$

$$|(f + g)(x_1) - (f + g)(x_2)| \leq \sup_{x_1, x_2 \in \Delta_r} |f(x_1) - f(x_2)| + \sup_{x_1, x_2, \in \Delta_r} |g(x_1) - g(x_2)|$$

or,
$$\sup_{x_1, x_2, \in \Delta_r} |(f + g)(x_1) - (f + g)(x_2)| \leq \sup_{x_1, x_2 \in \Delta_r} |f(x_1) - f(x_2)| + \sup_{x_1, x_2, \in \Delta_r} |g(x_1) - g(x_2)|$$

hence by Lemma 3. 9.2,

$$M_r - m_r \leq (M_r^1 - m_r^1) + (M_r^2 - m_r^2)$$

Multiplying by δ through out above equation and summing over $r = 1, 2, ...n$ we get,

$$\sum_{r=1}^n (M_r - m_r)\delta_r \leq \sum_{r=1}^n (M_r^1 - m_r^1)\delta_r + \sum_{r=1}^n (M_r^2 - m_r^2)\delta_r$$

That is, $U(P, f + g) - L(P, f + g) \leq [U(P, f) - L(P, f)] + [U(P, g) - L(P, g)]$ $\hspace{1em}$ (3.9.3.1)

Above equation is valid for any partition of $[a, b]$. Now as f & $g \in R[a, b]$ for $\varepsilon > 0$ there exists partitions P_1 and P_2 and (by necessary and sufficient condition) of $[a, b]$ such that,

$$U(P_1, f) - L(P_1, f) < \frac{\varepsilon}{2} \ \& \ U(P_2, g) - L(P_2, g) < \frac{\varepsilon}{2}.$$

Let P be the common refinement of P_1 and P_2 then from equation (3.9.3.1) we get,

$$U(P, f + g) - L(P, f + g) < \frac{\varepsilon}{2} + \frac{\varepsilon}{2} = \varepsilon$$

This proves that $f + g \in R[a, b]$. Similarly we can prove that $f - g \in R[a, b]$ by same argument choosing $-g = h$.

Now we prove $\int_a^b [f(x) \pm g(x)]\,dx = \int_a^b f(x)\,dx \pm \int_a^b g(x)\,dx$. For that we see that it is enough if we prove

$$\int_a^b [f(x) + g(x)]dx = \int_a^b f(x)\,dx + \int_a^b g(x)\,dx \text{ as}$$

$$\int_a^b [f(x) - g(x)]dx = \int_a^b f(x)dx + \int_a^b (-)\,g(x)dx$$

$$= \int_a^b f(x)\,dx + (-1)\int_a^b g(x)dx \qquad \text{(by Theorem 3.9.1)}$$

$$= \int_a^b f(x)\,dx - \int_a^b g(x)\,dx$$

For any partition P of $[a, b]$ we have,

$$L(P,f) + L(P, g) \le L(P, f + g) \le U(P, f + g) \le U(P, f) + U(P, g) \qquad (3.9.3.2)$$

Also we have,

$$\int_a^b \{f(x) + g(x)\}dx \le U(P, f + g) \le U(P, f) + U(P, g) \qquad (3.9.3.3)$$

Now for any $\varepsilon > 0$ there exists partitions P_1 and P_2 such that,

$$U(P_1, f) < L(P_1, f) + \frac{\varepsilon}{2} \le \int_a^b f(x)\,dx + \frac{\varepsilon}{2}$$

$$U(P_2, g) < L(P_2, g) + \frac{\varepsilon}{2} \le \int_a^b g(x)\,dx + \frac{\varepsilon}{2}.$$

Now if P be the common refinement of P_1 and P_2 we have,

$$U(P, f) < \int_a^b f(x)\,dx + \frac{\varepsilon}{2} \quad \text{and} \quad U(P, g) < \int_a^b g(x)\,dx + \frac{\varepsilon}{2}.$$

Now from (3.9.3.3) we have

$$\int_a^b \{f(x) + g(x)\}dx < \int_a^b f(x)dx + \int_a^b g(x)dx + \varepsilon \qquad (3.9.3.4)$$

As $\varepsilon > 0$ is arbitrary we have,

$$\int_a^b \{f(x) + g(x)\}\,dx \le \int_a^b f(x)\,dx + \int_a^b g(x)\,dx.$$

Now replacing $f(x)$ and $g(x)$ by $-f(x)$ and $-g(x)$ we get,

$$\int_a^b \{-f(x) - g(x)\}\,dx \le \int_a^b -f(x)\,dx + \int_a^b -g(x)\,dx \quad \text{using Theorem 3.9.1}$$

$$-\int_a^b \{f(x) + g(x)\}\,dx \le -\int_a^b f(x)\,dx - \int_a^b g(x)\,dx$$

or $\qquad \displaystyle\int_a^b \{f(x) + g(x)\}dx \geq \int_a^b f(x)\, dx + \int_a^b g(x)\, dx$ $\qquad\qquad$ (3.9.3.5)

From (3.9.3.4) & (3.9.3.5) we get the result.

Remark:

The outcome of Theorem 3.9.1 and Theorem 3.9.2 is very important as it tells us that the space $R[a, b]$ is a vector space over $\mathbb{R}$ and also the map $\int : R[a, b] \to \mathbb{R}$ is a linear transformation.

We have observed already that $R[a, b]$ is closed under the operation '+' ('–' can be considered as the inverse operation of '+') now we will see that it is closed under another operation which is pointwise multiplication.

Theorem 3.9.4

The product of two functions f and g in R[a, b] is also in R[a, b].

Proof: Let P be a partition of $[a, b]$ which is applied to the functions $f \cdot g, f$ and g. Let (M_r, m_r) (M_r^1, m_r^1), and (M_r^2, m_r^2) be (sup, inf) of the functions $f \cdot g, f,$ and g respectively on the interval Δ_r of P, $r = 1, 2, ...n$. Now for $x_1, x_2, \in \Delta_r$ we have,

$$|f(x_1) \cdot g(x_1) - f(x_2) \cdot g(x_2)\,| = |f(x_1)\,(g(x_1) - g(x_2)) - g(x_2)\,(f(x_1) - f(x_2))|$$

$$\leq |f(x_1)|\,|g(x_1) - g(x_2)| + |g(x_2)|\,|f(x_1) - f(x_2)|$$

$$\leq M|g(x_1) - g(x_2)| + M|f(x_1) - f(x_2)| \qquad\qquad (3.9.4.1)$$

Where, $M > 0$ is such that $|f(x) \leq M$ &. $|g(x) \leq M$. As (3.9.4.1) is true for any $x_1, x_2 \in \Delta_r$, taking sup over all $x_1, x_2 \in \Delta_r$ on left had side of (3.9.4.1) we get

$$|f(x_1) \cdot g(x_1) - f(x_2) \cdot g(x_2)| \leq M \mathop{\mathrm{Sup}}_{x_1, x_2 \in \Delta_r} |g(x_1) - g(x_2)| + M \mathop{\mathrm{Sup}}_{x_1, x_2, \in \Delta_r} |f(x_1) - f(x_2)| \qquad (3.9.4.2)$$

Now taking sup over all $x_1, x_2 \in \Delta_r$ on right had side of (3.9.4.2) we get

$$\mathop{\mathrm{Sup}}_{x_1, x_2, \in \Delta_r} |f(x_1), g(x_1) - f(x_2) \cdot g(x_2)| \leq M \mathop{\mathrm{Sup}}_{x_1, x_2, \in \Delta_r} |g(x_1) - g(x_2)| + M \mathop{\mathrm{Sup}}_{x_1, x_2, \in \Delta_r} |f(x_1) - f(x_2)|$$

or,

$$M_r - m_r = M(M_r^1 - m_r^1) + M(M_r^2 - m_r^2) \quad \text{(Using Lemma 3.9.2)} \qquad\qquad (3.9.4.3)$$

Multiplying through out equation (3.9.4.3) by δ_r and summing over $r = 1, 2, ...n$ we get,

$$\sum_{r=1}^{n} (M_r - m_r)\, \delta_r \leq M \sum_{r=1}^{n} (M_r^1 - m_r^1)\, \delta_r + M \sum_{r=1}^{n} (M_r^2 - m_r^2)\, \delta_r \text{ that is,}$$

$$U(P, fg) - L(P, fg) \leq M\{U(P, f) - L(P, f)\} + M\{U(P, g) - L(P, g)\} \qquad\qquad (3.9.4.4)$$

Now as f & $g \in R[a, b]$ there exists partitions P_1 & P_2 such that $U(P_1, f) - L(P_1, f) < \dfrac{\varepsilon}{2M}$ & $U(P_2,$ $g) - L(P_2, g) < \dfrac{\varepsilon}{2M}.$ We choose P to be the common refinement of P_1 & P_2 then from (3.9.4.4) we get,

$$U(P, fg) - L(P, fg) < \frac{\varepsilon}{2M} + M\,\frac{\varepsilon}{2M} = \frac{\varepsilon}{2} + \frac{\varepsilon}{2} = \varepsilon.$$

This proves the theorem.

We now prove another important result regarding modulus of a R-Integrable function. It says that if a function is R-Integrable function then its modulus is also R-Integrable.

Before proceeding to the next theorem we first define positive and negative parts of a function.

Definition   Let f be a real-valued function defined on a certain subset of real number, then the positive and negative parts of the function f denoted by f^+ and f^- respectively is defined as follows:

$$f^+(x) = \begin{cases} f(x) & \text{if } (x) \leq 0 \\ 0 & \text{if } f(x) < 0 \end{cases} \qquad f^-(x) = \begin{cases} 0 & \text{if } f(x) \geq 0 \\ f(x) & \text{if } f(x) < 0 \end{cases}$$

It follows from the definition that $f(x) = f^+(x) - f^-(x)$ and $|f(x)| = f^+(x) + f^-(x)$

Theorem 3.9.5

If $f \in R[a, b]$ then f^+ and f^- are also in $R[a, b]$.

Proof: We first note that as $f^- = g^+$ where $g(x) = -f(x)$ hence it suffices to prove that $f^+ \in R[a, b]$. Let P be a partition of $[a, b]$ and let (M_r, m_r) and (M_r^+, m_r^+) be (sup, inf) of the functions f and f^+ on the interval Δ_r of P, $r = 1, 2, \ldots n$. Now we can have the following cases:

(i) $M_r \geq 0$, $m_r \geq 0$, then we have $M_r^+ = M_r$ and $m_r^+ = m_r$ so $M_r - m_r = M_r^+ - m_r^+$

(ii) $M_r \geq 0$, $m_r \leq 0$, then we have $M_r^+ = M_r$ and $m_r^+ \geq m_r$ so $M_r - m_r \geq M_r^+ - m_r^+$

(iii) $M_r \leq 0$, $m_r \leq 0$, then we have $M_r^+ = 0 = m_r^+$ hence $M_r - m_r \geq M_r^+ - m_r^+ = 0$

So in any case we have,

$$M_r^+ - m_r^+ \leq M_r - m_r,\ r = 1, 2, \ldots n.$$

Multiplying above inequality with δ_r and summing over $r = 1, 2, \ldots n$ we get,

$$\sum_{r=1}^{n} (M_r^+ - m_r^+)\,\delta_r \leq \sum_{r=1}^{n} (M_r - m_r)\,\delta_r$$

that is ,

$$U(P, f^+) - L(P, f^+) \leq U(P, f) - L(P, f) \tag{3.9.6.1}$$

Now as $f \in R[a, b]$, for $\varepsilon > 0$ there exists a partition P of $[a, b]$ such that,

$$U(P, f) - L(P, f) < \varepsilon$$

Using (3.9.6.1) we get $U(P, f^+) - L(P, f^+) \leq U(P, f) - L(P, f) < \varepsilon$

Hence $f^+ \in R[a, b]$.

Theorem 3.9.6

If $f \in R[a, b]$ then $|f|$ is also in $R[a, b]$. Moreover,

$$\left| \int_a^b f(x)\,dx \right| \leq \int_a^b |f(x)|\,dx$$

Proof: Let P be a partition of $[a, b]$ which is applied to the functions f & $|f|$. Let (M_r, m_r) (M_r^1, m_r^1) be (sup, inf) of f & $|f|$ respectively on the interval Δ_r of P, $r = 1, 2, ...n$. Now for $x_1, x_2 \in \Delta_r$ we have, by triangular inequality,

$$||f(x_1)| - |f(x_2)|| \leq |f(x_1) - f(x_2)| \tag{3.9.5.1}$$

By taking sup first to the left hand side and then to the right hand side of (3.9.5.1) we get,

$$\underset{x_1, x_2 \in \Delta_r}{\text{Sup}} \ ||f(x_1)| - |(x_2)|| \leq \underset{x_1, x_2 \in \Delta_r}{\text{Sup}} \ |f(x_1) - f(x_2)|$$

that is,

$$M_r^1 - m_r^1 \leq M - m \tag{3.9.5.2}$$

Multiplying through out equation (3.9.5.2) by δ_r and summing over $r = 1, 2, ...n$ we get,

$$\sum_{r=1}^{n} (M_r^1 - m_r^1) \, \delta_r \leq \sum_{r=1}^{n} (M - m) \, \delta_r$$

or $\qquad\qquad U(P, |f|) - L(P, |f|) \leq U(P, f) - L(P, f), \tag{3.9.5.3}$

As $f \leq R[a, b]$ there exists a partition P such that

$$U(P, f) - L(P, f) < \varepsilon \text{ for any } \varepsilon > 0$$

then by equation (3.9.5.3) we have

$$U(P, |f|) - L(P, |f| \leq U(P, f) - L(P, f) < \varepsilon$$

This proves the first part of the theorem.

For the second part of the theorem we know by definition,

$$\int_a^b f(x) \, dx = \lim_{\delta \to 0} \ \sum_{r=1}^{n} f(\xi_r) \, \delta_r.$$

Now,

$$\left| \sum_{r=1}^{n} f(\xi_r) \, \delta_r \right| \leq \sum_{r=1}^{n} |f(\xi_r)| \, \delta_r \tag{3.9.5.4}$$

Taking limit $\delta \to 0$ to both sides of (3.9.5.4), and using continuity of $||$ we get,

$$\lim_{\delta \to 0} \ \left| \sum_{r=1}^{n} f(\xi_r) \, \delta_r \right| \leq \lim_{\delta \to 0} \ \sum_{r=1}^{n} |f(\xi_r)| \, \delta_r$$

or, $\qquad\qquad \left| \lim_{\delta \to 0} \sum_{r=1}^{n} f(\xi_r) \, \delta_r \right| \leq \lim_{\delta \to 0} \sum_{r=1}^{n} |f(\xi_r)| \, \delta_r$ That is,

$$\left| \int_a^b f(x) \, dx \right| \leq \int_a^b |f(x)| \, dx \text{ by definition.}$$

This completes the theorem.

Remark:

One can enquire about the converse of first part of the above theorem? That is, is it true that if $|f|$ is in $R[a, b]$ then f necessarily be in $R[a, b]$? Answer to this is in negative as shown by the following example:

Example 1.9   Let f be defined on $[0, 1]$ as follows,

$$f(x) = \begin{cases} 1 & \text{if } x \text{ is rational} \\ -1 & \text{if } x \text{ is irrational} \end{cases}$$

Then we know that f is not R-Integrable but see that as $|f(x)| = 1 \ \forall x \in [0, 1]$, $|f|$ is R-Integrable.

Next we ask the following question if a function f is R-Integrable in $[a, b]$ then is it R-Integrable in any closed subinterval contained in $[a, b]$? Also is the converse of above statement true? Intuitively, answers to both questions should be in affirmative which is supported by the following theorem.

Theorem 3.9.7

A function f is in $R[a, b]$ if and only if it is in $R[c, d]$ whenever $a \le c < d \le b$.

Proof: First suppose that $f \in R[a, b]$, we will show $f \in R[c, d]$ where $a \le c < d \le b$. Now for $\varepsilon > 0$, there exists a partition P of $[a, b]$ such that $U(P, f) - L(P, f) < \varepsilon$. We can assume that c & d are points of the partition otherwise we will conceder the refinement of P introducing points c & d and call that refinement partition as P. Now let P_1 be the partition induced by P on $[c, d]$. We see that $U(P, f) - L(P, f)$ contains all terms of $U(P_1, f) - L(P_1, f)$ plus some nonnegative terms. Hence,

$$U(P_1, f) - L(P_1, f) \le U(P, f) - L(P, f) < \varepsilon.$$

That is $f \in R[c, d]$.

Convesely, if $f \in R[c, d]$ for every c, d satisfying $a \le c < d \le b$ in particular choosing $a = c$ & $d = b$ we get $f \in R[a, b]$.

This proves the theorem.

Theorem 3.9.8

If $f \in R[a, b]$ then for any c such that $a < c < b$ we have,

$$\int_a^b f(x)\, dx = \int_a^c f(x)\, dx + \int_c^d f(x)\, dx \tag{3.9.8.1}$$

Proof: First observe that as $f \in R[a, b]$, if we choose $a = c < d \le b$ then from above theorem $f \in R[a, b]$ for any d satisfying $a < d < b$. Similarly choosing $a < c < d = b$ we can conclude $f \in R[c, b]$ for any c satisfying $a < c < b$. Hence if $a < c < b$ we conclude in particular $f \in R[a, c]$ and $f \in R[c, b]$.

As $f \in R[a, c]$ for $\varepsilon > 0$ there exists $\delta_1 > 0$ such that for any partition P_1 of $[a, b]$ with $\|P_1\| < \delta_1$ we have,

$$\left| \sum_{r=1}^{n_1} f(\xi_r)\, \delta r - \int_a^c f(x)\, dx \right| < \frac{\varepsilon}{4} \tag{3.9.8.2}$$

for every possible choice of ξ_r in Δ_r, $r = 1, 2 \dots n_1$.

Similarly as $f \in R[c, b]$ for $\varepsilon > 0$ there exists $\delta_2 > 0$ such that for any partition P_2 of $[a, b]$ with $\|P_1\| < \delta_1$ we have,

$$\left| \sum_{r=1}^{n_2} f(\xi_r') \, \delta_r - \int_c^b f(x) \, dx \right| < \frac{\varepsilon}{4} \tag{3.9.8.3}$$

for every possible choice of ξ_r in Δ_r, $r = 1, 2,...n$.

Now let $\delta < \min\left(\delta_1, \delta_2, \dfrac{\varepsilon}{4M}\right)$ where $M > 0$ is such that $|f(x)| < M \ \forall \ x$ and concider a partition P of $[a, b]$ with $\|P\| < \delta$.

Case 1: If c is a point of the partition P

In this case P induces partitions P_1 & P_2 of $[a, c]$ & $[c, b]$ respectively given by,

$$P_1 : x_0 = a < x_1 < x_2 < ... < x_k = c$$

$$P_2 : x_k = c < x_{k+1} < x_{k+2} < ... < x_n = b$$

also we note that as $\|P\| < \delta$ $\|P_1\| < \delta < \delta_1$ & $\|P_2\| < \delta < \delta_1$.

Now,

$$\left| \sum_{r=1}^n f(\xi_r) \, \delta_t - \int_a^c f(x) \, dx - \int_c^b f(x) \, dx \right| = \left| \sum_{r=1}^k f(\xi_r) \, \delta_r - \int_a^c f(x) \, dx + \sum_{r=k=1}^n f(\xi_r) \, \delta_r - \int_c^b f(x) \, dx \right|$$

$$< \left| \sum_{r=1}^k f(\xi_r) \, \delta_r - \int_a^c f(x) \, dx + \sum_{r=k=1}^n f(\xi_r) \, \delta_r - \int_c^b f(x) \, dx \right|$$

$$< \frac{\varepsilon}{4} + \frac{\varepsilon}{4} = \frac{\varepsilon}{2} < \varepsilon.$$

for every choice of ξ_r in Δ_r. This shows that,

$$\int_a^b f(x) \, dx = \lim_{\|P\| \to 0} \sum_{r=1}^n f(\xi_r) \, \delta_r = \int_a^c f(x) \, dx + \int_c^b f(x) \, dx$$

Case 2: c is not a point of the partition P

In this case say $x_{r-1} < c < x_r$ for some r. If $x_{r-1} \le \xi_r \le c$ then,

We concider the partition P_1 & P_2 of $[a, c]$ & $[c, b]$ as follows,

$$P_1 : x_0 = a < x_1 < x_2 < ... < x_{r-1} < c$$

$$P_2 : c < x_{r+1} < x_{r+2} < ... x_n = b$$

Again as case1 $\|P_1\| < \delta < \delta_1$ & $\|P_2\| < \delta < \delta_1$

Then we have,

$$\left| \sum_{r=1}^n f(\xi_r) \, \delta_r - \int_a^c f(x) \, dx - \int_c^b f(x) \, dx \right|$$

$$= \left| \left(\sum_{k=1}^{r=1} f(\xi_k) \, \delta_k + f(\xi_r)\,(c - x_{r-1}) - \int_a^b f(x) \, dx \right) + \left(\sum_{k=r+1}^n f(\xi_k) \, \delta_k + f(\eta)\,(x_r - c) - \int_c^b f(x) \, dx \right) \right.$$

$$\left. + f(\xi_r) \, \delta_r - f(\xi_r)\,(c - x_{r-1}) - f(\eta)\,(x_r - c) \right|$$

$$\leq \left| \sum_{k=1}^{r=1} f(\xi_k)\, \delta_k + f(\xi_r)\, (c - x_{r-1}) - \int_a^c f(x)\, dx \right| + \left| \sum_{k=r+1}^{n} f(\xi_k)\, \delta_k + f(\eta)\, (x_r - c) - \int_c^b f(x)\, dx \right|$$

$$+\, f(\xi_r)\, \delta_r - f(\xi_r)\, (c - x_{r-1}) - f(\eta)\, (x_r - c)| \qquad \text{for any } \eta \in [c, x_r])$$

$$\leq \frac{\varepsilon}{4} + \frac{\varepsilon}{4} + |f(\xi_r) - f(\eta)|\, \delta$$

$$\leq \frac{\varepsilon}{2} + 2M \times \frac{\varepsilon}{4M} = \varepsilon$$

This shows that,

$$\int_a^b f(x)\, dx = \lim_{\|P\| \to 0} \sum_{r=1}^{n} f(\xi_r)\, \delta_r = \int_a^c f(x)\, dx + \int_c^b f(x)\, dx$$

Similarly we can dispose of the case if $c < \xi_r \leq x_r$

Remark:

We define for $a < b$,

$$\int_a^a f(x)\, dx = 0,\ \int_b^a f(x)\, dx = -\int_a^b f(x)\, dx.$$

Then we can extend equation (3.9.8.1) to any arbitrary a, b, c. For example, if $a \leq c \leq b$, then

$$\int_a^b f(x)\, dx = \int_a^c f(x)\, dx + \int_c^b f(x)\, dx = \int_a^c f(x)\, dx - \int_b^c f(x)\, dx.$$

In the next section will now see how process of integration is connected to differentiation, which is called Fundamental Theorem of Integral Calculus and also Mean Value Theorems of Integral Calculus.

3.10   FUNDAMENTAL THEOREM AND MEAN VALUE THEOREMS OF INTEGRAL CALCULUS

We start with a very basic inequality,

Theorem 3.10.1

For a bounded function f in [a, b],

$$m(b - a) \leq \underline{\int_a^b} f(x)\, dx \leq \overline{\int_a^b} f(x)\, dx \leq M(b - a).$$

Proof: We know for any partition P,

$$m \leq m_r \leq M_r \leq M$$

hence,

$$\sum_{r=1}^{n} m\, \delta_r \leq \sum_{r=1}^{n} m_r\, \delta_r \leq \sum_{r=1}^{n} M_r\, \delta_r \leq \sum_{r=1}^{n} M\delta_r$$

or,

$$m(b - a) \leq L(P, f) \leq U(P, f) \leq M(b - a)$$

Now as $\|P\| \to 0$ we get

$$m(b - a) \leq \underline{\int_a^b} f(x)\, dx \leq \overline{\int_a^b} f(x)\, dx \leq M(b - a)$$

Theorem 3.10.2

If $f \in R[a, b]$ then,

$$\int_a^b f(x)\, dx = \lambda(b-a) \quad \text{where } m \le \lambda \le M, \tag{3.10.2.1}$$

Moreover, if f is continuous then there must exist some $\xi \in [a, b]$ such that,

$$\int_a^b f(x)\, dx = f(\xi)\,(b-a) \tag{3.10.2.2}$$

Proof: As $f \in R[a, b]$

$$\int_{\underline{a}}^b f(x)\, dx = \int_a^{\overline{b}} f(x)\, dx = \int_a^b f(x)\, dx,$$

Hence it follows from Proposition 3.10.1 that

$$\int_a^b f(x)\, dx = \lambda\,(b-a) \quad \text{where } m \le \lambda \le M,$$

Now if f is continuous then by Intermediate Value Theorem for continuous functions there must exist $\xi \in [a, b]$ such that $f(\xi) = \lambda$ hence in that case,

$$\int_a^b f(x)\, dx = f(\xi)\,(b-a).$$

Let us now see some consequences of Theorem 3.10.2 as corollaries.

Corollary 3.10.3

If $f \in R[a, b]$ and $f(x) \ge 0 \ \forall\ x \in [a, b]$ then,

$$\int_a^b f(x)\, dx \ge 0$$

Proof: For $f \in R[a, b]$, by Theorem 3.10.2

$$\int_a^b f(x)\, dx = \lambda(b-a), \quad m \le \lambda \le M.$$

As $f(x) \ge 0 \ \forall\ x \in [a, b]$ we have $m, M \ge 0$, so $\lambda \ge 0$ hence,

$$\int_a^b f(x)\, dx \ge 0 .$$

Corollary 3.10.4

If f & $g \in R[a, b]$ and $f(x) \ge g(x) \ \forall\ x \in [a, b]$ then

$$\int_a^b f(x)\, dx \ge \int_a^b g(x)\, dx.$$

Proof: Follows from Theorem 3.9.3 and Corollary 3.10.3 by taking $\phi(x) = f(x) - g(x) \ge 0$.

Corollary 3.10.5

If $f \in R[a, b]$ is never negative on $[a, b]$ and is continuous at $c \in [a, b]$ with $f(c) > 0$ then $\int_a^b f(x)\, dx > 0$.

Proof: First suppose that c is an interior point of $[a, b]$. Now as f is continuous at c by neighborhood property of continuous function there exists an interval $(c - \delta, c + \delta)$ where, $a < c - \delta < c + \delta < b$ such that $f(x) > k,\ 0\ \forall\ x \in (c - \delta, c + \delta)$ for some $k > 0$. Now $f(x) \geq 0$ in $[a, c - \delta]$ so $\int_a^{c - \delta} f(x)\, dx \geq 0$.

Similarly as $f(x) \geq 0$ in $[c + \delta, b]$, $\int_{c+\delta}^b f(x)\, dx \geq 0$.

Now as $f(x) > k,\ 0\ \forall\ x \in (c - \delta, c + \delta)$

$$\int_{c-\delta}^{c+\delta} f(x)\, dx \geq \int_{c-\delta}^{c+\delta} k\, dx \geq 2k\delta > 0.$$

Adding we get $\int_a^b f(x)\, dx > 0$.

Corollary 3.10.6

Let $f \in R[a, b]$, is never negative and is continuous on $[a, b]$. Suppose further $\int_a^b f(x)\, dx = 0$ then, $f(x) = 0\ \forall\ x \in [a, b]$.

Proof: Follows from Corollary 3.10.5.

Next, let $f \in R[a, b]$ then if we consider $F(x) = \int_a^x f(x)\, dx$ for $a \leq x \leq b$ then F is a well defined function from $[a, b]$ to R. We have the following theorem

Theorem 3.10.7

If f is bounded and Integrable on $[a, b]$ and if,

$$F(x) = \int_a^x f(x)\, dx$$

then, F is a continuous function of x on $[a, b]$. If f is continuous on $[a, b]$ then F is differentiable at every point of $[a, b]$ and moreover, $F'(x) = f(x)$ for $a \leq x \leq b$.

Proof: For $0 < h < b - a$ we have

$$F(x + h) - F(x) = \int_a^{x+h} f(t)\, dt - \int_a^x f(t)\, dt = \int_x^{x+h} f(t)\, dt \qquad (3.10.7.1)$$

Now using Proposition 3.10.2 we must have,

$$\int_x^{x+h} f(x)\, dx = \lambda h, \text{ for some } \lambda \text{ in } m' \leq \lambda \leq M' \qquad (3.10.7.2)$$

where m' and M' are supremum and infimum of f in $[x, x + h]$. Also as f is bounded we have some $K > 0$ such that, $|f| \leq K$. Now from (3.10.7.1) and (3.10.7.2) we get,

$$|F(x + h) - F(x)| = \left| \int_{x}^{x+h} f(t)\, dt \right| = |\lambda h| \leq K|h|, \qquad (3.10.7.3)$$

where $|\lambda| \leq K$. Now given any $\varepsilon > 0$ we can choose $|h| < \dfrac{\varepsilon}{K} = \delta$ so that whenever $|h| < \delta$ we have,

$$|F(x + h) - F(x)| < \varepsilon \quad \text{(from (3.10.7.3))}$$

That is F is a continuous function of x on $[a, b]$. In fact we have proved that F is uniformly on $[a, b]$. Suppose now f is continuous on $[a, b]$. Then from [3.10.7.1] and using Theorem 3.10.2 we have,

$$F(x + h) - F(x) = \int_{x}^{x+h} f(t)\, dt = hf(\xi)\ x \leq \xi \leq x + h$$

Now as f is continuous,

$$\lim_{h \to 0} \frac{F(x + h) - F(x)}{h} = \lim_{h \to 0} f(\xi) = f(x)$$

That is for $a \leq x \leq b\ F'(x) = f(x)$.

Corollary 3.10.8

At every point of continuity of f we have

$$\frac{d}{dx} \int_{x}^{b} f(x)\, dx = -f(x)$$

Proof: We just observe that by definition $\displaystyle\int_{x}^{b} f(t)\, dt = -\int_{b}^{x} f(t)\, dt$ and we get the required result.

Definition   A differentiable function $F(x)$ whose derivative is equal to $f(x)$ at every point $x \in [a, b]$ is called primitive of $f(x)$ on the interval.

Remark:

We observe that if f is a continuous function on $[a, b]$ then it immediately follows from last theorem that $F(x) = \displaystyle\int_{a}^{x} f(x)\, dx$ is a primitive of $f(x)$. We now ask whether it is possible for a function to have more than one primitive. The answer is yes as if $F(x)$ is a primitive of f then we note that $F(x) + c$ is also a primitive of f where c any arbitrary constant. In fact, we know from differential calculus that for two differentiable functions will have equal derivatives at every point if and only if they differ by a constant.

Theorem 3.10.9 (Fundamental Theorem of Integral Calculus) (Assuming Continuity of the function)

For a continuous function f if $F(x)$ denotes it's primitive at the point x on the interval $[a, b]$ then,

$$\int_{a}^{b} f(x)\, dx = F(b) - F(a)$$

Proof: Let us define a function G on $[a, b]$ as follows,

$$G(x) = \int_a^x f(x)\, dx.$$

The as we have noted in the remark following Corollary 3.10.8 $G(x)$ and $F(x)$ can differ by a constant. Hence,

$$F(x) = G(x) + c.$$

Now by definition of an integral $G(a) = 0$ so we get $F(a) = c$. That is,

$$F(x) = F(a) + \int_a^x f(x)\, dx \qquad a \le x \le b \tag{3.10.9.1}$$

In particular, putting $x = a$ in (3.10.9.1) we get

$$\int_a^b f(x)\, dx = F(b) - F(a)$$

Remark:

We give one example to show that a function may have a primitive but is not integrable

Example 1.10   Let $f(x)$ on $[-2, 2]$ be defined as follows

$$f(x) = \begin{cases} x^2 \sin\left(\dfrac{1}{x^2}\right) & \text{if } x \ne 0 \\ 0 & \text{if } x = 0 \end{cases}$$

Then,

$$f'(x) = \begin{cases} 2x \sin\left(\dfrac{1}{x^2}\right) - \dfrac{2}{x} \cos\left(\dfrac{1}{x^2}\right) & \text{if } x \ne 0, \\ 0 & \text{if } x = 0 \end{cases}$$

Since $f'(x)$ is not bounded on any neighbourhood of origin hence is not integrable in $[-2, 2$ but indeed it has $f(x)$ as its primitive.

However, if at all $f'(x)$ is integrable it must give back $f(x)$. This is essentially the next theorem.

Theorem 3.10.10 (Fundamental Theorem of Integral Calculus)

Suppose f is bounded and Integrable on $[a, b]$ and suppose there exists a differentiable function $F(x)$ such that $F'(x) = f(x)$ for $x \in [a, b]$, then

$$\int_a^b f(x)\, dx = F(b) - F(a)$$

Proof: Consider a partition P of $[a, b]$ then,

$$F(b) - F(a) = \sum_{r=1}^{n} \{F(x_r) - F(x_{r-1})\} = \sum_{r=1}^{n} F'(\xi_r)\, \delta_r.$$

where $\xi_r \in [x_{r-1}, x_r]$ by using Mean Value Theorem of Differential Calculus. Now if M_r and m_r denote supremum and infimum of f on Δ. Then as $m_r \le f(\xi_r) = F'(x) \le M_r$, we have

$$L(P, f) = \sum_{r=1}^{n} m_r\, \delta_r \le F(b) - F(a) \le \sum_{r=1}^{n} M_r\, \delta_r.$$

As above equation is true for every partition P we have

$$\int_a^b f(x)\, dx \le F(b) - F(a) \le \overline{\int_a^b} f(x)\, dx$$

And since f is Integrable we conclude

$$\int_a^b f(x)\, dx = F(b) - F(a).$$

We solve one example using Fundamental Theorem

Example 1.11 Evaluate $\displaystyle \lim_{x \to 0} \frac{\displaystyle\int_0^x t \sin t^2\, dt}{xe^x}$

Sol: Let $\phi(x) = \displaystyle\int_0^x t \sin t^2\, dt$ and $\psi(x) = xe^x$ then we see that $\phi(x) = \Psi(x) = 0$. As $t \sin t^2$ is continuous by fundamental theorem $\phi(x)$ is differentiable and $f'(x) = x \sin x^2$. Hence we apply L'Hospitals Rule to evaluate the above limit and obtain,

$$\lim_{x \to 0} \frac{\displaystyle\int_0^x t \sin t^2\, dt}{xe^x} = \lim_{x \to 0} \frac{x \sin x^2}{xe^x + e^x} = \frac{0}{1} = 0$$

Next we turn to Mean Value Theorems of Integral Calculus.

Theorem 3.10.11 (First Mean Value Theorem)

Let $f(x)$ and $f(x)$ be two functions in $R[a,b]$, suppose $g(x)$ maintains same sign on $[a, b]$ then,

$$\int_a^b f(x)\, g(x)\, dx = \lambda \int_a^b g(x)\, dx \qquad (3.10.11.1)$$

where $m \le \lambda \le M$, M & m being supremum and infimum of f on $[a, b]$.

Moreover, if we assume that f is continuous on $[a, b]$ then there must exist some $\xi \in [a, b]$ such that,

$$\int_a^b f(x)\, g(x)\, dx = f(\xi) \int_a^b g(x)\, dx$$

Proof: We first assume that $g(x) \ge 0$. We know,

$$m \le f(x) \le M$$

or,
$$mg(x) \le f(x)\, g(x) \le Mg(x)\ (g(x) \ge 0).$$

Now as, $mg(x), f(x)\, g(x)$ and $Mg(x)$ are R-Integrable

$$\int_a^b mg(x)\, dx \le \int_a^b f(x)\, g(x)\, dx \le \int_a^b Mg(x)\, dx$$

or,
$$m \int_a^b g(x)\, dx \le \int_a^b f(x)\, g(x)dx \le M \int_a^b g(x)dx.$$

So,

$$\int_a^b f(x)\, g(x)\, dx = \lambda \int_a^b g(x)\, dx, \text{ where } m \le \lambda \le M \tag{3.10.11.2}$$

Observe that if $g(x) \le 0$ then, $-g(x) \ge 0$. We get required result in this case by putting $-g(x)$ in place of $g(x)$ in (3.10.11.2).

Now if f is continuous on $[a, b]$ then by Intermediate Value Theorem for continuous functions, there must exist $\xi \in [a, b]$ such that $f(\xi) = \lambda$ hence in that case,

$$\int_a^b f(x)\, g(x)\, dx = f(\xi) \int_a^b g(x)\,dx.$$

Remarks

1. Observe that we get Proposition 3.10.2 from Theorem 3.10.11 by just putting $g(x) = 1$.
2. We can write (3.10.5.1) as,

$$\lambda = \frac{\displaystyle\int_a^b g(x)\, f(x)\, dx}{\displaystyle\int_a^b g(x)\, dx}$$

If $g(x) \ge 0$ then the ratio $\lambda = \dfrac{\displaystyle\int_a^b g(x)\, f(x)\, dx}{\displaystyle\int_a^b g(x)\, dx}$ can be interpreted as weighted average of $f(x)$ with on $[a, b]$ weight $g(x)$.

To prove the next theorem we need an inequality known as Abel's Inequality which is proved as a lemma,

Lemma 3.10.12 (Abel's Inequality)

Let $a_1, a_2, \ldots a_n$ be a sequence of non-increasing sequence of n positive numbers, and $b_1, b_2, \ldots b_n$ be any set of numbers satisfying, $1 < b_1 + b_2 + \ldots + b_k < L$, $1 \le k \le n$ for some real numbers l and L. Then,

$$a_1 l < a_1 b_1 + a_2 b_2 + \ldots + a_n b_n < a_1 L.$$

Proof: Let $S_k = b_1 + b_2 + \ldots + b_k$, $1 \le k \le n$. Then,

$$\sum_{r=1}^n a_r b_r = a_1 S_1 + a_2(S_2 - S_1) + \ldots a_r(S_r - S_{r-1}) + \ldots + a_n(S_n - S_{n-1})$$

$$= (a_1 - a_2)S_1 + (a_2 - a_3)S_2 + \ldots + (a_{n-1} - a_n)S_{n-1} + a_n S_n$$

Since $(a_1 - a_2), (a_2 - a_3), \ldots (a_{n-1} - a_n)$ are all non-negative and $l < S_k < L$, $1 \le k \le n$ by assumption we therefore have,

$$\sum_{r=1}^n a_r b_r < (a_1 - a_2)L + (a_2 - a_3)L + \ldots + (a_{n-1} - a_n)L + a_n L = a_1 L \tag{3.10.12.1}$$

$$> (a_1 - a_2)l + (a_2 - a_3)l + \ldots + (a_{n-1} - a_n)l + a_n l = a_1 l \tag{3.10.12.2}$$

Combining [3.10.12.1] and [3.10.12.2] we get the lemma.

We now give Second Mean Value Theorem which has two forms:

Theorem 3.10.13 (Second Mean Value Theorem) (Bonnet's Form)

Let $f(x)$ be a bounded monotone non-increasing and never negative on $[a, b]$ and $g(x)$ be in $R[a, b]$. Then there exists a value $\xi \in [a, b]$, such that,

$$\int_a^b f(x)\, g(x)\, dx = f(a) \int_a^\xi g(x)\, dx$$

(Weierstrass's Form)

Let $f(x)$ be a bounded monotonic on $[a, b]$ and $g(x)$ be in $R[a, b]$. Then there exists a value $\xi \in [a, b]$, such that,

$$\int_a^b f(x)\, g(x)\, dx = f(a) \int_a^\xi g(x)\, dx + f(b) \int_\xi^b g(x)\, dx$$

Proof: (Bonnet's Form) With usual notations let P be a partition of $[a, b]$ and let M_r and m_r be supremum and infimum of $g(x)$ on Δ_r, $r = 1, 2 \ldots n$. Let $\xi_1 = a$ and ξ_r, $r \neq 1$ be any point of Δ_r the rth subinterval of P. Then,

$$m_r \delta_r \leq \int_{x_{r-1}}^{x_r} g(x)\, dx \leq M_r \delta_r$$

summing above equation for $r = 1, 2, \ldots p$ where $p \leq n$ we get,

$$\sum_{r=1}^p m_r \delta_r \leq \int_a^{x_r} g(x)\, dx \leq \sum_{r=1}^p M_r \delta_r \tag{3.10.13.1}$$

Also we have,

$$m_r \delta_r \leq g(\xi_r)\, \delta_r \leq M_r \delta_r$$

again adding above equation for $r = 1, 2, \ldots p$ we obtain

$$\sum_{r=1}^p m_r \delta_r \leq \sum_{r=1}^p g(\xi_r)\, \delta_r \leq \sum_{r=1}^p M_r \delta_r \tag{3.10.13.2}$$

From (3.10.13.1) and (3.10.13.2) we have,

$$\left| \sum_{r=1}^p g(\xi_r)\, \delta_r - \int_a^{x_p} g(x)\, dx \right| \leq \sum_{r=1}^p (M_r - m_r)\, \delta_r \leq \sum_{r=1}^n (M_r - m_r)\, \delta_r$$

or,

$$\int_a^{x_p} g(x)\, dx - \sum_{r=1}^n (M_r - m_r)\, \delta_r \leq \sum_{r=1}^p g(\xi_r)\, \delta_r \leq \int_a^{x_p} g(x)\, dx + \sum_{r=1}^n (M_r - m_r)\, \delta_r$$

We note that as $g(x)$ is in $R[a, b]$ by Fundamental Theorem of Integral Calculus, $\int_a^x g(t)\, dt$ is a continuous function of x and so it must have maximum and minimum on $[a, b]$. Let M and m be maximum and minimum of $\int_a^x g(t)\, dt$ on $[a, b]$. Then we have

$$m - \sum_{r=1}^{n} (M_r - m_r)\, \delta_r \leq \sum_{r=1}^{p} g(\xi_r)\, \delta_r \leq M + \sum_{r=1}^{n} (M_r - m_r)\, \delta_r.$$

Now we use Abel's Inequality with $a_r = f(\xi_r)$, $b_r = g(\xi_r)$,

$$l = m - \sum_{r=1}^{n} (M_r - m_r)\, \delta_r \quad \text{and}$$

$$L = M + \sum_{r=1}^{n} (M_r - m_r)\, \delta_r \quad \text{and we get,}$$

$$f(a)\left\{ m - \sum_{r=1}^{n} (M_r - m_r)\, \delta_r \right\} \leq \sum_{r=1}^{n} f(\xi_r)\, g(\xi_r)\, \delta_r \leq f(a)\left\{ M + \sum_{r=1}^{n} (M_r - m_r)\, \delta_r \right\}.$$ If now the norm of the

partition P tends to zero then we know $\sum_{r=1}^{n} (M_r - m_r)\, \delta_r \to 0$ so that,

$$mf(a) \leq \int_a^b f(x)\, g(x)\, dx \leq Mf(a)$$

hence,

$$\int_a^b f(x)\, g(x)\, dx = \lambda f(a) \quad \text{where } m \leq \lambda \leq M.$$

We know that by Intermediate Value Property of continuous functions every value between m and M are attained hence there must exist some $\xi \in [a, b]$ such that $\lambda = \int_a^b g(x)\, dx$ that is,

$$\int_a^b f(x)\, g(x)\, dx = f(a) \int_a^\xi g(x)\, dx.$$

(Weierstrass's Form)

In this case let us first assume that $f(x)$ be monotone non-increasing. Then $f(x) - f(b)$ is non-negative and monotone non-increasing. Hence by previous form we get, for $\xi \in [a, b]$

$$\int_a^b \{f(x) - f(b)\} g(x)\, dx = \{f(a) - f(b)\} \int_a^\xi g(x)\, dx$$

or,

$$\int_a^b f(x)\, g(x)\, dx = f(a) \int_a^\xi g(x)\, dx + f(b) \int_\xi^b g(x)\, dx.$$

Similarly we get same result if $f(x)$ is monotone non-decreasing in that case we use the function $f(b) - f(x)$ which is non-negative and monotone non-increasing. Again by previous form we get for $\xi \in [a, b]$

$$\int_a^b \{f(b) - f(x)\} g(x)\, dx = \{f(b) - f(a)\} \int_a^\xi g(x)\, dx$$

or,

$$\int_a^b f(x)\, g(x)\, dx = f(a) \int_a^\xi g(x)\, dx + f(b) \int_\xi^b g(x)\, dx.$$

This proves the theorem.

Example 1.12 Show that $\left| \int_a^b \dfrac{\sin \lambda x}{x}\, dx \right| < \dfrac{4}{\lambda a},\ \lambda, a > 0$

Sol: Using Second Mean Value Theorem (Weirestrass Form) with $\phi(x) = \sin \lambda x$ (integrable) and $\psi(x) = \dfrac{1}{x}$ (monotone decreasing) we get,

$$\int_a^b \frac{\sin \lambda x}{x}\, dx = \frac{1}{a} \int_a^\xi \sin \lambda x\, dx + \frac{1}{b} \int_\xi^b \sin \lambda x\, dx$$

$$= \frac{1}{\lambda a}(\cos \lambda a - \cos \lambda \xi) + \frac{1}{\lambda b}(\cos \lambda \xi - \cos \lambda b)$$

Hence,

$$\left| \int_a^b \frac{\sin \lambda x}{x}\, dx \right| \le \frac{2}{\lambda a} + \frac{2}{\lambda b} < \frac{4}{\lambda a} \quad (\text{as } b > a)$$

Next we turn our attention to two very important techniques applied in the process of integration.

3.11 METHOD OF SUBSTITUTION IN AN INTEGRAL

Theorem 3.11.1

Let $f \in R[a, b]$ and let g be a strictly monotone differentiable function on $[\alpha, \beta]$ such that

$$g(\alpha) = a,\ g(\beta) = b \text{ with } f \circ g \text{ and } g \in R[\alpha, \beta] \text{ then,}$$

$$\int_a^b f(x)\, dx = \int_\alpha^\beta f\{g(t)\}\, g'(t)\, dt$$

Proof: For definiteness sake we take g strictly monotone increasing. Let

$P(\alpha = t_0, t_1, \ldots t_{r-1}, t_r, \ldots t_n = \beta)$ be a partition of $[\alpha, \beta]$ and let,

$P'(a = x_0, x_1, \ldots x_{r-1}, x_r, \ldots x_n = b)$ be the corresponding partition of $[a, b]$

where $x_r = g(t_r)$. Using mean-value theorem of differential calculus,

$$x_r - x_{r-1} = g(t_r) - g(t_{r-1}) = (t_r - t_{r-1})g'(\xi_r), \quad t_{r-1} < \xi_r < t_r$$

Let $g(\xi_r) = \eta_r$. Then,

$$g(t_{r-1}) < \eta_r < g(t_r)$$

that is,
$$x_{r-1} < \eta_r < x_r$$

Now,

$$\sum_{r=1}^n f(\eta_r)(x_r - x_{r-1}) = \sum_{r=1}^n f\{g(\xi_r)\}g'(\xi_r)(t_r - t_{r-1})$$

If norm of $P \to 0$, then the norm of $P' \to 0$ and using assumptions (itegrability of f, g' and $f \circ g$) we have,

$$\int_a^b f(x)\, dx = \int_\alpha^\beta f\{g(t)\}\, g'(t)\, dt.$$

Hence the theorem is true if we take g monotone increasing. By suitably modifying the above argument we can prove the theorem for g monotone decreasing.

3.12   INTEGRATION BY PARTS

Theorem 3.12.1

Suppose f and g are differentiable functions on $[a, b]$ with f' and $g' \in R[a, b]$ then,

$$\int_a^b f(x)\, g'(x)\, dx = [f(x)\, g(x)]_a^b - \int_a^b f'(x)\, g(x)\, dx$$

Proof: We observe that $[f(x)\, g(x)]$ is integrable and

$$[f(x)\, g(x)] = f'(x)\, g(x) + f(x)\, g'(x)$$

Rest follows from Fundamental Theorem of Integral Calculus.

Next we will see some examples:

Example 1.13   Suppose f is a bounded function in $[a, b]$ such that $f(x) = 0$ except finitely many points there, then show that $f \in R[a, b]$ and $\int_a^b f(x)\, dx = 0$.

Sol: As per given condition f is continuous except for finitely many points in $[a, b]$ hence by Theorem it is R-Integrable. Also it is easy to see that for every partition P, $L(P, f) = 0$ hence $J = 0$. As $f \in R[a, b]$ it follows that $\int_a^b f(x)\, dx = 0$.

Example 1.14   A function f is defined on $[0, 1]$ as follows,

$$f(x) = \begin{cases} x & \text{if } x \text{ is rational} \\ -x & \text{if } x \text{ is irrational} \end{cases}$$

Show that f is not Riemann Integrable.

Sol: We consider any partition P of $[0, 1]$. With usual notation consider $\Delta_r \equiv [x_{r-1}, x_r]$. Now if x_r is rational then $M_r = x_r$, if x_r is irrational then there exists infinite numbers of rational numbers are there in the neighbourhood of x_r then supremum of $f(x)$ is given by $\lim_{x \to x_r} x = x_r$ i.e. $M_r = x_r$ where the limit is taken over rationals. So in any case $M_r = x_r$. By similar arguments we can see that $m_r = -x_r$. Hence the upper integral is given by,

$$J = \lim_{\|P\| \to 0} \sum_{r=1}^n x_r \delta_r = \int_0^1 x\, dx = \frac{1}{2}$$

Similarly the lower integral,

$$I = \lim_{\|P\| \to 0} \sum_{r=1}^n -x_r \delta_r = \int_0^1 x\, dx = -\frac{1}{2}$$

Example 1.15   If f is a continuous function on $[0, 1]$ then show that,

$$\lim_{n \to \infty} \int_0^1 \frac{n f(x)}{1 + n^2 x^2}\, dx = \frac{\pi}{2} f(0)$$

Sol: Consider $\displaystyle\int_0^1 \frac{nf(x)}{1+n^2x^2}\,dx = \int_0^{\frac{1}{\sqrt{n}}} \frac{nf(x)}{1+n^2x^2}\,dx + \int_{\frac{1}{\sqrt{n}}}^1 \frac{nf(x)}{1+n^2x^2}\,dx$

Now we observe that as f is a continuous function and $g(x) = \dfrac{n}{1+n^2x^2}$ is non negative and monotone

non increasing in $[0, 1]$ so using first mean value theorem of integral calculus we get,

$$\int_0^{\frac{1}{\sqrt{n}}} \frac{nf(x)}{1+n^2x^2}\,dx = f(\xi)\int_0^{\frac{1}{\sqrt{n}}} \frac{n}{1+n^2x^2}\,dx, \quad 0 \le \xi \le \frac{1}{\sqrt{n}}$$

$$= f(\xi)\tan^{-1}(\sqrt{n}) \qquad 0 \le \xi \le \frac{1}{\sqrt{n}}$$

Now as, $n \to \infty\ f(\xi) \to f(0)$ (since f is continuous) and $\tan^{-1}(\sqrt{n}) \to \dfrac{\pi}{2}$.

That is as, $n \to \infty\ \displaystyle\int_0^{\frac{1}{\sqrt{n}}} \frac{nf(x)}{1+n^2x^2}\,dx \to \frac{\pi}{2}f(0)$ $(*)$

Again we see that,

$$\left|\int_{\frac{1}{\sqrt{n}}}^1 \frac{nf(x)}{1+n^2x^2}\,dx\right| \le M\left|\int_{\frac{1}{\sqrt{n}}}^1 \frac{n}{1+n^2x^2}\,dx\right| = M\int_{\frac{1}{\sqrt{n}}}^1 \frac{n}{1+n^2x^2}\,dx, \quad \text{where } |f(x)| \le M$$

$$= M\left[\tan^{-1}(n) - \tan^{-1}(\sqrt{n})\right]$$

$$\to M\left(\frac{\pi}{2} - \frac{\pi}{2}\right) = 0 \text{ as } n \to \infty \quad (**)$$

Combining $(*)$ & $(**)$ we get,

$$\lim_{n \to \infty} \int_0^1 \frac{nf(x)}{1+n^2x^2}\,dx = \frac{\pi}{2}f(0)$$

$$f \in R[a, b]$$

3.13 CHARACTERIZATION OF RIEMANN INTEGRABILITY

Recall that in Section 8 we remarked that functions which are Riemann Integrable are 'close' to continuous functions in this section we will justify this assertion. We first define what are called *negligible sets* of $\mathbb{R}$.

Definition A subset J of real number is called negligible if for any $\varepsilon > 0$ we can find a sequence of open intervals say, $\{I_n\}$ such that $J \subset \displaystyle\bigcup_{n=1}^{\infty} I_n$ and $\displaystyle\sum_{n=1}^{\infty} l(I_n) < \varepsilon$

Example 1.16　Let $S = \{x_1, x_2, \ldots x_n, \ldots\}$ any countable subset of real numbers. For any $\varepsilon > 0$ we see that,

$$x_n \in I_n = \left(x_n - \frac{1}{2^{n+2}}, \, x_n + \frac{1}{2^{n+2}}\right) \quad \text{and}$$

$$l(I_n) = \frac{1}{2^{n+1}}.$$

So that $S \subseteq \bigcup_{n=1}^{\infty} I_n$ and $\sum_{n=1}^{\infty} l(I_n) = \sum_{n=1}^{\infty} \left(\frac{\varepsilon}{2^{n+1}}\right) = \frac{\varepsilon}{2} < \varepsilon.$

This shows every countable subset of real number is negligible.

Lemma 3.13.1

Countable union of negligible sets are negligible.

Proof:　Left as an exercise.

Next we prove two lemmas which will be useful for characterizing Riemann Integrals.

Lemma 3.13.2

Suppose f is a bounded function defined on the interval $[a, b]$ and if for $x \in [a, b]$ we define,

$\omega(f, x) = \inf \, [\sup| f(x_1) - f(x_2)| \, x_1, x_2 \in U \colon U$ is an open neighbourhood of $x]$ *then f is continuous at x if and only if $\omega(f, x) = 0$.*

Proof:　Suppose $\omega(f, x) = 0$. Then given any $\varepsilon > 0$ there exists an open neighbourhood U of x such that $\sup |f(x_1) - f(x_2)| < \varepsilon \; \forall \; x_1, x_2 \in U$. In particular we have,

$$|f(x) - f(x_2)| < \varepsilon \; \forall \; x_1 \in U$$

In other words f is continuous at x.

Conversely suppose f is continuous at x then for any $\varepsilon > 0$ there exists an open neighbourhood U of x such that $|f(x) - f(x_1)| < \frac{\varepsilon}{2} \; \forall \; x_1 \in U$. Now for any $x_1, x_2 \in U$ we have,

$$|f(x_1) - f(x_2)| \le |f(x) - f(x_1)| + |f(x) - f(x_2)| < \frac{\varepsilon}{2} + \frac{\varepsilon}{2} = \varepsilon$$

Hence,

$$\sup |f(x_1) - f(x_2)| < \varepsilon \; \forall \; x_1, x_2 \in U$$

That is,

$\omega(f, x) = \inf \, [\sup |f(x_1) - f(x_2)| \, x_1, x_2 \in U \colon U$ is an open neighbourhood of $x] < \varepsilon$ which implies $\omega(f, x) = 0$.

Lemma 3.13.3

Suppose $I = [a, b] \subseteq \bigcup_{i=1}^{m} I_i$ where I_i are open intervals, then we can get a partition P of $[a, b]$ say

$P \colon a = x_0 < x_1 < x_2 < \ldots < x_n = b$ *such that for every $i \in \{1, 2, \ldots, n\}$ there exists a $j \in \{1, 2, \ldots, m\}$ such that $[x_{i-1}, x_i] \subseteq I_j.$*

Proof: As $a \in I$, there exists an open interval $I_{k_1} = (a_1, b_1)$ say such that $a \in I_{k_1}$ where $a_1 < a < b_1$. If $b_1 > b$ then, if we choose $x_0 = a$ & $x_1 = b$ then we obtain $[x_0, x_1] \subseteq (a_1, b_1)$ and we are done. Otherwise if $b_1 \le b$ then again as $b_1 \in I$, there exists an open interval $I_{k_2} = (a_2, b_2)$ say such that $a_2 < b_1 < b_2$ and we choose x_1 such that $a_2 < x_1 < b_1$ then, $[x_0, x_1] \subseteq I_{k_1}$. If $b_2 > b$ we are done. Otherwise using same argument as above we will obtain interval $I_{k_3} = (a_3, b_3,$ such that $a_3 < b_2 < b_3$ with $b_3 \le b$ otherwise we are done. If we choose x_2 such that $a_3 < x_2 < b_2$ then in that case we have $[x_1, x_2] \subseteq I_{k_2}$. We continue the process which must terminate as there are only finite number of intervals $I_1, I_2, ..., I_m$ covering interval $[a, b]$. But the process terminates only when $a_n < x_{n-1} < b_{n-1} < b_n$ & $b_n \ge b$ at the nth stage and we choose x_n in that case and obtain $[x_{n-1}, x_n] \subseteq I_{k_n} = [a_n, b_n]$ which proves the lemma.

Following theorem gives the characterization of Riemann-Integrable functions:

Theorem 3.13.4

If D_f denotes the set of discontinuities of a bounded function f defined on a closed interval $[a, b]$then $f \in R[a, b]$ if and only if D_f is a negligible set.

Proof: First we observe that by Lemma 3.13.1,

$$D_f = \{x \in [a, b] : \omega(f, x) > 0\}$$

$$= \left\{ \bigcup_{n=1}^{\infty} x \in [a, b] : \omega(f, x) \ge \frac{1}{n} \right\} = \bigcup_{n=1}^{\infty} D_n$$

where
$$D_n = \left\{ x \in [a, b] : \omega(f, x) \ge \frac{1}{n} \right\}$$

Now suppose that $f \in R[a, b]$. For a fixed $j \in \{1, 2, 3, ...n\}$, we will show that for D_j is a negligible set. Now for $\varepsilon > 0$ using necessary condition of Riemann integrability, there exists a partition P of $[a, b]$

$$P : x_0 = a < x_1 < x_2 < ... < x_m = b \text{ such that,}$$

$$\sum_{r=1}^{n} (M_r - m_r)\, \delta_r < \frac{\varepsilon}{2j} \tag{3.13.4.1}$$

Now, if $x \in D_i \cap (x_{r-1}, x_r)$ for some $r \in \{1, 2, 3, ...m\}$ then there exists a $\delta > 0$ such that,

$$S = (x - \delta, x + \delta) \subset (x_{r-1}, x_r) \subset [x_{r-1}, x_r]$$

Hence, for $x_1, x_2 \in S$ we have,

$$\sup_{x_1, x_2 \in S} |f(x_1) - f(x_2)| \le \sup_{x_1, x_2 \in [x_{r-1}, x_r]} |f(x_1) - f(x_2)| = M_r - m_r$$

So that,

$$\frac{1}{j} \le \omega(f, x) \le M_r - m_r \tag{3.13.4.2}$$

Let $A = \{r : D_j \cap (x_{r-1}, x_r) \ne f\}$ then from (3.13.3.2) we have,

$$\frac{1}{j} \sum_{r \in A} \delta_r \le \sum_{r \in A} (M_r - m_r)\, \delta_r < \frac{\varepsilon}{2j}$$

or,
$$\sum_{r \in A} l(I_r) \le \frac{\varepsilon}{2} \tag{3.13.4.3}$$

where
$$I_r = (x_{r-1}, x_r)$$

Also note that the points of D_j which does not intersect any of the n intervals $[x_{r-1}, x]$, $r = 1, 2, 3,$ ...n must be the points of the finite set $C = \{x_0, x_1, ...x_n\}$. Hence we can always get finite number of open intervals $I'_1, I'_2, I'_3, ...I'_n$ such that,

$$C \subset \bigcup_{i=1}^{n} I'_i \quad \text{and} \quad \sum_{i=1}^{n} l(I'_i) < \frac{\varepsilon}{2} \tag{3.13.4.4}$$

We now observe that,

$$D_j \subset \bigcup_{r \in A} I_r \cup \bigcup_{i=1}^{n} I'_i$$

and,
$$\sum_{r \in A} l(I_r) + \sum_{i=1}^{n} l(I'_i) < \frac{\varepsilon}{2} + \frac{\varepsilon}{2} = \varepsilon \quad \text{using (3.13.4.3) \& (3.13.4.4)}$$

Hence we conclude that D_j is negligible. By Lemma 3.13.3 it follows D_j is negligible.

Conversely, suppose that D_j is negligible. Then for $\varepsilon > 0$ there exists a sequence of open intervals $\{I_n\}$ such that $D_j \bigcup_{n=1}^{\infty} I_n$ and $\sum_{n=1}^{\infty} i(I_n) < \frac{\varepsilon}{4M}$ where $M = \sup_{x \in [a, b]} f(x)$. Let $y \in [a, b]$ then if $y \notin D_j$ that f is if f is continuous at y then there exists an open interval say $I(y)$ such that,

$$|f(y) - f(y')| < \frac{\varepsilon}{4(b-a)} \quad \forall\, y' \in I(y) \tag{3.13.4.5}$$

if $y \in D_j$ then $y \in I_k$ for some k, we denote this by I_k by $I(y)$. We see that the family of open intervals $\{I(y)\}_{y \in [a, b]}$ forms a cover of $[a, b]$. By Hiene-Borel theorem there exists finitely many intervals $I(y_1)$, $I(y_2)$, ... $I(y_m)$ such that

$$[a, b] \subset \bigcup_{i=1}^{m} I(y_i).$$

Using Lemma 3.13.3 we will obtain a partition P of $[a, b]$

$$P : x_0 = a < x_1 < x_2 < ... < x_n = b \text{ such that,}$$

for every $i \in \{1, 2, 3, ...n\}$ there exists a $j \in \{1, 2, 3, ...m\}$ satisfying,

$$[x_{i-1}, x_i] \subset I(y_j)$$

Let $A = \{i : \Delta_i \subset I(y_j) \,\&\, y_i \notin D_f\}$. If $i \in A$ then for $y_1, y_2 \in \Delta_i$ we have,

$$|f(x_1) - f(x_2)| \leq |f(x_j) - f(x_1)| + |f(x_j) - f(x_2)|$$

$$< \frac{\varepsilon}{4(b-a)} + \frac{\varepsilon}{4(b-a)} = \frac{\varepsilon}{2(b-a)} \quad \text{by using} \tag{3.13.4.5}$$

or,
$$M_i - m_i < \frac{\varepsilon}{2(b-a)} \Rightarrow \sum_{i \in A} (M_i - m_i)\,\delta_i < \frac{\varepsilon}{2} \tag{3.13.4.6}$$

Again if $i \notin A$ then, $\Delta_i \subset I_k$ for k some so that,

$$\sum_{i \in A^c} (M_i - m_i)\,\delta_i \leq 2M \sum_{i \in A^c} \delta_i < 2M \frac{\varepsilon}{4M} \tag{3.13.4.7}$$

Now,

$$U(P,f) - L(P,f) = \sum_{r=1}^{n} (M_r - m_r)\, \delta_r$$

$$= \sum_{i \in A} (M_i - m_i)\, \delta_i + \sum_{i \in A^c} (M_i - m_i)\, \delta_i$$

$$< \frac{\varepsilon}{2} + \frac{\varepsilon}{2} = \varepsilon \quad \text{using (3.13.4.6) \& (3.13.4.7)}$$

This shows that $f \in R[a, b]$.

PROBLEMS

1. Suppose $f \in R[a, b]$ then show that,

$$\int_{a}^{b} f(x)\, dx = \lim_{n \to \infty} \left(\frac{b-a}{n}\right) \sum_{r=1}^{n} f\left(a + \frac{b-a}{n}\, r\right).$$

   Hence show that $\int_{0}^{t} \sin x\, dx = 1 - \cos t$.

2. Suppose $f \in R[a, b]$ and let F be an infinite subset of $[a, b]$ having finite numbers of limit points. Let g be a function defined on $[a, b]$ such that $f(x) = g(x)$ except the points of then show that $g \in R[a, b]$.

3. Let f be a bounded function on $[a, b]$ and suppose there is sequence of partitions $\{P_n\}$ of $[a, b]$ such that $U(P_n, f) - L(P_n, f) \to 0$ as $n \to \infty$ then show that $f \in R[a, b]$ and

$$\lim_{n \to \infty} L(P_n, f) = \int_{a}^{b} f(x)\, dx = \lim_{n \to \infty} U(P_n, f)$$

4. Prove that if $f \in R[a, b]$ then $f^2 \in R[a, b]$ hence show that product of Riemann Integrable functions are Riemann Integrable. [Hint: $4fg = (f+g)^2 - (f-g)^2$]

5. If for a non-negative function, f, $\int_{a}^{b} f(x)\, dx = 0$ then show that $f = 0$ if $f \in C[a, b]$. Is the converse true?

6. If the function $f \in C[a, b]$ satisfies,

$$\int_{a}^{b} f(x)\, g(x)\, dx = 0$$

   for any $g \in c[a, b]$ then $f \equiv 0$.

7. If f is defined in $[0,1]$ as follows,

$$f(x) = \begin{cases} x & \text{if } x \text{ is rational} \\ 1 + x & \text{if } x \text{ is rational} \end{cases}$$

For any partition P of $[a, b]$ find $U(P, f)$ & $L(P, f)$ and hence show that f is not Riemann Integrable.

8. Show that the following functions a re integrable in their respective domains and find their integrals:

   (a) $f(x) = (x + [x]^2)$ in the interval $[-3, 2]$.

(b) $f(x) = \begin{cases} \dfrac{|x| + [x]}{x} & \text{if } x \neq 0 \\ 0 & \text{if } x = 0 \end{cases}$ in the interval $[-4, 4]$.

9. Show that if $f, g \in R[a, b]$ then the functions h & k defined by,

$$h(x) = \max\,(f(x), g(x))$$

$$k(x) = \min\,(f(x), g(x))$$

    are Riemann Integrable.

10. Let $f \in R[a, b]$ and suppose m & M are infimum and supremum of f in $[a, b]$. If $g \in C[m, M]$ then $g \circ f \in R[a, b]$. Is the above result true if we relax the condition $g \in C[m, M]$ by $g \in R[m, M]$?

11. Show that the functions & defined in $[0, 1]$ as follows

$$f(x) = \begin{cases} 0 & \text{if } x = 0 \\ (-1)^n & \text{if } \dfrac{1}{2n+1} < x \leq \dfrac{1}{2^n} \quad n = 0, 1, 2... \end{cases}$$

$$g(x) = \begin{cases} 0 & \text{if } x = 0 \\ \dfrac{1}{a^{r-1}} & \text{if } \dfrac{1}{2r} < x \leq \dfrac{1}{2^{r-1}} \quad r = 1, 2... \\ \text{where is } a \text{ is an integer greater than one.} \end{cases}$$

    are Riemann integrable and find their Integrals.

12. Using only definition to show that the function f defined as,

$$f(x) = \begin{cases} \cos\left(\dfrac{1}{x}\right) & \text{if } x \neq 0 \\ 0 & \text{if } x = 0 \end{cases}$$

    is not Riemann Integrable in $[-1, 1]$.

13. Show that second part of Theorem 3.10.7 may fail if f is not assumed to be continuous.

14. Suppose $f \in R[a, b]$ and let $F(x) \int_a^x f(t)\, dt$ then show that there exists some $\beta > 0$ such that

$|F(x) - F(y)| \leq \beta |x - y|\ \forall x, y \in [a, b]$ that is, F satisfies *Lipschitz condition* of first order). Hence prove that F is uniformly continuous.

15. Give an example of a function f on $[a, b]$ which is not continuous but Riemann Integrable such that the function F defined by $F(x) = \int_a^x f(t)\, dt$ is differentiable on $[a, b]$.

16. If f is strictly monotone increasing, continuous function defined from $[0, 1]$ to $[0, 1]$ show that

$$\int_0^1 f(x)\, dx + \int_0^1 f^{-1}(x)\, dx = 1$$

where f^{-1} is the inverse of the function.

17. Suppose f is a continuously differentiable function defined on the open interval (a, b) such that $f'(x) \neq 0$ for all $x \in (a, b)$ then show that,

$$\int_c^d f(x)\, dx + cf(c) = df(d) - \int_{f(c)}^{f(d)} f^{-1}(x)\, dx$$

18. Evaluate the following limits:

(a) $\displaystyle \lim_{x \to 0} \frac{\int_0^{\pi} t^2 \sin t\, dt}{x^3 e^{4x}}$

(b) $\displaystyle \lim_{x \to 4} \frac{1}{x^2 - 6x + 8} \int_4^{\pi} te^{t^2}\, dt$

19. Without evaluating the integral find $\phi'(x)$ where,

$$\phi(x) = \int_x^{x^2} \frac{t^5 e^{3t}}{t^3 + et}\, dt$$

20. Applying suitable Mean Value theorem of Integral Calculus show that,

$$\frac{\pi}{6} \leq \int_0^{\frac{1}{2}} \frac{dx}{\sqrt{(1 - x^2)(1 - 3x^2)}} \leq \frac{\pi}{3}.$$

21. Suppose $\phi(x)$ is a polynomial of n degree defined $[0, 1]$ on then show that

(i) $\displaystyle \int_0^1 \phi(x)\, dt = \frac{1}{2}[\phi(0) + \phi(1)]$, if $n \leq 1$

(ii) $\displaystyle \int_0^1 \phi(x)\, dt = \frac{1}{6}\left[\phi(0) + 4\phi\left(\frac{1}{2}\right) + \phi(1)\right]$, if $n \leq 2$

22. Let $f(x)$ be a function defined on the interval $[a, b]$ with $f^{(n)}(x)$ exists and continuous. Use integration by parts to deduce Taylor's Theorem and apply Mean value Theorem to obtain remainders after n terms.

23. (Differentiation of Integrals): Suppose $f \in C[a, b]$ and let $h(x), g(x)$ be two functions defined on the interval $[c, d]$ such that both of their range is contained the interval $[a, b]$. Define the functionas $G(x)$ follows:

$$G(x) = \int_{h(x)}^{g(x)} f(t)\, dt, \quad x \in [a, b]$$

Show that,

(i) $G(x)$ for $x \in [a, b]$

(ii) $G'(x) = f(g(x))\, g'(x) - f(h(x))\, h'(x)$

24. (Cauchy-Schwartz Inequality for Integrals): Let f & $g \in R[a, b]$ then show that,

$$\int_a^b |f(x)\, g(x)\, dx| \leq \left\{\int_a^b |f(x)|^2\, dx \int_a^b |g(x)|^2\, dx\right\}^{\frac{1}{2}}$$

25. Prove Lemma 3.13.3

Chapter 4

Riemann-Stieltjes Integral

4.1 INTRODUCTION

The definition Riemann Integral can now be generalized to impart a meaning to the symbol $\int_a^b f(x)d\,\alpha(x)$.

If we closely look at the definition of Upper (Lower Sums) we will find that it is nothing but a weighted sum where the length of the intervals act as weights. In Riemann-Stieltjes integral, the idea is to modify the above definition by replacing the length of the interval by a monotonically increasing function α, which may not be continuous. This general definition of integral helps us to handle practical situations which arise, for example in physics when one deals with problems of mass distribution which are partly discrete and partly continuous or while dealing with probability distribution of random variables which are partly discrete and partly continuous. Also it is possible to represent any infinite sum as Riemann-Stieltjes Integral by appropriate choice of α. We also note that for $\alpha(x) = x$ we get back ordinary Riemann Integral of the function. We will see that theorems, lemmas and deductions from them are very similar to what we have got in the previous chapter.

4.2 NOTATIONS AND DEFINITIONS

We follow same type of notations as developed in the last chapter. For a given interval $[a, b]$ by a partition P of $[a, b]$ we mean collection of points $\{x_0, x_2, ..., x_r, ... x_n\}$ in $[a, b]$ such that $a = x_0 < x_2 < ... < x_r, < ... x_n = b$. The rth sub-interval of $P[x_{r-1}, x_r]$ will be denoted by Δ_r. For a bounded function f in $[a, b]$ we define $M, m, M_r,$ and m_r as follows, $M = \sup\limits_{x\in[a,b]} f(x), m = \inf\limits_{x\in[a,b]} f(x), M_r = \sup\limits_{x\in\Delta_r} f(x)$ and $m_r = \inf\limits_{x\in\Delta_r} f(x)$. Let α be a monotonically increasing function on $[a, b]$ corresponding to a partition P we write,

$$\delta\alpha_r = \alpha(x_r) - \alpha(x_{r-1}) = 1, 2, 3, ... n$$

For any partition P we denote by $\|\delta\alpha_r\|_P$ as $\max\limits_r \delta\alpha_r$. For two partitions P and P' of $[a, b]$ their *common refinement* as already defined in the last chapter is a partition of $[a, b]$ consisting of all points of and P, that is if P'' is common refinement of P and P' then $P'' = P \cup P'$. As usual, $\|P\|$ will denote the norm of the partition P.

For a bounded function f in $[a, b]$ and for a partition 'P' of $[a, b]$ we define $U(P, f, \alpha)$ and $L(P, f, \alpha)$ as,

$$U(P, f, \alpha) = \sum_{r=1}^{n} M_r \, \delta\alpha_r$$

$$L(P, f, \alpha) = \sum_{r=1}^{n} m_r \, \delta\alpha_r$$

which are called Upper Sum and Lower Stieltjes Sum respectively.

Note: In this chapter we assume f to be a bounded function and α a monotonic increasing function in $[a, b]$ unless otherwise mentioned.

4.3   SOME BASIC LEMMAS AND THEOREMS

We now some lemmas whose proof we omit as they can be deduced following same line of arguments given in last chapter.

Lemma 4.3.1

For any partition p of $[a, b]$ we have,

$$m\,(\alpha(b) - \alpha(a)) \leq L(P, f, \alpha) \leq U\,(P, f, \alpha) \leq M(\alpha(b) - \alpha(a))$$

Lemma 4.3.2

For partitions P and P' of $[a, b]$ if 'P' is a refinement of the partition P then,

$$U(P, f, \alpha) \geq U(P, f, \alpha)$$

$$L(P, f, \alpha) \leq L(P, f, \alpha)$$

Lemma 4.3.3

For partitions P and P' of $[a, b]$ if P' is a refinement of the partition P having at most k more points other than the points of P then,

$$U(P, f, \alpha) - U(P, f, \alpha) \leq k(M - m)\, \|\delta\alpha_r\|_P$$

$$L(P, f, \alpha) - L(P, f, \alpha) \leq k(M - m)\, \|\delta\alpha_r\|_P$$

We now define,

$$\int_{a}^{\overline{b}} f(x)d\alpha = \inf U(P, f, \alpha)$$

$$\int_{\underline{a}}^{b} f(x)d\alpha = \sup L(P, f, \alpha)$$

Following theorem is similar to Lemma 3.3.5 and the proof is left as an exercise.

Theorem 4.3.4

$$\int_{\underline{a}}^{b} f(x)d\alpha \leq \int_{a}^{\overline{b}} f(x)d\alpha$$

Definition   We say f defined on $[a, b]$ is Riemann-Stieltjes Integrable (R.S-Integrable) with respect to α denoted as $f \in R_\alpha\,[a, b]$ if,

$$\int_{\underline{a}}^{b} f(x)d\alpha = \int_{a}^{\overline{b}} f(x)d\alpha$$

The common value is denoted by $\int\limits_a^b fd\,\alpha$ and is called Riemann-Stieltjes Integral of f with respect to α.

Note: We have taken here one of the several accepted definition of Riemann-Stieltjes Integral which are not equivalent. For other definitions of Riemann-Stieltjes Integral one can see Apostol [1] and Ross [13].

Example 4.1   Let f be any bounded function defined on $[a,\ b]$ and let $\alpha(x) = c$ where is any real constant, then $f \in R_\alpha\,[a,\ b]$.

Sol: Clearly for any partition P of we have

$$U(P, f,\ \alpha) = \sum_{r=1}^{n} M_r\,\delta\alpha_r = 0$$

$$L(P, f,\ \alpha) = \sum_{r=1}^{n} m_r\,\delta\alpha_r = 0$$

as $\delta\alpha_r = 0$ for $r = 1,\ 2,\ ...n$. So we get,

$$\int\limits_{\underline{a}}^{b} f(x)\,d\alpha = 0 = \int\limits_a^{\overline{b}} f(x)\,d\alpha.$$

That is $f \in R_\alpha\,[a,\ b]$.

Example 4.2   Let f on $[0,\ 1]$ be defined as follows,

$$f(x) = \begin{cases} 1 & \text{if } x \text{ is rational} \\ 0 & \text{if } x \text{ is irrational} \end{cases}$$

then it is known that $f \notin R[0,\ 1]$ but if we choose $\alpha(x)$ as in Example 4.1 then $f \in R_\alpha\,[0,\ 1]$.

Next we note that as, $\|P\| \to 0$, $\|\delta\alpha_r\|_p$ may not tend to zero hence a theorem similar to Darboux's theorem of Riemann Integration is not true in Riemann-Stieltjes Integral in general but if α is continuous we have the following theorem similar to Darboux's theorem.

Theorem 4.3.4

Let α be continuous and f be a bounded function on $[a,\ b]$ then given $\varepsilon > 0$ there exists $\delta > 0$ such that for any partition P with $\|P\| < \delta$ we have,

$$U(P, f,\ \alpha) < \int\limits_a^{\overline{b}} f(x)\,d\alpha + \varepsilon$$

$$L(P, f,\ \alpha) > \int\limits_{\underline{a}}^{b} f(x)\,d\alpha - \varepsilon.$$

Proof: By definition of inf. there exists a partition P_1 such that for $\varepsilon > 0$

$$U(P_1, f,\ \alpha) < \int\limits_a^{\overline{b}} f(x)\,d\alpha + \frac{\varepsilon}{2}$$

Let P_1 has K points other than the end points a & b. We may assume that $K \geq 1$ if necessary by allowing refinement of P_1. Now as α is given to be continuous and hence uniformly continuous there exists $\delta > 0$ such that for any partition P with $\|P\| < \delta$ we have $\|\delta\alpha_r\|_p < \dfrac{\varepsilon}{2(M - m)K}$. We chose such a partition P with $\|P\| < \delta$. Let P_2 be the common refinement of P_1 and P. Let P_2 has r more points than P. We see that these points are points of P_1 and as P_1 has K points other than end points we have $0 \le r \le K$. Now we have from Lemma 4.3.3

$$U(P, f, \alpha) - U(P_2, f, \alpha) \le r(M - m)\, \|\delta\alpha_r\|_P \tag{4.4.1}$$

also we have,

$$U(P_2, f, \alpha) \le U(P_1, f, \alpha) < \int_a^{\bar{b}} f(x)\, d\alpha + \frac{\varepsilon}{2} \tag{4.4.2}$$

Hence combining we get,

$$U(P, f, \alpha) \le U(P_2, f, \alpha) + r\,(M - m)\, \|\delta\alpha_r\|_P$$

$$< \int_a^{\bar{b}} f(x)\, d\alpha + \frac{\varepsilon}{2} + K(M - m)\, \|\delta\alpha_r\|_P$$

$$< \int_a^{\bar{b}} f(x)\, d\alpha + \frac{\varepsilon}{2} + \frac{\varepsilon}{2}$$

$$< \int_a^{\bar{b}} f(x)\, d\alpha + \varepsilon$$

Similarly we can show for this same P we an have,

$$L(P, f, \alpha) > \int_{\underline{a}}^{b} f(x)\, d\alpha - \varepsilon$$

Remark:

As in the case of Riemann Integrals above theorem implies,

$$\lim_{\delta \to 0} U(P, f, \alpha) = \int_a^{\bar{b}} f(x)\, d\alpha \ \text{ and } \ \lim_{\delta \to 0} L(P, f, \alpha) = \int_{\underline{a}}^{b} f(x)\, d\alpha$$

provided α is continuous.

Theorem 4.3.5

A function $f \in R_\alpha [a, b]$ if and only if for every $\varepsilon > 0$ there exists a partition P such that,

$$U(P, f, \alpha) - L(P, f, \alpha) < \varepsilon$$

Proof: Suppose, $f \in R_\alpha [a, b]$, then $\int_{\underline{a}}^{b} f(x)\, d\alpha = \int_a^{\bar{b}} f(x)\, d\alpha = \int_a^{b} f(x)\, d\alpha$

By definition of supremum and infimum, for $\varepsilon > 0$ there exists a partition P_1 and P_2 such that,

$$U(P_1, f, \alpha) < \int_a^{\bar{b}} f(x)\, d\alpha + \frac{\varepsilon}{2} = \int_a^{b} fd\, \alpha + \frac{\varepsilon}{2} \tag{4.3.5.1}$$

$$L(P_2, f, \alpha) > \int_{\underline{a}}^{b} f(x)\, d\alpha - \frac{\varepsilon}{2} = \int_a^{b} fd\, \alpha - \frac{\varepsilon}{2} \tag{4.3.5.2}$$

We now take common refinement of P_1 and P_2 as P, then (4.5.1) and (4.5.2) yields,

$$U(P, f, \alpha) < \int_a^{\bar{b}} f d\alpha + \frac{\varepsilon}{2} \qquad (4.3.5.3)$$

$$L(P, f, \alpha) > \int_a^b f d\alpha - \frac{\varepsilon}{2} \qquad (4.3.5.4)$$

Subtracting (4.5.3) and (4.5.4) we get,

$$U(P, f, \alpha) - L(P, f, \alpha) < \varepsilon$$

Conversely, suppose that $U(P, f, \alpha) - L(P, f, \alpha) < \varepsilon$ for some partition P, then we have

$$L(P, f, \alpha) \le \int_{\underline{a}}^b f(x)\, d\alpha \le \int_a^{\bar{b}} f(x)\, d\alpha \le U(P, f, \alpha)$$

Hence,

$$\int_a^{\bar{b}} f(x)\, d\alpha - \int_{\underline{a}}^b f(x)\, d\alpha \le U(P, f, \alpha) - L(P, f, \alpha) < \varepsilon.$$

So we conclude that $f \in R_\alpha[a, b]$.

Example 4.3   Show from first principle that $\int_0^2 [x]\, d\,(x)^2 = 3$.

Sol: We consider the partition P of $[0, 2]$ as follows,

$P: \left\{ 0, \dfrac{1}{n}, \dfrac{2}{n}, \ldots \dfrac{n-1}{n}, \dfrac{n}{n} = 1, 1 + \dfrac{1}{n}, + \dfrac{2}{n}, \ldots 1 + \dfrac{n-1}{n}, 1 + \dfrac{n}{n} = 2 \right\}$. We then calculate the upper sum and lower sum as follows.

$$U(P, f, \alpha) = 0 \cdot \left(\left(\tfrac{1}{n}\right)^2 - 0^2 \right) + 0 \cdot \left(\left(\tfrac{2}{n}\right)^2 - \left(\tfrac{1}{n}\right)^2 \right) + \ldots + 1 \cdot \left(1 - \left(1 - \tfrac{1}{n} \right)^2 \right) + 1 \cdot \left(1 + \left(\tfrac{1}{n}\right)^2 - (1)^2 \right)$$

$$+ 1 \cdot \left(\left(1 + \tfrac{2}{n} \right)^2 - \left(1 + \tfrac{1}{n} \right)^2 \right) + \ldots + 1 \cdot \left(\left(1 + \tfrac{n-1}{n} \right)^2 - \left(1 + \tfrac{n-2}{n} \right)^2 \right) + 2 \cdot \left((2)^2 - \left(1 + \tfrac{n-1}{n} \right)^2 \right)$$

$$= 8 - \left(1 + \tfrac{n-1}{n} \right)^2 - \left(1 - \tfrac{1}{n} \right)^2$$

That is,

$$U(P, f, \alpha) = (8 - 4 - 1) + B(n) = 3 + B(n)$$

where $B(n) \to 0$ as $n \to \infty$. Now see that as $n \to \infty$ $\|P\| \to 0$ and as $\alpha(x) = x^2$ is continuous by using Theorem 4.4 we get,

$$\lim_{\delta \to 0} U(P, f, \alpha) = \int_a^{\bar{b}} f(x)\, d\alpha = 3$$

By similar calculation we can show that,

$$L(P, f, \alpha) = 3$$

Hence,

$$\lim_{\delta \to 0} L(P, f, \alpha) = \int_a^b f(x)\, d\alpha = 3. \text{ Hence we get } \int_0^2 [x]\, d(x^2) = 3.$$

Theorem 4.3.6

If f is continuous on $[a, b]$ then $f \in R_\alpha[a, b]$.

Proof: We here can assume that α is not constant as if α is a constant then by Example 4.1 $f \in R_\alpha[a, b]$. Now as f is continuous on $[a, b]$ it is uniformly continuous there hence for $\varepsilon > 0$, there exists $\delta > 0$ such that for $x, y \in [a, b]$ such that $|x - y| < \delta$, we have

$$|f(x) - f(y)| < \frac{\varepsilon}{\alpha(b) - \alpha(a)}.$$

We choose a partition P with $\|P\| < \delta$. As for continuous functions supremum and infimum are attained we get,

$$U(P, f, \alpha) - L(P, f, \alpha) = \sum_{r=1}^{n} (M_r - m_r)\, \delta\alpha_r$$

$$= \sum_{r=1}^{n} (f(\xi_r) - f(\eta_r))\, \delta\alpha_r, \quad \text{for some } \xi_r, \eta_r \in \Delta_r$$

$$< \frac{\varepsilon}{\alpha(b) - \alpha(b)} \sum_{r=1}^{n} \delta\alpha_r = \varepsilon$$

Hence $f \in R_\alpha[a, b]$.

Theorem 4.3.7

If f is monotone and α continuous on $[a, b]$ then $f \in R_a[a, b]$

Proof: Suppose that f is monotone increasing. Also we can suppose that $f(a) \neq f(b)$ as by previous theorem every constant function is R-S Integrable. Since α continuous on $[a, b]$ it is uniformly continuous there hence for $\varepsilon > 0$, there exists $\delta > 0$ such that for $x, y \in [a, b]$ such that $|x - y| < \delta$ we have,

$$|\alpha(x) - \alpha(y)| < \frac{\varepsilon}{f(b) - f(a)}$$

We choose a partition P with $\|P\| < \delta$ then by above observation we have

$\|\delta\alpha_r\|_P < \dfrac{\varepsilon}{f(b) - f(a)}$. Also as f is monotone increasing we have,

$$U(P, f, \alpha) - L(P, f, \alpha) = \sum_{r=1}^{n} (M_r - m_r)\, \delta\alpha_r$$

$$= \sum_{r=1}^{n} f(x_r) - f(x_{r-1}))\, \delta\alpha_r$$

$$< \|\delta\alpha_r\|_P \sum_{r=1}^{n} (f(x_r) - f(x_{r-1}))$$

$$< \frac{\varepsilon}{f(b) - f(a)} \sum_{r=1}^{n} (f(x_r) - f(x_{r-1}))$$

$$< \frac{\varepsilon}{f(b) - f(a)} \, (f(b) - f(a)) = \varepsilon$$

This proves that $f \in R_\alpha [a, b]$.

Next we will see how Riemann-Stieltjes Integral can be conceived as a limit of sums as we had seen in the case of Riemann Integral. First we have the following definition,

Definition   For a partition P of $[a, b]$ let $\xi_r \in \Delta_r$, $r = 1, 2 \ldots n$. We now consider the sum,

$$S(P, f, \alpha) = \sum_{r=1}^{n} f(\xi_r) \, \delta \alpha_r.$$

We say that above sums converge to A as norm of partitions tend to zero, if for every $\varepsilon > 0$, there exists $a\ \delta > 0$ such that whenever $\|P\| < \delta$ we have

$$|S(P, f, \alpha) - A| < \varepsilon$$

for every possible choice of $\xi_r \in \Delta_r$. In that case we write

$$\lim_{\|P\| \to 0} S(P, f, \alpha) = A.$$

Theorem 4.3.8

If $\displaystyle\lim_{\|P\| \to 0} S(P, f, \alpha)$ *exists then* $f \in R_\alpha [a, b]$ *and* $\displaystyle\lim_{\|P\| \to 0} S(P, f, \alpha) = \int_a^b f d\alpha$

Proof: The proof can be deduced from the sufficiency part of Theorem 3.6.1 with some modification and is left as an exercise.

However converse of this theorem is not true in general as is shown in the following example:

Example 4.4   Consider the functions f and α defined as follows:

$$f(x) = 1 \text{ if } 0 \le x < 1$$
$$= 0 \text{ if } 1 \le x \le 2$$

$$\alpha(x) = 0 \text{ if } 0 \le x \le 1$$
$$= 1 \text{ if } 1 < x \le 2$$

We will show that $f \in R_\alpha [a, b] \displaystyle\lim_{\|P\| \to 0} S(P, f, \alpha) \ne \int_0^2 f d\alpha.$

Sol: Let us consider a partition P of $[0, 2]$, if $1 \in P$ alright, otherwise we take a refinement of P' say P' with $1 \in P'$. Let Δ_r denotes the rth subinterval of P such that $x_r = 1$. Now,

$$U(P, f, \alpha) \ge \sum_{i=1}^{n} M_i \delta_i = 1(\alpha(1) - \alpha(x_{r-1})) = 0. \text{ So we have}$$

$$\overline{\int_0^2} f d\alpha \ge 0 \tag{1}$$

Similarly,

$$L(P, f, \alpha) \le L(P, f, \alpha) = \sum_{r=1}^{n} m_r \, \delta_r = 0 \, (\alpha (x_{r-1}) = 0. \text{ Hence}$$

$$\int_{0}^{2} f d\alpha \le 0 \tag{2}$$

Also we see that $U(P', f, \alpha) - L(P', f, \alpha) = 0 < \varepsilon$ for any $\varepsilon > 0$. Hence $f \in R_\alpha[0, 2]$. Now using (1) and (2) we conclude that $\int_{a}^{b} f d\alpha = 0$.

Now let $\varepsilon < 1$ be any real number and let $\delta < 0$ be any positive number. We choose any partition P with $\|P\| < \delta$ such that for some interval Δ_r we have,

$$x_{r-1} < 1 < x_r$$

Now let $\xi_i \in \Delta_i$ be such that $\xi_r < 1$ and $\xi_i \, i \ne r$ be any point. Now,

$$S(P, f, \alpha) = \sum_{i=1}^{n} f(\xi_i) \, (\alpha (x_i) - \alpha (x_{i-1}))$$

$$= 1(1 - 0) = 1 > \varepsilon$$

So we conclude that $\lim_{\|P\| \to 0} S(P, f, \alpha) \ne 0 = \int_{0}^{2} f d\alpha.$

In fact, Theorem 4.3.8 tells us that $\lim_{\|P\| \to 0} S(P, f, \alpha)$ does not exist.

But the converse is true in some special cases like,

Theorem 4.3.9

We have,

$$\lim_{\|P\| \to 0} S(P, f, \alpha) = \int_{a}^{b} f d\alpha$$

if any one of the following conditions holds:

(A) $f \in R_\alpha[a, b]$ *and* α *is continuous monotonic increasing on* $[a, b]$

(B) f *is continuous on* $[a, b]$ *and* α *is monotonic increasing on* $[a, b]$

Proof: Suppose $f \in R_\alpha [a, b]$ and α is continuous monotonic increasing on $[a, b]$ we will show $\lim_{\|P\| \to 0}$ $S(P, f, \alpha) = \int_{a}^{b} f d\alpha.$ For $\varepsilon > 0$ by Theorem 4.3.5 there exists $\delta > 0$ such that for all partition P with $\|P\| < \delta$ we have,

$$U(P, f, \alpha) < \int_{a}^{\overline{b}} f(x) \, d\alpha + \frac{\varepsilon}{2}$$

$$L(P, f, \alpha) > \int_{\underline{a}}^{b} f(x) \, d\alpha - \frac{\varepsilon}{2}$$

But as $f \in R_\alpha [a, b]$ we have $\overline{\int_a^b} f(x)\, d\alpha = \underline{\int_a^b} f(x)\, d\alpha = \int_a^b fd\alpha$ hence

$$U(P, f, \alpha) < \int_a^b fd\alpha + \frac{\varepsilon}{2} \qquad\qquad (4.3.9.1)$$

$$L(P, f, \alpha) > \int_a^b fd\alpha - \frac{\varepsilon}{2} \qquad\qquad (4.3.9.2)$$

We note that,

$$U(P, f, \alpha) \le S(P, f, \alpha) \le L(P, f, \alpha)$$

Now using (4.3.9.1) and (4.3.9.2) we get,

$$S(P, f, \alpha) < \overline{\int_a^b} f(x)\, d\alpha + \frac{\varepsilon}{2}$$

$$S(P, f, \alpha) > \underline{\int_a^b} f(x)\, d\alpha - \frac{\varepsilon}{2}$$

Hence for all partition P with $\|P\| < \delta$ we have,

$$\left| S(P, f, \alpha) - \int_a^b fd\alpha \right| < \varepsilon.$$

This proves (A).

(B) is left as an exercise.

We list some properties of Riemann-Stieltjes Integral as a theorem without proof as these can be proved following same line of arguments as in ordinary Riemann Integrals.

Theorem 4.3.10

If f & $g \in R_\alpha [a, b]$ and α monotonic increasing function on $[a, b]$ then,

(i) $|f| \in R_\alpha [a, b]$ & $\left| \int_a^b fd\alpha \right| \le \int_a^b |f|\, d\alpha$

(ii) $kf \in R_\alpha [a, b]$ & $\int_a^b kfd\alpha = k \int_a^b fd\alpha$ *for any real number k*

(iii) $f \pm g \in R_\alpha [a, b]$ & $\int_a^b (f \pm g)\, d_\alpha = \int_a^b fd\alpha \pm \int_a^b gd\alpha$

Next we prove the following theorem,

Theorem 4.3.11

If α & β be two monotone increasing functions on $[a, b]$ such that $f \in R_\alpha [a, b]$ & $f \in R_\beta [a, b]$ then $f \in R_{\alpha + \beta} [a, b]$ and in that case we have,

$$\int_a^b fd(\alpha + \beta) = \int_a^b fd\alpha + \int_a^b fd\beta.$$

Proof: As $f \in R_\alpha [a, b]$ & $f \in R_\beta [a, b]$, for $\varepsilon > 0$ using refinement we can find a partition P of $[a, b]$ such that,

$$U(P, f, \alpha) - L(P, f, \alpha) < \frac{\varepsilon}{2} \tag{4.3.11.1}$$

$$\& \ U(P, f, \beta) - L(P, f, \beta) < \frac{\varepsilon}{2} \tag{4.3.11.2}$$

Let $\gamma = \alpha + \beta$ then γ is monotone increasing.

Now,

$$U(P, f, \gamma) = \sum_{r=1}^{n} M_i \, (\gamma(x_i) - \gamma(x_{i-1}))$$

$$= \sum_{r=1}^{n} M_r (\alpha(x_r) - \alpha(x_{r-1})) + \sum_{r=1}^{n} M_r (\beta(x_r) - \beta(x_{r-1}))$$

$$= U(P, f, \alpha) + U(P, f, \beta) \tag{4.3.11.3}$$

Similarly,

$$L(P, f, \gamma) = \sum_{r=1}^{n} m_i \, (\gamma(x_i) - \gamma(x_{i-1}))$$

$$= L(P, f, \alpha) + L(P, f, \beta) \tag{4.3.11.4}$$

Hence,

$$U(P, f, \gamma) - L(P, f, \gamma) = [U(P, f, \alpha) + U(P, f, \beta)] - [L(P, f, \alpha) + L(P, f, \beta)]$$

$$= [U(P, f, \alpha) - L(P, f, \alpha)] + [U(P, f, \beta) - L(P, f, \beta)]$$

$$< \frac{\varepsilon}{2} + \frac{\varepsilon}{2} = \varepsilon$$

Using (4.3.11.1), (4.3.11.2), (4.3.10.3) & (4.3.10.4).

Hence by necessary and sufficient condition $f \in R_{\alpha + \beta} [a, b]$.

Again we see that,

$$\int_a^b f d\gamma = \inf \, [U(P, f, \gamma)]$$

$$= \inf \, [U(P, f, \alpha) + U(P, f, \beta)]$$

$$\geq \inf \, [U(P, f, \alpha)] + \inf \, [U(P, f, \beta)]$$

$$= \int_a^{\overline{b}} f d\alpha + \int_a^b f d\beta$$

$$= \int_a^b f d\alpha + \int_a^b f d\beta \tag{4.3.11.5}$$

Similarly using $\int_a^b f d\gamma = \sup \, [L(P, f, \gamma)]$ we can show,

$$\int_a^b f d\gamma \leq \int_a^b f d\alpha + \int_a^b f d\beta \tag{4.3.11.6}$$

Using (4.3.11.5) & (4.3.11.6) we conclude that,

$$\int_a^b fd(\alpha + \beta) = \int_a^b fd\alpha + \int_a^b fd\beta.$$

Theorem 4.3.12

Suppose f is a bounded function on [a, b] and α be a monotone increasing differentiable function on [a, b] such that f $\alpha \in R[a, b]$ then if $f \in R_\alpha[a, b]$ we have,

$$\int_a^b fd\alpha = \int_a^b f(x)\, \alpha'(x)\, dx \qquad (4.3.12.1)$$

Proof: Note that as f is given to be a bounded function on $[a, b]$ and α is monotone increasing function then if $f \in R_\alpha[a, b]$ then by Theorem $\lim_{\|P\| \to 0} S(P, f, \alpha)$ exists and is equal to $\int_a^b fd\alpha$. Let be a partition of $[a, b]$ with usual notations. Now as α is differentiable on $[a, I]$ using Lagrange's Mean Value Theorem in Δ_r we get,

$$\alpha(x_r) - \alpha(x_{r-1}) = (x_r - x_{r-1})\, \alpha(\xi_r) \quad \text{where } x_r \le \xi_r \le x_{r-1} \qquad (4.3.12.2)$$

Again,

$$S(P, f, \alpha) = \sum_{r=1}^n f(\xi_r)\, (\alpha(x_r) - \alpha(x_{r-1}))$$

$$= \sum_{r=1}^n f(\xi_r)\, \alpha(\xi_r)\, (x_r - x_{r-1}) = S(P, g) \qquad (4.3.12.3)$$

Where $S(P, g)$ is the Riemann Approximating Sum of the function $g(x) = f(x)\, \alpha(x)$ corresponding to the partition P. As $\|P\| \to 0$ left-hand side of (4.3.12.3) tends to $\int_a^b fd\alpha$ and right hand side tends to

$$\int_a^b f(x)\, \alpha(x)\, dx \text{ as } \int_a^b f(x)\, \alpha(x)\, dx$$

Hence we get (4.3.12.1).

Example 4.5 Evaluate $\int_0^3 xd(e^x)$

Sol: We note that the function e^x is monotone increasing and differentiable in the interval $[0, 3]$ also as the function x is continuous hence it is R-S integrable. Using the above formula we get,

$$\int_0^3 xd(e^x) = \int_0^3 xe^x\, dx = [xe^x]_0^3 - \int_0^3 e^x\, dx$$

$$= 3e^3 - [e^x]_0^3 = 3e^3 - (e^3 - 1)$$

$$= 2e^3 + 1$$

4.4 FUNCTIONS OF BOUNDED VARIATIONS AND RIEMANN-STIELTJES INTEGRAL

Before proceeding further we need to know some facts about a special kind of functions known as *Function of Bounded Variations.*

Definition Let f is a function defined on $[a, b]$. For a partition P of $[a, b]$ let,

$$V(P, f) = \sum_{r=1}^{n} |f(x_i) - f(x_{i-1})|$$

We say f is a function of bounded variation on $[a, b]$ if

$$V(f) = \sup_{P} V(P, f)$$

is finite (where above supremum is taken over all partitions P of $[a, b]$).

$V(f)$ is called the total variation of f on $[a, b]$.

We denote by $BV[a, b]$ the set of all functions f defined on $[a, b]$ which is of bounded variations.

Now we note that as for a function of bounded variation f on $[a, b]$ as,

$$|f(x) - f(a)| \le V(f)$$

f is bounded on $[a, b]$. However the converse is not true.

Example 4.6 Let f on $[0, 1]$ is defined as follows:

$$f(x) = \begin{cases} x \cos\left(\dfrac{\pi}{2x}\right) & \text{if } x \ne 0 \\ 0 & \text{if } x = 0 \end{cases}$$

Then f is continuous on $[0, 1]$ and hence bounded there. We choose the partition

$$P = \left\{ 0, \frac{1}{2n}, \frac{1}{2n-1}, \; \cdots \; \frac{1}{3}, \frac{1}{2}, 1 \right\}$$

then,

$$V(P, f) = \sum_{r=1}^{n} |f(x_i) - f(x_{i-1})| = \frac{1}{2n} + \frac{1}{2n} + \frac{1}{2n-2} + \frac{1}{2n-2} + \cdots \frac{1}{2} + \frac{1}{2}$$

$$= 1 + \frac{1}{2} + \frac{1}{3} + \cdots + \frac{1}{n}.$$

As $\sum_{n=1}^{\infty} \frac{1}{n}$ is divergent, it follows that $V(f)$ is not finite hence f is not a function of bounded variation.

We also note that this example also shows that continuous functions may not be a function of bounded variation.

Example 4.7 It is easy to see that every monotone function f on $[a, b]$ is a function of bounded variation with $V(f) = |f(b) - f(a)|$.

Example 4.8 Suppose that f is differentiable on $[a, b]$ with f bounded there then we see that for any partition P of $[a, b]$,

$$V(P, f) = \sum_{r=1}^{n} |f(x_i) - f(x_{i-1})| = \sum_{r=1}^{n} |f'(\xi_i)|\,(x_i - x_{i-1}) \leq M \sum_{r=1}^{n} (x_i - x_{i-1}) = M(b - a)$$

Using Mean Value theorem of Differential Calculus, where $x_{i-1} \leq \xi_i \leq x_i$ and $|f(\xi_i)| \leq M$, $i = 1, 2, ...n$. So $V(f)$ finite.

Example 4.9 Suppose f is defined on $[0, 1]$ as follows,

$$f(x) = \begin{cases} x^2 \cos\left(\dfrac{1}{x}\right) & \text{if } x \neq 0 \\ 0 & \text{if } x = 0 \end{cases}$$

Then f' exists and is bounded there in fact $f'(0) = 0$ and if $x \neq 0$

$$f'(x) = \sin\left(\frac{1}{x}\right) + 2x \cos\left(\frac{1}{x}\right) \text{ if } x \neq 0 \text{ so that, } |f'(x)| \leq 3. \text{ Hence } f \text{ is a function of bounded variation.}$$

Theorem 4.4.1

If f and g belong to $BV[a, b]$ then $f \pm g$ and fg are also in $BV[a, b]$.

Proof: For any partition P of $[a, b]$ we have,

$$V(P, f \pm g) = \sum_{r=1}^{n} |(f(x_i) + g(x_i)) - (f(x_{i-1}) \pm g(x_{i-1}))|$$

$$\leq \sum_{r=1}^{n} |f(x_i) - f(x_{i-1})| + \sum_{r=1}^{n} |g(x_i) - g(x_{i-1})|$$

$$\leq V(f) + V(g).$$

So $V(f \pm g)$ is finite so $f \pm g \in BV[a, b]$.

Similarly we can show that,

$$V(fg) \leq MV(f) + MV(g)$$

Where $M > 0$ is a constant such that,

$$|f(x) < M,\ |g(x)| < M\ \forall\ x \in [a, b],.$$

So that we get $fg \in BV[a, b]$.

Corollary 4.4.2

As an immediate corollary of the above theorem we get if f and g are monotonically increasing function on $[a, b]$ then $f - g \in BV[a, b]$.

Remark:

Converse of above theorem is also true which we write as a theorem the proof is omitted for the proof interested reader is referred to Rudin [12] or Apostol [1].

Next we again return to the discussion of Riemann-Stieltjes Integral. Here we wish to extend the notion of R-S integral to impart a meaning to the symbol $\int_a^b fdg$ where g is any function not necessarily a monotone increasing function. For this we define Riemann-Stieltjes Integral with the help of $S(P, f, \alpha)$.

Definition   For any two functions f & g defined on $[a, b]$ we say $\int_a^b fdg$ exists provided $\lim\limits_{\|P\| \to 0} S(P, f, \alpha)$ exists (in the sense defined earlier) in that case we define,

$$\int_a^b fdg = \lim\limits_{\|P\| \to 0} S(P, f, \alpha)$$

We will denote by $R_g^*[a, b]$ the class of functions f which are integrable with respect to g.

We note that if g is continuous and monotone then $R_g[a, b] = R_g^*[a, b]$

Next we give a list of very elementary properties of the integral as a theorem which are very simple to prove and hence the proofs are kept as exercises.

Theorem 4.4.3

If $f_1, f_2, \in R_g^*[a, b]$ and c be any real number then, $cf_1, f_1 \pm f_2 \in R_g^*[a, b]$ and

$$\int_a^b (cf_1)\, dg = c \int_a^b f_1\, dg = \int_a^b f_1 d(cg).$$

$$\int_a^b (f_1 \pm f_2)\, dg = \int_a^b f_1 dg \pm \int_a^b f_2\, dg$$

Theorem 4.4.4

If then $f \in R_g^*[a, b]$ & $f \in R_h^*[a, b]$ then $f \in R_{g+h}^*[a, b]$ and

$$\int_a^b fd(g + h) = \int_a^b fdg + \int_a^b fdh$$

In view of Remark and above Theorem we have the following corollary:

Corollary 4.4.5

If $f \in C[a, b]$ and $g \in BV[a, b]$ then $f \in R_g^*[a, b]$.

Proof: If $g \in BV[a, b]$ then we can write $g = g_1 - g_2$ where g_1 & g_2 monotone increasing functions are. Now as $f \in C[a, b]$ we have,

$$f \in R_{g_1}^*[a, b] \ \& \ f \in R_{g_2}^*[a, b].$$

Now as,

$$\int_a^b fdg_1 - \int_a^b fdg_2 = \int_a^b fdg_1 + \int_a^b fd(-g_2)$$

$$\int_a^b fd(g_1 - g_2) = \int_a^b fdg.$$

As left hand side of the above equation is finite we conclude $f \in R_g^*[a, b]$.

Next theorem shows that there exists an interesting relationship between the spaces $R_f^*[a, b]$ & $R_g^*[a, b]$ in case $f \in R_g^*[a, b]$.

Theorem 4.4.6

If $f \in R_g^*[a, b]$ then $g \in R_g^*[a, b]$ and in that case,

$$\int_a^b f dg = [f(x)\, g(x)]_a^b - \int_a^b g df$$

where, $\qquad\qquad [f(x)\, g(x)]_a^b = f(b)\, g(b) - f(a)\, g(a).$

Proof: Let $P:\{a = x_0 \le x_1 \le ... \le x_{r-1} \le x_r \le ... \le x_{n-1} \le x_n = b\}$ and

$$P' = \{a = t_0 \le t_1 \le t_2 \le ... t_{r-1} \le t_r \le ... t_{n-1} \le t_n = b\} \text{ be}$$

partitions of $[a, b]$ such that,

$$x_{r-1} \le t_r \le x_r, r = 1, 2, ...n$$

Now we see that,

$$S(P, f, g) = \sum_{r=1}^{n} f(t_r)\, [g(x_r) - g(x_{r-1})]$$

$$= g(b)\, f(b) - g(a)\, f(a) - \sum_{r=1}^{n+1} g(x_{r-1})\, [f(t_i) - f(t_{i-1})]$$

$$= g(b)\, f(b) - g(a)\, f(a) - S(P', g, f) \qquad\qquad (4.4.5.1)$$

As $\|P\| \to 0$ we have $\|P'\| \to 0$ as $x_{r-1} \le t_r \le x_r$.

Hence as $\|P\| \to 0$ we have $S(P, f, g) \to \int_a^b f dg$ since $f \in R_g^*[a, b]$. From (4.4.5.1) it is clear that

$S(P, g, f) \to g(b)\, f(b) - g(a)\, f(a) - \int_a^b f dg$. Hence it follows from definition that,

$$\int_a^b f dg = [f(x)\, g(x)]_a^b - \int_a^b g df.$$

In particular we have the following theorem,

Theorem 4.4.6

If $f \in C[a, b]$ and $\alpha(x)$ be a monotonic increasing function then

$$\int_a^b f d\alpha = [f(x)\, \alpha(x)]_a^b - \int_a^b \alpha df$$

where, $[f(x)\, \alpha(x)]_a^b = f(b)\, \alpha(b) - f(a)\, \alpha(a).$

Example 4.10   Evaluate $\int_0^2 x^2 d(x - [x])$

Sol: We first note that the functions $\alpha(x) = x$ & $\beta(x) = [x]$ are monotone functions and hence are of bounded variations so their difference is again of bounded variations and since the function x^2 is continuous hence it belongs to $R_g^*[a, b]$ where $g(x) = x - [x]$. Now,

$$\int_0^2 x^2 d(x - [x]) = \int_0^2 x^2 dx - \int_0^2 x^2 d[x].$$

$$= \left[\frac{x^3}{3}\right]_0^2 - (x^2[x])_0^2 + \int_0^2 [x]\, dx^2 \quad \text{(Using previous theorem)}$$

$$= \frac{16}{3} - 8 + \int_0^2 2[x]x\, dx = -\frac{8}{3} + \int_1^2 2x\, dx$$

$$= -\frac{8}{3} + [x^2]_1^2 = -\frac{8}{3} + 3 = \frac{1}{3}$$

4.5   SOME MISCELLANEOUS PROBLEMS

Example 4.11   Show that $\int_0^3 x^2 d[x] = 14$

Sol: Note that as $f(x) = x^2$ is continuous and $\alpha(x) = [x]$ is monotone we can use Theorem to get,

$$\int_0^3 x^2\, d[x] = [x^2[x]]_0^3 - \int_0^3 [x]\, dx^2$$

$$= 27 - \int_0^3 2x[x]\, dx = 27 - \int_0^1 2x.0\, dx - \int_1^2 2x.1\, dx - \int_2^3 2x.2\, dx$$

$$= 27 - 3 - 2(9 - 4) = 27 - 13 = 14$$

Alternatively

As $f(x) = x^2$ is continuous and $\alpha(x) = [x]$ is monotone we know from Theorem that $\lim_{\|P\| \to 0} S(P, f, \alpha)$ exists and is equal to $\int_0^3 x^2 d[x]$. We will find $\lim_{\|P\| \to 0} S(P, f, \alpha)$. As usual let us start with partition P of $[0, 3]$ as

$$P : \left\{0, \frac{1}{n}, \frac{2}{n}, \dots \frac{n}{n} = 1, \frac{n+1}{n}, \dots, \frac{2n}{n} = 2, \frac{2n+1}{n}, \dots \frac{3n}{n} = 3\right\}$$

Here we have $\|P\| = \frac{1}{n}$ hence $\|P\| \to 0$ as $n \to \infty$. Now,

$$\lim_{\|P\| \to 0} S(P, f, \alpha) = \lim_{n \to \infty} \sum_{k=1}^{3n} \left(f\left(\frac{k}{n}\right)\left[\alpha\left(\frac{k}{n}\right) - \alpha\left(\frac{k-1}{n}\right)\right]\right)$$

$$= \lim_{n \to \infty} \left(f(1)\left(\alpha(1) - \alpha\left(\frac{n-1}{n}\right)\right) + f(2)\left(\alpha(2) - \alpha\left(\frac{2n-1}{n}\right)\right) + f(3)\left(\alpha(3) - \alpha\left(\frac{3n-1}{n}\right)\right) \right)$$

(as all other terms vanishes)

$$= (1^2(1-0) + 2^2(2-1) + 3^2(3-2)) = 14$$

Hence we have $\int\limits_0^3 x^2 d[x] = 14$.

Example 4.12 Show that $\lim\limits_{\|P\| \to 0} S(P, f, \alpha) = [x]$ where $f(t) = 1$ & $a(x) = [t]$ in the interval $[0, [x]$ and hence conclude that $\int\limits_0^x d[t] = [x]$.

Sol: We note that as $f(x)$ is continuous and $\alpha(x)$ is monotone $\lim\limits_{\|P\| \to 0} S(P, f, \alpha)$ exists in the above interval. We take several cases,

Case (I) $0 < x < 1$

For a fixed positive integer n let $\Delta = \dfrac{x}{n}$ and consider the partition P_n of $[0, x]$ as,

$P_n : \{x_0 = 0, x_1 = \Delta, x_2 = 2\Delta, ..., x_n = n\Delta = x\}$. We note that $\|P_n\| = \dfrac{x}{n} \to 0$ as $n \to \infty$

Now,

$$U(P_n, f, \alpha) = \sum_{r=1}^{n} 1(\alpha(x_r) - \alpha(x_{r-1}))$$

$$= \sum_{r=0}^{n} ([x_r] - [x_{r-1}]) = 0 = L(P_n, f, \alpha).$$

As $f(x)$ is continuous using Theorem we get,

$$\lim_{n \to \infty} U(P_n, f, \alpha) = J = 0 = I = \lim_{n \to \infty} L(P_n, f, \alpha)$$

$$\int\limits_0^x d[t] = [x] = 0.$$

Case (II) $x \geq 1$ is not an integer.

For a fixed positive integer n let $\Delta = \dfrac{x - [x]}{n}$ and consider the partition P_n of $[0, x]$ as,

$P_n = \left\{ 0, \dfrac{1}{n}, \dfrac{2}{n}, ..., \dfrac{n}{n} = 1, 1 + \dfrac{1}{n}, ...1 + \dfrac{n}{n} = 2, ..., ([x] - 1) + \dfrac{n}{n} = [x], [x] + \Delta, ..., [x] + n\Delta = x \right\}$. Note that

$$\|P\| = \max\left\{ \dfrac{1}{n}, \dfrac{x - [x]}{n} \right\} = \dfrac{1}{n} \to 0 \text{ as } n \to \infty .$$

$$U(P_n, f, \alpha) = \sum_{r=1}^{n} 1\left(\left[\frac{r}{n}\right] - \left[\frac{r-1}{n}\right]\right) + \sum_{r=1}^{n} 1\left(\left[1 + \frac{r}{n}\right] - \left[1 + \frac{r-1}{n}\right]\right) + \ldots \sum_{r=1}^{n} 1\left(\left[([x]-1) + \frac{r}{n}\right]\right)$$

$$-\left[([x]-1) + \frac{r-1}{n}\right] \sum_{r=1}^{n} 1([x+r\Delta] - [x+(r-1)\Delta])$$

$$= \underbrace{1 + 1 + \ldots + 1}_{[x] \text{ times}} + 0 \; [x] \text{ times} = [x] = L(P_n, f, \alpha)$$

Again using Theorem we get,

$$\lim_{n \to \infty} U(P_n, f, \alpha) = J + [x] = I = \lim_{n \to \infty} L(P_n, f, \alpha)$$

$$\int_{0}^{x} d[t] = [x].$$

Case (III) x is an integer.

This case is easy to dispose.

Example 4.13   Suppose that α is monotonically increasing on $[a, b]$ and $a \leq c \leq b$ with α continuous at c. Let f be a function defined on $[a, b]$ such that $f(x) = 1$ if $x = c$ and $f(x) = 0$ if $x \neq c$. Then show that $f \in R_\alpha[a, b]$ and $\int_{a}^{b} f d\alpha = 0..$

Sol: First we take $a < c < b$. As α is continuous at c, for $\varepsilon > 0$ there exists a $\delta > 0$ such that whenever $|x - c| < \delta$ we have $|\alpha(x) - \alpha(c)| < \varepsilon$. Let P be a partition of $[a, b]$ with $\|P\| < \delta$ without loss of generality we may assume c is a point of the partition (otherwise we will consider the refinement) and suppose $[c, x_i]$ be the ith subinterval of the partition. Then for this partition we have,

$$U(P, f, \alpha) = \alpha(x_i) - \alpha(c) < \varepsilon \tag{1}$$

and also note that

$$L(P, f, \alpha) = 0 \text{ for any partition } P.$$

Hence we see that for any $\varepsilon > 0$ there exists a partition p such that,

$$U(P, f, \alpha) - L(P, f, \alpha) < \varepsilon$$

that is $f \in R_\alpha[a, b]$. Also from (1) we see that for any $\varepsilon > 0$ there exists a partition for which

$$U(P, f, \alpha) < \varepsilon$$

Now as $\int_{a}^{b} f d\alpha = \inf U(P, f, \alpha)$ it follows that $\int_{a}^{b} f d\alpha = 0$.

Example 4.14   Suppose $\beta(x)$ on $[-1, 1]$ is defined as follows:

$$\beta(x) = \begin{cases} 1 & \text{if } x > 0 \\ \dfrac{1}{2} & \text{if } x = 0 \\ 0 & \text{if } x < 0 \end{cases}$$

Let f be any bounded function on $[-1, 1]$ then prove that $f \in R_\beta[-1, 1]$ if and only if f is continuous at $x = 0$.

Sol: Let $\varepsilon > 0$ be any real number and suppose f is continuous at $x = 0$ then there exists a $\delta > 0$ such that whenever $|x| < \delta$ we have. $|f(x) - f(0)| < \dfrac{\varepsilon}{2}$. Let P be a partition (using usual notations) with $\|P\| = \delta_1 < \delta$. Without loss of generality we assume 0 is a point of the partition otherwise we will use refinement. Then we have,

$$U(P, f, \beta) - L(P, f, \beta) = (M_i - m_i)(\beta(0) - \beta(x_i)) + (M_{i+1} - m_{i+1})(\beta(x_{i+1}) - \beta(0))$$

$$= \frac{1}{2}(M_i - m_i) + \frac{1}{2}(M_{i+1} - M_{i+1}) \tag{1}$$

Now observe that as norm of P is less than δ_1 we have $(M_i - m_i) < \varepsilon$ & $(M_{i+1} - M_{i+1}) < \varepsilon$

(Recall that) $(M_i - m_i) = \sup\limits_{x, x \in \Delta_r} |f(x) - f(x)|)$. Hence from (1) we conclude that

$$U(P, f, \beta) - L(P, f, \beta) < \varepsilon$$

That is $f \in R_\beta[-1, 1]$.

Conversely, suppose that $f \in R_\beta[-1, 1]$. Then for $\varepsilon > 0$ there is a partition P such that $U(P, f, \beta) - L(P, f, \beta) < \varepsilon$. Let $0 \in \Delta_i$ then we get,

$$U(P, f, \beta) - L(P, f, \beta) = (M_i - m_i)(\beta(x_i) - \beta(x_{i-1}))$$

$$= (M_i - m_i)(1 - 0) < \varepsilon.$$

Hence we see that for $x \in \Delta_r, |f(x) - f(0)| \le M_i - m_i < \varepsilon$. that is f is continuous at $x = 0$.

PROBLEMS

1. Evaluate from first principle the following integrals:

   (i) $\displaystyle\int_0^3 [x]\, dx^2$

   (ii) $\displaystyle\int_0^1 |x|\, d(e^x)$

2. Prove Lemma 4.3.1, Lemma 4.3.2 and Lemma 4.3.3

3. Prove Theorem 4.3.4

4. Prove Theorem 4.3.8

5. Suppose $\beta(x)$ is defined on $[0, 1]$ as follows:

$$\beta(x) = \begin{cases} c_1 & \text{if } 0 \le x \le \dfrac{1}{2} \\[2mm] c_2 & \text{if } x = \dfrac{1}{2} \\[2mm] c_3 & \text{if } \dfrac{1}{2} < x \le 1 \end{cases}$$

where $c_1 \leq c_2 \leq c_3$. Show that $\int_0^1 fd\beta$ exists and its value is independent of c_2.

6. If $f(x)$ is continuous and $\alpha(x)$ is monotone on $[a, b]$ then show that

$$\lim_{\|P\| \to 0} S(P, f, \alpha) = \int_a^b fd\alpha.$$

7. Prove Theorem 4.3.10

8. Evaluate:

   (i) $\displaystyle\int_0^3 (x^2 + 1)\, d[x]$

   (ii) $\displaystyle\int_0^\pi x^2 d(\cos x)$

   (iii) $\displaystyle\int_{-1}^1 e^x d|x|$

   (iv) $\displaystyle\int_0^2 x^2 d(x - [x])$

9. Prove Theorem 4.4.3 and Theorem 4.4.3

10. Show that the following function are of bounded variation in the respective intervals they are defined :

    (i) $x^2 - [x], 0 \leq x \leq 1$

    (ii) $\cos x, 0 \leq x \leq \dfrac{\pi}{2}$

    (iii) $3x^2 - 2x, 0 \leq x \leq 2$.

11. Prove Theorem 4.3.12.

12. Suppose $f \in C[a, b]$. For a finite sequence of points $\{k_i\}_{i=1}^n$ on $[a, b]$ satisfying $a < k_1 < k_2 < ... < k_n < b$ define the function $g(x)$ on $[a, b]$ as follows:

$$g(x) = \begin{cases} 1 & \text{if } x \neq k_i\ i = 1, 2, ...n \\ P_i & \text{if } x = k_i \end{cases}$$

where P_i' are some real constants. Show that $f \in R^*[a, b]$ and $\int_a^b fdg = 0$.

What happens if we take $a \leq k_1 < k_2 < ... < k_n \leq b$.

13. Let $f(x)$, $g(x)$ be defined on $[0, 1]$ as follows:

$$f(x) = \begin{cases} x & \text{if } 0 \leq x < \dfrac{1}{2} \\[2mm] 1 & \text{if } x = \dfrac{1}{2} \\[2mm] 1 + x & \text{if } \dfrac{1}{2} < x \leq 1 \end{cases} \qquad g(x) = \begin{cases} \dfrac{1}{2} & \text{if } 0 \leq x < \dfrac{1}{2} \\[2mm] \dfrac{1}{3} & \text{if } x = \dfrac{1}{2} \\[2mm] 1 & \text{if } \dfrac{1}{2} < x \leq 1 \end{cases}$$

Show that $f \notin R_g^* [a, b]$.

14. Let $f(x)$, $g(x)$ be defined on $[0, 1]$ as follows:

$$f(x) = \begin{cases} 1 & \text{if } x \text{ is rational} \\ 0 & \text{if } x \text{ is irrational} \end{cases} \quad \text{and}$$

$g(x) = x^2$. Show that $f \notin R_g^* [a, b]$.

15. (Mean Value Theorem) Let $f(x)$ be continuous and $\alpha(x)$ be monotone increasing in the on the interval $[a, b]$. Show that there exists $\xi \in [a, b]$ such that,

$$\int_a^b f\, d\alpha = f(\xi) \int_a^b d\alpha.$$

Chapter 5

Improper Integrals

5.1 INTRODUCTION

In the last chapter for the existence of proper Riemann Integrals we have assumed that the function should be defined on a closed interval and should be bounded. In this chapter we will extend the notion of integrals for functions which are defined on an infinite intervals and can have infinite discontinuities such integrals will be called '*improper integrals*'. Again we know proper integrals represents area of a 'closed region' in case of improper integrals it represents area of some 'open regions'. The following examples will give some idea how naturally we can extend the definition of a proper integral to a improper integral.

Example 5.1 Consider the function $f(x) = \dfrac{1}{\sqrt{x}}$ defined on the interval I = $\{0 < x \le 1\}$. We see that the function is unbounded near 0. Let c be any point of *I* then,

$$\int_{c}^{1} f(x)\, dx = 2 - 2\sqrt{c}$$

That is the area of the closed region bounded by the curve $f(x)$ and ordinates $x = 1$ & $x = c$ is $2 - 2\sqrt{c}$.

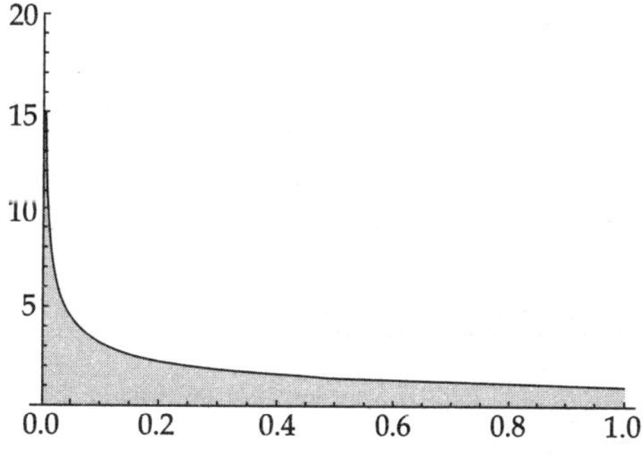

Fig. 5.1

Now we will define the area of the open shaded region in Fig. 5.1 naturally by,

$$\lim_{c \to 0^+} \int_c^1 f(x)\,dx \quad \text{if it is finite}$$

Now see that as $c \to 0^+$, $2 - 2\sqrt{c} \to 2$ (finite)

Hence in this case we will say the value of the integral $\int_0^1 f(x)\,dx$ is 2.

Consider the next example,

Example 5.2   Let $f(x) = \dfrac{1}{x}$ defined on the interval $I = \{0 < x \leq 1\}$. Again we see that the function is unbounded near 0. For any point c in I we have,

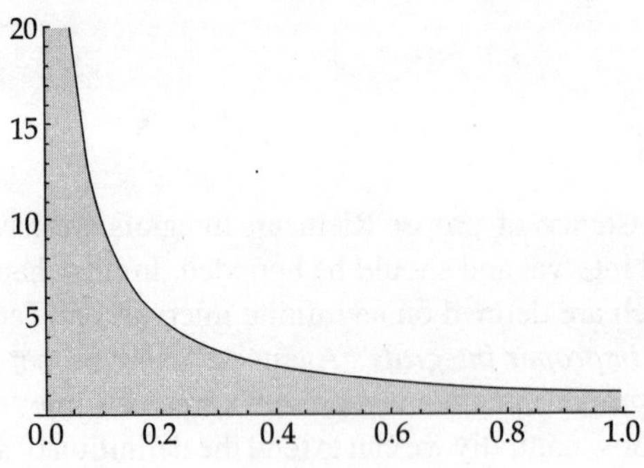

Fig. 5.2

$$f(x) = \frac{1}{x}$$

$$\int_c^1 f(x)\,dx = -\log c$$

hence,
$$\int_c^1 f(x)\,dx \to +\infty \quad \text{as} \quad c \to 0^+.$$

This means that the area of the shaded region of Fig. 5.2 is not finite and hence this case we will say the value of the integral $\int_0^1 f(x)\,dx$.

Compare Fig. 5.1 and Fig. 5.2 and notice the gap between two functions and Y-axis.

We now proceed to classify the types of improper integrals.

5.2   TYPES OF IMPROPER INTEGRALS

We get two kinds of Improper Integrals, Type I and Type II Integrals depending on which assumption of the proper Riemann Integral being relaxed.

Type I: This type of improper integral occurs when we assume that the interval is infinite but the function is bounded.

Type II: We get this type of improper integrals when the interval is bounded but the function has infinite discontinuities at finite number of points of the interval.

5.3 TYPE I INTEGRALS

There can be three possible unbounded intervals giving rise to three types of
Type I integrals:

$$\text{(i)} \int_a^\infty f(x)\,dx \qquad\qquad \text{(ii)} \int_{-\infty}^a f(x)\,dx \text{ and} \qquad\qquad \text{(iii)} \int_{-\infty}^\infty f(x)\,dx$$

We will now define what we mean by these above symbols.

(i) Suppose f be bounded and integrable in the interval $[a, M]$ for any $M > a$ then we say that exists

$$\int_a^\infty f(x)\,dx \text{ if } \lim_{M\to\infty} \int_a^M f(x)\,dx \text{ exists } \textbf{\textit{finitely}} \text{ and in that case we write } \int_a^\infty f(x)\,dx = \lim_{M\to\infty} \int_a^M f(x)\,dx$$

(ii) Suppose f be bounded and integrable in the interval $[A, a]$ for any $A < a$ then we say that $\int_{-\infty}^a f(x)\,dx$

exists if $\lim\limits_{A\to\infty} \int_A^a f(x)\,dx$ exists $\textbf{\textit{finitely}}$ and in that case we write $\int_{-\infty}^a f(x)\,dx = \lim\limits_{A\to-\infty} \int_A^a f(x)\,dx$

(iii) Suppose a be any real number and suppose f be bounded and integrable in the intervals $[a, M]$

for any $M > a$ and $[A, a]$ for any $A < a$ then we say that $\int_{-\infty}^\infty f(x)\,dx$ exists if both $\int_a^\infty f(x)\,dx$ and

$\int_{-\infty}^a f(x)\,dx$ exist in the sense defined above and in that case we write $\int_{-\infty}^\infty f(x)\,dx = \int_{-\infty}^a f(x)\,dx$

$$+ \int_a^\infty f(x)\,dx.$$

Remarks:

1. See that existence of $\int_{-\infty}^\infty f(x)\,dx$ demands separate existence of both the limits

$$\lim_{M\to\infty} \int_{-a}^\infty f(x)\,dx \text{ and } \lim_{M\to-\infty} \int_M^a f(x)\,dx \text{ which is stronger than the existence of}$$

$$\lim_{M\to\infty} \int_{-M}^M f(x)\,dx. \text{ (For example take } f(x) = x)$$

2. Note that definition of $\int_{-\infty}^\infty f(x)\,dx$ is independent of the choice of a. That is, if b is any real

number and if $\int_{-\infty}^\infty f(x)\,dx$ exists in above sense then $\int_{-\infty}^\infty f(x)\,dx = \int_{-\infty}^b f(x)\,dx + \int_b^\infty f(x)\,dx$ (Prove!)

3. We must see that there is a remarkable similarity between infinite series and Type I improper integrals. Remember that sum of an infinite series $\sum\limits_{n=1}^{\infty} U_n$ is defined by $\lim\limits_{n \to \infty} S_n$ where $S_n = \sum\limits_{i=1}^{n} U_i$ is the nth 'partial sum' provided the limit exits that is sum of an infinite series is based on the notion of finite sum. In case of Type I improper integral of the form $\int\limits_{a}^{\infty} f(x)\, dx$ n^{th} 'partial sum' is replaced by so called 'partial integrals' denoted by $\int\limits_{a}^{M} f(x)\, dx$ and the value of the integral is by the limit of the partial sums as $M \to \infty$ provided the limit exists hence existence of improper integral is based on the notion of proper Riemann integral which is the analogue of finite sum in case of infinite sum.

Example 5.3: $\int\limits_{0}^{\infty} e^{-x}\, dx$ converges, as for any $a > 0$,

$$\lim_{a \to \infty} \int\limits_{0}^{a} e^{-x}\, dx = \lim_{a \to \infty} [1 - e^{-a}] = 1$$

Example 5.4 $\int\limits_{0}^{\infty} xe^{-x^2}\, dx$ converges, as for any $a > 0$

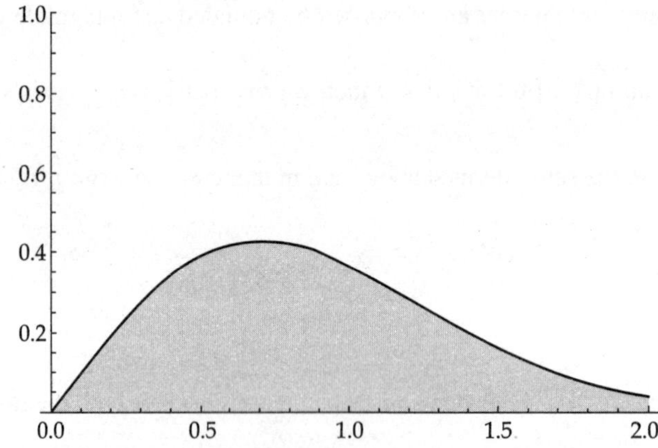

Fig. 5.3

$$\lim_{a \to \infty} \int\limits_{0}^{a} xe^{-x^2}\, dx = \frac{1}{2} \lim_{a \to \infty} (1 - e^{-a^2}) = \frac{1}{2}$$

Example 5.5 $\int\limits_{0}^{\infty} \cos x\, dx$ diverges as, as for any $a > 0$ $\lim\limits_{a \to \infty} \sin x$ does not exist.

Next example is an important example which may be taken as analogue of *p*-series in case of infinite series which is some time known as p-integral and will see later that it is a useful comparison integral.

Example 5.6 $\displaystyle\int_a^\infty \frac{dx}{x^p}\ (a > 0)$ exists if $p > 1$ and does not exist if $p \le 1$.

First suppose that $p = 1$ then,

$$\lim_{M \to \infty} \int_a^M \frac{dx}{x} = \lim_{M \to \infty} \ [\log |x|]_a^M = \lim_{M \to \infty} \ \log M - \log a = \infty$$

Hence $\displaystyle\int_a^\infty \frac{dx}{x}$ does not exist. Suppose now $p \ne 1$ then,

$$\lim_{M \to \infty} \int_a^M \frac{dx}{x^p} = \lim_{M \to \infty} \left[\frac{1}{1-p} x^{1-p}\right]_a^M = \lim_{M \to \infty} \frac{1}{1-p} [M^{1-p} - a^{1-p}]$$

$$= \begin{cases} \dfrac{1}{p-1} a^{1-p} & \text{if } p > 1 \\ \infty & \text{if } p < 1 \end{cases}$$

Hence we can conclude that the integral exists only when $p > 1$.

We now give analytic definition of Type I integrals.

Definition   Let f be bounded integrable in $a \le x \le B$ for every $B > a$. The integral $\displaystyle\int_a^\infty f(x)\ dx$ is said to converge and has the value I if given any $\varepsilon > 0$ there exists a positive number M depending on ε such that

$$\left| I - \int_a^B f(x)\ dx \right| < \varepsilon \text{ tor } B > M$$

Using the fact that $\displaystyle\int_a^B f(x)\ dx$ can be treated as a function of its upper limit and using the Cauchy Criterion for the existence of limit of functions we get the following theorem:

Theorem 5.3.1

A necessary and sufficient condition for the convergence of the integral $\displaystyle\int_a^\infty f(x)\ dx$ is that for any positive number ε there exists a positive number M depending on ε such that for al x_1, x_2 such that $x_2 > x_1 > M$ we have,

$$\left| \int_{x_1}^{x_2} f(x)\ dx \right| < \varepsilon$$

Note: We can similarly define analytically the existence of the integral $\displaystyle\int_{-\infty}^b f(x)\ dx$ and obtain similar necessary and sufficient condition for its convergence.

5.4   ABSOLUTE CONVERGENCE

The integral $\displaystyle\int_a^\infty f(x)\ dx$ is said to converge absolutely if the integral $\displaystyle\int_a^\infty |f(x)|\ dx$ is convergent. Similarly we define absolute convergence of $\displaystyle\int_{-\infty}^b f(x)\ dx$ and $\displaystyle\int_\infty^{-\infty} f(x)\ dx$.

As we know that for any x_1, x_2 $\left| \int_{x_1}^{x_2} f(x)\, dx \right| \le \int_{x_1}^{x_2} |f(x)|\, dx$ and using Theorem 5.3.1 we get the following obvious theorem which shows absolute convergence is stronger than ordinary convergence.

Theorem 5.4.1

Absolute convergence implies ordinary convergence.

We will now show by an example that converse of the above Theorem is not true.

Example 5.5   Consider the improper integral $\int_0^\infty \dfrac{\sin x}{x}\, dx$.

First we observe that as $\lim\limits_{x \to 0} \dfrac{\sin x}{x} = 0$ there is no infinite discontinuity at $x = 0$. Now let x_1, x_2 be any to positive real numbers then by using second mean value theorem it is easy to see that

$$\int_{x_1}^{x_2} \frac{\sin x}{x}\, dx = \frac{1}{x_1} \int_{x_1}^{\xi} \sin x\, dx + \frac{1}{x_2} \int_{\xi}^{x_2} \sin x\, dx,\ x_1 \le \xi \le x_2$$

Again we see that $\left| \int_{x_1}^{\xi} \sin x\, dx \right| \le 2$ and also $\int_{\xi}^{x_2} \sin |x\, dx| \le 2$ hence,

$$\left| \int_{x_1}^{x_2} \frac{\sin x}{x}\, dx \right| \le 2\left(\frac{1}{x_1} + \frac{1}{x_2} \right) < \frac{4}{x_1} \to 0 \text{ as } x_1 \to \infty \text{ this show that } \int_0^\infty \frac{\sin x}{x}\, dx \text{ is convergent. Now we proceed}$$

to show that the above integral is not absolutely convergent that is we will show that $\int_0^\infty \dfrac{|\sin x|}{x}\, dx$ diverges. For any positive integer n we see that,

$$\int_0^{n\pi} \frac{|\sin x|}{x}\, dx = \sum_{r=1}^{n} \int_{(r-1)\pi}^{r\pi} \frac{|\sin x|}{x}\, dx$$

Now $\int_{(r-1)\pi}^{r\pi} \dfrac{|\sin x|}{x}\, dx = \int_0^{\pi} \dfrac{\sin t}{(r-1)\pi + t}\, dx$ where we substitute $x = (r-1)\pi + t$ and using the fact that

$|\sin\{(r-1)\pi + t\}| = |(-1)^n \sin t| = |\sin t| = \sin t$ as $0 \le t \le \pi$. Also we see that

$(r-1)\pi + t \le r\pi$ for $0 \le t \le \pi$.

Hence $\int_{(r-1)\pi}^{r\pi} \dfrac{|\sin x|}{x}\, dx \ge \dfrac{1}{r\pi} \int_0^{\pi} \sin t\, dt = \dfrac{2}{r\pi}$. As a result we get,

$\int_0^{n\pi} \dfrac{|\sin x|}{x}\, dx \ge \dfrac{2}{\pi} \sum_{r=1}^{n} \dfrac{1}{r}$. Now as $\sum_{r=1}^{n} \dfrac{1}{r}$ is the partial sum of the divergent series $\sum_{r=1}^{\infty} \dfrac{1}{r}$ we have

$\int\limits_{0}^{r\pi} \dfrac{|\sin x|}{x}\, dx \to \infty$ as $n \to \infty$. This in turn implies that $\int\limits_{0}^{X} \dfrac{|\sin x|}{x}\, dx$ can be made arbitrarily large for

sufficiently large X or in other words $\int\limits_{0}^{\infty} \dfrac{|\sin x|}{x}\, dx$ diverges.

The integral $\int\limits_{a}^{\infty} f(x)\, dx$ is said to converge ***conditionally*** if $\int\limits_{0}^{\infty} f(x)\, dx$ converges but $\int\limits_{0}^{\infty} |f(x)|\, dx$ does not converge

We will now device certain tools as we have done in case of infinite series which we call 'tests' to ascertain whether an improper integral of first kind is convergent or not. As in case of infinite series these tests only can say whether an improper integral is convergent or not.

5.5 TEST FOR TYPE I INTEGRALS

We know that, if f is a monotone increasing function in $[a, \infty]$ then $\lim\limits_{x \to \infty} f(x)$ exists finitely if and only if f is bounded above. This fact gives the following theorem:

Theorem 5.5.1

If f is nonnegative function in $[a, \infty]$ such that f is integrable in $a \leq x \leq M$

$M \geq a$ then $\int\limits_{a}^{\infty} f(x)\, dx$ converges if and only if the set $\left\{ \int\limits_{a}^{M} f(x)\, dx, M > a \right\}$ is bounded above

Theorem 5.5.2 (Comparison Test)

Suppose f and g be two nonnegative functions in $[a, \infty]$ such that $f \leq g$ and both are integrable in $a \leq x \leq M$ for every $M > a$. Let

$$F = \int\limits_{a}^{\infty} f(x)\, dx \quad and \quad G = \int\limits_{a}^{\infty} g(x)\, dx$$

then,

    (i) F converges if G converges

    (ii) G diverges if F diverges.

Proof of above theorem is left as an exercise which can be proved by using Theorem 5.5.1

Theorem 5.5.3 (Limit Test)

Suppose f and g be two functions in $[a, \infty]$ which are integrable in $a \leq x \leq M$ for every $M > a$. Let F and G be defined as in Theorem 5.6.1. Then if g be positive then,

    (i) *If $\lim\limits_{x \to \infty} \dfrac{f(x)}{g(x)} = \lambda \,(\neq 0)$ the integrals F and G converge absolutely or both diverge.*

    (ii) *If $\lim\limits_{x \to \infty} \dfrac{f(x)}{g(x)} = 0$ and G converges then F converges.*

    (iii) *if $\lim\limits_{x \to \infty} \dfrac{f(x)}{g(x)} = \pm\, \infty$ and G diverges then F diverges.*

Proof: (i) Suppose $\lim\limits_{x \to \infty} \dfrac{f(x)}{g(x)} = \lambda\,(\neq 0)$. Now as $\int\limits_a^\infty f(x)\,dx$ and $\int\limits_a^\infty -(f(x))\,dx$ converge diverge simultaneously

we may assume without loss of generality that $\lambda > 0$. Then for $0 < \varepsilon < \lambda$ there exists a $M > a$ such that,

$$\left| \frac{f(x)}{g(x)} - \lambda \right| < \varepsilon \text{ whenever } x \geq M$$

or
$$(\lambda - \varepsilon)\, g(x) < f(x) < (\lambda + \varepsilon)\, g(x) \text{ for } x \geq M.$$

Hence we see that f is positive (as it lies between positive numbers) for $x \geq M$.

Now suppose that G converge. Then by using Theorem 5.6.2 and right side of above inequality we

conclude $\int\limits_M^\infty f(x)\,dx$ converge. But we see that, $\int\limits_a^\infty |f(x)|\,dx = \int\limits_a^M |f(x)|\,dx + \int\limits_M^\infty f(x)\,dx$ (as $f \geq 0$ for $x \geq M$)

Now as $\int\limits_a^M |f(x)|\,dx$ is proper, F converges absolutely as $\int\limits_M^\infty f(x)\,dx$ converges.

(Divergence part is left as an exercise)

(ii) Let $\lim\limits_{x \to \infty} \dfrac{f(x)}{g(x)} = 0$, then for $\varepsilon > 0$ there exists $M > a$ such that,

$$|f(x)| < \varepsilon g(x) \text{ whenever } x \geq M.$$

Suppose G converges. Then using above inequality and Theorem 5.6.2 we conclude $\int\limits_M^\infty |f(x)|\,dx$

converges but as before $\int\limits_a^M |f(x)\,dx$ is proper hence we have F converges absolutely

(iii) Suppose $\lim\limits_{x \to \infty} \dfrac{f(x)}{g(x)} = +\infty$. Then for any arbitrary large number L there exists $M > a$ such that

$f(x) > Lg(x)$ whenever $x \geq M$. Now it is clear that if G diverges then $\int\limits_M^\infty f(x)\,dx$ diverges and again as
$\int\limits_a^M f(x)\,dx$ is proper we get F diverges.

The theorem next will show a connection between an infinite series and a Type I integral. This theorem is often referred to as 'Integral Test' of infinite series.

Theorem 5.5.4 (Integral Test)

Let f be a positive and monotonically decreasing function in $[1, \infty]$. Then the series $\sum\limits_{n=1}^\infty f(n)$ is convergent if and only if $\int\limits_1^\infty f(x)\,dx$ converges.

Proof: We first note that as f is monotonically decreasing function in $[1, \infty]$ by Theorem we know f is integrable in $1 \leq x \leq M$ for any $M > 1$. Let n be any integer greater than 1. Now for all integers k such that $1 \leq k \leq n + 1$ and for all such that $k \leq x \leq k + 1$ we have

$$f(k + 1) \leq f(x) \leq f(k)$$

By integrating the above inequality in $[k, k+1]$ we get,

$$f(k+1) \leq \int_{k}^{k+1} f(x)\, dx \leq f(k)$$

Putting $k = 1, 2, \ldots n-1$ in the above inequality and adding we get

$$\sum_{k=2}^{n} f(k) \leq \int_{1}^{n} f(x)\, dx \leq \sum_{k=1}^{n-1} f(k)$$

Now if $\sum_{n=1}^{\infty} f(n)$ is convergent and converges to S then,

$$\int_{1}^{n} f(x)\, dx \leq \sum_{k=1}^{n-1} f(k) \leq S$$

and as n is arbitrary it implies that $\int_{1}^{\infty} f(x)\, dx$ converges.

Suppose now that $\sum_{n=1}^{\infty} f(n)$ is divergent. Then for any large G, there exists an integer N such that for

$n > N \sum_{k=2}^{n} f(k) > G$. Hence we have,

$$\int_{1}^{n} f(x)\, dx \geq \sum_{k=2}^{n} f(k) > G \text{ for all } n > N.$$

Now choose $X > N$ then we have $\int_{1}^{x} f(x)\, dx > G$. Which shows that $\int_{1}^{\infty} f(x)\, dx$ is divergent.

5.6 TYPE II INTEGRALS

Here we assume the interval of integration is bounded say $[a, b]$ and the integrand f has infinite discontinuity at some point of the interval. As the integrand can have infinite discontinuity at two end points of the interval or at any intermediate point of the interval we have three kinds of Type II integrals depending whether the integrant has infinite discontinuity at upper end point, lower end point or at any intermediate point of the interval. Before we proceed to give the formal definitions of these three kinds of Type II integrals we observe the following: Suppose f is a function having infinite discontinuity at $x = b$

that is, $\lim_{x \to b^-} f(x) = \infty$ then by putting $x = b - \dfrac{1}{t}$ in the integral $\int_{a}^{b} f(x)\, dx$, we get $\int_{a}^{b} f(x)\, dx = \int_{\frac{1}{b-a}}^{\infty} \dfrac{1}{t^2} f\left(b - \dfrac{1}{t}\right) dt$

the integral on the right side is an improper integral of first kind. Similarly if f has infinite discontinuity at

$x = a$ we can transform $\int_{a}^{b} f(x)\, dx$ to $\int_{\frac{1}{b-a}}^{\infty} \dfrac{1}{t^2} f\left(a + \dfrac{1}{t}\right) dt$ by substituting $x = a + \dfrac{1}{t}$. Hence we conclude that

convergence of Type II integrals can be discussed in terms of Type I integrals. Next we give the formal definitions of Type II integrals as follows:

(i) Suppose f has a infinite discontinuity at $x = b$ then we say that $\int_{a}^{b} f(x)\, dx$ converges if,

$$\lim_{\varepsilon \to 0^+} \int_a^{b-\varepsilon} f(x)\, dx$$

exists finitely and in that case the value of the integral is given by the limit. (We will refer such integrals as improper integrals having infinite discontinuity at upper end point)

Example 5.7 The improper integral $\int_0^1 \frac{1}{x-1}$ does not converges since

$$\lim_{\varepsilon \to 0^+} \int_0^{1-\varepsilon} \frac{1}{x-1}\, dx = \lim_{\varepsilon \to 0^+} [\log |x-1|]_0^{1-\varepsilon} = \lim_{\varepsilon \to 0^+} [\log \varepsilon - \log 1] = \lim_{\varepsilon \to 0^+} \log \varepsilon$$

which is certainly not finite (why?)

Example 5.8 The improper integral $\int_0^1 \frac{1}{\sqrt{1-x^2}}\, dx$ converges as,

$$\lim_{\varepsilon \to 0^+} \int_0^{1-\varepsilon} \frac{1}{\sqrt{1-x^2}}\, dx = \lim_{\varepsilon \to 0^+} [\sin^{-1} x]_0^{1-\varepsilon} = \lim_{\varepsilon \to 0^+} [\sin^{-1}(1-\varepsilon) - \sin^{-1}(0)] = \frac{\pi}{2} \text{ is finite.}$$

(ii) Suppose f has a infinite discontinuity at $x = a$ then we say that $\int_a^b f(x)\, dx$ converges if

$$\lim_{\varepsilon \to 0^+} \int_{a+\varepsilon}^b f(x)\, dx \text{ exists finitely and in that case the value of the integral is given by the limit.}$$

(We will refer these integrals as improper integrals having infinite discontinuity at lower end point)

Example 5.9 The improper integral $\int_0^1 \frac{1}{\sqrt{x}}\, dx$ converges as $\lim_{\varepsilon \to 0^+} \int_\varepsilon^1 \frac{1}{\sqrt{x}}\, dx = \lim_{\varepsilon \to 0^+} [2\sqrt{x}]_\varepsilon^1 = 2$ finite.

Example 5.10 The improper integral $\int_0^1 \frac{1}{x}\, dx$ does not converge as $\lim_{\varepsilon \to 0^+} \int_\varepsilon^1 \frac{1}{x}\, dx = \lim_{\varepsilon \to 0^+} [\log x]_\varepsilon^1 = - \lim_{\varepsilon \to 0^+}$

$(\log \varepsilon)$ is not finite. (why?)

(iii) Suppose f has an infinite discontinuity at $x = c$ where $a < c < b$ then we say that $\int_a^b f(x)\, dx$ converges

if both $\lim_{\varepsilon \to 0^+} \int_a^{c-\varepsilon} f(x)\, dx$ & $\lim_{\delta \to 0^+} \int_{c+\delta}^b f(x)\, dx$ exists finitely and in that case the value of the integral is

given by $\lim_{\varepsilon \to 0^+} \int_a^{c-\varepsilon} f(x)\, dx + \lim_{\delta \to 0^+} \int_{c+\delta}^b f(x)\, dx.$

Example 5.11   The improper integral $\int\limits_{-1}^{1} \dfrac{1}{x^3}\, dx$ does not converge. We see that the integrant has a infinite discontinuity at $x = 0$. Consider,

$$\lim_{\varepsilon \to 0^+} \int\limits_{-1}^{-\varepsilon} \frac{1}{x^3}\, dx = \lim_{\varepsilon \to 0^+} \frac{1}{2} \left\{ -\left[\frac{1}{x^2}\right]_{-1}^{-\varepsilon} \right\} = \frac{1}{2} - \lim_{\varepsilon \to 0^+} \left(\frac{1}{2\varepsilon^2}\right)$$

which is not finite.

Similarly, it is easy to see that $\lim\limits_{\delta \to 0^+} \int\limits_{\delta}^{1} \dfrac{1}{x^3}\, dx = \lim\limits_{\delta \to 0^+} \left(\dfrac{1}{2\delta^2}\right) - \dfrac{1}{2}$ is also not finite.

Example 5.12   The improper integral $\int\limits_{-1}^{1} \dfrac{1}{\sqrt{|x|}}\, dx$ converges. Clearly this is a modification of example 7.3.

See that $\lim\limits_{\varepsilon \to 0^+} \int\limits_{-1}^{-\varepsilon} \dfrac{1}{\sqrt{|x|}}\, dx = \lim\limits_{\varepsilon \to 0^+} \int\limits_{\varepsilon}^{1} \dfrac{1}{\sqrt{y}}\, dy$ (putting $y = -x$) $= 2$.

Note: It may happen that may have f infinite discontinuity at both the end points in that case we will choose any point c such that $a < c < b$ and will write the integral as $\int\limits_{a}^{b} f(x)\, dx = \int\limits_{a}^{c} f(x)\, dx + \int\limits_{c}^{b} f(x)\, dx$ so that the integrals in the right hand side is improper at only one end point of the interval. Now we will say that $\int\limits_{a}^{b} f(x)\, dx$ converges if both $\int\limits_{a}^{c} f(x)\, dx$ & $\int\limits_{c}^{b} f(x)\, dx$ converges as described above.

Example 5.13   The improper integral $\int\limits_{0}^{2} \dfrac{\log x}{\sqrt{2-x}}\, dx$ has infinite discontinuities at both the end points later we will write as $I = \int\limits_{0}^{2} \dfrac{\log x}{\sqrt{2-x}}\, dx + \int\limits_{0}^{1} \dfrac{\log x}{\sqrt{2-x}}\, dx + \int\limits_{1}^{2} \dfrac{\log x}{\sqrt{2-x}}\, dx$. Later we will see that both integrals on the right hand side converges (Example 5.21) and hence the given integral converges.

We again look Example 5.11 although the integral does not converge in usual sense but we see that,

$$\lim_{\varepsilon \to 0^+} \left[\int\limits_{-0}^{-\varepsilon} \frac{1}{x^2}\, dx + \int\limits_{0+\varepsilon}^{1} \frac{1}{x^2}\, dx \right] = \left[\left(\frac{1}{2} - \frac{1}{2\varepsilon^2}\right) + \left(\frac{1}{2\varepsilon^2} - \frac{1}{2}\right) \right] = 0.$$

This motivates us to give the definition of ***Cauchy Principal Value*** of an improper integral.

Definition   Suppose f has an infinite discontinuity at $x = c$ where $a < c < b$ then the ***Cauchy Principal Value*** of $\int\limits_{a}^{b} f(x)\, dx$ is defined as $\lim\limits_{\varepsilon \to 0^+} \left[\int\limits_{a}^{c-\varepsilon} f(x)\, dx + \int\limits_{c+\varepsilon}^{b} f(x)\, dx \right]$ provided the limit exists finitely.

Clearly if the improper integral is convergent then Cauchy Principal Value definitely exists and in that case two are equal. However Example 5.11 and discussion preceding the definition shows that the improper integral may not be convergent but the Cauchy Principal Value may exist. Of course there exist examples where neither the improper integral is convergent nor the Cauchy Principal Value exists. (Find one such example).

Next we give an important example which will later serve as a useful comparison integral for Type II integrals.

Example 5.14 $\displaystyle\int_a^b \frac{dx}{(x-a)^p}$ exists if $p < 1$ and does not exist if $p \geq 1$. This is an easy exercise proceed as example 5.3.4

Note this integral will be useful when we will compare with an integral having infinite discontinuity at lower end point)

Following example will be useful when we will compare with an integral having infinite discontinuity at upper end point.

Example 5.15 $\displaystyle\int_a^b \frac{dx}{(b-x)^p}$ exists if $p < 1$ and does not exist if $p \geq 1$

Now as in Type I integral we give the analytical definition of Type II integral then followed by necessary sufficient condition for its existence. As usual we will define Type I integral when the integrant has infinite singularity only at lower end point of the bounded interval as the definition for Type I integral when the integrant has infinite singularity only at lower end point will be similar and when the integrant has infinite singularity at any intermediate point we divide the interval into two parts and apply above definitions in two parts.

Definition   Suppose the function f has only point of infinite discontinuity at $x = a$ and is bounded and integrable in $a < x \leq b$. We say the improper integral $\displaystyle\int_a^b f(x)\,dx$ is convergent and has value I if given any $\varepsilon > 0$ there exists $\delta > 0$ $(< b - a)$ such that whenever $0 < \varepsilon_1 < \delta$ we have,

$$\left| I - \int_{a+\varepsilon_1}^b f(x)\,dx \right| < \varepsilon$$

Theorem 5.6.1

A necessary and sufficient condition for the convergence of the improper integral $\displaystyle\int_a^b f(x)\,dx$ *where function f has only point of infinite discontinuity at $x = a$ is that for any $\varepsilon > 0$ there exists $0 < \delta < b - a$ such that whenever $0 < \varepsilon_1 < \varepsilon_2 < \delta$ we have*

$$\left| \int_{a+\varepsilon_1}^{a+\varepsilon_2} f(x)\,dx \right| < \varepsilon$$

Proof: Immediate consequence of Cauchy criterion for the existence of limit of a function.

5.7 ABSOLUTE CONVERGENCE

The improper integral $\int_a^b f(x)\, dx$ where function f has only point of infinite discontinuity at $x = a$ is said to converge absolutely if $\int_a^b |f(x)|\, dx$ converges.

As in the case of Type *I* integrals we have a similar theorem like Theorem .5.1.

Theorem 5.7.1

If the improper integral $\int_a^b f(x)\, dx$ where function f has only point of infinite discontinuity at $x = a$ is absolutely convergent then it is convergent but the converse is not necessarily true.

Proof: Suppose the improper integral $\int_a^b f(x)\, dx$ is absolutely convergent then by definition $\int_a^b |f(x)|\, dx$ converges. Let now $\varepsilon > 0$, then by using the necessary part of Theorem. 8.1 we must have $0 < \delta < b - a$ such that whenever $0 < \varepsilon_1 < \varepsilon_2 < \delta$ we have

$$\left| \int_{a + \varepsilon_1}^{a + \varepsilon_2} |f(x)|\, dx \right| < \varepsilon$$

But we also have $\left| \int_{a + \varepsilon_1}^{a + \varepsilon_2} f(x)\, dx \right| \le \int_{a + \varepsilon_1}^{a + \varepsilon_2} |f(x)|\, dx < \varepsilon$ whenever $0 > \varepsilon_1 < \varepsilon_2 < \delta$

Hence using the sufficient condition of Theorem 5.8.1 we conclude that $\int_a^b f(x)\, dx$ is convergent.

We now show with help of an example that converse is not true. Consider $\int_0^1 \dfrac{\sin \frac{1}{x}}{x}\, dx$ and substitute $x = \dfrac{1}{t}$ the integral transform to $\int_1^\infty \dfrac{\sin t}{t}\, dx$ which we have already shown to be convergent but ***not*** absolutely convergent.

Definition The improper integral $\int_a^b f(x)\, dx$ is said to converge conditionally if $\int_a^b f(x)\, dx$ converges but $\int_a^b |f(x)|\, dx$ does not converge.

Now as in Type I integrals we will formulate certain tests which will help to determine whether a Type II integral is convergent or not. We will only give tests assuming the integrant has only point of infinite discontinuity at lower end point of the interval as one can easily formulate analogous test for the case where integrant has only point of infinite discontinuity at upper end point of the interval. For the case where integrant has only point of infinite discontinuity at any intermediate point one has to divide the interval into two parts and apply previous two tests separately in these interval.

5.8   TEST FOR TYPE II INTEGRALS

We get almost similar theorems like those obtained for Type I integrals proofs are also similar and left as an exercise for the reader. In the following theorems we assume that the functions f & g defined in the interval $a < x \leq b$ has only point of infinite discontinuity at $x = a$. We define F & G as $F = \int_a^b f(x)\, dx$ and

$$G = \int_a^b f(x) = dx$$

Theorem 5.8.1

Let f be a non negative function integrable in $[a + \varepsilon, b]$ for $0 < \varepsilon < b - a$ and then F converges if and only if the set $\left| \int_{a+\varepsilon}^b f(x)\, dx, 0 < \varepsilon < b - a \right|$ *is bounded above.*

Theorem 5.8.2 (Comparison Test)

Let f & g be two functions satisfying $0 \leq f \leq g$ in the interval $a < x \leq b$ then

    (i) If G converges F also converge.

    (ii) If F diverges then G also diverges.

Theorem 5.9.3 (Limit Comparison Test)

Let f & g be two functions defined in the interval $a < x \leq b$ with $g \geq 0$ there in then,

    (i) *If* $\lim\limits_{x \to a^+} \dfrac{f(x)}{g(x)} = \lambda\,(\neq 0)$ *then F & G either both converge absolutely or both diverge.*

    (ii) *If* $\lim\limits_{x \to a^+} \dfrac{f(x)}{g(x)} = 0$ *and G converges then F converges absolutely.*

    (iii) *If* $\lim\limits_{x \to a^+} \dfrac{f(x)}{g(x)} = \pm\, \infty$ *and G diverges then F also diverge.*

5.9   ABSOLUTE AND ORDINARY CONVERGENCE OF INTEGRANT AS PRODUCT OF FUNCTIONS

Now we consider Type I integrals where the integrant is a product of two functions and formulate tests to see they are convergent or not. First test is for absolute convergence followed by two which are to determine ordinary convergence.

Theorem 5.9.1   Suppose that f & g be two functions defined for $x \geq a$ such that $f(x)$ is bounded and $\int_a^\infty |g(x)|\, dx$ is convergent then the integral $\int_a^\infty f(x)\, g(x)\, dx$ converges absolutely.

Proof: As $\int_a^\infty |g(x)|\, dx$ is convergent using necessary part of Theorem 5.6.1 we can find $M_1 > 0$ such that, $\int_a^x |g(x)|\, dx < M_1$ for all $X \geq a$. Also as $f(x)$ is bounded there exists $M_2 > 0$ so that, $|f(x)| < M_2$ for $x \geq a$.

Now $\int\limits_{a}^{x} |f(x)\,g(x)|\,dx < M_2 \int\limits_{a}^{x} |g(x)|\,dx < M_1 M_2$ for $X \geq a$ hence using sufficient part of Theorem 5.6.1 we

conclude that $\int\limits_{a}^{\infty} f(x)\,g(x)\,dx$ converges absolutely.

Theorem 5.9.2 (Abel's Theorem)

Let f & g be two functions defined for $x \geq a$ such that $f(x)$ is bounded and monotonic if further $\int\limits_{a}^{\infty} g(x)\,dx$ is convergent then $\int\limits_{a}^{\infty} f(x)\,g(x)\,dx$ converges.

Proof: We apply second mean value of integral calculus to the functions f & g in the interval $[x_1, x_2]$ and get,

$$\int\limits_{x_1}^{x_2} f(x)\,g(x)\,dx = f(x_1) \int\limits_{x_1}^{\xi} g(x)\,dx + f(x_2) \int\limits_{\xi}^{x_2} g(x)\,dx \text{ for } a < x_1 \leq \xi \, x_2 \tag{5.9.2.1}$$

Hence,

$$\left| \int\limits_{x_1}^{x_2} f(x)\,g(x)\,dx \right| = |f(x_1)| \left| \int\limits_{x_1}^{\xi} g(x)\,dx \right| + |f(x_2)| \left| \int\limits_{\xi}^{x_2} g(x)\,dx \right| \text{ for } a < x_1 \leq \xi - x_2 \tag{5.9.2.2}$$

As $f(x)$ is bounded there exists $G > 0$ such that $|f(x)| < G$ for all $x \geq a$ hence we have,

$$|f(x_1)| < G \ \& \ |f(x_2)| < G \tag{5.9.2.3}$$

Now again as $\int\limits_{a}^{\infty} g(x)\,dx$ is convergent for $\varepsilon > 0$ we can get $A(\varepsilon) > a$ such that whenever $x_2 > x_1 > A(\varepsilon)$

we have $\left| \int\limits_{x_1}^{x_2} g(x)\,dx \right| < \dfrac{\varepsilon}{2G}$. Hence,

$$\int\limits_{x_1}^{\xi} g(x)\,dx < \dfrac{\varepsilon}{2G} \ \& \ \left| \int\limits_{\xi}^{x_2} g(x)\,dx \right| < \dfrac{\varepsilon}{2G}, \text{ if we choose } x_2 > x_1 > A(\varepsilon) \tag{5.9.2.4}$$

Combining equations (10.2.2), (10.2.3) & (10.2.4) we get for

$$x_2 > x_1 > A(\varepsilon),$$

$$\left| \int\limits_{x_1}^{x_2} f(x)\,g(x)\,dx \right| < G \cdot \dfrac{\varepsilon}{2G} + G \cdot \dfrac{\varepsilon}{2G} = \varepsilon \tag{5.9.2.5}$$

Hence $\int\limits_{a}^{\infty} f(x)\,g(x)\,dx$ converges.

If we closely look at the proof of above theorem we see that we have used the conditions of the theorem to make right of equation (5.9.2.2) arbitrarily small. Now we observe that if we can make $|f(x)|$

arbitrarily small and integral the $\int_{x_1}^{x_2} g(x)\,dx$ remains bounded for any $x_2 > x_1 > a$ then again we can achieve

right of equation (5.9.2.2) arbitrarily small this is precisely the conditions of next theorem which is known as Dirichlet's Theorem.

Theorem 5.9.3 (Dirichlet's Theorem)

Suppose that f & g be two functions defined for $x \geq a$ such that $f(x)$ is bounded, monotonic and $f(x) \to 0$ as $x \to \infty$ and also let the set $\left\{ \int_{a}^{X} g(x)\,dx,\ X > a \right\}$ is bounded then, $\int_{a}^{\infty} f(x)\,g(x)\,dx$ converges.

Proof: As in the previous theorem using second mean value theorem we again get equation the following

$$\left| \int_{x_1}^{x_2} f(x)\,g(x)\,dx \right| = |f(x_1)| \left| \int_{x_1}^{\xi} g(x)\,dx \right| + |f(x_2)| \left| \int_{\xi}^{x_2} g(x)\,dx \right| \text{ for, } a < x_1 \leq \xi \leq x_2 \tag{5.9.3.1}$$

Now as $\left\{ \int_{a}^{X} g(x)\,dx,\ X > a \right\}$ is bounded there exists $G > 0$ such that,

$$\left| \int_{a}^{X} g(x)\,dx \right| < G \text{ for } X > a \tag{5.9.3.2}$$

Now for $x_2 > \xi < x_1 > a$ we have $\int_{x_1}^{\xi} g(x)\,dx = \int_{a}^{\xi} g(x)\,dx - \int_{a}^{x_1} g(x)\,dx$

Hence we get,

$$\left| \int_{x_1}^{\xi} g(x)\,dx \right| = \left| \int_{a}^{\xi} g(x)\,dx \right| + \left| \int_{a}^{x_1} g(x)\,dx \right| < 2G \tag{5.9.3.3}$$

Similarly,

$$\int_{\xi}^{x_2} g(x)\,dx < 2G \tag{5.9.3.4}$$

Since $f(x) \to 0$ as $x \to \infty$ for $\varepsilon > 0$ there exists $A(\varepsilon) > a$ such that whenever $x_2 > x_1 > A(\varepsilon)$ we must have,

$$|f(x_1)| < \frac{\varepsilon}{2G} \text{ and } |f(x_1)| < \frac{\varepsilon}{2G} \tag{5.9.3.5}$$

Now using (5.10.3.1), (5.10.3.3), (5.10.3.4) & (5.10.3.5) we get that whenever $x_2 > x_1 > A(\varepsilon)$

$$\left| \int_{x_1}^{x_2} f(x)\,g(x)\,dx \right| < G.\frac{\varepsilon}{2G} + G.\frac{\varepsilon}{2G} = \varepsilon.$$

Hence $\int_{a}^{\varepsilon} f(x)\,g(x)\,dx$ converges.

Remark:

Analogous theorems for absolute convergence and convergence of above form for Type II integrals can easily be formulated.

5.10 SOME MISCELLANEOUS PROBLEMS

We will now work out various examples so that the readers get to know how to use various tests we have formulated to see whether a given improper integral is convergent or divergent. We must point out here that many a times reader will come across improper integrals which may fall into both Type I and Type II then in that case one has to write the improper integral as sum of two or more integrals where each of them will be only one category either Type I or Type II and then apply tests on each of them. The given integral in this case will be convergent if each of the integrals is convergent. Finally readers must note that the solution to problems given here is just one of the various possible methods available and they must try out other methods to have better grasp on technique of solving these problems.

Example 5.17 Show that $\displaystyle\int_{1}^{\infty} \frac{dx}{x\sqrt{x^2+1}}$ converges.

See that,

$$\sqrt{x^2+1} > \sqrt{x^2} = x$$

hence $\dfrac{1}{\sqrt{x^2+1}} < \dfrac{1}{x}$ or, $g(x) = \dfrac{1}{x\sqrt{x^2+1}} < \dfrac{1}{x^2} = f(x)$. But $F = \displaystyle\int_{1}^{\infty} \frac{1}{x^2}\,dx$ converges by Example 5.6 Now using comparison test (Theorem 5.5.2) we conclude above integral converges.

Example 5.18 Examine the convergence of $\displaystyle\int_{0}^{\infty} \frac{dx}{e^x+1}$

Sol: Let $f(x) = \dfrac{1}{e^x+1}$ & $g(x) = \dfrac{1}{x^2}$

then, $\displaystyle\lim_{x\to\infty} \frac{f(x)}{g(x)} = \lim_{x\to\infty} \frac{x^2}{e^x+1} = 0.$

Again as observed in previous example $F = \displaystyle\int_{1}^{\infty} \frac{1}{x^2}\,dx$ converges. We now use Theorem 5.5.3 part (ii) to conclude that given integral converges.

Example 5.19 Examine the convergence of $\displaystyle\int_{2}^{\infty} \frac{dx}{\log x}$.

Sol: Let $f(x) = \dfrac{1}{\log x}$ & $g(x) = \dfrac{1}{x}$

then, $\displaystyle\lim_{x\to\infty} \frac{f(x)}{g(x)} = \lim_{x\to\infty} \frac{x}{\log x} = \infty$ (using L'Hospital's rule).

Now, by Example 5.6 $F = \int_1^\infty \frac{1}{x}\, dx$ diverges and by using Theorem 5.5.3 part (iii) we conclude the given integral is divergent.

Example 5.20 Examine the convergence of $\int_0^1 \log x\, dx$

Sol: This is a Type II integral where the integrant has an infinite singularity at $x = 0$. Let $f(x) = \log x$ & $g(x) = \frac{1}{\sqrt{x}}$. Then $\lim\limits_{x \to 0} \frac{f(x)}{g(x)} = \lim\limits_{x \to 0} \sqrt{x} \log x = 0$. By Example 5.9 $\int_0^1 \frac{1}{\sqrt{x}}\, dx$ is convergent and by Theorem 5.6.3 part (iii) given integral convergent absolutely.

Example 5.21 Examine the convergence of $\int_0^2 \frac{\log x}{\sqrt{2 - x}}\, dx$

Sol: Let $I = \int_2^0 \log \frac{x}{\sqrt{2 - x}}\, dx$ & $f(x) = \log \frac{x}{\sqrt{2 - x}}$ &. Notice that g has infinite discontinuities at both end points of the intervals. Hence we write $I = \int_0^2 \frac{\log x}{\sqrt{2 - x}}\, dx = \int_0^1 \frac{\log x}{\sqrt{2 - x}}\, dx + \int_1^2 \frac{\log x}{\sqrt{2 - x}}\, dx$.

Let $I_1 = \int_0^1 \frac{\log x}{\sqrt{2 - x}}\, dx$ & $I_2 = \int_1^2 \frac{\log x}{\sqrt{2 - x}}\, dx$.

Convergence of I_1

Let $g_1(x) = \log x$, then $\lim\limits_{x \to 0^+} \frac{f(x)}{g_1(x)} = \lim\limits_{x \to 0^+} \frac{1}{\sqrt{2 - x}} = \frac{1}{\sqrt{2}}$ (finite non-zero)

By Theorem 5.6.3 part (ii) $\int_0^1 \log x\, dx$ and I_1 will converge or diverge simultaneously but by Example 5.21 $\int_0^1 \log x\, dx$ converge hence I_1 converges.

Convergence of I_2

Note that I_2 has an infinite discontinuity at upper end point. Let $g_2(x) = \frac{1}{\sqrt{2 - x}}$ then,

$$\lim\limits_{x \to 2^-} \frac{f(x)}{g_2(x)} = \lim\limits_{x \to 2^-} \log x = \log 2 \text{ (finite non-zero)}.$$

By modified form of Theorem 5.6.3 part (ii) $\int_1^2 g_2(x)\, dx$ & I_2 will converge or diverge simultaneously but by Example 5.4,

$$\int_1^2 g_2(x)\, dx \text{ converges as } p = \frac{1}{2} < 1$$

Hence I_2 converges.

So as both I_1 & I_2 converge we conclude that I converges.

Example 5.22   Show that $\int_0^{\frac{\pi}{2}} \log \sin x \, dx$ is convergent and find it's value.

Sol: We write $I = \int_0^{\frac{\pi}{2}} \log \sin x \, dx$ then,

$$I = \int_0^{\frac{\pi}{2}} \left(\log \frac{\sin x}{x} + \log x \right) dx = \int_0^{\frac{\pi}{2}} \log \left(\frac{\sin x}{x} \right) dx + \int_0^{\frac{\pi}{2}} \log x \, dx = I_1 + I_2$$

As, $\displaystyle\lim_{x \to 0^+} \log \left(\frac{\sin x}{x} \right) = \log 1 = 0$, I_1 is proper and we have already seen I_2 is convergent (by Example 5.21)

hence, I is convergent. Now we evaluate the integral,

$$I = \int_0^{\frac{\pi}{2}} \log \sin x \, dx = \int_0^{\frac{\pi}{2}} \log \sin \left(\frac{\pi}{2} - x \right) dx = \int_0^{\frac{\pi}{2}} \log \cos x \, dx$$

or,
$$2I = \int_0^{\frac{\pi}{2}} \log (\sin x \cos x) \, dx = \int_0^{\frac{\pi}{2}} \log \left(\frac{\sin 2x}{2} \right) dx$$

$$= -\frac{\pi}{2} \log 2 + \int_0^{\frac{\pi}{2}} \log \sin 2x \, dx \qquad (1)$$

Now in $\int_0^{\frac{\pi}{2}} \log \sin 2x \, dx$ we put $2x = t$ then,

$$\int_0^{\frac{\pi}{2}} \log \sin 2x \, dx = \frac{1}{2} \int_0^{\pi} \log \sin t \, dt$$

$$= \frac{1}{2} \cdot 2 \int_0^{\frac{\pi}{2}} \log \sin t \, dt \ (\text{as } \sin (\pi - t) = \sin t) = I$$

From (1)
$$2I = -\frac{\pi}{2} \log 2 + I$$

Hence,
$$I = -\frac{\pi}{2} \log 2.$$

Example 5.23   Examine the convergence of $\displaystyle\int_{1}^{\infty} \dfrac{dx}{(x+1)^{\frac{1}{2}}\, x^{\frac{1}{2}}\, (x-1)^{\frac{1}{3}}}$

Sol: Let $f(x) = (x+1)^{\frac{1}{2}}\, x^{\frac{1}{2}}\, (x-1)^{\frac{1}{3}}$ & $g(x) = \dfrac{1}{x^{1+\frac{1}{3}}} = \dfrac{1}{x^{4/3}}$. Then, $\displaystyle\lim_{x\to\infty} \dfrac{f(x)}{g(x)} = 1$.

Now $\displaystyle\int_{1}^{\infty} g(x)\, dx$ converges by Example 5.6 and by Theorem 5.5.3 part (i) $\displaystyle\int_{1}^{\infty} f(x)\, dx$ converges.

Example 5.24   Examine the convergence of $\displaystyle\int_{0}^{\infty} \dfrac{\cos h\,(bx)}{\cos h\,(ax)}\, dx \quad a>0, b>0$

Sol: We see that,

$$\frac{\cos h\,(bx)}{\cos h\,(ax)} = \frac{e^{bx}+e^{-bx}}{e^{ax}+e^{-ax}} < \frac{e^{bx}+e^{-bx}}{e^{ax}} < e^{(b-a)x} + e^{-(b+a)x} \quad \text{if } b<a \tag{1}$$

As,

$$(a+b) > (a-b)$$

$$\Rightarrow\quad -(a+b) < -(a-b)$$

$$\Rightarrow\quad e^{-(a+b)} < e^{b-a}$$

From (1) we get,

$$\frac{\cos h\,(bx)}{\cos h\,(ax)} < 2e^{(b-a)x} = 2e^{-(a-b)x} \tag{2}$$

It is easy to check that,

$$\int_{a}^{\infty} e^{-bx}\, dx \text{ converges if } b>0 \text{ and diverges if } b<0 \tag{3}$$

Using (2) and comparison test we conclude that $\displaystyle\int_{0}^{\infty} \dfrac{\cos h\,(bx)}{\cos h\,(ax)}\, dx$ converges if $b<a$.

Now if $b>a$

$$\frac{\cos h\,(bx)}{\cos h\,(ax)} = \frac{e^{bx}+e^{-bx}}{e^{ax}+e^{-ax}} > \frac{e^{bx}}{e^{ax}+e^{-ax}} > \frac{e^{bx}}{e^{ax}+e^{ax}} = \frac{1}{2}\, e^{(b-a)x} \ \left(\text{as } e^{ax} > e^{-ax}\right)$$

Now using (3) we get that $\displaystyle\int_{0}^{\infty} e^{(b-a)x}\, dx$ diverges hence by comparison test we get $\displaystyle\int_{0}^{\infty} \dfrac{\cos h\,(bx)}{\cos h\,(ax)}\, dx$

diverges in this case.

Example 5.25   Examine the convergence of $\displaystyle\int_{0}^{\pi/2} \dfrac{\sin^{m} x}{x^{n}}\, dx$

Sol: Let $I = \displaystyle\int_{0}^{\pi/2} \dfrac{\sin^{m} x}{x^{n}}\, dx$ & $f(x) = \dfrac{\sin^{m} x}{x^{n}} = \left(\dfrac{\sin x}{x}\right)^{m} \cdot \dfrac{1}{x^{n-m}}$

When $n \le m$ the integral is proper as,

$$\lim_{x \to 0} \left\{ \left(\frac{\sin x}{x} \right)^m \cdot \frac{1}{x^{n-m}} \right\} = \lim_{x \to 0} \left(\frac{\sin x}{x} \right)^m \cdot \lim_{x \to 0} x^{m-n} = 0.$$

When $n > m$, let $g(x) = \dfrac{1}{x^{n-m}}$

Then $\lim\limits_{x \to 0^+} \dfrac{f(x)}{g(x)} = 1$. Now $\int\limits_0^{\pi/2} g(x)\,dx$ converges if $0 < n - m < 1$ that is if $m < n < m + 1$. Now using

Theorem 5.6.3 (i) we conclude $\int\limits_0^{\pi/2} \dfrac{\sin^m x}{x^n}\,dx$ converges absolutely if $n < m + 1$.

Example 5.26   Examine the convergence of $\int\limits_0^{\infty} \dfrac{\sin x^m}{x^n}\,dx$.

Sol: Let $f(x) = \dfrac{\sin x^m}{x^n}$ & $I = \int\limits_0^{\infty} \dfrac{\sin x^m}{x^n}\,dx$.

$$I = \int\limits_0^1 f(x)\,dx + \int\limits_1^{\infty} f(x)\,dx = I_1 + I_2$$

Case (i)   If $m = 0$

$$I_1 = \int\limits_0^1 \frac{\sin 1}{x^n}\,dx \quad \text{converges if } n < 1 \text{ \&}$$

$$I_2 = \int\limits_1^{\infty} \frac{\sin 1}{x^n}\,dx \quad \text{converges if } n > 1$$

Hence I does not converge in this case.

Case (ii)   If $m > 0$

Substituting $x^m = y$ in I we get

$$I = \frac{1}{m} \int\limits_0^{\infty} \frac{\sin y}{y^{\frac{n-1}{m}+1}}\,dx$$

$$= \frac{1}{m} \left(\int\limits_0^1 h(y)\,dy + \int\limits_1^{\infty} h(y)\,dy \right), \quad \text{where } h(y) = \frac{\sin y}{y^{\frac{n-1}{m}+1}}$$

$$= \frac{1}{m}(J_1 + J_2).\ I \text{ will converge if both } J_1 \text{ \& } J_2 \text{ converge.}$$

For convergence of J_1

Let $g(y) = \dfrac{1}{y^{\frac{n-1}{m}}}$ then $\lim\limits_{x \to 0^+} \dfrac{f(y)}{g(y)} = 1$ and $\int\limits_0^1 g(y)\, dy$ converges if $\dfrac{n-1}{m} < 1$ that is if $n < m + 1$

Again using Theorem 5.6.3 (i) we conclude that J_1 converges absolutely if $n < m + 1$ (1)

For convergence of J_2

Let $k(y) = \dfrac{1}{y^{\frac{n-1}{m}} + 1}$. Then $k(y)$ is monotone decreasing function and $k(y) \to 0$ as $y \to \infty$ provided

$n + m - 1 > 0$ that is $n > 1 - m$.

Also we see that, $\left| \int\limits_1^B \sin y\, dy \right| \le 2$ for any $B > 1$. Hence by Dirichlet's test J_2 converges if $n > 1 - m$ (2)

Combining we get if $m > 0$ I converges if $m - 1 < n < m + 1$.

Case (iii)   If $m < 0$

Suppose $m = -p,\, p > 0$, then,

$$I = \int\limits_1^\infty \frac{\sin y^p}{y^{2-n}}\, dy \text{ by substituting } \frac{1}{x} = y \text{ now this case can now be disposed as case (ii)}$$

Example 5.27   Examine the convergence of $\int\limits_0^\infty \dfrac{\sin x}{x}\, dx$

Sol: First we note that $I = \int\limits_0^1 \dfrac{\sin x}{x}\, dx + \int\limits_1^\infty \dfrac{\sin x}{x}\, dx = I_1 + I_2$. As, $\lim\limits_{x \to 0^+} \dfrac{\sin x}{x} = 1,\, I_1,$ is proper.

In I_2 let $\phi(x) = \dfrac{1}{x}$ then $\phi(x)$ is bounded, monotone decreasing in $[1, \infty]$ and tens to zero as $x \to \infty$.

Also, $\left| \int\limits_1^B \sin x\, dx \right| \le 2$, for $B > 1$ hence by Dirichlet's test $\int\limits_0^\infty \dfrac{\sin x}{x}\, dx$ is convergent.

Example 5.28   Examine the convergence of $\int\limits_0^\infty e^{-px} \left(\dfrac{\sin x}{x} \right) dx,\, p \ge 0,$

Sol: Here we see that e^{-px} is bounded, monotone decreasing and $e^{-px} \to 0$. Also we know from Example 5.27 that,

$$\int\limits_0^\infty \frac{\sin x}{x}\, dx$$

is convergent. Now use Abel's Test to conclude that $\int\limits_0^\infty e^{-px} \left(\dfrac{\sin x}{x} \right) dx$ convergent.

Example 5.29   Examine the convergence of $\int\limits_{0}^{\infty} \sin x^2 \, dx$

Sol: $\int\limits_{0}^{\infty} \sin x^2 \, dx = \int\limits_{0}^{1} \sin x^2 \, dx + \int\limits_{1}^{\infty} \sin x^2 \, dx = I_1 + I_2.$

I_1 is proper. For convergence of I_2 we see that

$$\sin x^2 = \left(\frac{1}{2x}\right)(2x \sin x^2) = g(x)\, k(x)$$

$g(x)$ is bounded, monotone decreasing in $[1, \infty]$ & tends to zero as $x \to \infty$.

Also $\left|\int\limits_{0}^{B} h(x)\, dx\right| = \left|[\cos x^2]_B^1\right| = |\cos 1 - \cos B^2| \le 2$ now using Dirichlet's test I_2 converges.

Hence $\int\limits_{0}^{\infty} \sin x^2 \, dx$ converges.

Example 5.30   Examine the convergence of $\int\limits_{0}^{\pi} \dfrac{\sqrt{x}}{\sin x} \, dx$

Sol: We see that the integrant has infinite discontinuities at both the end points hence have to split the integral.

Let $f(x) = \dfrac{\sqrt{x}}{\sin x} = \dfrac{1}{\sqrt{x}}\left(\dfrac{x}{\sin x}\right)$ & $I = \int\limits_{0}^{\pi/2} \dfrac{\sqrt{x}}{\sin x} \, dx + \int\limits_{\pi/2}^{0} \dfrac{\sqrt{x}}{\sin x} \, dx + = I_1 + I_2.$

For convergence of I_1 we put $g(x) = \dfrac{1}{\sqrt{x}}$. Then $\lim\limits_{x \to \infty} \dfrac{f(x)}{g(x)} = 1$. Again we note that $\int\limits_{0}^{\pi/2} g(x)\, dx$ converges hence I_1 converges. For I_2 we put

$$k(x) = \frac{1}{\sin x}$$

then
$$\lim\limits_{x \to \infty} \frac{f(x)}{k(x)} = \sqrt{\pi}.$$

But we see that $\int\limits_{\pi/2}^{\pi} \dfrac{dx}{\sin x}$ does not converge as,

$$\int\limits_{\pi/2}^{\pi} \frac{dx}{\sin x} = \lim\limits_{\varepsilon \to 0^+}\left[\log \tan \frac{x}{2}\right]_{\pi/2}^{\pi-\varepsilon} = \lim\limits_{\varepsilon \to 0^+}\left[\log \tan \frac{\pi-\varepsilon}{2} - \log \tan \frac{\pi}{4}\right]$$

$$\lim\limits_{\varepsilon \to 0^+} \log \tan \frac{\pi-\varepsilon}{2} - 0 = \infty$$

Application of Theorem 5.6.3 shows that I_2 does not converge. Hence we conclude that $\int\limits_{0}^{\pi} \dfrac{\sqrt{x}}{\sin x} \, dx$ does not converge.

Example 5.31   (Gamma Function) Examine the convergence of $\int_0^\infty e^{-x} x^{n-1}\, dx$.

Sol: Let $f(x) = e^{-x} x^{n-1}$. We note that $f(x)$ will have infinite discontinuity if $n < 1$. Hence we split the integral as

$$I = \int_0^\infty e^{-x} x^{n-1}\, dx = \int_0^1 e^{-x} x^{n-1}\, dx + \int_1^\infty e^{-x} x^{n-1}\, dx = I_1 + I_2$$

Now I_1 will be proper if $n \geq 1$. In case $n < 1$ let $g(x) = \dfrac{1}{x^{1-n}}$ then $\lim\limits_{x \to 0^+} \dfrac{f(x)}{g(x)} = \lim\limits_{x \to 0^+} e^{-x} = 1$

Now $\int_0^1 g(x)\, dx$ will converge if $0 < 1 - n < 1$ or, $0 < n < 1$.

Hence I_1 converges absolutely if $0 < n < 1$ by Theorem 5.6.3 (i)   (1)

For convergence of I_2 we see that if,

$$k(x) = \frac{1}{x^2}$$

then $\lim\limits_{x \to \infty} \dfrac{f(x)}{g(x)} = \lim\limits_{x \to \infty} (x^2 e^{-x} x^{n-1}) = \lim\limits_{x \to \infty} (e^{-x} x^{n+1}) = 0.$

Also note that $\int_0^1 g(x)\, dx$ converges.

Hence by Theorem 5.5.3 (ii) I_2 converges absolutely for all values of n.   (2)

From (1) & (2) we conclude I converges for $n > 0$.

We see that the value of I when exists is a function of n. When I exists we call it ***Gamma Function*** and denote it by $\Gamma(n)$. Thus,

$$\Gamma(n) = \int_0^\infty e^{-x} x^{n-1}\, dx \quad n > 0$$

Example 5.32   *(Beta Function)* Examine the convergence of

$$\int_0^1 x^{m-1}(1-x)^{n-1}\, dx.$$

Sol: Let $I = \int_0^1 x^{m-1}(1-x)^{n-1}\, dx$ & $f(x) = x^{m-1}(1-x)^{n-1}$. We note that if $m \geq 1$ & $n \geq 1$ then I is proper.

If either $n < 1$ or $m < 1$ then $f(x)$ may have infinite discontinuities at both end points. Hence we write the integral as

$$I = \int_0^{1/2} x^{m-1}(1-x)^{n-1}\, dx + \int_{1/2}^1 x^{m-1}(1-x)^{n-1}\, dx = I_1 + I_2$$

For convergence of I_1, let $g(x) = \dfrac{1}{x^{1-n}}$ then $\lim\limits_{x \to 0^+} \dfrac{f(x)}{g(x)} = \lim\limits_{x \to 0^+} \left[x^{1-m} x^{m-1} (1-x)^{n-1} \right] = 1$. Also note

that $\int_0^{1/2} g(x)\, dx$ converges if $0 < 1 - m < 1$ that is if $0 < m < 1$. Hence I_1 converges for $0 < m < 1$ and all values of n. It is also easy to show that I_1 does not converge if $m < 0$. Now for convergence of I_2 if we substitute $t = 1 - x$ then $I_2 = \int_0^{1/2} t^{n-1}(1-t)^{m-1}\, dt$ observe that this is I_1 with m & n interchanged. Hence we conclude that I_2 converges for $0 < n < 1$ and all values of m. Combining we conclude that I converges if $m > 0, n > 0$.

Here we observe that value of I if exists is a function of m & n. When I exists we call it **Beta Function** and is denote as $B(m, n)$. Thus,

$$B(m, n) = \int_0^1 x^{m-1}(1-x)^{n-1}\, dx \qquad m > 0, n > 0$$

Example 5.33 Show that the integral $\int_1^\infty \dfrac{\sin x}{x^p}\, dx$ converges absolutely for $p > 1$.

Sol: Observe $\left| \int_1^\infty \dfrac{\sin x}{x^p}\, dx \right| \leq \int_1^\infty \left| \dfrac{\sin x}{x^p} \right| dx = \int_1^\infty \dfrac{dx}{x^p}$. Also $\int_1^\infty \dfrac{dx}{x^p}$ converges if $p > 1$. Hence the conclusion follows by comparison test.

Example 5.34 Show that the integral $\int_1^\infty \dfrac{\sin x}{x^p}\, dx$ converges for $p > 0$

Sol: The result follows if we apply Dirichlet's Test. We take $f(x) = \dfrac{1}{x^p}$ which is bounded monotone decreasing and tends to 0 as $x \to \infty$ and $g(x) = \sin x$ so

that
$$\left| \int_1^B x\, dx \right| \leq 2.$$

(Compare Example 5.33 & Example 5.34)

Example 5.35 Examine the convergence of $\int_0^1 x^{m-1}(1-x)^{n-1} \log\left(\dfrac{1}{x}\right) dx$

Sol: Let $f(x) = x^{m-1}(1-x)^{n-1}\log\left(\dfrac{1}{x}\right) = -x^{m-1}(1-x)^{n-1}\log x$ and $I = \int_0^1 x^{m-1}(1-x)^{n-1}\log x\, dx$. Note that if I converges then the given integral converges so we only discuss convergence of I instead of the given integral. Now $f(x)$ can have infinite discontinuity at both the end points hence we split the integral and write,

$$I = \int_0^1 x^{m-1}(1-x)^{n-1}\log x\, dx = \int_0^{1/2} x^{m-1}(1-x)^{n-1}\log x\, dx + \int_{1/2}^1 x^{m-1}(1-x)^{n-1}\log x\, dx = I^1 + I_2$$

For convergence of I_1 we put $g(x) = x^{m-1}\log x$. Then as $\lim\limits_{x \to 0^+} \dfrac{f(x)}{g(x)} = 1$. By Theorem 5.9.3 (i) I_1 converges if $J = \int_0^{1/2} x^{m-1}\log x\, dx$ converges. In J we put $z = -\log x$ then we get

$$J = \int\limits_{\log 2}^{\infty} z e^{-mz}\, dz$$

$$= \frac{1}{m^2} \int\limits_{m \log 2}^{\infty} t e^{-t}\, dt, \text{ by substituting } mz = t.$$

Now comparing with Gamma Function we conclude that J converges. Hence I_1 converges.

For convergence of I_2 we put $k(x) = (1-x)^{n-1} \log x$. Then $\lim\limits_{x \to 1^-} \dfrac{f(x)}{k(x)} = 1$.

Hence I_2 converges if $J_1 = \int\limits_{1/2}^{1} (1-x)^{n-1} \log x\, dx$ converge. Again we know

$$-\log x < \frac{(1-x)}{x}, x \in (0,1) \text{ hence,}$$

$$-J_2 \leq \int\limits_{1/2}^{1} \frac{(1-x)^n}{x}\, dx \text{ hence } J_2 \text{ converges if } n > 0.$$

If $n < 0$ let $n = -k, k > 0$.

$$-J_2 \leq \int\limits_{1/2}^{1} \frac{(1-x)^{-k}}{x}\, dx = \int\limits_{1/2}^{1} \frac{dx}{(1-x)^k\, x}. \text{ Extreme right integral converges if } 0 < k < 1 \text{ that is } n > -1 \text{ hence}$$

J_2 converges if $n > -1$. So the given integral converges if $m > 0$ and $n > -1$.

Example 5.36   Examine the convergence of $\int\limits_{0}^{\infty} \left(\dfrac{1}{x} - \dfrac{1}{\sin hx} \right) \dfrac{dx}{x}$.

Sol: Let $I = \int\limits_{0}^{\infty} \left(\dfrac{1}{x} - \dfrac{1}{\sin hx} \right) \dfrac{dx}{x}$ & $f(x) = \left(\dfrac{1}{x} - \dfrac{1}{\sin hx} \right) \dfrac{1}{x} = \dfrac{\sin hx - x}{x^2 \sin hx}$.

Now $\sinh x - x = \left[x + \dfrac{x^3}{3!} + \dfrac{x^5}{5!} + \dots \right] - x = \dfrac{x^3}{3!} + \dfrac{x^5}{5!} + \dots$

Hence, $\lim\limits_{x \to 0^+} f(x) = \lim\limits_{x \to 0^+} \dfrac{\sin hx - x}{x^2 \sin hx} = \lim\limits_{x \to 0^+} \dfrac{\dfrac{x^3}{3!} + \dfrac{x^5}{5!} + \dots}{x^3} \lim\limits_{x \to 0^+} \dfrac{x}{\sin hx} = \dfrac{1}{3!}$ \hfill (1)

We write $I = \int\limits_{0}^{1} f(x)\, dx + \int\limits_{1}^{\infty} f(x)\, dx = I_1 + I_2$. By (1) I_1 is proper.

For I_2 we observe that $f(x) = \dfrac{\sin hx - x}{x^2 \sin hx} = \dfrac{1}{x^2} - \dfrac{2e^{-x}}{x(1 - e^{2x})}$.

Let $g(x) = \dfrac{1}{x^2}$ then $\lim\limits_{x \to \infty} \dfrac{f(x)}{g(x)} = 1$. Now as $\int\limits_{1}^{\infty} g(x)\, dx$ converges, I also converge.

Example 5.37   Show that $\displaystyle\int_0^1 x^n \log x \, dx$ converges when $n > 0$

When $n = 1$ $$\int_0^1 \log x \, dx = \lim_{x \to 0^+} \varepsilon(1 - \log \varepsilon) - 1 = -1.$$

If $n > 1$ $\displaystyle\int_0^1 x^n \log x \, dx$ is proper.

Lastly if $1 - n > 0$ we take $p > 1 - n$ and

$$g(x) = \frac{1}{x^p} \text{ also } \lim_{x \to 0^+} \frac{f(x)}{g(x)} = \lim_{x \to 0^+} x^{p+n-1} \log x = 0.$$

Since $\displaystyle\int_0^1 g(x) \, dx$ convergent the given integral is convergent in this case.

Combining we get $\displaystyle\int_0^1 x^n \log x \, dx$ when $n > 0$.

PROBLEMS

1. Let $f(x)$ be a function defined on the interval $[a, M]$ for $M \geq a$. Then show that

    (i) The improper integral $\displaystyle\int_a^\infty f(x) \, dx$ converges absolutely if $\displaystyle\lim_{x \to \infty} x^a f(x) = b$ (finite) $a > 1$

    (ii) The improper integral $\displaystyle\int_a^\infty f(x) \, dx$ diverges if $\displaystyle\lim_{x \to \infty} x^a f(x) = b \ (\neq 0)$ or $\pm \infty$ and $a \leq 1$.

2. Prove an analogous result as problem 1 for Type II integrals.

3. Show that the following integrals converge

    (i) $\displaystyle\int_1^\infty \frac{dx}{(x + 1)^{1/2} \, x^{1/2}(x + 3)^{1/2}}$

    (ii) $\displaystyle\int_1^\infty \frac{\log x \, dx}{x^3}$

    (iii) $\displaystyle\int_0^\infty \frac{dx}{e^{3x} + 1}$

    (iv) $\displaystyle\int_0^1 \frac{dx}{x^{1/2}(1 - x)^{1/2}}$

    (v) $\displaystyle\int_0^1 \log x \, dx$

4. Show that the following integrals diverge:

(i) $\displaystyle\int_{-1}^{1} \frac{dx}{(2-x)\,(1-x^2)^{1/2}}$

(ii) $\displaystyle\int_{1}^{3} \frac{x^2\,dx}{\sqrt{(x-1)\,(3-x)}}$

(iii) $\displaystyle\int_{0}^{\infty} \frac{x^{\frac{3}{2}}}{b^2 x^2 + c^2}\,dx$

(iv) $\displaystyle\int_{a}^{b} \frac{dx}{(x-a)\sqrt{b-x}}$

(v) $\displaystyle\int_{2}^{\infty} \frac{dx}{\log x}$

5. Prove that $\displaystyle\int_{0}^{\frac{\pi}{2}} \log(\sin x)\,dx$ converges and show that its value is equal to $-\dfrac{\pi}{2}\log 2$.

6. Show that $\displaystyle\int_{2}^{\infty} \frac{\cos x}{\log x}\,dx$ is conditionally convergent.

7. Discuss the convergence and divergence of the following integrals:

(i) $\displaystyle\int_{0}^{\infty} \left(\frac{1}{x} - \frac{1}{\sin hx}\right)\frac{dx}{x}$

(ii) $\displaystyle\int_{0}^{\infty} \frac{\cos h\,(bx)}{\cos h\,(ax)}\,dx \quad a>0\ b>0$

(iii) $\displaystyle\int_{0}^{1} \frac{x^{a-1}}{1-x}\,dx$

(iv) $\displaystyle\int_{0}^{1} x^{m-1}\,(1-x)^{n-1}\log\left(\frac{1}{x}\right)dx$

(v) $\displaystyle\int_{0}^{2\pi} \frac{1}{\sqrt{x}}\,\sin h\left(\frac{1}{x}\right)dx$

8. Show that:

(i) $\displaystyle\int_{0}^{\infty} \frac{t^{l-1}}{1+t^{m}}\,dt$ is convergent if and only if $0<l<m$

(ii) $\displaystyle\int_0^\infty \frac{t^{l-1} - t^{s-1}}{1-x}\, dt$ is convergent if and only if $0 < l < 1$ and $0 < t < 1$

9. Use Dirichlet's Test to show that following integrals are convergent:

(i) $\displaystyle\int_0^1 \frac{1}{x^{\frac{3}{2}}} \sin\left(\frac{1}{x}\right) dx$

(ii) $\displaystyle\int_0^\infty \sin x^2\, dx$

<h2>Chapter 6</h2>

Beta and Gamma Functions

6.1 INTRODUCTION

Beta and Gamma functions are used widely as tools not only in mathematics but also in other subjects like Physics, Probability, Statistics to name a few. The purpose of this chapter is to explore some important properties of these functions and relation between them.

6.2 BETA FUNCTION

The integral

$$\int_0^1 x^{m-1}(1-x)^{n-1}\,dx$$

converges for $m > 0$ & $n > 0$ (as shown in Chapter 5) and is clearly a function of m and n. The above integral when treated as a function of m and n $m > 0$ & $n > 0$ is called *Beta function* and is denoted by $B(m, n)$. That is,

$$B(m, n) = \int_0^1 x^{m-1}(1-x)^{n-1}\,dx \quad m > 0 \ \& \ n > 0$$

6.3 PROPERTIES OF BETA FUNCTION

Following theorem gives some of the important properties Beta function.

Theorem 6.3.1

If $B(m, n)$ $m >$ & $n > 0$ denotes Beta function then,

    (i) $B(m, n) = B(n, m)$

    (ii) $B(m, n) = \int_0^\infty \dfrac{x^{m-1}dx}{(1+x)^{m+n}} = \int_0^\infty \dfrac{x^{n-1}\,dx}{(1+x)^{m+n}}$

        (The integrals on the right are called Beta Integrals of Second Kind)

(iii) $B(m, n) = 2 \int\limits_{0}^{\frac{\pi}{2}} \sin^{2m-1} \theta \cos^{2n-1} \theta \, d\theta$

Proof: (i) For $\varepsilon > 0$ & $\delta > 0$ consider,

$$\int\limits_{\varepsilon}^{1-\delta} x^{m-1} (1 - x)^{n-1} \, dx \tag{6.3.1.1}$$

Substituting in the above integral $t = 1 - x$ we obtain,

$$\int\limits_{\varepsilon}^{1-\delta} x^{m-1}(1 - x)^{n-1} \, dx = \int\limits_{1-\varepsilon}^{\delta} t^{n-1} (1 - t)^{m-1} \, (-dt) \tag{6.3.1.2}$$

Taking limit as $\varepsilon \to 0^+$ and $\delta \to 0^+$ (6.3.1.2) adjusting the limit with minus sign we get,

$$B(m, n) = B(n, m)$$

(ii) Substituting $x = \dfrac{1}{1 + t}$ in (6.3.1.1) we get,

$$\int\limits_{\varepsilon}^{1-\delta} x^{m-1} (1 - x)^{n-1} \, dx = \int\limits_{\frac{1}{\varepsilon}-1}^{\frac{\delta}{1-\delta}} \frac{1}{(1 + t)^{m-1}} \frac{t^{n-1}}{(1 + t)^{n-1}} \left\{ -\frac{1}{(1 + t)^2} \right\} dt \tag{6.3.1.3}$$

Now again taking limits as $\varepsilon \to 0^+$ and $\delta \to 0^+$ one obtains,

$$B(m, n) = \int\limits_{0}^{\infty} \frac{x^{n-1} \, dx}{(1 + x)^{m+n}} \tag{6.3.1.4}$$

Hence interchanging m & n in (6.3.1.4) we get

$$B(n, m) = \int\limits_{0}^{\infty} \frac{x^{m-1} \, dx}{(1 + x)^{m+n}} \tag{6.3.1.5}$$

But from (i) $B(m, n) = B(n, m)$ hence using (6.3.1.4) and (6.3.1.5) we conclude,

(iii) Substituting $x = \sin^2 \theta$ in (6.3.1.1) we get,

$$\int\limits_{\varepsilon}^{1-\delta} x^{m-1} (1 - x)^{n-1} dx = \int\limits_{\sin^{-1}\sqrt{\varepsilon}}^{\sin^{-1}\sqrt{1-\delta}} \sin^{2m-1} \theta \cos^{2n-1} \theta (2 \sin \theta \cos \theta) \, d\theta \tag{6.3.1.6}$$

Now taking limit as $\varepsilon \to 0^+$ and $\delta \to 0^+$ in both sides of (6.3.1.6) we get,

$$B(m, n) = 2 \int\limits_{0}^{\frac{\pi}{2}} \sin^{2m-1} \theta \cos^{2n-1} \theta \, d\theta$$

Noting that $\sin^{-1} \sqrt{1 - \delta} \to \dfrac{\pi}{2}$ as $\delta \to 0$ and $\sin^{-1} \varepsilon \to 0$ as $\varepsilon \to 0$.

6.4 GAMMA FUNCTION

The integral $\int\limits_{0}^{\infty} e^{-x} x^{n-1} dx$ converges for $n > 0$ (as shown in Chapter 5). When treated as a function of n the above integral is called *Gamma Function* and is denoted by $\Gamma(n)$. That is,

$$\Gamma(n) = \int\limits_{0}^{\infty} e^{-x} x^{n-1} dx$$

6.5 PROPERTIES OF GAMMA FUNCTION

Next theorem gives some important properties of Gamma function.

Theorem 6.5.1

For the Gamma Function $\Gamma(n)$ $n > 0$ we have,

    (i) $\Gamma(n) > 0$ and $\Gamma(n) \to \infty$ as $n \to 0^+$

    (ii) $\Gamma(n + 1) = n\Gamma(n)$ $n > 0$

    (iii) $\Gamma(n + 1) = n!$ if n is a positive integer

Proof: (i) Let $\varepsilon > 0$ and M (≥ 1) be arbitrary small and large numbers respectively. We see that,

$$\int\limits_{\varepsilon}^{M} e^{-x} x^{n-1} dx \geq \int\limits_{\varepsilon}^{1} e^{-x} x^{n-1} dx \text{ (as the integrant is positive)}$$

$$\geq e^{-1} \int\limits_{\varepsilon}^{1} x^{n-1} dx$$

Now as $\varepsilon \to 0^+$ and $M \to \infty$ we get,

$$\Gamma(n) > \frac{1}{en} > 0 \tag{6.5.1.1}$$

From (6.5.1.1) it is clear that $\Gamma(n) \to \infty$ as $n \to 0^+$.

    (ii) Let ε & M be chosen as in part (i) and consider,

$$\int\limits_{\varepsilon}^{M} e^{-x} x^{n-1} dx \tag{6.5.1.2}$$

Integrating by parts the above integral we get,

$$\int\limits_{e}^{M} e^{-x} x^{n-1} dx = \left[e^{-x} \frac{x^{n}}{n} \right]_{\varepsilon}^{M} + \frac{1}{n} \int\limits_{e}^{M} e^{-x} x^{n} dx$$

$$= e^{-M} \frac{M^{n}}{n} - e^{-\varepsilon} \frac{\varepsilon^{n}}{n} + \frac{1}{n} \int\limits_{\varepsilon}^{M} e^{-x} x^{n} dx \tag{6.5.1.3}$$

If now $\varepsilon \to 0^+$ and $M \to \infty$, $e^{-M} \dfrac{M^{n}}{n}$ and $e^{-\varepsilon} \dfrac{\varepsilon^{n}}{n}$ both tends to zero hence from (6.5.1.3) we obtain,

$$\Gamma(n + 1) = n\Gamma(n)$$

(iii) Note that,

$$\Gamma(1) = \lim_{M \to \infty} \int_0^M e^{-x}\, dx = 1$$

Now if n is a positive integer then repeated use of (ii) yields,

$$\Gamma(n + 1) = n(n - 1)\,(n - 2)\, \ldots \Gamma(1) = n!$$

6.6 RELATION BETWEEN BETA & GAMMA FUNCTION

There is a nice relation between Beta and Gamma function which we give as a theorem whose proof is given in Chapter 10 as an application of Double integral.

Theorem 6.6.1

For $m > 0$ & $n > 0$ we have,

$$B(m, n) = \frac{\Gamma(m)\,\Gamma(n)}{\Gamma(m + n)} \tag{6.6.1.1}$$

Corollary 6.6.2

For $p, q > -1$

$$\int_0^{\frac{\pi}{2}} \sin^p \theta \cos^q \theta\, d\theta = \frac{1}{2} B\left(\frac{p + 1}{2}, \frac{q + 1}{2}\right) = \frac{1}{2}\, \frac{\Gamma\left(\frac{p + 1}{2}\right)\Gamma\left(\frac{q + 1}{2}\right)}{\Gamma\left(\frac{p + q + 1}{2}\right)} \tag{6.6.2.1}$$

Proof: Follows immediately from part (iii) of Theorem 6.3.1 and Theorem 6.6.1.

Corollary 6.6.3

$$\Gamma\left(\frac{1}{2}\right) = \sqrt{\pi}$$

Proof: If we put $p = q = 0$ in (6.6.2.1) we get,

$$\frac{1}{2}\, \frac{\Gamma\left(\frac{1}{2}\right)\Gamma\left(\frac{1}{2}\right)}{\Gamma(1)} = \int_0^{\frac{\pi}{2}} \sin^0 \theta \cos^0 \theta\, d\theta = \frac{\pi}{2}$$

$$\Rightarrow \qquad \left(\Gamma\left(\frac{1}{2}\right)\right)^2 = \pi \qquad (\text{As } \Gamma(1) = 1)$$

i.e. $\Gamma\left(\frac{1}{2}\right) = \sqrt{\pi}$ (Since $\Gamma(m) > 0$ for $m > 0$).

Corollary 6.6.4 (Duplication Formula)

For any real number $m > 0$ we have,

$$\sqrt{\pi}\,\Gamma(2m) = 2^{2m - 1}\,\Gamma(m)\,\Gamma\left(m + \frac{1}{2}\right)$$

Proof: In the equation (6.6.2.1) if we put p = q = 2m – 1 we get.

$$\frac{\Gamma(m)\,\Gamma(m)}{\Gamma(2m)} = 2 \int_0^{\frac{\pi}{2}} \sin^{2m-1}\theta \cos^{2m-1}\theta \, d\theta$$

$$= \frac{2}{2^{2m-1}} \int_0^{\frac{\pi}{2}} (\sin 2\theta)^{2m-1}\, d\theta$$

$$= \frac{1}{2^{2m-1}} \int_0^{\pi} (\sin \phi)^{2m-1}\, d\phi \qquad \text{(Putting } 2\theta = \phi)$$

$$= \frac{2}{2^{2m-1}} \int_0^{\frac{\pi}{2}} (\sin \phi)^{2m-1}\, d\phi \qquad (\because \sin^{2m-1}(\pi - \theta) = \sin^{2m-1}\theta)$$

$$= \frac{2}{2^{2m-1}} \frac{\Gamma(m)\,\Gamma\left(\frac{1}{2}\right)}{2\Gamma\left(m + \frac{1}{2}\right)} \qquad \text{(Using (6.6.2.1)} \; p = 2m - 1 \; \& \; q = 0). \qquad (6.6.4.1)$$

Hence we get $\sqrt{\pi}\ \Gamma(2m) = 2^{2m-1}\ \Gamma(m)\ \Gamma\left(m + \dfrac{1}{2}\right)$ canceling $\Gamma(m)\ (> 0)$ from both sides of equation

(6.6.4.1) and noting that $\Gamma\left(\dfrac{1}{2}\right) = \sqrt{\pi}$.

Corollary 6.6.5

If $0 < p < 1$ then,

$$\Gamma(p)\,\Gamma(1 - p) = \frac{\pi}{\sin px}$$

Proof: To prove this result we will assume the following result,

$$\sum_{n=0}^{\infty} (-1)^n \left(\frac{1}{n+p} + \frac{1}{n+1-p}\right) = \frac{\pi}{\sin p\pi}, \; 0 < p < 1 \tag{6.6.5.1}$$

Above result can be proved using standard Fourier Series technique.

Now using Beta Gamma relation and Part (ii) of Theorem 6.3.1 we get,

$$\frac{\Gamma(p)\,\Gamma(1 - p)}{\Gamma(1)} = B(p,\, 1 - p) = \int_0^{\infty} \frac{x^{p-1}\, dx}{(1 + x)}$$

Hence,

$$\Gamma(p)\,\Gamma(1 - p) = \int_0^{\infty} \frac{x^{p-1}\, dx}{(1 + x)} \qquad \text{(as } \Gamma(1) = 1) \tag{6.6.5.2}$$

Now, $\displaystyle \int_0^{\infty} \frac{x^{p-1}\, dx}{(1 + x)} = \int_0^{1} \frac{x^{p-1}\, dx}{(1 + x)} + \int_1^{\infty} \frac{x^{p-1}\, dx}{(1 + \pi)}$

(Note that both the integrals converge on the right in the respective intervals)

$$= \int_0^1 \frac{x^{p-1}dx}{(1+x)} + \int_0^1 \frac{t^{-p}dt}{(1+t)} \quad \left(\text{putting } x = \frac{1}{t} \text{ in the second integral}\right)$$

$$= \int_0^1 \frac{(x^{p-1} + x^{-p})\, dx}{(1+x)} \quad \text{(Changing the dummy variable } t \text{ to } x) \qquad (6.6.5.3)$$

Again, $(x^{p-1} + x^{-p})(1+x)^{-1} = (x^{p-1} + x^{-p})\left(1 - \dfrac{x}{1+x}\right)$

$$= (x^{p-1} + x^{-p}) - (x^{p} + x^{1-p})(1+x)^{-1} \qquad (6.6.5.4)$$

It is easy to check that.

$$\int_0^1 (x^{p-1} + x^{-p})\, dx = \frac{1}{p} + \frac{1}{1-p} \qquad (6.6.5.5)$$

Now we see that,

$$(x^{p} + x^{1-p})(1+x)^{-1} = (x^{p} + x^{1-p})(1 - x + x^2 - \ldots\ldots)$$

$$= (x^{p} + x^{1-p}) - x(x^{p} + x^{1-p}) + x^2(x^{p} + x^{1-p}) - \ldots\ldots$$

$$= (x^{p} + x^{1-p}) - (x^{p+1} + x^{2-p}) + (x^{p+2} + x^{3-p}) - \ldots\ldots$$

and the series on the right is uniformly convergent in $0 \le x < 1$ and hence integrating term by term we get,

$$\int_0^t (x^{p} + x^{1-p})(1+x)^{-1}dx = \left[\frac{x^{p+1}}{p+1} + \frac{x^{2-p}}{2-p}\right]_0^t - \left[\frac{x^{p+2}}{p+2} + \frac{x^{3-p}}{3-p}\right]_0^t + \left[\frac{x^{p+3}}{p+3} + \frac{x^{4-p}}{4-p}\right]_0^t - \ldots\ldots$$

$$= \frac{t^{p+1}}{p+1} + \frac{t^{2-p}}{2-p} - \frac{t^{p+2}}{p+2} - \frac{t^{3-p}}{3-p} + \frac{t^{p+3}}{p+3} + \frac{t^{4-p}}{4-p} - \ldots\ldots,$$

$0 \le t < 1$

Now note that the series,

$$\frac{1}{p+1} + \frac{1}{2-p} - \frac{1}{p+2} - \frac{1}{3-p} + \frac{1}{p+3} + \frac{1}{4-p} - \ldots\ldots$$

is convergent. Hence using Abel's Theorem we get,

$$\int_0^1 (x^{p} + x^{1-p})(1+x)^{-1}dx = \frac{1}{p+1} + \frac{1}{2-p} - \frac{1}{p+2} - \frac{1}{3-p} + \frac{1}{p+3} + \frac{1}{4-p} - \ldots\ldots \qquad (6.6.5.6)$$

Hence using (6.6.5.2), (6.6.5.5) and (6.6.5.6) we get,

$$\Gamma(p)\,\Gamma(1-p) = \frac{1}{p} + \frac{1}{1-p} - \frac{1}{p+1} - \frac{1}{2-p} + \frac{1}{p+2} + \frac{1}{3-p} - \frac{1}{p+3} - \frac{1}{4-p} + \ldots\ldots$$

$$= \sum_{n=0}^{\infty} (-1)^n \left(\frac{1}{n+p} + \frac{1}{n+1-p}\right) = \frac{\pi}{\sin p\pi} \quad \text{(using 6.6.5.1)}$$

6.7 SOME MISCELLANEOUS PROBLEMS

Example 1   Show that $\displaystyle\int_0^\infty a^{-x^2}\, dx = \dfrac{\sqrt{\pi}}{2\sqrt{\log a}}, \ a < 0$

Sol: We see that,

$$\int_0^\infty a^{-x^2}\, dx = \int_0^\infty e^{-x^2 \log a}\, dx$$

$$= \int_0^\infty e^{-z}\ \frac{dz}{2\,\dfrac{\sqrt{z}}{\sqrt{\log a}}\,\sqrt{\log a}} \qquad \text{(putting } z = x^2 \log a)$$

$$= \frac{1}{2\sqrt{\log a}} \int_0^\infty e^{-z}\, z^{\frac{1}{2}}\, dz = \frac{1}{2\sqrt{\log a}}\, \Gamma\!\left(\frac{1}{2}\right) = \frac{\sqrt{\pi}}{2\sqrt{\log a}}$$

Example 2   Evaluate $\displaystyle\int_a^b (x-a)^{\alpha-1}(b-x)^{\beta-1}\, dx,\ \alpha,\ \beta > 0$

Sol: If we put $x = a\cos^2\theta + b\sin^2\theta$ in the above integral then,

$\theta \to \dfrac{\pi}{2}$ as $x \to b$ and $\theta \to 0$ as $x \to a$ so that we get,

$$\int_a^b (x-a)^{\alpha-1}(b-x)^{\beta-1}dx$$

$$= \int_0^{\frac{\pi}{2}} (b-a)^{\alpha-1}\sin^{2\alpha-2}(b-a)^{\beta-1}\sin^{2\beta-2}(-2a\cos\theta\sin\theta + 2a\sin\theta\cos\theta)\, d\theta$$

$$= \int_0^{\frac{\pi}{2}} 2(b-a)^{\alpha+\beta-1}\sin^{2\alpha-1}\cos^{2\beta-1}d\theta$$

$$= 2(b-a)^{\alpha+\beta-1}\int_0^{\pi/2}\sin^{2\alpha-1}\cos^{2\beta-1}d\theta$$

$$= (b-a)^{\alpha+\beta-1}\, B(\alpha,\ \beta)$$

Example 3   Show that $\displaystyle\int_0^{\pi/2}\frac{\sin^{2\alpha-1}\theta\,\cos^{2\beta-1}\theta\, d\theta}{(a\sin^2\theta + b\cos^2\theta)^{\alpha+\beta}} = \frac{\Gamma(\alpha)\,\Gamma(\beta)}{2a^\alpha\, b^\beta\, \Gamma(\alpha+\beta)},\ a,\ b,\ \alpha,\ \beta > 0,$

Sol: In the above integral if we put $bx = a\tan^2\theta$ then,

$$\int_0^{\pi/2}\frac{\sin^{2\alpha-1}\theta\,\cos^{2\beta-1}\theta\, d\theta}{(a\sin^2\theta + b\cos^2\theta)^{\alpha+\beta}} = \int_0^\infty \frac{\sin^{2\alpha-1}\theta\,\cos^{2\beta-1}\theta\, d\theta}{(1+x)^{\alpha+\beta}\, b^{\alpha+\beta}\cos^{2(\alpha+\beta)}\theta}\ \frac{b\cos^2\theta}{2a\sin\theta}\, dx$$

$$= \frac{1}{2a^\alpha b^\beta} \int_0^\infty \frac{x^{\alpha-1} dx}{(1-x)^{\alpha+\beta}} = \frac{1}{2a^\alpha b^\beta} B(\alpha, \beta) = \frac{1}{2a^\alpha b^\beta} \frac{\Gamma(\alpha)\,\Gamma(\beta)}{\Gamma(\alpha+\beta)}$$

Example 4   Show that $\displaystyle\int_0^{\frac{\pi}{2}} \frac{d\theta}{\sqrt{1 - \frac{1}{2}\sin^2\theta}} = \frac{\left(\Gamma\left(\frac{1}{4}\right)\right)^2}{4\sqrt{\pi}}$

Sol: If we put $\sin\theta = \sqrt{2}\,\sin\left(\dfrac{\phi}{2}\right)$ in the above integral then we have,

$$\int_0^{\frac{\pi}{2}} \frac{d\theta}{\sqrt{1 - \frac{1}{2}\sin^2\theta}} = \frac{1}{\sqrt{2}} \int_0^{\frac{\pi}{2}} \frac{\cos\frac{\phi}{2}\, d\phi}{\cos\frac{\phi}{2}\,\sqrt{1 - 2\sin^2\frac{\phi}{2}}}$$

$$= \frac{1}{\sqrt{2}} \int_0^{\frac{\pi}{2}} (\cos\phi)^{\frac{1}{2}}\, d\phi$$

$$= \frac{1}{2\sqrt{2}} B\left(\frac{1}{2}, \frac{1}{4}\right) = \frac{1}{2\sqrt{2}} \frac{\Gamma\left(\frac{1}{2}\right)\Gamma\left(\frac{1}{4}\right)}{\Gamma\left(\frac{3}{4}\right)}$$

$$= \frac{1}{2\sqrt{2}} \frac{\sqrt{\pi}\,\Gamma\left(\frac{1}{4}\right)\Gamma\left(\frac{1}{4}\right)}{\Gamma\left(\frac{3}{4}\right)\Gamma\left(\frac{1}{4}\right)} = \frac{1}{2\sqrt{2}} \frac{\sqrt{\pi}\left(\Gamma\left(\frac{1}{4}\right)\right)^2}{\pi\,\mathrm{cosec}\,\frac{\pi}{4}}$$

$$= \frac{\left(\Gamma\left(\frac{1}{4}\right)\right)^2}{4\sqrt{\pi}} \qquad \left(\text{Using } \Gamma(p)\,\Gamma(1-p) = \frac{\pi}{\sin p\pi}\right)$$

Example 5   Evaluate $\displaystyle\int_0^{\frac{\pi}{2}} \sqrt{\tan x}\, dx$

Sol: $\displaystyle\int_0^{\frac{\pi}{2}} \sqrt{\tan x}\, dx = \int_0^{\frac{\pi}{2}} \sin^{\frac{1}{2}}\theta \cos^{\frac{1}{2}} dx$

$$= \frac{1}{2} B\left(\frac{3}{4}, \frac{1}{4}\right) = \frac{1}{2} \frac{\Gamma\left(\frac{3}{4}\right)\Gamma\left(\frac{1}{4}\right)}{\Gamma(1)} = \frac{1}{2}\,\pi\,\mathrm{cosec}\,\frac{\pi}{4} = \frac{\pi}{\sqrt{2}} \qquad \left(\text{Using } \Gamma(p)\,\Gamma(1-p) = \frac{\pi}{\sin p\pi}\right)$$

Example 6   If $0 < \alpha < 1$ show that

$$\int_0^{\frac{\pi}{2}} \tan^\alpha x\, dx = \frac{1}{2}\,\pi \sec\left(\alpha\,\frac{\pi}{2}\right)$$

Sol:

$$\int_0^{\frac{\pi}{2}} \tan^\alpha x\, dx = \int_0^{\frac{\pi}{2}} \sin^\alpha x \cos^{-\alpha} x\, dx$$

$$= \frac{1}{2}\, B\left(\frac{\alpha+1}{2}, \frac{1-\alpha}{2}\right)$$

$$= \frac{1}{2}\, \frac{\Gamma\left(\frac{\alpha+1}{2}\right)\Gamma\left(\frac{1-\alpha}{2}\right)}{\Gamma(1)}$$

$$= \frac{1}{2}\, \Gamma\left(\frac{\alpha+1}{2}\right)\Gamma\left(1 - \frac{\alpha+1}{2}\right)$$

$$= \frac{1}{2}\, \frac{\pi}{\cosec\left(\frac{\alpha+1}{2}\,\pi\right)} = \frac{1}{2}\,\pi \sec\left(\alpha\,\frac{\pi}{2}\right)$$

PROBLEMS

1. Prove that $\displaystyle\int_0^1 \frac{x\, dx}{\sqrt{1-x^5}} = \frac{1}{5}\, B\left(\frac{2}{5}, \frac{1}{2}\right)$

2. $\displaystyle\int_0^1 \frac{x^{\alpha-1}(1-x)^{\beta-1}\, dx}{(b+cx)^{\alpha+\beta}} = \frac{B(\alpha, \beta)}{b^\beta(b+c)^\alpha},\ \alpha, \beta, > 0$

3. Evaluate $\displaystyle\int_0^1 \frac{dx}{\sqrt{\log\left(\frac{1}{x}\right)}}$

4. Evaluate $\displaystyle\int_0^\infty e^{-ax^b}\, x^c\, dx,\ a, b, c > 0$

5. Show that $\displaystyle\int_0^a x^4 \sqrt{a^2 - x^2}\, dx = \frac{\pi a^5}{16},\ a > 0$

6. Show that $\displaystyle\int_0^\pi \left(1 - \frac{x}{n}\right)^n x^{\alpha-1}\, dx = n^\alpha B(x, n+1),\ n, \alpha > 0,$

7. Show that

   (i) $B(\alpha, \alpha)\, B\left(\alpha + \dfrac{1}{2}, \alpha + \dfrac{1}{2}\right) = \left(\dfrac{\pi}{\alpha}\right) 2^{1-4\alpha}$, $\alpha > 0$,

   (ii) $\alpha B(\alpha, \beta) = (\beta - 1)\, B(\alpha + 1, \beta - 1)$, $\alpha, \beta > 0$

8. Evaluate $\displaystyle\int_a^b (x - a)^\alpha\, (b - x)^\beta\, dx$ where $\alpha, \beta > -1$ & $b > a$ in terms of Beta function.

9. Show that $\displaystyle\int_0^1 \dfrac{dx}{\sqrt{1 - x^n}} = \dfrac{2^{\frac{2}{n}-1}\left[\Gamma\left(\frac{1}{n}\right)\right]^2}{n\Gamma\left(\frac{2}{n}\right)}$

10. Show that $\displaystyle\lim_{n \to \infty} \dfrac{\Gamma(n + k)}{n^k\,\Gamma(n)} = 1$, k being a positive integer.

Chapter **7**

Sequence and Series of Functions

7.1 INTRODUCTION

We have studied sequences of real numbers in Chapter 1 where our primary objective was to study their convergence and which ultimately led us to the study of infinite series of real numbers. In this chapter we are going to study the concept convergence and limit of a sequence of real valued functions defined on a subset of real numbers (generally on an interval). We will also investigate some interesting questions like, 'If all the members of the sequence are continuous (differentiable) is the limit of this sequence continuous (differentiable) in case the such a limit exists' Similar questions can be asked about integrability of limit of sequence of functions when all members of the sequence are integrable. These questions actually will lead us to the study of uniform convergence.

Finally, we will study notion of sum of series of functions and theorems leading to tem by tem integration, differentiation of such series.

7.2 DEFINITIONS AND EXAMPLES

Let E be a subset of real numbers $\mathbb{R}$ and suppose $\mathfrak{I}$ denotes set of all real functions from E to $\mathbb{R}$. If there exist a one to one function ϕ from the set of natural numbers N to $\mathfrak{I}$ then, $\{\phi(n)\}_{n \in \mathbb{N}}$ is called a sequence of real valued functions defined on E. It is customary to denote $\phi(n)$ the nth term of the sequence by f_n, where f_n is a real valued functions defined on E hence we will denote such sequence as $\{f_n\}_{n=1}^{\infty}$ or simply by $\{f_n\}$.

Example 7.1 Let $E = [0, 1]$ & if

$$f_n(x) = x^n \quad n \in \mathbb{N}$$

then $\{f_n\}$ is a sequence of functions defined on the closed interval $[0, 1]$.

Example 7.2 Let $E = [0, 1]$ & $Q = \{r_1, r_2, r_3, ..., r_n, ...\}$ be the enumeration of rationals of $[0, 1]$. For any $n \in \mathbb{N}$, we define $f_n: [0, 1] \to \mathbb{R}$

$$f_n(x) = 1, \text{ for } x = r_1, r_2, ..., r_n$$

$$= 0, \text{ otherwise.}$$

Then $\{f_n\}$ is a sequence of functions defined on the closed interval $[0, 1]$

7.3 POINTWISE CONVERGENCE AND UNIFORM CONVERGENCE

Suppose $\{f_n\}$ is a sequence of functions defined on certain interval I of R. Now for any $x \in I$, $\{f_n(x)\}$ is a sequence of real number. Let x_1 & x_2 be two points of I and consider the sequences $\{f(x_1)\}$ & $\{f(x_2)\}$, suppose that both of them converge with limits l_1 & l_2 that is $\lim_{n \to \infty} f_n(x_1) = l_1$ &. $\lim_{n \to \infty} f_n(x_2) = l_2$ Now there is no reason to believe that $l_1 = l_2$, in general they will not be equal. We now assume that $\{f_n(x)\}$ converges for every $x \in I$ then by what we have just now observed we can say $\lim_{n \to \infty} f_n(x)$ depends on x.

As we know that limit of a convergent sequence is unique the correspondence $x \to \lim_{n \to \infty} f_n(x)$ from I to R is well defined and hence defines one function, which is called *pointwise limit* of the sequence $\{f_n\}$ which is denote by f and we say $\{f_n\}$ *converges pointwise* to f. In symbol we write, $\lim_{n \to \infty} f_n(x) = f(x)$ or simply, $f_n \to f$.

Again using the definition of limit of a sequence we say that f is pointwise limit of the sequence $\{f_n\}$ defined on the interval I, if for a given any $\epsilon > 0$ and a fixed $x \in I$ there exits a positive integer $n_1(\epsilon, n)$ such that, $|f_n(x) - f(x)| < \epsilon$ whenever $n > n_1$.

The important point to be noted here is that the positive integer n_1 which we obtain depends not only on given ϵ but also on the fixed x. We exhibit this with following examples:

Example 7.3 Let $\{f_n\}$ be a sequence of functions defined on $[0, \infty)$ as,

$$f_n(x) = \frac{nx}{n + x}; n = 1, 2, 3...$$

First we guess the point wise limit $f(x)$ by taking any fixed $x \in [0, \infty)$ and then find $\lim_{n \to \infty} f_n(x)$ in this case we see that if $x = 0$ then $f_n(x) = 0$ for all n hence $f(0) = 0$. Now for any $x > 0$ we have

$$f(x) = \lim_{n \to \infty} f_n(x) = \lim_{n \to \infty} \frac{nx}{n + x}.$$

$$= \lim_{n \to \infty} \frac{x}{1 + \dfrac{x}{n}} = x$$

So we get the point wise limit as $f(x) = x \; \forall \; x \geq 0$. Now to prove formally that, $f(x) = x \; \forall \; x \geq 0$ is the point wise limit. For that we start with any $\epsilon > 0$ and a fixed $x \geq 0$ and see that,

$$|f_n(x) - f(x)| = \left| \frac{nx}{n + x} - x \right| = \left| \frac{-x^2}{n + x} \right| = \frac{x^2}{n + x}$$

So $$|f_n(x) - f(x)| < \varepsilon \text{ if } \frac{x^2}{n+x} < \varepsilon, \text{ that is if,}$$

$$n > \frac{x(x + \varepsilon)}{\varepsilon} \tag{1}$$

If we choose the positive integer n_1 as $n_1 = \left[\dfrac{x(x+\varepsilon)}{\varepsilon}\right] + 1$ ([.] is the greatest integer function) then $\forall \, n \geq n_1$ we get $|f_n(x) - f(x)| < \varepsilon$. Hence, $f(x) = x \, \forall \, x \geq 0$ is the pointwise limit of the above sequence. Also note that n_1 depends both on ε & x.

Example 7.4 Let $\{f_n\}$ be a sequence of functions defined on [1, 2] as,

$$f_n(x) = \frac{\sin nx}{nx^2}, n = 1, 2, 3 \ldots$$

As in the previous example we first find $\lim\limits_{n \to \infty} f_n(x) \, \dfrac{\sin nx}{nx^2}$ for a fixed $x \in [1, 2]$. We see that,

$$\left|\frac{\sin nx}{nx^2}\right| \leq \frac{1}{nx^2} \to 0, \text{ as } n \to \infty \, \forall \, x \in [1, 2]. \text{ Hence, the pointwise limit is given by}$$

$f(x) = 0 \, \forall \, x \in [1, 2]$. We now give the formal proof that $f(x) = 0 \, \forall \, x \in [1, 2]$ is the pointwise limit.

As usual let $\varepsilon > 0$ be any number and $x \in [1, 2]$ then, $|f_n(x) - f(x)| = \left|\dfrac{\sin nx}{nx^2} - 0\right| < \dfrac{1}{nx^2} < \varepsilon$ if, $n > \dfrac{1}{\varepsilon x^2}$. So

in this case if we choose $n_1 = \left[\dfrac{1}{\varepsilon x^2}\right] + 1$ we get $|f_n(x) - f(x)| < \varepsilon \, \forall \, n \geq n_1$. This proves that, $f(x) = 0 \, \forall \, x$

$\in [1, 2]$ is the pointwise limit of the above sequence of functions. Again we observe that n_1 is a function of both ε & x.

Now suppose f is the pointwise limit of a sequence of functions $\{f_n\}$ defined on the interval I, then a natural question arises, whether the positive integer n_1 we obtain in the definition of pointwise limit of a sequence of function can be made independent of fixed x? In other words, we want to investigate if it is possible to obtain a positive integer n_1 such that $|f_n(x) - f(x)| < \varepsilon$ whenever $n > n_1$ and for all x in the domain of definition of the sequence of function. Which means that this n_1 so obtained works uniformly for all x. Not always such positive integer exits, but when it does exists we say that the sequence $\{f_n\}$ converges uniformly to f. We now give the formal definition of uniform convergence.

Definition A sequence of function $\{f_n\}$ defined on an interval I is said to converge uniformly to a function f there, if given any $\varepsilon > 0$, there exists a positive integer n_1 depending on ε only, such that $|f_n(x) - f(x)| < \varepsilon$ whenever $n > n_1$ for all $x \in I$.

We try to see the geometrical significance of uniform convergence, suppose $\{f_n\}$ defined on an interval $I = [a, b]$ is uniformly convergent to a function f on I then for any $\varepsilon > 0$ we can find a positive integer n_1 such that graph of any function f_n where $n > n_1$ will be contained within the graph of the functions $f(x) + \varepsilon$ & $f(x) - \varepsilon$ (see Fig. 7.1)

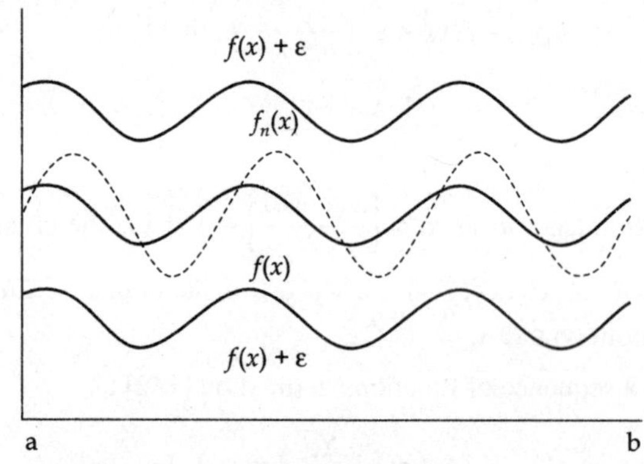

Fig. 7.1

If $\{f_n\}$ converges uniformly to f we will denote it by $f_n \Rightarrow f$ and f will be called the uniform limit of the sequence. It is evident that if $f_n \Rightarrow f$ on an interval I then $\{f_n\}$ converges pointwise to f on I. Since if $f_n \Rightarrow f$ then for any $\varepsilon > 0$ there exists a positive integer n_1 depending on ε such that $|f_n(x) - f(x)| < \varepsilon$ whenever $n > n_1$ & for all x in I and hence for any particular x in I that is, f is point wise limit of $\{f_n\}$ or in other words $\{f_n\}$ converges pointwise to f.

We will now show by examples that uniform convergence is a stronger than just pointwise convergence that is, pointwise convergence does not imply uniform convergence.

Example 7.5   Let $\{f_n\}$ be a sequence of functions defined on [0, 1] as follows,

$$f_n(x) = x^n$$

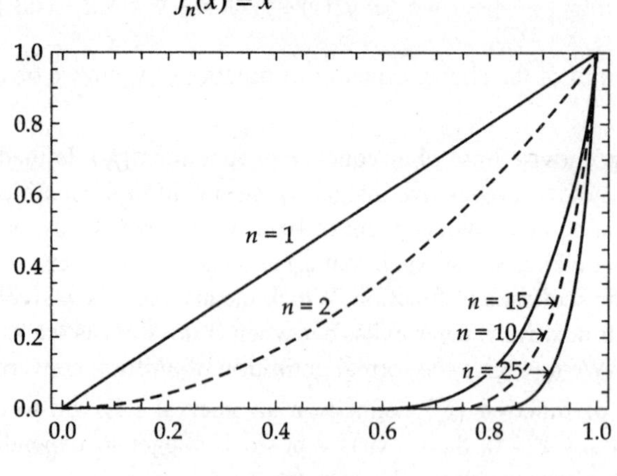

Fig. 7.2

We know

$$\lim_{n \to \infty} x^n = 0 \text{ if } 0 \le x < 1$$

$$= 1 \text{ if } x = 1$$

Hence $\{f_n\}$ converges pointwise to the limit f given by,

$$f(x) = 0 \text{ if } 0 \le x < 1$$

$$= 1 \text{ if } x = 1$$

Let $0 < x_1 < 1$, then,

$$|f_n(x_1) - f(x_1)| = x_1^n < \varepsilon$$

or, $$n \log x_1 < \log \varepsilon$$

or, $$-n \log x_1 > -\log \varepsilon$$

or, $$n > \frac{\log (1/\varepsilon)}{\log (1/x_1)} \tag{1}$$

Now if we choose $n_1 = \left[\dfrac{\log (1/\varepsilon)}{\log (1/x_1)} \right] + 1$ then for all $n > n_1$ we have $|f_n(x_1) - f(x_1)| < \varepsilon$ but clearly depends on the choice of x_1. As from right hand side of (1) it is clear that as $x_1 \to 1$ $n_1 \to \infty$

Hence it is not possible to choose a positive integer n_1 which works uniformly for all x in $[0, 1]$. In other words $\{f_n\}$ converge pointwise to f but not uniformly.

Example 7.6   We again consider Example 3 where we have,

$$f_n(x) = \frac{nx}{n + x}, x \in [0, \infty) \ \& \ n = 1, 2, 3...$$

We have already seen that $\{f_n\}$ converges pointwise to $f(x) = x$, $x \in [0, \infty)$. Now from equation (1) of Example 7.3 it is clear that for fixed $\varepsilon > 0$, as $x \to \infty$, $n \to \infty$ hence there can not exist any positive integer n_1 for which $|f_n(x) - f(x)| < \varepsilon$ holds uniformly for all $x \in [0, \infty)$ hence $\{f_n\}$ does not converge uniformly.

We summarize observations made in this paragraph in form of the following theorem,

Theorem 7.3.1

Suppose $\{f_n\}$ is a sequence of functions converging to a function uniformly on I, then $\{f_n\}$ converges pointwise to f. Converse of the above statement is not true.

We next note another point, same sequence of functions may be uniformly convergent in one domain where, as it may not so in other domain. In other words, for uniform convergence domain of definition of functions plays a very important and crucial role. We illustrate this by following examples:

Example 7.7   Let $\{f_n\}$ be a sequence of functions defined on $[0, 1]$ by,

$$f_n(x) = \frac{nx}{n + x}, \quad n = 1, 2, 3...$$

Let $f(x) = x$, $x \in [0, 1]$ then we will show that $f_n \Rightarrow f$. To show this we proceed as in Example 3 and we get from equation (3.1),

$$n > \frac{x(x + \varepsilon)}{\varepsilon}.$$

It is clear that, $\dfrac{x(x + \varepsilon)}{\varepsilon}$ is an increasing function of x and the maximum value of is attained at $x = 1$.

Hence if we choose $n_1 = \left[\dfrac{1(1 + \varepsilon)}{\varepsilon} \right] + 1$ then for $n > n_1$ $|f_n(x) - f(x)| < \varepsilon$ holds uniformly for all $f_n \Rightarrow fx \in$ $[0, 1]$. So $f_n \Rightarrow f$.

We note that same sequence of functions is not uniformly convergent in $[0, \infty]$ (by Example 7.6)

Example 7.8 Let $\{f_n\}$ be a sequence of functions defined on $[0, a]$, $a < 1$ as follows,

$$f_n(x) = x^n, \quad n = 1, 2, 3\ldots$$

It is easy to see that pointwise limit of this sequence is given by,

$$f(x) = 0, x \in [0, a]\ 0 < a < 1.$$

We now show that $f_n \Rightarrow f$. For any real number $\varepsilon > 0$ and any fixed $x \in [0, a]$,

Now, $|f_n(x) - f(x)| < \varepsilon$, if $|x^n| < \varepsilon$, that is if,

$$n > \frac{\log (1/\varepsilon)}{\log (1/x)} \tag{1}$$

We observe that $\dfrac{\log (1/\varepsilon)}{\log (1/x)}$ is an increasing function of x, hence if we choose $n_0 = \left[\dfrac{\log (1/\varepsilon)}{\log (1/a)}\right] + 1$ then

from (8.1) $|f_n(x) - f(x)| < \varepsilon$ for all $n > n_0$ and for all $x \in [0, a]$.

Which shows that $f_n \Rightarrow f$.

Note that in the previous example we have shown that same sequence is not uniformly convergent in $[0, \infty]$.

We now give a theorem which is a necessary and sufficient condition for a sequence to be uniformly convergent. Readers are advised to master this theorem as this is a useful tool to test whether a sequence of function is uniformly convergent or not and will be used quite frequently.

Theorem 7.3.3 (M_n-Test)

Suppose $\{f_n\}$ be a sequence of functions defined on an interval I converging point wise to a functionand if M_n be defined as,

$$M_n = \sup_{x \in I} |f_n(x) - f(x)|$$

Then $f_n \Rightarrow f$ if and only if $M_n \to 0$ as $n \to \infty$.

Proof: First suppose that $M_n \to 0$ as $n \to \infty$. Then for any given $\varepsilon > 0$ there exists a positive integer n_1 depending on ε only such that $M_n < \varepsilon$ whenever $n > n_1$ that is $\sup\limits_{x \in I} |f_n(x) - f(x)| < \varepsilon$ whenever $n > n_1$. This

means that, $|f_n(x) - f(x)| < \varepsilon$ for $n > n_1$ and for $x \in I$ or in other words $f_n \Rightarrow f$.

Conversely, suppose $f_n \Rightarrow f$ and let $\varepsilon > 0$, then by definition of uniform convergence there exists a positive integer n_1 depending on ε only such that $|f_n(x) - f(x)| < \varepsilon$ for $n > n_1$ and for $x \in I$ that is $\sup\limits_{x \in I}$ $|f_n(x) - f(x)| < \varepsilon$ for $n > n_1$. Hence $M_n \to 0$ as $n \to \infty$.

Example 7.9 Show that the sequence $\{f_n\}$ defined by,

$$f_n(x) = \frac{1}{n} x \quad x \in [0, 1]$$

is uniformly convergent.

Sol: Here we first try to find the pointwise limit of the above sequence. For this we choose any $x \in [0, 1]$ fix it and then find $\lim\limits_{n \to \infty} f_n(x)$. But $\lim\limits_{n \to \infty} \dfrac{1}{n} x = 0$ for any $x \in [0, 1]$. Hence pointwise limit in this case is given by,

$$f(x) = 0 \ \forall\ x \in [0, 1].$$

Now let us consider,

$$M_n = \sup_{x \in [0,\, 1]} |f_n(x) - f(x)|$$

$$= \sup_{x \in [0,\, 1]} \left| \frac{1}{n} x \right| = \frac{1}{n}$$

$$= \sup_{x \in [0,\, 1]} \left(\frac{1}{n} x \right)$$

$$= \max_{x \in [0,\, 1]} \left(\frac{1}{n} x \right) \qquad \left(\text{as } \frac{1}{n} x \text{ is a continuous function of } x \right)$$

$$= \frac{1}{n} \to 0 \text{ as } n \to \infty$$

Hence by previous theorem we get $\{f_n\}$ converges uniformly to f in the interval $[0, 1]$.

Example 7.10 The sequence of functions $\{f_n\}$ defined by,

$$f_n(x) = (nx)e^{-nx^2} \quad x \geq 0$$

is not uniformly convergent in $[0, k]$ for any $k > 0$.

Sol: As before first we find the pointwise limit of the sequence, it is easy to see (Using L'Hospital's Rule) that,

$$\lim_{x \to \infty} (nx)e^{-nx^2} = 0 \; x \geq 0.$$

That is pointwise limit of this sequence is given by,

$$f(x) = 0 \; x \geq 0$$

Now,

$$M_n = \sup_{x \in [0,\, k]} |f_n(x) - f(x)|$$

That is,

$$M_n = \max_{x \in [0,\, k]} (nx)e^{-nx^2} \quad (\text{as } f_n\text{'s are continuous})$$

We now find the maximum value of f_n for fixed value of n. For that let $y = (nx)e^{-nx^2}$, then

$$\frac{dy}{dx} = ne^{-nx^2} - n^2 2x^2 e^{-nx^2} = ne^{-nx^2}(1 - 2nx^2)$$

For maximum we put $\dfrac{dy}{dx} = 0$ so that we get $x = \pm \dfrac{1}{\sqrt{2n}}$. Now we see that.

$$\frac{d^2y}{dx^2} = -2n^2 xe^{-nx^2}(1 - 2nx^2) - 4n^2 xe^{-nx^2}$$

$$= -6n^2 xe^{-nx^2} + 4n^3 x^3 e^{-nx^2}$$

At $x = \dfrac{1}{\sqrt{2n}}$,

$$\frac{d^2y}{dx^2} = 4n^3 \left(\frac{1}{\sqrt{2n}}\right) e^{-nx^2}3 - 6n^2 \left(\frac{1}{\sqrt{2n}}\right) e^{-nx^2}$$

$$e^{-nx^2} \frac{1}{\sqrt{2n}} (2n^2 - 6n^2) = -\frac{1}{\sqrt{2n}} 4n^2 e^{-nx^2} < 0$$

Similar calculation will show that at $x = -\dfrac{1}{\sqrt{2n}}, \dfrac{d^2y}{dx^2} > 0$. Hence we conclude that y attains maximum

at $x = \dfrac{1}{\sqrt{2n}}$ and the value of y at that point is given by $y = n \dfrac{1}{\sqrt{2n}} e^{-1/2} = \sqrt{\dfrac{n}{2e}}$.

Hence, $M_n = \sqrt{\dfrac{n}{2e}} \to \infty$ as $n \to \infty$. By above theorem we conclude that the sequence $\{f_n\}$ does not converge uniformly in the interval $[0, k]$

Example 7.11   Show that the sequence of function $\{f_n\}$ defined by,

$$f_n(x) = \frac{nx}{(1 + n^2x^2)}$$

does not converge uniformly in any interval containing 0 as an interior point.

Sol: Let I be any interval containing 0 as an interior point. It is easy to see that pointwise limit of this sequence is given by $f(x) = \lim\limits_{n \to \infty} f_n(x) = 0 \ x \in I$. As before,

$$M_n = \sup_{x \in I} |f_n(x) - f(x)|$$

$$= \sup_{x \in I} \left|\frac{nx}{(1 + n^2x^2)}\right| = \max_{x \in I} \left|\frac{nx}{(1 + n^2x^2)}\right| \quad \text{(As each } f_n\text{'s are continuous)}$$

Let $y = \dfrac{nx}{(1 + n^2x^2)}$ then $\dfrac{dy}{dx} = \dfrac{n(1 - n^2x^2)}{(1 + n^2x^2)^2}$. Hence $\dfrac{dy}{dx} = 0$ at $x = \pm \dfrac{1}{n}$. Then it is easy to show that $\dfrac{d^2y}{dx^2} < 0$

when $x = \dfrac{1}{n}$ and $\dfrac{d^2y}{dx^2} > 0$ when $x = -\dfrac{1}{n}$ so y attains maximum at $x = \dfrac{1}{n}$ and the maximum value of y is

$\dfrac{1}{2}$. So we get $M_n = \dfrac{1}{2}$ which does not tend to zero as $n \to \infty$. So we conclude that $\{f_n\}$ does not converge uniformly on I.

In the above argument where did we use the fact that I contains zero as interior point? See that the

maximum of y is attained at $x = \dfrac{1}{n}$ so that supremum of $\left|\dfrac{nx}{(1 + n^2x^2)}\right|$ is also attained at $x = \dfrac{1}{n}$ which $\to 0$

as $n \to \infty$. Now suppose $0 \notin I$ then $\dfrac{1}{n}$ will lie outside I for all but finitely many n hence $M_n \leq \dfrac{1}{2}$ for all but

finitely many n. Hence we cannot say for sure whether M_n tends to 0 or not hence we cannot conclude whether $\{f_n\}$ converges uniformly or not I this case.

Example 7. 12   Show that the sequence of function $\{f_n\}$ defined by

$$f_n(x) = \frac{x}{1 + nx^2}, \; x \in \mathbb{R},$$

converges uniformly on any closed interval I.

Sol: Again as before in the previous examples it is easy to see that the pointwise limit of this sequence is given by $f(x) = \lim_{n \to \infty} f_n(x) = 0$. Now here,

$$M_n = \sup_{x \in I} |f_n(x) - f(x)|$$

$$= \sup_{x \in I} \left| \frac{x}{1 + nx^2} \right|$$

$$= \max_{x \in I} \left| \frac{x}{1 + nx^2} \right| \qquad \text{(As each } f_n\text{'s are continuous and } I \text{ is closed)}$$

Again if we put $y = \dfrac{x}{1 + nx^2}$ then by use of differentiation we can show that maximum of y occurs at $x = \dfrac{1}{\sqrt{n}}$ and the maximum value of y is given by $\dfrac{1}{2\sqrt{n}}$ (Note that maximum of f and $|f|$ are attained at the same point). Hence,

$$M_n = \frac{1}{2\sqrt{n}} \to 0 \text{ as } n \to \infty.$$ Hence we conclude by previous theorem that $\{f_n\}$ converges uniformly on any closed interval I.

In the next section we formulate another important necessary and sufficient condition for uniform convergence of a sequence of functions

7.4   THE CAUCHY CRITERION FOR UNIFORM CONVERGENCE OF A SEQUENCE OF FUNCTIONS

Theorem 7.4.1 (Cauchy Criterion)

A sequence of functions $\{f_n\}$ defined on an interval I is uniformly convergent there if and only if given $\varepsilon > 0$ there exists a positive integer n_0 depending on ε only such that $|f_n(x) - f_m(x)| < \varepsilon$ for all $m, n > n_0$ and for all $x \in I$.

Proof: First we prove the necessity part. Suppose $f_n \Rightarrow f$ and let $\varepsilon > 0$ be any given number. Then by definition of uniform convergence there exists a positive integer n_0 depending on ε only such that whenever $n, m > n_0$ we have $|f_n(x) - f(x)| < \dfrac{\varepsilon}{2}$ and $|f_m(x) - f(x)| < \dfrac{\varepsilon}{2}$ for all $x \in I$. Now we see that,

$$|f_m(x) - f_n(x)| = |f_m(x) - f(x)| + |f_n(x) - f(x)| < \frac{\varepsilon}{2} + \frac{\varepsilon}{2} = \varepsilon.$$

This proves that the condition is necessary.

To prove the sufficiency part let us assume that the condition of the theorem holds. Hence given any $\varepsilon > 0$ there must exist a positive integer n_0 depending on ε only such that for $n, m > n_0$ we must have

$|f_n(x) - f_m(x)| < \dfrac{\varepsilon}{2}$ for all $x \in I$. Now for a fixed $x \in I$ and using above condition we see that the sequence

of real numbers $\{f_n(x)\}$ is a cauchy sequence and hence converges. Since this is true for every $x \in I$, we can define the function $f(x)$ as $f(x) = \lim\limits_{n \to \infty} f_n(x)$. Hence f is the pointwise limit of the sequence $\{f_n\}$. We

now show that $f_n \to f$ on I. From the definition of pointwise convergence, for given $\varepsilon > 0$ and for any $x \in I$ we get a positive integer $n(x) > n_0$ such that $|f_m(x) - f(x)| < \dfrac{\varepsilon}{2}$ whenever $m > n_0 > n(x)$. We choose such an.

Now we see that for $n > n_0$,

$$|f_n(x) - f(x)| < |f_n(x) - f_m(x)| + |f_m(x) - f(x)| < \frac{\varepsilon}{2} + \frac{\varepsilon}{2} = \varepsilon.$$

This shows $f_n \Rightarrow f$ and hence the condition is sufficient.

7.5 SERIES OF FUNCTIONS AND UNIFORM CONVERGENCE

In Chapter 1 as we have defined convergence of an infinite series of real numbers via the notion of sequence of partial sum in this section we will take same route to define uniform convergence of an infinite series of functions with the help of concept of uniform convergence of sequence of functions.

To begin with, let $\{f_n\}$ is a sequence of functions defined on an interval I. Then the symbol $\sum\limits_{n=1}^{\infty} f_n(x)$ will

denote an infinite series of functions defined on I with n^{th} term as $f_n(x)$ and the sequence $\{s_n(x)\}$ defined by $s_n(x) = \sum\limits_{i=1}^{n} f_n(x)$ will be called the sequence of partial sum function of the above series.

Definition   The series $\sum\limits_{n=1}^{\infty} f_n(x)$ is said to converge uniformly to the sum function $f(x)$ if the sequence of

partial sum functions $\{s_n(x)\}$ converges uniformly to the function $f(x)$.

Using above definition and Theorem 7.4.1 one can easily obtain what we call Cauchy Criterion for uniform convergence of an infinite series.

Theorem 7.5.1 (Cauchy Criterion)

A necessary and sufficient condition for a series of function $\sum\limits_{n=1}^{\infty} f_n(x)$ *defined on an interval I is, for any given $\varepsilon > 0$ there must exist a positive integer n_0 depending on ε only such that whenever $n > m > n_0$ we have,*

$$|f_{m+1}(x) + f_{m+2}(x) + \ldots + f_n(x)| < \varepsilon \; \forall \; x \in I.$$

Proof: We note that by definition $\sum\limits_{n=1}^{\infty} f_n(x)$ is uniformly convergent if the sequence of partial sum

functions $\{s_n(x)\}$ converges uniformly. Now apply Theorem 7.4.1 to the sequence $\{s_n(x)\}$ to obtain above theorem.

As an immediate corollary of the above theorem we get the following:

Corollary 7.5.2

If the infinite series of functions $\sum_{n=1}^{\infty} f_n(x)$ defined on I is uniformly convergent then the sequence $\{f_n(x)\}$ converges uniformly to 0 ($f_n \Rightarrow 0$) on I.

Proof: Suppose $\sum_{n=1}^{\infty} f_n(x)$ converges to f uniformly on I and let $\{s_n(x)\}$ denote the sequence of partial sums of given series then $s_n(x) \Rightarrow f$ on I. Hence by Theorem 7.5.1, given any $\varepsilon > 0$ there exists a positive integer n_0 depending on ε only such that whenever $n > m > n_0$ we have,

$$|f_{m+1}(x) + f_{m+2}(x) + \dots + f_n(x)| < \varepsilon \ \forall \ x \in I \text{ that is,}$$

$$|s_n(x) - s_m(x)| < \varepsilon \ \forall \ x \in I \tag{7.5.2.1}$$

Now we take $N = n_0 + 1$ then for $n > N$, we have $n > n - 1 > n_0$, hence form (7.5.2.1) we get,

$$|f_n(x)| = |s_n(x) - s_{n-1}(x)| < \varepsilon \ \forall \ x \in I.$$

Hence $f_n \Rightarrow 0$ on I.

Corollary 7.5.3

If the infinite series of functions $\sum_{n=1}^{\infty} f_n(x)$ defined on I is uniformly convergent then,

$$\lim_{n \to \infty} \left(\sup_{x \in I} |f_n(x)| \right) = 0$$

Proof: By Corollary 7.5.2 $\{f_n(x)\}$ converges uniformly to 0 Now using necessity part of M_n, – Test we get Corollary 7.5.3.

Although Theorem 7.5.1 gives a necessary and sufficient condition of uniform convergence of a series but it is difficult to apply in practical situations, whereas the following theorem which provides a sufficient condition will be very useful.

Theorem 7.5.3 (Weierstrass' M- Test)

Let $\sum_{n=1}^{\infty} f_n(x)$ be a series of functions defined on an interval I and if $\{M_n\}$ be a sequence of positive constants satisfying,

(i) $|f_n(x)| \le M_n \ \forall \ x \in I$ and

(ii) $\sum_{n=1}^{n} M_n$ *is convergent then* $\sum_{n=1}^{\infty} f_n(x)$ *converges uniformly and absolutely on I.*

Proof: As $\sum_{n=1}^{\infty} M_n$ converges, for $\varepsilon > 0$ there exists a positive integer n_0 such that whenever $n > m > n_0$ we have,

$$M_{m+1} + M_{m+2} + \dots + M_n < \varepsilon \quad \text{(by Theorem 1.8.1)} \tag{7.5.3.1}$$

Note that n_0 depends on ε only. Now if $\{s_n(x)\}$ denotes the sequence of partial sum functions of the given series, then for $n > m > n_0$ we have,

$$|s_n(x) - s_m(x)| = |f_{m+1}(x) + f_{m+2}(x) + \ldots + f_n(x)|$$

$$< |f_{m+1}(x) + |f_{m+2}(x)| + \ldots + |f_n(x)|$$

$$< M_{m+1} + M_{m+2} + \ldots + M_n < \varepsilon \; \forall \, x \in I \qquad \text{(by (7.5.3.1))}$$

This shows that $\{s_n(x)\}$ converge uniformly on I and hence same conclusion holds for the series $\sum\limits_{n=1}^{\infty} f_n(x)$.

It is also clear from (i) and (ii) that $\sum\limits_{n=1}^{\infty} |f_n(x)| < \infty$ or each $x \in I$ which proves that the series $\sum\limits_{n=1}^{\infty} f_n(x)$ is absolutely convergent on I.

Example 7.13 Show that the series $\sum\limits_{n=1}^{\infty} \dfrac{x}{1 + n^2 x^2}$ does not converge uniformly in $[0, 1]$ whereas it converges uniformly on $[a, 1]\, a > 0$.

Sol: Let $f_n(x) = \dfrac{x}{1 + n^2 x^2}$ then it is clear $f_n(x) \to 0$ pointwise. Now for $\varepsilon > 0$,

$\dfrac{x}{1 + n^2 x^2} < \varepsilon$ if $n^2 > \dfrac{1}{\varepsilon x} - \dfrac{1}{x^2}$. We see that $\left(\dfrac{1}{\varepsilon x} - \dfrac{1}{x^2} \right)$ increases as x decreases to 0 hence there cannot

exist any positive integer n_0 independent of x such that $\dfrac{x}{1 + n^2 x^2} < \varepsilon$ for $n \geq n_0$. In other words $f_n(x)$ does

not converge uniformly in $[0, 1]$ by Corollary 7.5.2.

Again
$$|f_n(x)| \leq \frac{1}{1 + n^2 a^2} \quad \text{for, } a \leq x \leq 1$$

$$\leq \frac{1}{a^2 n^2} = M_n$$

As $\sum M_n < \infty$, $\sum\limits_{n=1}^{\infty} \dfrac{x}{1 + n^2 x^2}$ converges uniformly in $[a, 1]\, a > 0$

Example 7.14 Show that $\sum\limits_{n=1}^{\infty} \sin \dfrac{nx}{n^p}$ converges uniformly in $(-\infty, \infty)$ for $p > 1$.

Sol: Observe that $\left| \dfrac{\sin nx}{n^p} \right| \leq \dfrac{1}{n^p} = M_n \; \forall \, x \in (-\infty, \infty)$ ($|\sin nx| \leq 1$)

As $\sum\limits_{n=1}^{\infty} M_n < \infty$ (being a p series with $n > 1$) we conclude using above theorem that $\sum\limits_{n=1}^{\infty} \dfrac{\sin nx}{n^p}$ converges

uniformly in $(-\infty, \infty)$ for $p > 1$.

Example 7.15   Show that the series $\dfrac{2x}{1+x^2} + \dfrac{4x^3}{1+x^4} + \dfrac{8x^7}{1+x^8} + \ldots$

converges uniformly in $\left[-\dfrac{1}{2}, \dfrac{1}{2}\right]$.

Sol: We note here the nth term of the above series is given by,

$$f_n(x) = \frac{2^n x^{2^n-1}}{1+x^{2^n}}.$$

Now we see that,

$$\frac{2^n x^{2^n-1}}{1+x^{2^n}} \le |2^n x^{2^n-1}| \le 2^n \lambda^{2^n-1} \qquad \text{where } |x| \le \lambda < \frac{1}{2}$$

We put $M_n = 2^n \lambda^{2^n-1}$ then by using Cauchy's Root test we can show $\displaystyle\sum_{n=1}^{\infty} M_n < \infty$. Hence the series

converges $\forall\, x \in \left[-\dfrac{1}{2}, \dfrac{1}{2}\right]$.

Example 7.16   Show that the series $\displaystyle\sum_{n=1}^{\infty} \frac{x}{n(1+nx^2)}$ is uniformly convergent in any finite interval closed

interval I.

Sol: We here try to estimate the nth term of the above series so let $y = \dfrac{x}{n(1+nx^2)}$ then, $y = \dfrac{1-nx^2}{n(1+nx^2)^2}$,

for maximum of y we put $y = 0$ and we obtain $x = \pm\dfrac{1}{\sqrt{n}}$. Now as

$$y'' = \frac{-2nx(1+nx^2)^2 - (1-nx^2)\dfrac{d}{dx}\{(1-nx^2)^2\}}{n(1+nx^2)^4}$$

$$y'' < 0 \qquad \text{if } x = \frac{1}{\sqrt{n}} \text{ and } y'' > 0 \text{ if } x = -\frac{1}{\sqrt{n}}$$

Hence we conclude y attains maximum at $x = \dfrac{1}{\sqrt{n}}$ where the value of $y = \dfrac{\dfrac{1}{\sqrt{n}}}{n\left(1+n\cdot\dfrac{1}{n}\right)} = \dfrac{1}{2n^{3/2}}$. So we

get $\left|\dfrac{x}{n(1+nx^2)}\right| \le \dfrac{1}{2n^{3/2}} = M_n$ (say) $\forall\, x \in I$. Again we see that $\displaystyle\sum_{n=1}^{\infty} M_n = \sum_{n=1}^{\infty} \dfrac{1}{2n^{3/2}} < \infty$ being a p-series with

$p > 1$. So by above theorem the series $\displaystyle\sum_{n=1}^{\infty} \frac{x}{n(1+nx^2)}$ is uniformly convergent in any finite interval closed

interval I.

Example 7.17   Show that the series $\sum\limits_{n=1}^{\infty} \dfrac{x}{n^p + x^2 n^q}$ converges uniformly on any closed interval I when

   (i) $q > 0$ & $p > 1$ and

   (ii) $0 < p \le 1$ & $p + q > 2$

Sol: Let $I = [a, b]$ and suppose $q > 0$ & $p > 1$ then if $f_n(x)$ denotes the nth term of the above series we have,

$$|f_n(x)| = \frac{x}{n^p + x^2 n^q} \le \frac{|x|}{n^p} \le \frac{\alpha}{n^p} = M_n \ \forall \ x \in I \quad \text{where } \alpha = \max \{|a|, |b|\}.$$

Now $\sum M_n = \sum\limits_{n=1}^{\infty} \dfrac{\alpha}{n^p} < \infty$ as $p > 1$.

Hence in this case above series converges uniformly on I.

Next suppose $0 < p \le 1$ & $p + q > 2$ and let,

$$y = \frac{x}{n^p + x^2 n^q} \text{ then } y = \frac{(n^p + x^2 n^q) - x(2xn^q)}{(n^p + x^2 n^q)^2} = \frac{n^p - x^2 n^q}{(n^p + x^2 n^q)^2}.$$

For maximum value of y we put $y = 0$ to obtain $x = \sqrt{\dfrac{n^p}{n^q}}$.

By second derivative test we can show at this point y attains maximum. The maximum value of y is

given by $y_{\max} = \dfrac{1}{2n^{1/2\,(p+q)}}$. Hence $|f_n(x)| = \left| \dfrac{x}{n^p + x^2 n^q} \right| \le \dfrac{1}{2n^{1/2\,(p+q)}} = M_n \ \forall \ x \in I$. Now as $\sum M_n$ converges

if $p + q > 2$ so above series converges uniformly on I in this case.

Example 7.18   Show that the series $\sum\limits_{n=1}^{\infty} (xe^{-x})^n$ is uniformly convergent on the interval $[0, 1]$.

Sol: Let $f_n(x) = (xe^{-x})^n$ then we observe that $f_n(x)$ attains maximum when the function $y = xe^{-x}$ attains maximum. Now again using derivatives we can show that y attains maximum at $x = 1$ hence we get $|f_n(x)| \le e^n = M_n \ \forall x \in I$. Obviously $\sum M_n$ converges as it is a geometric series with common ratio $e^{-1} < 1$ hence above series converges.

Next we will now study whether under uniform convergence the limit function retains same properties of sequence or series. Precisely we want to investigate the following type of questions: Suppose $\{f_n(x)\}$ be a sequence of continuous (integrable/differentiable) functions converging pointwise to the limit function $f(x)$ then does it always imply that $f(x)$ is continuous (integrable/differentiable)? What happens if we assume that the convergence is uniform? Similar corresponding questions can be asked about a series of functions $\sum\limits_{n=1}^{\infty} f_n(x)$ and its sum function $f(x)$.

7.6   UNIFORM CONVERGENCE AND CONTINUITY

First consider the following example,

   Let $\{f_n(x)\}$ be a sequence of functions defined as,

$$f_n(x) = x^n \ x \in [0, 1]$$

We observe that each $f_n(x)$ is a continuous function. Also we have already seen that this sequence converge pointwise to the function $f(x)$ defined as,

$$f(x) = \begin{cases} 0 & \text{if } 0 \le x < 1 \\ 1 & \text{if } x = 1 \end{cases}$$

Clearly $f(x)$ is discontinuous. Hence pointwise limit of a sequence of continuous functions need not be continuous. We have also seen that above sequence is not a uniformly convergent sequence.

This phenomenon naturally raises the question what happens if we assume a more stronger mode of convergence that is uniform convergence? We establish the following theorem:

Theorem 7.6.1

Suppose $\{f_n(x)\}$ be a sequence of continuous functions defined on an interval I (that is each f_n continuous on I) converging uniformly to $f(x)$ then $f(x)$ is continuous on I.

Proof: Let $x \in I$ be any point and $\varepsilon > 0$ be any arbitrary number. As $f_n \Rightarrow f$ there must exist a positive integer n_0 depending on ε such that,

$$|f_n(x) - f(x)| < \frac{\varepsilon}{3} \text{ for all } n > n_0 \ \& \ \forall \ x \varepsilon I \tag{7.6.1.1}$$

We choose an $n_1 > n_0$ and as f_{n1} is a continuous function on I, in particular it is continuous at x'. Hence there must exist $a \ \delta > 0$ such that,

$$|f_{n1}(x) - f_{n1}(x')| < \frac{\varepsilon}{3} \quad \text{for all } x \text{ satisfying } |x - x'| < \delta \tag{7.6.1.2}$$

Now for any x satisfying $|x - x'| < \delta$ we get,

$$|f(x) - f(x')| = |f(x) - f_{n1}(x) + f_{n1}(x) - f_{n1}(x') + f_{n1}(x') - f(x')|$$

$$< |f(x) - f_{n1}(x)| + |f_{n1}(x) - f_{n1}(x') + fn_1(x) - f(x')|$$

$$< \frac{\varepsilon}{3} + \frac{\varepsilon}{3} + \frac{\varepsilon}{3} = \varepsilon.$$

This shows that the limit function $f(x)$ is continuous at x' and as x' was chosen arbitrary we conclude $f(x)$ is continuous on I.

The corresponding theorem for an infinite series of function easily follows from above theorem and we just give its statement below.

Theorem 7.6.2

Let $\displaystyle\sum_{n=1}^{\infty} f_n(x)$ be an infinite series of continuous functions defined on an interval I (that is each f_n continuous on I)

converging uniformly to the sum function $f(x)$. Then $f(x)$ is continuous on I.

Proof: This follows as an easy corollary to Theorem 7.6.1, we only observe that if we conceder the sequence of partial sum functions $\{s_n(x)\}$ then each $s_n(x)$ being finite sum of continuous functions is continuous. Now by definition $s_n \Rightarrow f$, hence by Theorem 7.6.1 $f(x)$ is continuous on I.

Remarks

1. The Theorem 7.6.1 says that in case of uniform convergence we will have $\displaystyle\lim_{n \to \infty} \left(\lim_{x \to x_0} f_n(x) \right)$

    $= \displaystyle\lim_{x \to x_0} \lim_{n \to \infty} f_n(x))$ for any $x_0 \in I$. That is we can interchange $\displaystyle\lim_{n \to \infty}$ & $\displaystyle\lim_{x \to x_0}$.

2. See that uniform convergence is a sufficient condition for the limit function (sum function) to be continuous in case each term of sequence (series) are continuous no way it is necessary. That is if the limit function (sum function) of a sequence (series) of continuous functions is continuous then the convergence may not be uniform. We show this by the following example:

Example 7.19   Consider the sequence of functions $\{f_n\}$ defined by,

$$f_n(x) = (nx)e^{-nx^2} \ x \geq 0$$

in $[0, k] \ k > 0$. Then we have seen before that $\{f_n\}$ converges pointwise to $f(x) = 0$ on $[0, k]$ which is a continuous function but the sequence does not converge uniformly in the above interval.

3. The contra-positive statement of above theorem is some time useful. That is if the limit function (sum function) of a sequence (series) of continuous functions is not continuous then the convergence cannot be uniform and one can make use of this statement as a technique to show some sequence (series) not convergent.

We have seen by Remark 1 a sequence of continuous functions can converge to a continuous point function even when the convergence is not uniform. But this raises a natural question: Does there exists some particular condition when continuous limit function of a sequence of continuous functions will ensure the convergence is uniform? The answer to this question is given in form of a theorem popularly known as Dini's Theorem.

Theorem 7.6.3 (Dini)

If $\{f_n\}$ is a sequence of continuous functions defined on the interval $[a, b]$ converges pointwise to a continuous function f and if $f_n(x) \geq f_{n+1}(x)$ for each x in $[a, b]$ and for $n = 1, 2, \dots$ then $f_n \Rightarrow f$.

Proof: Let us define $g_n(x)$ as,

$$g_n(x) = f_n(x) - f(x)$$

then $g_n(x) \geq g_{n+1}(x)$ and $g_n \to 0$. Let $x \in [a, b]$ and $\varepsilon > 0$ be any real number then there exists a positive integer $n(x')$ depending on x such that,

$$|g_n(x')| < \frac{\varepsilon}{2} \ \forall \ n \geq n(x') \tag{7.6.2.1}$$

Again as each g_n is continuous there exists a open neighbourhood of x
(i.e. an open interval containing x) say $U(x')$ such that,

$$|g_n(x) - g_n(x')| < \frac{\varepsilon}{2} \ \forall \ x \in U(x') \ \& \ \forall \ n \geq 1 \tag{7.6.2.2}$$

Hence combining (7.6.2.1) & (7.6.2.2) and monotone decreasing nature of the sequence we get,

$$|g_n(x)| < |g_n(x) - g_n(x')| + |g_n(x')| < \frac{\varepsilon}{2} + \frac{\varepsilon}{2} = \varepsilon \ \forall \ n \geq n(x') \ \& \ \forall \ x \in U(x') \tag{7.6.2.3}$$

Now we observe that $\{U(x') : x' \in [a, b]\}$ is an open cover of $[a, b]$ hence by Hine-Borel Theorem there exists finite subcover of $[a, b]$ say $U(x_1), U(x_2), U(x_3), \dots, U(x_n)$. We choose $n_0 = \max \{n(x_1), n(x_2), \dots, n(x_n)\}$. Then using (7.6.2.3) we see that,

$$|g_n(x)| < \varepsilon \ \ \forall \ x \in [a, b] \text{ and } \forall \ n \geq n_0.$$

That is $g_n \Rightarrow 0$ which implies that $f_n \Rightarrow f$.

Again the corresponding theorem for series of functions is a easy consequence of above theorem and we just give its statement below

Theorem 7.6.4

If $\sum\limits_{n=1}^{\infty} f_n(x)$ be a series of continuous functions defined on a closed interval $[a, b]$ converges point wise to a sum function which is continuous and if $f_n(x) \geq f_{n+1}(x)$ for each x in $[a, b]$ and for $n = 1, 2, \ldots$ then the above series converges uniformly to f.

Remarks

1. The above theorems remains true if we assume $f_n(x) \leq f_{n+1}(x)$.

2. Note that in the proof we have used the fact that $[a, b]$ is compact. So the theorem holds if we replace $[a, b]$ by a compact set E. We now show by an example that the condition of compactness can not be dropped.

Example 7.20 Consider the sequence of functions $\{f_n\}$ defined by,

$$f_n(x) = \frac{1}{(nx + 1)} \; x \in (0, 1)$$

Then each f_n is continuous and the sequence converges pointwise to $f \equiv 0$ on $(0, 1)$. So the pointwise limit is also a continuous function. For $\varepsilon > 0$ we have,

$$\frac{1}{(nx + 1)} < \varepsilon \text{ if } n > \frac{1}{x}\left(\frac{1}{\varepsilon} - 1\right).$$

Now we see that $\frac{1}{x}\left(\frac{1}{\varepsilon} - 1\right)$ increases to ∞ when $x \to 0^+$ hence it is not possible to find a positive integer n_0 such that $f_n(x) < \varepsilon$ for all $n > n_0$ and for all $x \in (0, 1)$. That is $\{f_n\}$ does not converge uniformly. This shows that even if sequence of continuous functions converge pointwise to a continuous function the convergence may not be uniform unless we assume the domain of the functions a compact set, observe here that $(0, 1)$ is not compact.

7.7 UNIFORM CONVERGENCE AND INTEGRATION

We now want to study two very natural questions:

1. If we have a sequence of functions $\{f_n\}$ defined on a closed interval $[a, b]$ each of which is R-Integrable converging point wise to a function f then does it imply f is R-Integrable?

2. If a sequence of R-Integrable functions defined on a closed interval $[a, b]$ converges pointwise to a R-Integrable function f then does it follow that $\lim\limits_{n \to \infty} \int\limits_{a}^{b} f_n(x) \, dx = \int\limits_{a}^{b} f(x) \, dx$?

We give two examples to show that answers to both questions above are in negative.

Example 7.21 Let $\{r_i\}_{r=1}^{\infty}$ be an enumeration of rational numbers in $[0, 1]$ also suppose $Q_n = \{r_1, r_2, \ldots, r_n\}$. We define a sequence of functions $\{f_n\}$ on $[0, 1]$ as follows,

$$f_n(x) = 1 \quad \text{if } x \in Q_n$$

$$= 0 \quad \text{if } x \in Q_n^c \cap [0, 1]$$

Then we see that since each f_n is continuous except for a finite number of points it is R-Integrable. It is clear that $f_n \to f$ pointwise where the function f is defined as,

$$f(x) = 1 \quad \text{if } x \text{ is a rational number in } [0, 1]$$

$$= 0 \quad \text{if } x \text{ is a irrational number in } [0, 1]$$

But we know that f is not R-Integrable. Hence pointwise limit of a sequence of R-Integrable functions may not be R-Integrable.

Example 7.22   Consider the sequence of functions $\{f_n\}$ on $[0, 1]$ as follows.

$$f_n(x) = \frac{n}{1 + n^2 x^2}, \; x \in [0, 1].$$

Then each f_n is R-Integrable and $f_n \to f$ pointwise where the function f is defined as,

$$f(x) = 0 \; \forall \, x \in [0, 1]$$

Hence f is also R-Integrable. Now,

$$\lim_{n \to \infty} \int_0^1 \frac{n}{1 + n^2 x^2} \, dx = \lim_{n \to \infty} [\tan^{-1}(n) - \tan^{-1}(0)] = \frac{\pi}{2}. \text{ But } \int_0^1 f(x) \, dx = 0. \text{ Hence in this case,}$$

$$\lim_{n \to \infty} \int_a^b f_n(x) \, dx \neq \int_a^b f(x) \, dx.$$

Uniform convergence turns out to be a sufficient condition which will ensure that limit function of a sequence of R-Integrable functions will be R-Integrable and also in that case we have limit of integrals is integral of the limit $\left(\text{i.e. } \lim_{n \to \infty} \int_a^b f_n(x) \, dx = \int_a^b f(x) \, dx \right)$. This is precisely the next theorem,

Theorem 7.7.1

Let $\{f_n\}$ be a sequence of R-Integrable functions defined on the interval $[a, b]$ converging uniformly to f then $f \in R[a, b]$. Moreover in that case we have,

$$\lim_{n \to \infty} \int_a^b f_n(x) \, dx = \int_a^b f(x) \, dx.$$

Proof: As $f_n \Rightarrow f$ for $\varepsilon > 0$ there exists a positive integer N_0 such that whenever $n > N_0$ we have,

$$|f_n(x) - f(x)| < \frac{\varepsilon}{3(b - a)} \; \forall \, x \in [a, b] \tag{7.7.1.1}$$

We choose such a n and keep it fixed. Now as each $f_n \in R[a, b]$ f_n is bounded so there exists $M > 0$ such that $|f_n(x)| < M$. Using (7.7.1.1) we get,

$$|f(x)| < |f_n(x)| + |f_n(x) - f(x)| < M + \frac{\varepsilon}{3(b - a)} \; \forall \, x \in [a, b]$$

Hence f is bounded. Using notations of Chapter-3 and using Theorem 3.7.1 we can find a partition P such that,

$$U(P, f_n) - L(P, f_n) < \frac{\varepsilon}{3} \tag{7.7.1.2}$$

Again using (7.7.1.1) we see that,

$$f_n(x) - \frac{\varepsilon}{3(b-a)} < f(x) < f_n(x) + \frac{\varepsilon}{3(b-a)} \tag{7.7.1.3}$$

For above partition P let us now define,

$$M_r = \sup_{x \in \delta_r} f(x) \qquad\qquad m_r = \inf_{x \in \delta_r} f(x)$$

$$M_r^* = \sup_{x \in \delta_r} f_n(x) \qquad\qquad m_r^* = \inf_{x \in \delta_r} f_n(x)$$

Hence for $x \in \delta_r$, using (7.7.1.3) we have,

$$m_r^* - \frac{\varepsilon}{3(b-a)} < m_r \leq M_r < M_r^* + \frac{\varepsilon}{3(b-a)} \tag{7.7.1.4}$$

Hence we get,

$$U(P, f) - L(P, f) = \sum_{r=1}^{n} (M_r - m_r)\Delta_r \leq \sum_{r=1}^{n} \left(M_r^* - m_r^* + \frac{2\varepsilon}{3(b-a)} \right) \Delta_r$$

$$= \sum_{r=1}^{n} (M_r^* - m_r^*)\Delta_r + \frac{2\varepsilon}{3}$$

So that,

$$U(P, f_n) - L(P, f_n) = \frac{2\varepsilon}{3} < \frac{\varepsilon}{3} + \frac{2\varepsilon}{3} = \varepsilon$$

Hence $f \in R[a, b]$.

Again we see that for $n > N_0$

$$\left| \int_a^b f_n(x)\, dx - \int_a^b f(x)\, dx \right| < \int_a^b |f_n(x) - f(x)|\, dx$$

$$< \int_a^b \frac{\varepsilon}{3(b-a)}\, dx \quad \text{(Using (7.7.1.1))}$$

$$= \frac{\varepsilon}{3} < \varepsilon.$$

In other words $\displaystyle\lim_{n \to \infty} \int_a^b f_n(x)\, dx = \int_a^b f(x)\, dx$.

The corresponding theorem for series is a result which tells us that in case a series of functions where each term is R-Integrable converges uniformly to a sum function then the sum function is R-Integrable and sum of integral is integral of the sum.

Theorem 7.7.2

Let $\sum\limits_{n=1}^{\infty} f_n(x)$ be a series of functions defined on $[a, b]$ converging to $f(x)$ uniformly. Suppose $f_n \in R[a, b]$ for all n then f is R-Integrable and moreover in that case we get,

$$\sum_{n=1}^{\infty} \int_a^b f_n(x)\, dx = \int_a^b f(x)\, dx \tag{7.7.2.1}$$

Proof: We just observe that here the sequence of partial sum functions $\{s_n(x)\}$ is a sequence of R-Integrable functions as,

$$s_n(x) = f_1(x) + f_2(x) + \ldots + f_n(x)$$

Is a finite sum of R-Integrable functions. Hence by using Theorem 7.7.1 on the sequence $\{s_n(x)\}$ we get $f \in R[a, b]$ and also we have,

$$\lim_{n \to \infty} \int_a^b s_n(x)\, dx = \int_a^b f(x)\, dx$$

But $\int_a^b s_n(x)\, dx = \int_a^b \sum_{r=1}^{n} f_i(x) = \sum_{r=1}^{n} \int_a^b f_i(x)\, dx$ (Using Theorem 3.9.3). Hence,

$$\lim_{n \to \infty} \sum_{r=1}^{n} \int_a^b f_i(x)\, dx = \int_a^b f(x)\, dx$$

That is,

$$\sum_{n=1}^{\infty} \int_a^b f_n(x)\, dx = \int_a^b f(x)\, dx$$

Remarks

1. If for a series of functions equation (7.7.2.1) is satisfied we say term by term integration is possible for the series. In particular by above theorem we can conclude that if a series of R-Integrable functions converge uniformly then term by term integration is possible for the series.

2. Theorem 7.7.1 says that in case of uniform convergence we can interchange $\lim\limits_{n \to \infty}$ & $\int_a^b$ for a sequence of R-Integrable functions and Theorem 7.7.2 tells us that for a series of R-Integrable functions under uniform convergence we can interchange $\sum$ & $\int$ signs.

3. We must realize that uniform convergence is just a sufficient condition in the above theorems and no way it is necessary. We give an example below to illustrate this.

Example 7.23   Let us consider the sequence $\{f_n\}$ defined as,

$$f_n(x) = x^n \; x \in [0, 1]$$

Then we know $f_n \to f$ pointwise and not uniformly where f is defined as,

$$f(x) = \begin{cases} 0 & \text{if } 0 \leq x < 1 \\ 1 & \text{if } x = 1. \end{cases}$$

Now $\int_0^1 f_n(x)\,dx = \dfrac{1}{n+1} \to 0$ as $n \to \infty$ and it clear that f is R-Integrable and $\int_0^1 f(x)\,dx = 0$. So in this case we have $\lim\limits_{n \to \infty} \int_a^b f(x)\,dx$ even when the sequence is not uniformly convergent.

4. We can use the contra-positive statement of above theorems to prove a sequence or a series to be non-uniformly convergent. Suppose $\{f_n\}$ be a sequence of R-Integrable functions defined on the interval $[a, b]$ converging pointwise to f which is R-Integrable and suppose that $\lim\limits_{n \to \infty} \int_a^b f_n(x)\,dx \neq \int_a^b f(x)\,dx$ then we will conclude that the convergence here cannot be uniform. Similarly if $\sum\limits_{n=1}^{\infty} f_n(x)$ be a series of R-Integrable functions defined on $[a, b]$ converging to $f(x)$ which is R-Integrable and if $\sum\limits_{r=1}^{\infty} \int_a^b f_i(x)\,dx \neq \int_a^b f(x)\,dx$ then we will conclude that the series does not converge uniformly on $[a, b]$

7.8 UNIFORM CONVERGENCE AND DIFFERENTIATION

Let us examine the following example:

Example 7.24 Let us consider the sequence $\{f_n\}$ defined as

$$f_n(x) = \frac{\sin nx}{\sqrt{n}} \quad x \in (-\infty, \infty)$$

We see that $f_n \to f = 0$. Also $|f_n(x)| = \dfrac{\sin nx}{\sqrt{n}} \leq \dfrac{1}{\sqrt{n}} \to 0$ hence f_n converges uniformly to 0. Further each f_n is differentiable and $f'_n(x) = \sqrt{n}\,\cos nx$ hence $\lim\limits_{n \to \infty} f'_n(0) = \lim\limits_{n \to \infty} \sqrt{n} = \infty$. Although $f'(0) = 0$. So in this case $\lim\limits_{n \to \infty} \dfrac{d}{dx} f_n(0) \neq \dfrac{d}{dx} \lim\limits_{n \to \infty} f_n(0)$.

From above example it is clear that even if a sequence of differentiable uniformly convergent to a differentiable function we may not be able to interchange $\lim\limits_{n \to \infty}$ & $\dfrac{d}{dx}$. So to ensure this interchange we must assume some thing more that just uniform convergence of the given sequence in fact here we have to assume uniform convergence of the derived sequence. Precisely we have the following theorem:

Theorem 7.8.1

Let $\{f_n\}$ be a sequence of functions defined on a closed interval $[a, b]$ such that,

   (i)   *Each f_n is differentiable on $[a, b]$*

   (ii) *There exists at least one point $p \in [a, b]$ such that $\{f_n(p)\}$ converges*

   (iii) *The derived sequence $\{f'_n\}$ is uniformly convergent to a function $g(x)$*

Then we have the following

(a) *The sequence $\{f_n\}$ converges uniformly to a function f (say) on $[a, b]$*

(b) $g(x) = \lim\limits_{n\to\infty} f'_n(x) = f'(x) \left(\text{i.e. } \lim\limits_{n\to\infty} f'_n(x) \right) = \dfrac{d}{dx}\left(\lim\limits_{n\to\infty} \dfrac{d}{dx} (f_n(x)) \right) \forall\, x \in [a, b]$

Proof: Let $\varepsilon > 0$ be any given number using conditions (2) & (3) we can get a positive integer n_0 such that whenever $m, n > n_0$ we have,

$$|f_n(p) - f_m(p)| < \frac{\varepsilon}{2} \tag{7.8.1.1}$$

$$|f'_n(x) - f'_m(x)| < \frac{\varepsilon}{2(b-a)} \quad \text{(Using Cauchy Criterion for uniform convergence)} \tag{7.8.1.2}$$

Now let $x, y \in [a, b]$ the function $f_n(x) - f_m(x)$ satisfies all conditions of Lagrange's Mean Value Theorem on the interval $[x, y]$ hence using it we get,

$$|(f_n(y) - f_m(y)) - (f_n(x) - f_m(x))| < |x - y|\, |f'n(t) - f'_m(t) - f'm(t)| \quad \text{for } t \in (x, y)$$

$$< |x - y| \frac{\varepsilon}{2(b-a)} \quad\quad \text{(by 7.8.1.2)) (7.8.1.3)}$$

$$< \frac{\varepsilon}{2} \tag{7.8.1.4}$$

Now for $m, n > n_0$ and for any $x \in [a, b]$ we have,

$$|f_n(x) - f_m(x)| = |f_n(x) - f_n(p) + f_n(p) - f_m(p) + f_m(p) - f_m(x)|$$

$$< |(f_n(x) - f_m(x)) - (f_n(p) - f_m(p)| + |f_n(p) - f_m(p)| \quad\quad \text{(by (7.8.1.1) \& (7.8.1.4))}$$

$$< \frac{\varepsilon}{2} + \frac{\varepsilon}{2} = \varepsilon.$$

This shows that $\{f_n\}$ converges uniformly. Let f denote the limit function.

Next let us choose any $x \in [a, b]$ and keep it fixed. Define the functions φ_n and φ as follows,

$$\varphi_n(y) = \frac{f_n(y) - f_n(x)}{y - x}\, y \in [a, b]\, y \neq x\, n \in \mathbb{N} \tag{7.8.1.5}$$

$$\varphi(y) = \frac{f(y) - f(x)}{y - x}\, y \in [a, b]\, y \neq x \tag{7.8.1.6}$$

$$\text{Now } \lim_{y\to x} \varphi_n(y) = \lim_{y\to x} \frac{f_n(y) - f_n(x)}{y - x} = f'_n(x) \quad \text{(as each } f_n \text{ is differentiable on } [a, b]) \tag{7.8.1.7}$$

Hence $\{\varphi_n\}$ is a sequence of continuous function. We will now show that the sequence $\{\varphi_n\}$ a sequence is uniformly convergent. For that if we choose $m, n > n_0$ we have,

$$\left|\varphi_n(y) - \varphi_m(y)\right| = \left|\frac{1}{y - x}\right| |(f_n(y) - f_n(x)) - (f_m(y) - f_m(x))|$$

$$< \frac{\varepsilon}{2(b-a)} < \varepsilon \tag{7.8.1.8}$$

Hence $\{\varphi_n\}$ is uniformly convergent. Also we observe that,

$$\lim_{n\to\infty} \varphi_n(y) = \lim_{n\to\infty} \frac{f_n(y)-f_n(x)}{y-x} = \frac{f(y)-f(x)}{y-x} = \varphi(y) \; y \neq x \qquad (7.8.1.9)$$

(As $\{f_n\}$ converges uniformly to f)

Now as $\{\varphi_n\}$ is uniformly convergent sequence of continuous functions we have the following

$$\lim_{y\to x} \left(\lim_{n\to\infty} \varphi_n(y) \right) = \lim_{n\to\infty} \left(\lim_{y\to x} \varphi_n(y) \right)$$

i.e. $\displaystyle\lim_{y\to x} \frac{f(y)-f(x)}{y-x} = \lim_{n\to\infty} f_n'(x) = g(x)$ (Using (7.8.1.7), (7.8.1.9) & condition(3)). (7.8.1.10)

By definition, $\displaystyle\lim_{y\to x} \frac{f(y)-f(x)}{y-x} = f'(x)$

Hence using (7.8.1.10) we conclude that $f'(x)$ exists and is given by the equation,

$$f'(x) = g(x) \qquad (7.8.1.11)$$

The corresponding theorem for series is given as follows:

Theorem 7.8.2

Let $\displaystyle\sum_{n=1}^{\infty} f_n(x)$ be a series of functions defined on a closed interval $[a, b]$ such that,

(i) *Each $f_n(x)$ is differentiable on $[a, b]$*

(ii) *There exists at least one point $p \in [a, b]$ such that the series $\displaystyle\sum_{n=1}^{\infty} f_n(p)$ converges*

(iii) *The derived series $\displaystyle\sum_{n=1}^{\infty} f_n'(x)$ is uniformly convergent to a function $g(x)$*

Then we have the following

(a) *The series $\displaystyle\sum_{n=1}^{\infty} f_n(x)$ converges uniformly to a function f (say) on $[a, b]$*

(b) $g(x) = \displaystyle\sum_{n=1}^{\infty} f'n(x) = f'(x) \left(\text{i.e. } \sum_{n=1}^{\infty} \frac{d}{dx}(f_n(x)) \; \frac{d}{dx}\left(\sum_{n=1}^{\infty} f_n(x)\right) \right) \forall \, x \in [a, b]$ (7.8.2.1)

Proof: We apply Theorem 7.8.1 to the sequence of partial sum function $\{s_n(x)\}$ of the above series to get (a) & (b).

Remark

If for a series of functions equation (7.8.2.1) is satisfied we say term by term differentiation is possible for the series. Above theorem gives a sufficient condition under which term by term differentiation is possible for a series.

Example 7.24   Let $\{f_n(x)\}$ be a sequence of function defined as,

$$f_n(x) = e^{-nx^2} \quad x \in [-1, 1]$$

Does $\{f_n(x)\}$ converge uniformly in the above domain?

Sol: First we observe the point wise limit f of the sequence is given by

$$f(x) = \begin{cases} 0 & \text{if } x \neq 0 \in [-1, 1] \\ 1 & \text{if } x = 0 \end{cases}$$

Now each $f_n(x) = e^{-nx^2}$ is a continuous function whereas f is clearly discontinuous hence by remark 3 following Theorem 7.6.2 we conclude $\{f_n(x)\}$ does not converge uniformly in $[-1, 1]$.

Example 7.25   Let $\{f_n(x)\}$ be a sequence of function defined as,

$$f_n(x) = \tan^{-1}(nx), \, x \in [0, \infty)$$

Show that $\{f_n(x)\}$ does not converge uniformly in the above domain.

Sol: The pointwise limit f of the sequence is given by,

$$f(x) = \begin{cases} \pi/2 & \text{if } x \neq 0 \in [0, \infty) \\ 0 & \text{if } x = 0 \end{cases}$$

Clearly f is discontinuous whereas each $f_n(x)$ is continuous. Hence by remark 3 following Theorem 7.6.2 we conclude $\{f_n(x)\}$ does not converge uniformly in the above domain.

Example 7.26   If $\{f_n(x)\}$ be a sequence of function defined as,

$$f_n(x) = \frac{1}{nx + 1} \, x \in [0, b]$$

then show that above sequence does not converge uniformly in the above domain.

Sol: The pointwise limit f of the sequence is given by,

$$f(x) = \begin{cases} 0 & \text{if } x \neq 0 \in [0, b] \\ 1 & \text{if } x = 0 \end{cases}$$

Clearly f is discontinuous whereas each $f_n(x)$ is continuous. Again as preceding example $\{f_n(x)\}$ does not converge uniformly in the above domain.

Example 7.27   Show that the series $\displaystyle\sum_{n=0}^{\infty} \frac{x}{(1 + x)^n} \, x \in [0, 1]$ is not uniformly convergent.

Sol: Let us find the sum function of the series. For that consider the sequence of partial sum function $\{s_n(x)\}$ where, $s_n(x) = x + \dfrac{x}{(1 + x)} + \dfrac{x}{(1 + x)^2} + \dots + \dfrac{x}{(1 + x)^{n-1}}$

Hence,

$$s_n(x) = \begin{cases} 0 \text{ if } x = 0 \\ = \dfrac{x[1 - (1/1 + x)^n]}{1 - 1/1 + x} = (1 + x)\,[1 - (1/1 + x)^n] & \text{if } x \neq 0 \end{cases}$$

Hence the sum function $f(x) = \lim_{n \to \infty} s_n(x)$ is given by,

$$f(x) = \begin{cases} 0 & \text{if } x = 0 \\ 1 + x & \text{if } x \neq 0 \end{cases}$$

Now observe that each term of above series is continuous and the sum function is discontinuous hence the convergence cannot uniform by Remark 3 following Theorem 7.6.2.

Example 7.28 Show that the series $\displaystyle\sum_{n=1}^{\infty} \frac{\sin x}{[n \sin x + 1][(n-1)\sin x + 1]}$ is not uniformly convergent in $\left[0, \dfrac{\pi}{2}\right]$.

Sol: Again as before we will try to find the sum function. For that we see that the nth term of the above series can be written as,

$$f_n(x) = \frac{\sin x}{[n \sin x + 1][(n-1)\sin x + 1]}$$

$$= \frac{1}{(n-1)\sin x + 1} - \frac{1}{n \sin x + 1}$$

So that n^{th} term of the partial sum sequence

$$s_n(x) = \sum_{i=1}^{n}\left[\frac{1}{(i-1)\sin x + 1} - \frac{1}{i \sin x + 1}\right]$$

$$= 1 - \frac{1}{n \sin x + 1}$$

Hence the sum function $f(x)$ is given as,

$$f(x) = \lim_{n\to\infty} s_n(x) = \begin{cases} 1 & \text{if } x \neq 0 \in [0, \pi/2] \\ 0 & \text{if } x = 0 \end{cases}$$

Since each $f_n(x)$ is continuous and $f(x)$ is discontinuous the series cannot converge uniformly.

Example 7.29 Show that the sequence $\{f_n(x)\}$ defined by

$$f_n(x) = (x - \pi/2)\sin^n x, \quad x \in [0, \pi/2]$$

Is uniformly convergent in $[0, \pi/2]$.

Sol: Observe that each $f_n(x)$ is continuous. If f denotes the pointwise limit of the above sequence, then as,

$$f_n(x) = 0 \text{ for } x = 0, \frac{\pi}{2}$$

we have

$$f(x) = 0 \text{ for } x = 0, \frac{\pi}{2}$$

Also, for $0 < x < \dfrac{\pi}{2}$, $0 < \sin x < 1$ hence we have,

$$\sin^n x \to 0 \text{ as } n \to \infty$$

So that, $f(x) = 0$ for $0 < x < \dfrac{\pi}{2}$. That is $f(x) = 0 \ x \in \left[0, \dfrac{\pi}{2}\right]$.

Now let,

$$M_n = \sup |f_n(x) - f(x)|$$

$$= \sup |(x - \pi/2)\sin^n x|$$

But as $f_n(x) = (x - \pi/2)\sin^n x$ is continuous for each n there exists some $0 < c < 1$ $\left(c \neq 0, \dfrac{\pi}{2}\right)$ such that,

$$M_n = |(c - \pi/2)\sin^n c|$$

We have $M_n \to 0$ $n \to \infty$ (as $\sin^n c \to 0$ when $n \to \infty$)

Hence we conclude that above sequence converges uniformly

Example 7.30   For $n \geq 2$ we define the sequence $\{f_n(x)\}$ on $[0, 1]$ as,

$$f_n(x) = \begin{cases} n^2 x & \text{if } 0 \leq x \leq 1/n \\ n(2 - nx) & \text{if } 1/n < x \leq 2/n \\ 0 & \text{if } 2/n < x \leq 1 \end{cases}$$

Then show that $\{f_n(x)\}$ does not converge uniformly on $[0, 1]$.

Sol:

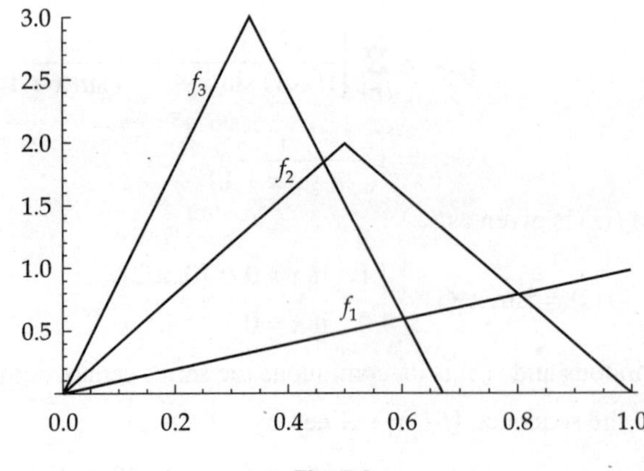

Fig. 7.3

We first show that the pointwise limit of $\{f_n(x)\}$ is given by $f(x) = 0$ $\forall\, x \in [0, 1]$. Observe that $f_n(x) = 0$ $\forall\, n$ if $x = 0$ hence $|f_n(0) - f(0)| = 0$ $\forall\, n$. Now for any $x \in (0, 1)$ we can find a natural number n_0 such that $x > \dfrac{2}{n_0}$. Hence by definition $f_n(x) = 0$ $\forall\, n > n_0$. So $|f_n(x) - f(x)| = 0$ $\forall\, n > n_0$ that is $\displaystyle\lim_{n \to \infty} f_n(x) = f(x)$ $\forall\, x \in [0, 1]$.

Next we see that as each f_n is continuous $f_n \in R[0, 1]$. From the figure it is clear $\displaystyle\int_0^1 f_n(x)\, dx$ is the area of the triangle with base of length $\dfrac{2}{n}$ and height n. Hence $\displaystyle\int_0^1 f_n(x)\, dx = 1$ $\forall\, n$.

So we have $\displaystyle\lim_{n \to \infty} \int_0^1 f_n(x)\, dx = 1$ but it is clear that $\displaystyle\int_0^1 f(x)\, dx = 0$. Hence $\displaystyle\lim_{n \to \infty} \int_0^1 f_n(x)\, dx = 1 \neq 0$ $= \displaystyle\int_0^1 f(x)\, dx$.

Hence from Remark 4 following Theorem 7.7.2 we conclude that the convergence is not uniform.

Example 7.31 Show that the sequence of functions $\{f_n(x)\}$ defined as,

$$f_n(x) = n^2 x^n (1 - x), x \in [0, 1]$$

is pointwise convergent but not uniformly convergent.

Sol: We show that the sequence converge pointwise to $f = 0$. If $x = 0, 1$ $f(x) = \lim_{x \to \infty} f_n(x) = 0$. Now if $0 < x < 1$ then,

$$x = \frac{1}{1 + p}, \text{ for some } p > 0.$$

See that $(1 + p)^n > n(n - 1)(n - 2)p^3$ hence we get

$$n^2 x^n = \frac{n^2}{(1 + p)^n} < \frac{n^2}{n(n - 1)(n - 2)p^3} \to 0 \text{ as the power of } n \text{ in the denominator is more than that in the}$$

numerator. Hence the sequence converge pointwise to $f \equiv 0$. Next see that as each $f_n(x)$ is a continuous function hence R-Integrable also,

$$\int_0^1 f_n(x)\, dx = \frac{n^2}{(n + 1)(n + 2)}$$

Hence,

$$\lim_{n \to \infty} \int_0^1 f_n(x)\, dx = \lim_{n \to \infty} \left[\frac{n^2}{(n + 1)(n + 2)} \right] = 1.$$

But $\int_0^1 f(x)\, dx = 0$. So as in the previous example we conclude that $\{f_n(x)\}$ does not converge uniformly.

Example 7.32 Show that the sequence of functions $\{f_n(x)\}$ defined as

$$f_n(x) = nx^n (1 - x), x \in [0, 1]$$

does not converge uniformly to its pointwise limit f, but $\lim_{n \to \infty} \int_0^1 f_n(x)\, dx = \int_0^1 f(x)\, dx.$

(This example again exhibits that the condition in Theorem 7.7.1 is not necessary.)

Sol: Using a similar argument as in previous example we can show that the sequence converge pointwise to $f \equiv 0$. To show $\{f_n(x)\}$ is not uniformly convergent we appeal to M_n – Test. Now we that if

$$y = nx^n (1 - x) \text{ we have } \frac{dy}{dx} = nx^{n-1}[n - (n + 1)x]$$

hence,

$$\frac{dy}{dx} = 0 \Rightarrow x = 0 \text{ or } x = \frac{n}{n + 1}.$$

Again,

$$\frac{d^2y}{dx^2} = nx^{n-1}[-(n + 1)] + n(n - 1)x^{n-2}[n - (n + 1)n]$$

$$\Rightarrow \text{ at } x = \frac{n}{n + 1}, \frac{d^2y}{dx^2} = -n(n + 1)\left(\frac{n}{n + 1}\right)^{n-1} < 0$$

So y attains maximum at $x = \dfrac{n}{n+1}$. Now

$$M_n = \sup_{x \in [0,\, 1]} |f_n(x) - f(x)|$$

$$= \max_{x \in [0,\, 1]} y = n \left(\frac{n}{n+1}\right)^n \left(1 - \frac{n}{n+1}\right) = \frac{1}{\left(1 + \frac{1}{n}\right)^n} \frac{1}{\left(1 + \frac{1}{n}\right)}$$

So that $M_n \to \dfrac{1}{e}$ as $n \to \infty$ which shows that $\{f_n(x)\}$ is not uniformly convergent. Next a simple

calculation shows that, $\displaystyle\lim_{n \to \infty} \int_0^1 f_n(x)\, dx = \lim_{n \to \infty} \frac{n}{n(n+1)(n+2)} = 0 = \int_0^1 f(x)\, dx.$

Example 7.33 Let $\{f_n(x)\}$ be a sequence of functions defined on $[a, b]$ converging uniformly to $f(x)$. If

further we assume that $f_n \in R[a, b] \ \forall\ n$ then show that the sequence $\left\{\displaystyle\int_a^x f_n(t)\, dt\right\}$ converging uniformly

to $\left(\displaystyle\int_a^x f(t)\, dt\right) \forall\ x \in [a, b].$

Sol: First observe that $f(x) \in R[a, b]$ (by Theorem 7.7.1) and hence $f(x) \in R[a, x] \ \forall\ x \in [a, b]$
(by Theorem). Now as $f_n \to f$ uniformly for $\varepsilon > 0$ there must exist a positive integer n_0 such that,

$$|f_n(x) - f(x)| < \frac{\varepsilon}{b-a} \ \forall\ n > n_0 \ \&\ \forall\ x \in [a, b] \quad (1)$$

Now for $n > n_0$,

$$\left|\int_a^x f_n(t)\, dt - \int_a^x f(t)\, dt\right| \leq \int_a^x |f_n(t) - f(t)|\, dt < \int_a^x \frac{\varepsilon}{b-a}\, dt = \frac{\varepsilon(x-a)}{b-a} < \varepsilon.$$

Hence we conclude $\displaystyle\int_a^x f_n(t)\, dt \to \int_a^x f(t)\, dt$ uniformly.

Example 7.34 If $f(x) = \displaystyle\sum_{n=1}^{\infty} \frac{e^{-nx}}{n^2}$ $x \in [0, 1]$ then show that,

$$\int_0^1 f(x)\, dx = \sum_{n=1}^{\infty} \left(\frac{1 - e^{-n}}{n^3}\right)$$

Sol: Consider the series $\displaystyle\sum_{n=1}^{\infty} \frac{e^{-nx}}{n^2}$ and note that $\dfrac{e^{-nx}}{n^2} \in R[a, b] \ \forall\ n$. As $e^{nx} \geq 1$ for $x \geq 0$ we have $e^{-nx} \leq 1$

for $x \geq 0$. Hence for $x \in [0, 1]$, $\dfrac{e^{-nx}}{n^2} \leq \dfrac{1}{n^2} = M_n$ (say). Also as $\sum M_n = \sum \dfrac{1}{n^2} < \infty$ the above series is

uniformly convergent by Weierstrass' M-Test and by Theorem 7.7.2 term by term integration is possible.
So we get,

$$\int_0^1 f(x)\,dx = \sum_{n=0}^{\infty} \int_0^1 \frac{e^{-nx}}{n^2}\,dx = \sum_{n=0}^{\infty} \frac{1}{n^2} \int_0^1 e^{-nx}\,dx = \sum_{n=0}^{\infty} \frac{1}{n^2}\left[\frac{e^{-nx}}{-n}\right]_0^1 = \sum_{n=1}^{\infty}\left(\frac{1-e^{-n}}{n^3}\right).$$

Example 7.35 If $\{f_n(x)\}$ is defined as,

$$f_n(x) = \frac{e^{-n^2x^2}}{n}\, x \in [0,\,1]$$

show that $\{f'_n\}$ does not uniformly in $[0,\,1]$ but,

$$\lim_{n\to\infty} \frac{d}{dx}\,(f_n(x)) = \frac{d}{dx}\left(\lim_{n\to\infty} f_n(x)\right) \forall\, x \in [0,\,1]$$

(This example again shows that conditions of Theorem 7.8.2 are sufficient but not necessary)

Sol: As usual first we note that pointwise limit of the above sequence is given by,

$$f(x) = 0$$

Also we observe that, $f'_n(x) = -2nxe^{-n^2x^2} \to 0$ as $n \to \infty$. Hence we get

$$\lim_{n\to\infty} \frac{d}{dx}\,(f_n(x)) = 0 = \frac{d}{dx}\,(f(x)) = \frac{d}{dx}\left(\lim_{n\to\infty} f_n(x)\right).$$

But we have seen by Example 10 we get that $\{nxe^{-n^2x^2}\}$ does not converge uniformly for $x \geq 0$ in particular in $[0,\,1]$. Hence we conclude $\{f'_n\}$ does not uniformly in $[0,\,1]$.

Example 7.36 Show that the series

$$\log(1-x) + \sum_{n=1}^{\infty} \log\left(1 + x^{2^{n-1}}\right) \tag{1}$$

converges in $-1 < x < 1$ and deduce that,

$$\frac{1}{1+x} + \frac{2x}{1+x^2} + \frac{4x^3}{1+x^4} + \dots = \frac{1}{1\,x} \text{ for } -1 < x < 1$$

Sol: The sequence of partial sum function $\{s_n(x)\}$ of the series $\log(1-x) + \sum_{n=1}^{\infty} \log(1 + x^{2^{n-1}})$ is given as $s_1(x) = \log(1-x)$ and

$$s_n(x) = \log(1-x) + \log(1+x) + \log(1+x^2) + \log(1+x^4) + \dots + \log(1+x^{2^{n-2}}),\, n \geq 2$$

$$= \log(1-x^2)\log(1+x^2)\dots\log(1+x^{2^{n-2}})$$

$$= \log(1-x^{2^{n-1}}) \to \log 1 = 0 \text{ as } n \to \infty \text{ for } -1 < x < 1$$

Hence the above series converges to the sum function $f(x) = 0$ in $(-1,\,1)$, in particular it converges for $x = 0$ also we can write

$$\log(1-x) + \log(1+x) + \log(1+x^2) + \log(1+x^4) + \dots = 0 \tag{2}$$

The series of derivatives obtained by ignoring the first two terms of above series is given as,

$$\frac{2x}{1+x^2} + \frac{4x^3}{1+x^4} + \dots + \frac{2^n x^{2^{n-1}}}{1+x^{2^n}} + \dots \tag{3}$$

Now $\left[\dfrac{2^n x^{2^{n}-1}}{1+x^{2^n}}\right] < 2^n \lambda^{2^{n}-1} = M_n$ for $|x| \le \lambda < 1$. As $\sum M_n < \infty$ series (3) converges uniformly by Weierstrass' M- Test in $-1 < x < 1$. Observe that the series (2) and the series of derivatives obtained from the series (1) differs by a finite number of terms hence we conclude that the series of derivatives obtained from the series (1) converges uniformly in $-1 < x < 1$. Hence we find that all conditions of Theorem 7.8.2 are satisfied by the series (1) so applying term by term differentiation on (2) we get,

$$-\frac{1}{1-x} + \frac{1}{1+x} + \frac{2x}{1+x^2} + \frac{4x^3}{1+x^4} + \dots = 0$$

Transferring sides we get as desired,

$$\frac{1}{1+x} + \frac{2x}{1+x^2} + \frac{4x^3}{1+x^4} + \dots = \frac{1}{1-x} \quad \text{for } -1 < x < 1.$$

Example 7.38   Let $\{f_n\}$ be a sequence of continuous functions defined on $[a, b]$ converging uniformly to f and suppose $\varphi : [c, d] \to [a, b]$ be another continuous function on then $f_n \circ \varphi \to f \circ \varphi$ uniformly on $[c, d]$.

Sol: As $f_n \to f$ uniformly we have by M_n test,

$$\sup_{x \in [a, b]} |f_n(x) - f(x)| \to 0 \text{ as } n \to \infty$$

Now as φ is continuous it is bounded and also as $\varphi([c, d]) \subseteq [a, b]$, hence we have,

$$M_n = \sup_{x \in [c, d]} |f_n \circ \varphi(x) - f \circ \varphi(x)|$$

$$= \sup_{x \in [c, d]} |(f_n - f) \circ \varphi(x)|$$

$$= \sup_{y \in \varphi[c, d]} |(f_n - f)(y)| \le \sup_{y \in [a, b]} |(f_n - f)(y)| = M_n \to 0 \text{ as } n \to \infty$$

As a result we conclude that $f_n \circ \varphi \to f \circ \varphi$ uniformly on $[c, d]$

7.9   UNIFORM CONVERGENCE OF SERIES OF PRODUCT OF TWO FUNCTIONS

Suppose $\{f_n\}$ & $\{g_n\}$ be two sequences of functions defined on an interval I. We will study in this section under what conditions the series $\sum\limits_{n=1}^{\infty} f_n(x) g_n(x)$ converges uniformly on I. We give two theorems in this regard known as Abel's Theorem and Dirichlet's theorem.

Theorem 7.9.1 (Abel's Theorem)

Suppose $\{f_n(x)\}$ & $\{g_n(x)\}$ be two sequence of functions defined on the closed interval $[a, b]$ such that,

(i) $\sum\limits_{n=1}^{\infty} f_n(x)$ converges uniformly on $[a, b]$

(ii) $\{g_n(x)\}$ is monotone sequence for each $x \in [a, b]$

(iii) $\{g_n(x)\}$ is uniformly bounded (that is there exists some $M > 0$ such that $|g_n(x)| < M$ for all $n \in \mathbb{N}$ & for all $x \in [a, b]$)

Then the series $\sum\limits_{n=1}^{\infty} f_n(x)\, g_n(x)$ converges uniformly on $[a, b]$.

Proof: Let $\{s_n(x)\}$ & $\{h_n(x)\}$ be the sequence of partial sum functions of the series $\sum\limits_{n=1}^{\infty} f_n(x)$

and $\sum\limits_{n=1}^{\infty} f_n(x)\, g_n(x)$ respectively. Now as $\sum\limits_{n=1}^{\infty} f_n(x)$ converges uniformly on $[a, b]$ for,. $\varepsilon > 0$, by using

Cauchy's Criterion there must exist a positive integer n_0 such that if $m > n > n_0$ we have,

$$\left| \sum_{i=n+1}^{m} f_i(x) \right| < \frac{\varepsilon}{3M} \tag{7.9.1.1}$$

We see that for $m > n > n_0$ and $\forall\, x \in [a, b]$,

$$h_m(x) - h_n(x) = \sum_{i=n+1}^{m} f_i(x)\, g_i(x)$$

$$= \sum_{i=n+1}^{m} [s_{i+1}(x) - s_i(x)]\, g_i(x)$$

$$= [s_{n+1}(x) - s_n(x)]g_{n+1}(x) + [s_{n+2}(x) - s_{n+1}(x)]g_{n+2}(x)$$

$$+ \dots [s_{m-1}(x) - s_{m-2}(x)]g_{m-1}(x) + [s_m(x) - s_{m-1}(x)]g_m(x)$$

$$= s_{n+1}(x)[g_{n+1}(x) - g_{n+2}(x)] + s_{n+2}(x)[g_{n+2}(x) - g_{n+3}(x)] + \dots$$

$$+ s_{m-1}(x)[g_{m-1}(x) - g_m(x)] + s_m(x)g_m(x)$$

$$= \sum_{k=n+1}^{m-1} s_k(x)[g_k(x) - g_{k+1}(x)] + s_m(x)\, g_m(x)$$

That is $|h_m(x) - h_n(x)| \le \sum\limits_{k=n+1}^{m-1} |s_k(x)|\, |g_k(x) - g_{k+1}(x)| + |s_m(x)|\, |g_m(x)|$

$$< \frac{\varepsilon}{3M} \sum_{k=n+1}^{m-1} |g_k(x) - g_{k+1}(x)| + \frac{\varepsilon}{3M} |g_m(x)| \qquad \text{(using (7.9.1.1)) (7.9.1.2)}$$

Now we see that as $\{g_n(x)\}$ is monotone $|g_k(x) - g_{k+1}(x)|$ keeps same sign for all k, also using given condition (c) from (7.9.1.2) we get,

$$|h_m(x) - h_n(x)| < \frac{\varepsilon}{3M} \sum_{k=n+1}^{m-1} (g_k(x) - g_{k+1}(x))| + \frac{\varepsilon}{3M} M$$

$$= \frac{\varepsilon}{3M}(g_{n+1}(x) - g_m(x)) + \frac{\varepsilon}{3}$$

$$\leq \frac{\varepsilon}{3M}\left(|g_{n+1}(x)| + |g_m(x)|\right) + \frac{\varepsilon}{3}$$

$$\leq \frac{\varepsilon}{3M} \cdot 2M + \frac{\varepsilon}{3} = \varepsilon$$

Which implies via Cauchy Criterion that the series $\sum\limits_{n=1}^{\infty} f_n(x)g_n(x)$ converges uniformly on $[a, b]$.

In the next theorem we replace the condition of uniform convergence of $\sum\limits_{n=1}^{\infty} f_n(x)$ by uniform boundedness of sequence of its partial sum function also we put a stricter condition on $\{g_n(x)\}$ to obtain the same result.

Theorem 7.9.2 (Dirichlet's Theorem)

Suppose $\{f_n(x)\}$ & $\{g_n(x)\}$ be two sequence of functions defined on the closed interval $[a, b]$ such that,

   (i)   *The sequence of partial sum function $\{s_n(x)\}$ of the series $\sum\limits_{n=1}^{\infty} f_n(x)$ is uniformly bounded.(that is*
         there exists $M > 0$ such that $|s_n(x)| < M$ for all $n \in \mathbb{N}$ and for all $x \in [a, b]$)
   (ii) *For any fixed $x \in [a, b]$ the sequence $\{g_n(x)\}$ is positive monotone decreasing sequence converging uniformly to 0.*

Then the series $\sum\limits_{n=1}^{\infty} f_n(x)g_n(x)$ converges uniformly on $[a, b]$.

Proof: As in the previous theorem let $\{h_n(x)\}$ be the sequence of partial sum functions of the series $\sum\limits_{n=1}^{\infty} f_n(x)\, g_n(x)$. We see that as $g_n(x) \to 0$ uniformly there exists a positive integer n_0 such that for $n > n_0$ and $\forall\, x \in [a, b]$,

$$|g_n(x)| < \frac{\varepsilon}{3M} \tag{7.9.2.1}$$

Now proceeding as in the previous theorem we see that for $m > n > n_0$ and $\forall\, x \in [a, b]$ we have,

$$|h_m(x) - h_n(x)| \leq \sum_{k=1}^{m-1} |s_k(x)|\,|g_k(x) - g_{k+1}(x)| + |s_m(x)|\,|g_m(x)|$$

$$< M \sum_{k=1}^{m-1} |g_k(x) - g_{k+1}(x)| + Mg_m(x)$$

(using given condition (i) and that $\{g_n(x)\}$ is positive (7.9.2.2)

Now as $\{g_n(x)\}$ is positive monotone decreasing sequence we have, $\sum\limits_{k=n+1}^{m-1} |g_k(x) - g_{k+1}(x)| = g_{n+1}(x) - g_m(x)$

Using this in (7.9.2.2) we get for $m > n > n_0$ and $\varepsilon\, x \in [a, b]$,

$$|h_m(x) - h_m(x)| = Mg_{n+1}(x) < \frac{\varepsilon}{M} = \varepsilon \tag{Using (7.9.2.1)}$$

That is the series $\sum\limits_{n=1}^{\infty} f_n(x)\, g_n(x)$ converges uniformly on $[a, b]$.

Example 7.37   Show that the series $\sum\limits_{n=1}^{\infty} \dfrac{(-1)^n (1 + |\sin^n x|)}{x^3 + n}$ is uniformly convergent in the interval $[0, \infty)$

Sol: We see that by Leibniz Test $\sum\limits_{n=1}^{\infty} \dfrac{(-1)^n}{x^3 + n}$ converges for $x \in [0, \infty)$. Let $\{s_n(x)\}$ be the sequence of

partial sum function of the series $\sum\limits_{n=1}^{\infty} \dfrac{(-1)^n}{x^3 + n}$. Then we see that,

$$s_{2m-1} - s_{2n-1}(x) = \sum_{k=2n}^{2m-1} \frac{(-1)^n}{x^3 + n}$$

$$= \sum_{k=n}^{m-1} \left(\frac{1}{(x^3 + 2k)} - \frac{1}{x^3 + 2k + 1} \right)$$

$$= \sum_{k=m}^{m-1} \frac{1}{(x^3 + 2k)(x^3 + 2k + 1)}$$

$$\leq \sum_{k=n}^{m-1} \frac{1}{(2k)(2k + 1)}$$

$$= \sum_{t=2n}^{2m-2} \frac{1}{(t)(t + 1)} = \sum_{t=2n}^{2m-2} \left[\frac{1}{t} - \frac{1}{t + 1} \right] = \frac{1}{2n} - \frac{1}{2m - 1}$$

Hence,

$$|s_{2m-1}(x) - s_{2n-1}(x)| < \frac{1}{2n}$$

Similarly,

$$|s_{2m-1}(x) - s_{2n}(x)| < \frac{1}{2n + 1}.$$

Similar calculations will show that ,

$|s_{2m}(x) - s_{2n-1}(x)| < \dfrac{1}{2n}$ & $|s_{2m}(x) - s_{2n-1}(x)| < \dfrac{1}{2n + 1}$. Hence given any $\varepsilon > 0$ we will get some positive

integer n_0 such that for $m > n > n_0$ and for $x \in [0, \infty)$,

$$|s_m(x) - s_n(x)| < \varepsilon$$

This shows $\sum\limits_{n=1}^{\infty} \dfrac{(-1)n}{x^3 + n}$ converges uniformly for $x \in [0, \infty)$ also it is clear that for $x \in [0, \infty)$, $(1 + |\sin^n x|)$

is monotone decreasing hence by Abel's Theorem the series $\sum\limits_{n=1}^{\infty} \dfrac{(-1)^n (1 + |\sin^n x|)}{x^3 + n}$ is uniformly convergent

in the interval $[0, \infty)$.

Example 7.38   Show that the series $\sum\limits_{n=0}^{\infty} (\cos^n x)\, x^n$ $x \in [a, b]$ $0 < a < b < 1$ is uniformly convergent.

Sol: Let $\{s_n(x)\}$ denote the sequence of partial sum of the series $\sum x^n$ then,

$$|s_n(x)| = \left|\sum_{i=1}^{n} x^n\right| = \left|\frac{1 - x^n}{1 - x}\right| \le \frac{2}{1 - b}.$$

Also $\cos^n x$ is monotone decreasing for fixed $x \in [a, b]$ and uniformly convergent to 0. Hence by Dirichlet's Theorem the conclusion follows.

7.10   WEIERSTRASS APPROXIMATION THEOREM

We end this chapter by a famous result which says that any continuous function in a closed interval $[a, b]$ can be uniformly approximated by a polynomial.

Theorem 7.10.1

Let $f(x)$ be a continuous function defined in a closed interval $[a, b]$ then there exists a sequence of polynomials $\{p_n(x)\}$ such that $p_n \to f$ uniformly in $[a, b]$.

Proof: Suppose the theorem is true in the interval $[0, 1]$. Let f be any continuous function in a closed interval $[a, b]$. We know that the function $\varphi(x) = \dfrac{x - a}{b - a}$ maps $[a, b]$ to $[0, 1]$ in one to one fashion and is continuous. We now consider the function $(f \circ \varphi - 1)(x)$ which denotes a continuous function on $[0, 1]$. Hence by supposition there exists a sequence of polynomials $\{q_n(x)\}$ such that $q_n \to (f \circ \varphi^{-1})$ uniformly on $[0, 1]$. Now by Example we conclude $q_n \circ \varphi \to f \circ \varphi - 1 \circ \varphi = f$ uniformly on $[a, b]$. For above observation it is enough if we prove the theorem for the interval $[0, 1]$.

For $f \in C[0, 1]$ we define the nth degree **Bernstein Polynomial** $B_n(x)$ associated with f as,

$$B_n(x) = \sum_{k=0}^{n} {}^n C_x f\left(\frac{k}{n}\right) x^k (1 - x)^{n-k}, \, x \in [0, 1]$$

We will show that $B_n(x)$ converges uniformly to $f(x)$ as $n \to \infty$.

Let us define for a fixed positive integer n and fixed $x \in [0, 1]$ the polynomials $b'(k, x)$ as $B'(k, x) = {}^n C_x x^k (1 - x)^{n-k}$ $0 \le k \le n$.

We immediately observe the following,

$$\sum_{k=0}^{n} B'(k, x) = \sum_{k=0}^{n} {}^n C_x x^k (1 - x)^{n-k} = (x + 1 - x)^n = 1 \tag{7.10.1.1}$$

and

$$B_n(x) = \sum_{k=0}^{n} {}^n C_x f\left(\frac{k}{n}\right) x^k (1 - x)^{n-k} = \sum_{k=0}^{n} f\left(\frac{k}{n}\right) B'(k, x) \tag{7.10.1.2}$$

Differentiating the identity (7.10.1.1) with respect to x we get,

$$\sum_{k=0}^{n} (k - xn)\, {}^n C_x x^k (1 - x)^{n-k} = 0 \tag{7.10.1.3}$$

Again differentiating the identity (7.10.1.3) with respect to x we obtain,

$$\sum_{k=0}^{n} (k - xn)^2 \, {}^nC_x x^k (1 - x)^{n-k} = nx(1 - x) \tag{7.10.1.4}$$

As $f \in C[0, 1]$, $f(x)$ is uniformly continuous on $[0, 1]$, hence for any $\varepsilon > 0$ there exists a $\delta > 0$ such that whenever $|x - y| < \delta$ $|f(x) - f(y)| < \dfrac{\varepsilon}{2}$. For a fixed $x \in [0, 1]$ consider the sets A & B defined as,

$$A = \left\{ k \in \{1, 2, ...n\} : \left| \frac{k}{n} - x \right| < \delta \right\}, \, B = \{1, 2, ...n\} - A$$

Now for fixed x and n

$$|f(x) - B_n(x)| = \left| \sum_{k=0}^{n} f(x) \, B'(k, x) - \sum_{k=0}^{n} f\left(\frac{k}{n}\right) B'(k, x) \right| \quad \text{(using (7.10.1.1))}$$

$$\leq \sum_{k=0}^{n} \left| f(x) - f\left(\frac{k}{n}\right) \right| B'(k, x)$$

$$= \sum_{k \in A} \left| f(x) - f\left(\frac{k}{n}\right) \right| B'(k, x) + \sum_{k \in B} \left| f(x) - f\left(\frac{k}{n}\right) \right| B'(k, x) \tag{7.10.1.5}$$

We now show that each sum on the right is less than $\varepsilon/2$

If $k \in A$ we have $\left| f(x) - f\dfrac{k}{n} \right| < \dfrac{\varepsilon}{2}$ hence,

$$\sum_{k \in A} \left| f(x) - f\left(\frac{k}{n}\right) \right| B'(k, x) < \frac{\varepsilon}{2} \sum_{k \in A} B'(k, x) < \frac{\varepsilon}{2} \sum_{k=1}^{n} B'(k, x) = \frac{\varepsilon}{2} \tag{7.10.1.6}$$

Now,

$$\sum_{k \in B} \left| f(x) - f\left(\frac{k}{n}\right) \right| B'(k, x) \leq 2M \sum_{k \in B} B'(k, x) \tag{7.10.1.7}$$

where $M = \sup_{x \in [0, 1]} |f(x)|$

Again from (7.10.1.4) we get,

$$nx(1 - x) = \sum_{k=0}^{n} (k - xn)^2 B'(k, x)$$

$$\geq \sum_{k \in R} (k - xn)^2 B'(k, x) \tag{7.10.8}$$

Now we observe that if $k \in B$, $\left| \dfrac{k}{n} - x \right| \geq \delta \Rightarrow (k - nx)^2 \geq (n\delta)^2$.

Hence from (7.12.8) we get

$$\sum_{k \in B} B'(k, x) \leq \frac{nx(1 - x)}{n^2 \delta^2}.$$

But as $\displaystyle \sup_{x \in [0,\,1]} [x(1-x)] = \frac{1}{4}$ we finally we obtain,

$$\sum_{k \in B} B'(k, x) \le \frac{1}{4n\delta^2}$$

Now, by (7.10.7) we get,

$$\sum_{k \in B} \left| f(x) - f\left(\frac{k}{n}\right) \right| B'(k, x) \le \frac{M}{2n\delta^2}$$

We see that $\dfrac{M}{2n\delta^2} \to 0$ as $n \to \infty$

Hence for $\varepsilon > 0$ there exists a positive integer $n_0(\varepsilon)$ such that whenever $n \ge n_0(\varepsilon)$ we get,

$$\sum_{k \in B} f(x) - f\left(\frac{k}{n}\right) B'(k, x) < \frac{\varepsilon}{2} \tag{7.10.9}$$

Thus by (7.10.5) we get,

$$|f(x) - B_n(x)| < \frac{\varepsilon}{2} + \frac{\varepsilon}{2} = \varepsilon \quad \text{for } \forall\, n \ge n_0(\varepsilon) \text{ and } x \in [0,\,1].$$

Hence the theorem.

Remark:

In the above Theorem see that we have actually proved that the ***Bernstein Polynomials*** $B_n(x)$ associated with f converges to the function f. Let us compute the first three Bernstein polynomials associated with the function with the simple function $f(x) = x^2$ in $[0,\,1]$ and see graphically how they actually converge to the actual function. Simple calculations show that,

$$B_1(x) = \frac{1}{3}x(1-x), \; B_2(x) = \frac{1}{2}x(1-x), \; B_3(x) = \frac{1}{3}x(1-2x).$$

We now look at their graph (Fig 7.4, 7.5, 7.6).

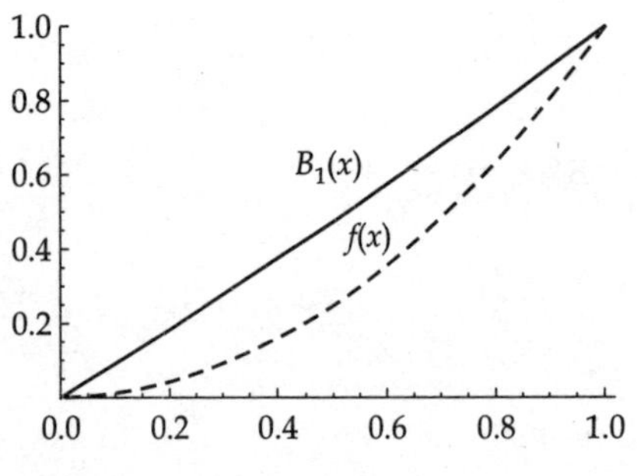

Fig. 7.4

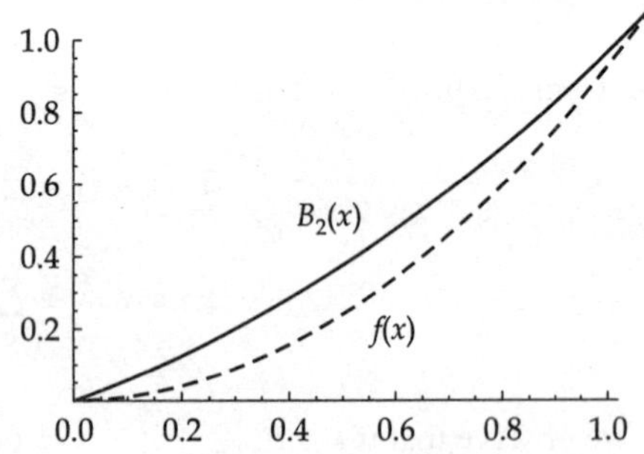

Fig. 7.5

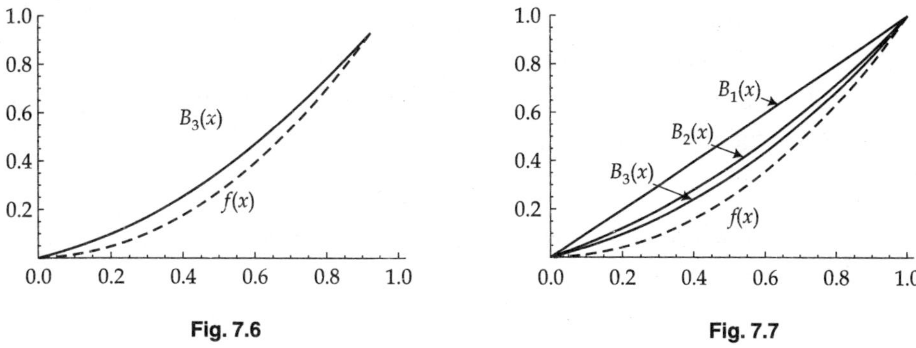

Fig. 7.6   **Fig. 7.7**

Figure 7.7 shows how the Bernstein polynomials approaches to the actual function.

PROBLEMS

1. Let $U[a, b]$ denote the space of sequences of real valued defined on $[a, b]$ that are uniformly convergent to a bounded limit function. If $\tilde{f}$ and $\tilde{g}$ are elements of $U[a, b]$ and c be any real number we define operations '+' and scalar multiplication on $U[a, b]$ as follows:

   $(\tilde{f} + \tilde{g})\,(x) = \langle f_n(x) + g_n(x)\rangle_{n=1}^{\infty}$ and $(c\tilde{g})\,(x) = \langle cg_n(x)\rangle_{n=1}^{\infty}$ where, $\tilde{f}(x) = \langle f_n(x)\rangle_{n=1}^{\infty}$ and $\tilde{g}(x) = \langle g_n(x)\rangle_{n=1}^{\infty}$. Show that $U[a, b]$ is closed under these operations and hence show that $U[a, b]$ is a real vector space.

2. On $U[a, b]$ as in Problem 1 define multiplication 'o' as follows:

   $(\tilde{f} \,\mathrm{o}\, \tilde{g})\,(x) = \langle f_n(x)\, g_n(x)\rangle_{n=1}^{\infty}$.

   Show that $U[a, b]$ is closed under 'o'. Does there exists an identity element of $(U[a, b], \mathrm{o})$? Is $(U[a, b], \mathrm{o})$ a group? We say an element $\tilde{f}$ of $U[a, b]$ bounded away from zero if $f_n(x) \neq 0 \; \forall \; x \in 0 \; \forall \, x \in [a, b]$ & $\forall n$ and if there exists $\delta > 0$ such that $f(x) \geq \delta \; \forall \, x \in [a, b]$ where

   $\tilde{f}(x) = \langle f_n(x)\rangle_{n=1}^{\infty}$ and $f_n(x) \to f(x)$ uniformly. Show that for every element which is of $U[a, b]$ which bounded away from zero there exists an inverse element in $U[a, b]$.

3. Let the sequence $\{f_n(x)\}$ on $[0, 1]$ be defined as follows

   $$f_n(x) = \begin{cases} \dfrac{2}{n^2} & \text{if } x \text{ is rational} \\ 0 & \text{otherwise} \end{cases}$$

   Show that for each $n\, f_n(x)$ is discontinuous at each point of $[0, 1]$ but $\{f_n(x)\}$ converges uniformly to identically zero function.

4. Show that the following functions are uniformly convergent in their respective intervals

5. Show that '0' is the point of non-uniform convergence for the sequence $\left\{ f_n(x) = \dfrac{n^2 x}{1 + n^4 x^2} \right\}$.

6. Let $\{f_n(x)\}$ be a sequence of uniformly continuous functions on (a, b) converging uniformly to a function $f(x)$ on (a, b). Show that $f(x)$ is niformly continuous.

7. If $\{f_n(x)\}$ converges uniformly to $f(x)$ on $[a, b]$ where $f_n \in C[a, b]$ for all n. If $\{x_n\}$ is a sequence in $[a, b]$ converging to a point x in $[a, b]$ then show that,

$$\lim_{n \to \infty} f_n(x_n) = f(x)$$

8. Find a sequence of functions $\{f_n(x)\}$ in an interval $[a, b]$ converging pointwise to a function $f(x)$ such that each $f_n(x) \in R[a, b]$ but $f(x) \notin R[a, b]$.

9. Show that $f_n(x) = nxe^{-nx^2}$ $0 \le x \le 1$ converges pointwise to $f(x) \equiv 0$ on $[0, 1]$ but

$$\lim_{n \to \infty} \int_0^1 f_n(x)\, dx \ne \int_0^1 f(x)\, dx.$$

Does $f_n(x) \to f(x)$ uniformly?

10. Let $\{f_n(x)\}$ be a sequence of functions defined on the interval $[0, 1]$ as follows,

$$f_n(x) = \begin{cases} n^2 x & 0 \le x \le \dfrac{1}{n} \\[2mm] -n^2 x + 2n & \dfrac{1}{n} \le x \le \dfrac{2}{n} \\[2mm] 0 & \dfrac{2}{n} \le x \le 1 \end{cases}$$

Show that $f_n(x) \to 0$ and $\displaystyle\lim_{n \to \infty} \int_0^1 f_n(x)\, dx = 1$. Hence conclude that $\{f_n(x)\}$ is not uniformly convergent.

11. Show that

(i) $\displaystyle\lim_{n \to \infty} \int_{\frac{\pi}{2}}^{\pi} \frac{\sin nx}{nx}\, dx = 0$

(ii) $\displaystyle\lim_{n \to \infty} \int_0^1 nx(1 - x^2)\, dx \ne \int_0^1 \lim_{n \to \infty} [nx(1 - x^2)]\, dx.$

12. Show that for the following series term by term integration is valid and hence find the integral of the series on the respective intervals:

(i) $\displaystyle\sum_{n=1}^{\infty} \frac{\sin nx}{n^2}$ in $[0, \pi]$

(ii) $\displaystyle\sum_{n=1}^{\infty} \frac{\log nx}{n^2}$ in $[1, 2]$

(iii) $\displaystyle\sum_{n=1}^{\infty} ne^{-nx}$ in $[\log 2, \log 3]$.

13. Show that term by term differentiation is valid for the following series and show that:

(i) $\dfrac{d}{dx}\left(\displaystyle\sum_{n=1}^{\infty}\dfrac{n}{x^n}\right)=\displaystyle\sum_{n=1}^{\infty} n^2/x^{n+1}$, for $|x|>1$

(ii) $\dfrac{d}{dx}\left(\displaystyle\sum_{n=1}^{\infty}\dfrac{1}{n^3+n^4x^2}\right)=-\displaystyle\sum_{n=1}^{\infty}\dfrac{2n^4x}{n^3+n^4x^2}$, $\forall\, x \in R$

(iii) $\dfrac{d}{dx}\left(\displaystyle\sum_{n=1}^{\infty}\dfrac{\cos nx}{n^4}\right)=-\displaystyle\sum_{n=1}^{\infty}\dfrac{\sin nx}{n^3}$, $\forall\, x \in R$

14. For $f \in C[0, 1]$ if

$$\int_0^1 x^n f(x)\, dx = 0,\ n = 0, 1, 2\ldots$$

Show that $f = 0$ in $[0, 1]$.

15. Let $B_n(x)$ is the nth degree Bernstein Polynomial associated with $f(x)$ defined on $[0, 1]$. If $B_n(f) \to f$ uniformly in $[0, 1]$ show that $f(x)$ is uniformly continuous on $[0, 1]$.

Chapter **8**

Power Series

8.1 INTRODUCTION

In this chapter we will be interested in studying convergence and related questions as in the previous chapter for a special series of the form $\sum\limits_{n=0}^{\infty} a_n x^n$ (where a_n's are real numbers independent of x), which we can see is a series involving power of x called in short power series. In the Chapter 1 we have seen that under certain conditions a function defined on an interval can be expanded in Maclaurin's series in that interval which is nothing but a power series. We in this chapter will see that under certain conditions a power series will represent a function in a certain neighborhood of whose Maclaurin's expansion is same as the given power series.

8.2 DEFINITION

An infinite series of the form $a_0 + a_1 x + a_2 x^2 + \ldots$ written in short $\sum\limits_{n=0}^{\infty} a_n x^n$ where $a_0, a_1, \ldots a_n, \ldots$ are real constants independent of x is called a power series about the point 0. More generally an infinite series of the form $\sum\limits_{n=0}^{\infty} a_n(x - a)^n$ where $x_0, a_0, a_1, \ldots a_n, \ldots$ are real constants independent of x will be called as a power series about the point a.

We note that any power series of the form $\sum\limits_{n=0}^{\infty} a_n(x - a)^n$ can be reduced to the power series of the form $\sum\limits_{n=0}^{\infty} a_n x^n$ by a simple change of variable $x' = x - a$. Hence no generality will be lost if we study power series about the point zero and hence forth in this chapter we will only take up power series of the form $\sum\limits_{n=0}^{\infty} a_n x^n$.

8.3   RADIUS OF CONVERGENCE

Given a power series $\sum\limits_{n=0}^{\infty} a_n\, x^n$ we want to we ask natural question for what values of real x above series

converges. We first note an interesting point, any power series converges at least for one point $x = 0$. For

$\sum\limits_{n=0}^{\infty} a_n$, $a^n = a_0$, a constant for $x = 0$ and hence convergent. Next we see following examples,

Example 8.1   The power series $\sum\limits_{n=0}^{\infty} n^n\, x^n$ converge only for any $x = 0$. As for any $a \neq 0$,

$$\frac{(n+1)^{n+1}\, a^{n+1}}{n^n\, a^n} = \left\{ (n+1)\left(1 + \frac{1}{n}\right)^n \right\} |a| \to \infty \text{ as } n \to \infty \text{ hence ultimately exceeds 1 so we conclude by}$$

Ratio Test that $\sum\limits_{n=0}^{\infty} nx^n$ does not converge for any $x \neq 0$.

Example 8.2   The power series $\sum\limits_{n=0}^{\infty} \frac{x_n}{n!}$ which represents e^x converges for all values x. This can be easily

verified by using Ratio Test.

Example 8.3   We know that the power series $\sum\limits_{n=0}^{\infty} x^n$ which represents the infinite geometric series

converges for $|x| < 1$ and diverges for $|x| > 1$ and oscillates for $x = \pm\, 1$.

If the power series does not converge for any other value of except x we say the power is ***nowhere convergent*** (Example 1). If the power series converges for all values of x we say the power series is ***everywhere convergent*** (Example 2). We now ask is it possible to have a power series which converges only at a single point other than 0? Following theorem tells us that this is not possible as, if a power series is convergent for at least one non zero value of x then there exists an interval containing 0 such that the power series converges for all values of x in that interval.

Theorem 8.3.1

If the power series $\sum\limits_{n=0}^{\infty} a_n\, x^n$ converges for $x = a\ (\neq 0)$ then the power series converges absolutely and

uniformly in $[-l, l]$ where $0 \leq l < |a|$. Moreover if the power series diverges for $x = b$ then the power series diverges for all x satisfying $|x| > |b|$.

Proof: As $\sum\limits_{n=0}^{\infty} a_n\, a^n$ converges, $\{a_n\, a^n\}$ is bounded so there must exists some $M > 0$ such that,

$$|a_n\, a^n| < M \ \forall n \tag{8.3.1.1}$$

Now for any $x \in [-\, l,\, l]$ we have,

$$|a_n\, x^n| = a_n\, a^n \left|\frac{x^n}{a^n}\right| = |a_n\, a^n| \left|\frac{x^n}{a^n}\right| < M \left|\frac{l}{a}\right|^n = M_n \text{ (say)} \tag{8.3.1.2}$$

Then as $\left|\dfrac{l}{a}\right| < 1$ the geometric series $\sum\limits_{n=1}^{\infty} \left|\dfrac{l}{a}\right|^n$ converges and hence $\sum M_n$ converges, now using

Theorem 7.5.3 we conclude that the series $\sum\limits_{n=0}^{\infty} a_n x^n$ converges absolutely and uniformly in $[-l, l]$ where $0 \le l < |a|$.

Next suppose $\sum\limits_{n=0}^{\infty} a_n b^n$ diverges and suppose x be any element such that $|x| > |b|$. If possible, let $\sum\limits_{n=0}^{\infty} a_n x^n$ be convergent then by what we have proved in the first part, it follows that $\sum\limits_{n=0}^{\infty} a_n b^n$ has to converge which is a contradiction. Hence $\sum\limits_{n=0}^{\infty} a_n x^n$ diverges. Since x is arbitrary we get the result.

Remark

Note that the first part of the theorem is true if we only assume that the sequence $\{a_n a^n\}$ is bounded

Using above theorem we get the following important result:

Theorem 8.3.2 (Existence of Radius of convergence)

For a power series $\sum\limits_{n=0}^{\infty} a_n x^n$ only one of the following holds:

   (i) *The power series does not converge for any value of x except $x = 0$*

   (ii) The power series converges for all values of x

   (iii) *There exists a unique real number $R > 0$ such that for all x satisfying $|x| < R$ the power series converges absolutely whereas for all x satisfying $|x| > R$ the power series diverges.*

Proof: Consider case (iii). Define,

$$R = \sup \left\{ |x| : \sum_{n=0}^{\infty} a_n x^n \text{ is convergent} \right\}$$

As by assumption $\sum\limits_{n=0}^{\infty} a_n x^n$ is convergent for values of x other than 0 we have $R > 0$. Let $|x| < R$ then by definition of supremum there exists some a satisfying $|x| < |a| < R$ such that $\sum\limits_{n=0}^{\infty} a_n a^n$ is convergent. Hence it follows from first part of Theorem 8.2.1 $\sum\limits_{n=0}^{\infty} a_n x^n$ converges absolutely. Again suppose that $|x| > R$ then if possible, let $\sum\limits_{n=0}^{\infty} a_n x^n$ be convergent then we have got an $|x|$ element bigger than R for which the power series converges this contradicts the definition of R hence $\sum\limits_{n=0}^{\infty} a_n x^n$ must diverge. We now show that R is unique. For this let R' be any other positive real number having same property of R. Then by definition of R we have,

$$R \le R' \tag{8.3.2.1}$$

again for any $\varepsilon > 0$ the power series converges for $x = R' - \varepsilon$ (by definition of R') hence $R' - \varepsilon \le R \ \forall \ \varepsilon > 0$ that is

$$R' \le R \tag{8.3.2.2}$$

From (8.3.2.1) & (8.3.2.2) we conclude R is unique and hence the proof of the theorem is complete.

The unique number R in case (c) above is called ***Radius of convergence*** of the power series. The radius of convergence for a nowhere convergent power series is defined to be 0 and the radius of convergence for everywhere convergent power series is defined to be ∞ The interval $(-R, R)$ is called the ***interval of convergence*** of the power series.

8.4 FORMULAE FOR FINDING RADIUS OF CONVERGENCE

In this section we will be interested to some useful formulae which will enable us to find radius of convergence of a given power series. We have basically two formulae the first one is based on ratio test of infinite series and the second one is based on the root test.

Theorem 8.4.1 Suppose $\sum\limits_{n=0}^{\infty} a_n x_n$ be a power series with $a_n \neq 0$ for large values of n and if ,

$$\lim_{n \to \infty} \left| \frac{a_{n+1}}{a_n} \right| = \lambda \quad \text{(may be infinite)}$$

then the radius of convergence R of the power series is given by ,

$$R = \begin{cases} 0 & \text{if } \lambda = \infty \\ \dfrac{1}{\lambda} & \text{if } \lambda \neq 0 \\ \infty & \text{if } \lambda = 0 \end{cases}$$

Proof: We use Ratio Test for the series $\sum\limits_{n=1}^{\infty} |a_n x^n|$. For fixed x let $u_n = |a_n x^n|$ then we see that,

$$\lim_{n \to \infty} \left(\frac{u_{n+1}}{u_n} \right) = \lim_{n \to \infty} \left(\left| \frac{a_{n+1}}{a_n} \right| |x| \right) = \lambda |x|$$

So by Ratio Test $\sum\limits_{n=0}^{\infty} a_n x^n$ converges absolutely if $\lambda |x| < 1$ that is $|x| < \dfrac{1}{\lambda}$ if $\lambda \neq 0$

$$\sum_{n=0}^{\infty} a_n x^n \text{ diverges if } \lambda |x| > 1 \text{ that is } |x| > \frac{1}{\lambda} \neq 0$$

Now using above observations and uniqueness of radius of convergence we conclude that the radius of convergence of the power series is given by

$$R = \frac{1}{\lambda} \text{ in this case.}$$

Now see that if $\lambda = \infty$ then for every x, $\left(\dfrac{u_{n+1}}{u_n} \right)$ exceeds 1 for large values of n hence the power series

diverges for any x that is the power series is nowhere convergent hence by definition $R = 0$ in this case. Again if $\lambda = 0$ we see that for any x, $\lambda |x| = 0 < 1$. Hence the power series converges for all x that is power series is everywhere convergent and hence by definition $R = \infty$ in this case.

Theorem 8.4.2 (Cauchy-Hadamard Formula)

For a power series $\sum\limits_{n=0}^{\infty} a_n x^n$ *let* $\lambda = \lim \sup (|a_n|)^{1/n}$. *Then the radius of convergence* R *of the power series is given by,*

$$R = \begin{cases} 0 & \text{if } \lambda = \infty \\[2mm] \dfrac{1}{\lambda} & \text{if } \lambda \neq 0 \\[2mm] \infty & \text{if } \lambda = 0 \end{cases}$$

(Note: If in particular $\lim\limits_{n \to \infty} (|a_n|)^{1/n}$ *exists then we have* $\lambda = \lim \sup (|a_n|)^{1/n} = \lim\limits_{n \to \infty} (|a_n|)^{1/n}$*)*

Proof: We just apply Cauchy's Root test on test on the series $\sum\limits_{n=0}^{\infty} |a_n x^n|$. For that we note,

$$\lim \sup (|a_n x^n|)^{1/n} = \lim \sup |a_n|^{1/n} |x| = \lambda |x|.$$

Now by applying similar argument as Theorem 8.3.1 we arrive at the theorem.

Example 8.4 Find the radius of convergence R of the following power series

(i) $\sum\limits_{n=1}^{\infty} \dfrac{7^{n+1}}{e^n} x^{n+4}$

(ii) $\sum\limits_{n=1}^{\infty} \dfrac{6^{3n}}{n} (x-2)^n$

(iii) $\sum\limits_{n=1}^{\infty} x^{n!}$

Sol: (i) Observe that if we write the above series as $\sum\limits_{n=0}^{\infty} a_n x^n$ then $a_n = 0$ for $n = 0, 1, 2, 3, 4$ and $a_n = \dfrac{7^{n-3}}{e^{n-4}}$ for $n \geq 5$. Then for

$$n \geq 5 \quad \lim_{n \to \infty} \left| \frac{a_{n+1}}{a_n} \right| = \lim_{n \to \infty} \left(\frac{7}{e} \right) = \frac{7}{e}. \text{ Then } R = \frac{e}{7} \quad \text{(by Theorem 8.4.1)}$$

(ii) Let $y = x - 2$ then if we write the above series as $\sum\limits_{n=0}^{\infty} a_n y^n$ then $a_0 = 0$ and $a_n = \dfrac{6^{3n}}{n} \geq 1$. Hence for

$$\lim_{n \to \infty} (|a_n|)^{1/n} = \lim_{n \to \infty} \left(\frac{6^{3n}}{n} \right)^{1/n} = \lim_{n \to \infty} \frac{6^3}{n^{1/n}} = 6^3. \text{ Therefore } R = \frac{1}{6^3} \text{ (by Theorem 8.4.2)}$$

(iii) Here if we represent the series as $\sum\limits_{n=0}^{\infty} a_n x^n$ then,

$$a_n = \begin{cases} 1 & \text{if } n = 1!, 2!, 3! \ldots \\ 0 & \text{otherwise} \end{cases}$$

So $\lim \sup (|a_n|)^{1/n} = 1$ and we have in this case $R = 1$. (Observe that here $\lim\limits_{n \to \infty} (|a_n|)^{1/n}$ does not exist)

Example 8.5 Find interval of convergence for the following power series,

$$1 + \frac{a \cdot b}{1 \cdot c} x + \frac{a \cdot (a+1) \cdot b \cdot (b+1)}{1 \cdot 2 \cdot c\,(c+1)} x^2 + \frac{a\,(a+1) \cdot (a+2) \cdot b \cdot (b+1) \cdot (b+2)}{1 \cdot 2 \cdot c \cdot (c+1) \cdot (c+2)} x^3 + \dots$$

a, b, c are positive real numbers

Sol: The above series can be represented as $\sum\limits_{n=0}^{\infty} a_n\, x^n$ where $a_0 = 1$ and

$$a_n = \frac{a \cdot (a+1) \cdot (a+2) \dots (a+n-1)\, b \cdot (b+1) \cdot (b+2) \dots (b+n-1)}{n!\, c \cdot (c+1) \cdot (c+2) \dots (c+n-1)}\; n \geq 1.$$

We get,

$$\lim_{n\to\infty} \left| \frac{a_{n+1}}{a_n} \right| = \lim_{n\to\infty} \frac{(a+n)\,(b+n)}{(n+1)\,(c+n)}$$

$$= \lim_{n\to\infty} = \frac{(a/n+1)\,(b/n+1)}{(c/n+1)\,(1/n+1)} = 1$$

so that the radius of convergence $R = 1$ hence interval of convergence is $(-1, 1)$

8.5 SIGNIFICANCE OF RADIUS OF CONVERGENCE: TERM BY TERM INTEGRATION & DIFFERENTIATION OF A POWER SERIES

In this section we will see what important role radius of convergence plays with regard to a given power series. We will find that a power series represents a well behaved function within it's radius of convergence. For example we will show that a power series is infinitely differentiable within the radius of convergence and moreover term by term differentiation and integration of the power series is possible inside the interval of convergence. We give the following important theorem:

Theorem 8.5.1

Let $\sum\limits_{n=0}^{\infty} a_n\, x^n$ be a power series with non-zero radius of convergence R and suppose $[a, b] \subset (-R, R)$ then,

(i) *$f(x) = \sum\limits_{n=0}^{\infty} a_n\, x^n\; x \in [a, b]$ represents a continuous function.*

(ii) *$\int\limits_{a}^{b} f(x)\, dx = \sum\limits_{n=1}^{\infty} \frac{a_n}{n+1} (b^{n+1} - a^{n+1})$ (Term by term integration is possible)*

Proof: (i) First observe that, as $x \in [a, b]$ we have, $|x| < R$ and hence, $\sum\limits_{n=0}^{\infty} a_n\, x^n$ converges, so it is justified to write,

$$f(x) = \sum_{n=0}^{\infty} a_n\, x^n \quad \text{for } x \in [a, b]$$

If $l = \max\{|a|, |b|\}$ then $0 < l < R$ so we can choose some c such that $0 < l < c < R$. Then by definition of R the power series converges absolutely for $x = c$. By Theorem 8.3.1 the power series converges absolutely in $[-l, l]$ and hence in $[a, b]$ as $[a, b] \subset [-l, l]$. Now observe that $a_n x^n$ is continuous for each n and as the series $\sum\limits_{n=0}^{\infty} a_n x^n$ is uniformly convergent in $[a, b]$ the sum function $f(x)$ is continuous there by Theorem 7.6.2.

(ii) As $f(x)$ is continuous it is integrable in $[a, b]$. Also we see that each $a_n x^n$ is continuous hence integrable in $[a, b]$ for each n and as seen in part (a) the series $\sum\limits_{n=0}^{\infty} a_n x^n$ is uniformly convergent there hence by Theorem 7.7.2 we get

$$\int\limits_a^b f(x)\, dx = \sum_{n=1}^{\infty} \int\limits_a^b a_n x^n\, dx = \sum_{n=1}^{\infty} \frac{a_n}{n+1}(b^{n+1} - a^{n+1}).$$

Hence the theorem.

Remarks

1. If we closely look at the first part of the theorem it states that for any $\varepsilon > 0$ the function $f(x) = \sum\limits_{n=0}^{\infty} a_n x^n$ is defined and continuous in $[-R + \varepsilon, R - \varepsilon]$. Also if $|c| < R$ be any element then we can find a suitable $\delta > 0$ such that $I = [c - \delta, c + \delta] \subset (-R, R)$ hence $f(x)$ is continuous in I and in particular at $x = c$. Hence we see that the function $f(x) = \sum\limits_{n=0}^{\infty} a_n x^n$ is defined and continuous in $(-R, R)$.

2. Consider the power series $\sum\limits_{n=0}^{\infty} \frac{a_n}{n+1} x^{n+1}$ which is obtained by term by term integration of the power series $\sum\limits_{n=0}^{\infty} a_n x^n$ with radius of convergence R. Now by using Cauchy-Hadamard formula we have,

$$\lim_{n \to \infty} \left| \frac{a_n}{n+1} \right|^{1/n} = \lim_{n \to \infty} |a_n|^{1/n} \lim_{n \to \infty} \left(\frac{1}{n+1} \right)^{1/n}$$

$$= R \left(\text{as } \lim_{n \to \infty} \left(\frac{1}{n+1} \right)^{1/n} = 1 \right)$$

Hence we conclude that integrated power series has same radius of convergence as the original power series.

Theorem 8.5.2

Let $\sum\limits_{n=0}^{\infty} a_n x^n$ be a power series with non-zero radius of convergence R. Then the function defined by $f(x) = \sum\limits_{n=0}^{\infty} a_n x^n$, $-R < x < R$, is a differentiable function and moreover,

$$f'(x) = \sum_{n=1}^{\infty} n a_n x^{n-1}, \quad -R < x < R.$$

Proof: First note that by Remark 1 above it is justified to write $f(x) = \sum_{n=0}^{\infty} a_n x^n$, $-R < x < R$ as the series converges for every x in that interval. We use Cauchy-Hadamard formula to find the radius of convergence R' of the series $\sum_{n=0}^{\infty} a_n x^{n-1}$ which is the differentiated series of the given power series. Now,

$$R' = \lim_{n \to \infty} |na_n|^{1/n} = \lim_{n \to \infty} n^{1/n} \lim_{n \to \infty} |a_n|^{1/n}$$

$$= \lim_{n \to \infty} |a_n|^{1/n} = R \ (\text{as } \lim_{n \to \infty} n^{1/n} = 1)$$

Hence we find that radius of convergence of the differentiated series is same as the original series so the series $\sum_{n=1}^{\infty} na_n x^{n-1}$ is uniformly convergent in the interval $(-R, R)$. We now use Theorem 7.8.1 to conclude the sum function $f(x)$ is differentiable $\forall x$ in $(-R, R)$ and

$$f'(x) = \sum_{n=1}^{\infty} na_n x^{n-1}, \quad -R < x < R$$

We obtain the following immediate corollaries:

Corollary 8.5.3

The function $f(x) = \sum_{n=0}^{\infty} a_n x^n$ represents an infinitely differentiable function in $(-R, R)$ where $\sum_{n=0}^{\infty} a_n x^n$ is a power series with radius of convergence R. Moreover, for any $k \in \mathbb{N}$,

$$f^{(k)}(x) = \sum_{n=k}^{\infty} \frac{a_n n!}{(n-k)!} x^{n-k}, \quad -R < x < R \tag{8.5.3.1}$$

In particular,

$$f^{(k)}(0) = a_k \, k! \ k \in \mathbb{N} \ \text{ and } a_0 = f^{\circ}(0) = f(0) \tag{8.5.3.2}$$

Proof: We have seen while proving the above theorem that if the radius of convergence of a power series $f(x) = \sum_{n=0}^{\infty} a_n x^n$ is R then the radius of convergence of the series $f'(x) = \sum_{n=1}^{\infty} na_n x^{n-1}$ is also R. We can now apply previous theorem on $f'(x)$ to show that it is differentiable and

$$f^{(2)}(x) = \sum_{n=2}^{\infty} \frac{a_n n!}{(n-2)!} x^{n-2}, \quad -R < x < R.$$

Using inductive logic we can prove (8.5.3.1) & (8.5.3.2) is an easy consequence of (8.5.3.1).

From the above corollary we get that a power series represents an infinitely differentiable function within the interval of convergence.

8.6 REAL ANALYTIC FUNCTIONS AND POWER SERIES

We first define what are called **Real Analytic function**. Let f be an infinitely differentiable function defined on an open interval I such that for every point $c \in I$ there is a real number $R > 0$ such that,

$$f(x) = \sum_{n=0}^{\infty} \frac{f^n(c)}{n!} (x-c)^n \ \forall x \in \{x \in I: |x-c| < R\}$$

then f is said to be a real analytic function.

From the definition it is clear that a function is real analytic if the Taylor's series of the function around any point in the domain of definition converges to the function in some neighbourhood of that point.

Now Corollary 8.5.3 tells us that any power series is an infinitely differentiable function in its interval of convergence so a natural questions arises does Maclaurin's Series expansion of this function exists? If yes, then how it looks like? The answer is given by the following theorem:

Theorem 8.6.1

The Maclaurin's series expansion of the power series $f(x) = \sum\limits_{n=0}^{\infty} a_n\, x^n$ *exists in* $(-R < R)$ *where R is the radius of convergence is equal to itself.*

Proof: From Corollary 8.5.4 we get $f^{(k)}(0) = a_k\, k!$ and $a_0 = f^0(0) = f(0)$. Hence we can write,

$$f(x) = \sum_{n=0}^{\infty} \frac{f^{(n)}(0)}{n!}\, x^n.$$

Also note that we can write $f(x) = \sum\limits_{k=0}^{n} a_k\, x^k + \sum\limits_{k=n}^{\infty} a_k\, x^k = \sum\limits_{k=0}^{n} a_k\, x^k + R_n$ where R_n is the remainder after n terms. Now as $\sum\limits_{n=0}^{\infty} a_n\, x^n$ converges for $x' (-R, R)$ we have $R_n \in 0$ as $n \to \infty$ that is, condition for Maclaurin's expansion is automatically satisfied.

In view of Theorem 8.6.1 we can conclude that any power series is a real analytic function.

8.7 ABEL'S THEOREM

We have seen that a given power series $\sum\limits_{n=0}^{\infty} a_n\, x^n$ with radius of convergence R converges absolutely for $x \in (-R, R)$ but we still have no conclusion for the points $x = \pm R$. The answer is provided by Abel's Theorem.

Theorem 8.5.1 (Abel's Theorem)

For a power series $\sum\limits_{n=0}^{\infty} a_n\, x^n$ *with radius of convergence R if we define*

$$f(x) = \sum_{n=0}^{\infty} a_n\, x^n, \ -R < x < R$$

and if $\sum\limits_{n=0}^{\infty} a_n R^n$ *is convergent then,* $\lim\limits_{x \to R^-} f(x) = \sum\limits_{n=0}^{\infty} a_n R^n$. *Further if we assume that* $\sum\limits_{n=0}^{\infty} a_n (-R)^n$ *is convergent then,* $\lim\limits_{x \to -R^+} f(x) = \sum\limits_{n=0}^{\infty} a_n (-R)^n$.

Proof: We give a proof using Theorem 7.10.1. For $n \in \mathbb{N}$ let $f_n(x) = a_n R^n$, $x \in [0, R]$ (constant function) then trivially $\sum\limits_{n=1}^{\infty} f_n(x)$ converges uniformly in $[0, R]$. We define $g_n(x) = \dfrac{x^n}{R^n}$, $x \in [0, R]$ then $\{g_n(x)\}$ is monotone sequence for each $x \in [0, R]$ and $|g_n(x)| < 2$ for all $x \in [0, R]$ hence $\{g_n(x)\}$ is uniformly bounded. Hence by

Theorem 7.10.1 the series $\sum_{n=0}^{\infty} f_n(x)\, g_n(x) = \sum_{n=0}^{\infty} a_n\, x^n$ converges uniformly on $[0, R]$. Now we use Theorem 7.6.2 to get $f(x)$ is continuous on $[0, R]$ and hence we write,

$$\lim_{x \to R^-} f(x) = \sum_{n=0}^{\infty} a_n\, R^n$$

For the next part we first observe that $\sum_{n=0}^{\infty} a_n\, x^n$ converges in $[-R, R]$ if and only if $\sum_{n=0}^{\infty} (-1)^n\, a_n x^n$ converges in $[0, R]$ then conclusion follows by a similar argument as the first part.

Example 8.6 Using series expansion of $(1 + x)^{-1}$ show that,

$$\log 2 = 1 - \frac{1}{2} + \frac{1}{3} - \frac{1}{4} \ldots$$

Sol: We know that,

$$(1 + x)^{-1} = 1 - x + x^2 - x^3 + x^4 - \ldots -1 < x < 1 \tag{1}$$

By using Cauchy-Hadamard formula we get radius of convergence of (1) is 1. Hence we can integrate (1) term by term in the interval $[0, t]$ where $0 < t < 1$ and obtain,

$$\int_0^t (1 + x)^{-1}\, dx = \sum_{n=0}^{\infty} \int_0^t (-1)^n\, x^n\, dx = \sum_{n=0}^{\infty} (-1)^n\, \frac{t^{n+1}}{n+1}$$

or,
$$\log (1 + t) = \sum_{n=0}^{\infty} (-1)^n \frac{t^{n+1}}{n+1} \quad 0 < t < 1 \tag{2}$$

similarly we get the same series (2) if we integrate series (1) in the interval $[t, 0]$ for $-1 < t < 0$. Hence we conclude that,

$$\log (1 + t) = \sum_{n=0}^{\infty} (-1)^n \frac{t^{n+1}}{n+1}\, |t| < 1 \tag{3}$$

We again observe that (3) is a power series with radius of convergence 1 (using Cauchy- Hadamard formula). The right hand series at $t = 1$ is given by $\sum_{n=0}^{\infty} \frac{(-1)^n}{n+1}$ which converges by Leibniz Test for alternating series so by Abel's Theorem we have,

$$\lim_{x \to 1^-} \log (1 + x) = \sum_{n=0}^{\infty} \frac{(-1)^n}{n+1}$$

or,
$$\log 2 = \sum_{n=0}^{\infty} \frac{(-1)^n}{n+1}$$

$$= 1 - \frac{1}{2} + \frac{1}{3} - \frac{1}{4} \ldots \text{ (using continuity of } \log x)$$

Example 8.7 Expand $\tan^{-1} x$ in power series and deduce that,

$$\frac{\pi}{4} = 1 - \frac{1}{3} + \frac{1}{5} - \frac{1}{7} + \ldots \text{ (Gregory's series)}$$

Sol: We start with the expansion of $(1 + x^2)^{-1}$ which is given as

$$(1 + x^2)^{-1} = x^2 + x^4 - x^6 + x^8 - \ldots -1 < x < 1 \tag{1}$$

Radius of convergence of power series (1) is 1 (verify). As in the previous example by integrating (1) term by term in the interval $[0, t]$ $0 < t < 1$ we obtain,

$$\tan^{-1} t = \int_0^t (1 + x^2)^{-1}\, dx = t - \frac{t^3}{3} + \frac{t^5}{5} - \frac{t^7}{7} + \ldots\ 0 < t < 1. \tag{2}$$

Again as in previous example by integrating (1) term by term in the interval $[t, 0]$ $-1 < t < 0$ we obtain same equation as (2). Hence we write,

$$\tan^{-1} t = \int_0^t (1 + x^2)^{-1}\, dx = t - \frac{t^3}{3} + \frac{t^5}{5} - \frac{t^7}{7} + \ldots\ |t| < 1. \tag{3}$$

At $t = 1$ the right-hand series is given by

$$1 - \frac{1}{3} + \frac{1}{5} - \frac{1}{7} + \ldots$$

which converges by Leibniz Test. Now by using Abel's Theorem,

$$\lim_{x \to 1^-} \tan^{-1} x = 1 - \frac{1}{3} + \frac{1}{5} - \frac{1}{7} + \ldots$$

or,
$$\frac{\pi}{4} = \tan^{-1} 1 = 1 - \frac{1}{3} + \frac{1}{5} - \frac{1}{7} + \ldots \qquad \text{(using continuity of } \tan^{-1} x)$$

PROBLEMS

1. Find radius of convergence of the following series:

(i) $\displaystyle\sum_{n=1}^{\infty} \frac{n^3 x^n}{n!}$

(ii) $\displaystyle\sum_{n=1}^{\infty} a_n x^n$ where $a_{2n} = \dfrac{1}{5^n}, n = 1, 2 \ldots$ & $a_{2n-1} = \dfrac{1}{4^{n+1}}, n = 1, 2 \ldots$

(iii) $\displaystyle\sum_{n=1}^{\infty} \frac{n x^n}{(1 + n)^2}$

(iv) $x + \dfrac{(1.\,2)^2}{4!} x^2 + \dfrac{(1.\,2.\,3)^2}{6!} x^3 + \ldots + \dfrac{(n!)^2}{2(n)!} x^n + \ldots$

2. Let $\displaystyle\sum_{n=0}^{\infty} a_n x^n$ be a power series such that $0 < a \le |a_n| \le b$. Find the radius of convergence of this series.

3. Define the function $f(x)$ as follows:

$$f(x) = 1 + \frac{1}{1!} x + \frac{1}{2!} x^2 + \frac{1}{3!} x^3 + \ldots - \infty < x < \infty.$$

Prove that $f(x)$ is well defined and show that $f'(x) = f(x)$.

4. Define $f(x)$ and $g(x)$ as follows:

$$f(x) = x - \frac{x^3}{3!} + \frac{x^5}{5!} - \cdots$$
$$g(x) = 1 - \frac{x^2}{2!} + \frac{x^4}{4!} - \cdots$$

$$-\infty < x < \infty$$

Show that

  (i) $f'(x) = g(x)$ and $g'(x) = f(x)$

 (ii) $(f^2(x) + g^2(x))' = 0$

(iii) $f^2(x) + g^2(x) = 1$

Function of Two Variables

9.1 INTRODUCTION

In Chapter 2 we have studied real valued functions of real variables or real valued functions of a single variable. In this chapter we propose to study functions of two independent variables or functions which are defined on a subset of $\mathbb{R}^2$ (or $\mathbb{R}^2$). Where $\mathbb{R}^2$ is the set of all ordered tuples of the form (a, b) a, b are real numbers in other words, $\mathbb{R}^2$ is Cartesian product $\mathbb{R} \times \mathbb{R}$. We will in general, study the same concepts of limit, continuity, differentiability etc as introduced in Chapter 2 and while doing so, it will be evident that although the definitions are very similar in nature but not same. For example, existence of limit of a function defined in a subset of $\mathbb{R}^2$ at a point is much stronger concept than existence of limit at a point of a function of one variable. To begin our study we start with some basic facts about $\mathbb{R}^2$.

9.2 BASIC FACTS OF $\mathbb{R}^2$

In this section we will discuss and define some basic concepts of $\mathbb{R}^2$ which will be needed to study functions of two variables. We begin by defining the elements of the set $\mathbb{R}^2$. An element x of $\mathbb{R}^2$ is an ordered tuple (x_1, x_2) where $x_1, x_2 \in \mathbb{R}$. It is clear that there is an one to one correspondence between the points of a plane and $\mathbb{R}^2$. That is given any point (x_1, x_2) of $\mathbb{R}^2$ we can plot the point in the co-ordinate plane with the cartesian coordinate as x_1 & x_2 taking two mutually perpendicular axes as X-axis and Y-axis and vice-versa. x_1 & x_2 are called X and Y *coordinates* of the point. Also observe that any point

$P(x_1, x_2)$ of $\mathbb{R}^2$ can be regarded as a *vector* with magnitude as $\sqrt{x_1^2 + x_2^2}$ and

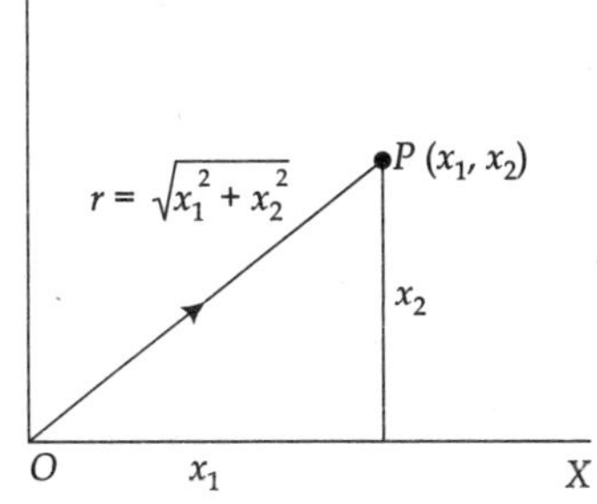

Fig. 9.1

direction as OP where O is the **origin** representing the point $\mathbf{0} = (0, 0)$ of $\mathbb{R}^2$ as shown in the figure (Fig. 9.1).

We want to remind our readers that although we will give relevant definitions in context of $\mathbb{R}^2$ only but all these concepts will have analogous definitions in $\mathbb{R}^n$ for any positive integer n. For example any element of $\mathbb{R}^n$ is a n- tuple $(x_1, x_2,.......,x_n)$ where $x_i \in \mathbb{R}$, $i = 1, 2...n$.

We define two operations in $\mathbb{R}^2$ as follows:

(i) If $\mathbf{x} = (x_1, x_2)$ & $\mathbf{y} = (y_1, y_2) \in \mathbb{R}^2$ then their **sum** denoted by $\mathbf{x} + \mathbf{y}$ is defined by the equation,

$$\mathbf{x} + \mathbf{y} = (x_1 + y_1, x_2 + y_2)$$

(ii) If $\mathbf{x} = (x_1, x_2) \in \mathbb{R}^2$ then **multiplication** of $\mathbf{x}$ with any real number a denoted by $a\mathbf{x}$ is defined by the equation,

$$a\mathbf{x} = (ax_1, ax_2)$$

Note that we for $\mathbf{x}, \mathbf{y} \in \mathbb{R}^2$ we will define $\mathbf{x} - \mathbf{y}$ as $\mathbf{x} + (-\mathbf{y})$.

The **Euclidean norm** or the **distance** of an element $\mathbf{x} = (x_1, x_2) \in \mathbb{R}^2$ from the origin $\mathbf{0} = (0, 0)$ denoted by $\|\mathbf{x}\|$ is defined by,

$\|\mathbf{x}\| = \sqrt{x_1^2 + x_2^2}$ Thus the distance between two elements $\mathbf{x} = (x_1, x_2)$ & $\mathbf{y} = (y_1, y_2) \in \mathbb{R}^2$ is defined by the equation,

$$\|\mathbf{x} - \mathbf{y}\| = \sqrt{(x_1 - y_1)^2 + (x_2 - y_2)^2}$$

For all $\mathbf{x}$ & $\mathbf{y} \in \mathbb{R}^2$ and $a \in \mathbb{R}$ Euclidean norm $\|\cdot\|$ satisfies following properties,

(i) $\|\mathbf{x}\| \geq 0$

(ii) $\|a\mathbf{x}\| = |a| \|\mathbf{x}\|$

(iii) $\|\mathbf{x} + \mathbf{y}\| \leq \|\mathbf{x}\| + \|\mathbf{y}\|$

The property (iii) is called **triangular inequality**.

An **open ball** (or **open disc**) in $\mathbb{R}^2$ with centre $\mathbf{a} = (a_1, a_2)$ and radius $\delta > 0$ denoted by, $B(\mathbf{a}, \delta)$ is defined by the equation,

$$B(\mathbf{a}, \delta) = \{\mathbf{x} \in \mathbb{R}^2 : \|\mathbf{x} - \mathbf{y}\| < \delta\}$$

A **closed ball** (or **closed disc**) in $\mathbb{R}^2$ with centre $\mathbf{a} = (a_1, a_2)$ and radius $\delta > 0$ denoted by, $\vec{B}(a, \delta)$ is defined by the equation,

$$\overline{B}(a, \delta) = \{\mathbf{x} \in \mathbb{R}^2 : \|\mathbf{x} - \mathbf{y}\| \leq \delta\}$$

Let $D \subset \mathbb{R}^2$ then a point $x \in D$ is called an **interior point** if there exists some $\delta > 0$ such that $B(x, \delta) \subset D$. A set $U \subset \mathbb{R}^2$ is called **open** if it's every point is an interior point. An open ball is trivially open. If (a, b) & (c, d) are two open intervals in $\mathbb{R}$ then t their cartesian product $(a, b) \times (c, d)$, is known as **open rectangle** . It is easy to see that an open rectangle is an open set in $\mathbb{R}^2$. By an **open neighbourhood** of a point $\mathbf{x}$ of $\mathbb{R}^2$ we mean an open set U such that $x \in U$. In particular, an open ball or an open rectangle containing the point are open neighbourhoods of $\mathbf{x}$ which are called **circular** and **rectangular** open neighbourhoods. Now observe that given any open ball we can find a sufficiently small open rectangle which lies completely inside the ball, similarly given any open rectangle it is possible to find an open ball lying inside the rectangle (Fig. 9.2 & Fig. 9.3) in this sense an open ball or an open rectangle are **equivalent**.

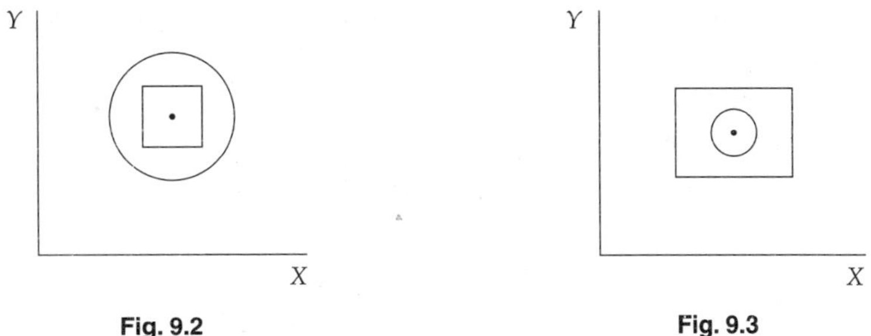

Fig. 9.2   **Fig. 9.3**

A set $F(\subset \mathbb{R}^2)$ is ***closed*** if it's complement in $\mathbb{R}^2$ is open. A closed ball is an example of a closed set. If $[a, b]$ & $]c, d]$ are two closed intervals in $\mathbb{R}$ then their cartesian product $[a, b] \times [c, d]$ known as ***closed rectangle*** is a closed set in $\mathbb{R}^2$.

A point $\mathbf{a} \in R^2$ is called a ***limit point*** of a set $D \subseteq \mathbb{R}^2$ if every open neighbourhood of $\mathbf{a}$ contains at least one point of D other than $\mathbf{a}$. Note that a limit point of a set of a set may or may not belong to the set. For example, consider the unit circle $C = \{(x, y): x^2 + y^2 = 1\}$ then it is easy to see that every point of C is the limit point of both the open and closed unit disc $\overline{D} = \{(x, y): x^2 + y^2 < 1\}$ and $\overline{D} = \{(x, y): x^2 + y^2 \leq 1\}$. Also see that every point of C belongs to $\overline{D}$ whereas no point of C belongs to D.

Let $D \subseteq \mathbb{R}^2$ then any mapping $f: \mathbb{R}^2 \to \mathbb{R}$ is called ***function of two variables*** and image of any point $\mathbf{x} = (x, y) \in D$ under f will be either denoted by $f(x, y)$ or $f(\mathbf{x})$.

9.3 LIMIT OF A FUNCTION

Let us recall the definition of limit of a function of single variable. We say that $f(x) \to l$ as $x \to a$ if the values of $f(x)$ are '*arbitrarily close*' to l as x approaches a from any *possible direction* which is only *two* in this case from right or from left of a. Now it is clear that a point in $\mathbb{R}^2$ can be approached in infinite ways for example if we take the point to be the **origin** $\mathbf{0} = (0, 0)$ then it can be approached through straight lines of the form $y = mx$ $(m \neq 0)$, parabolas of the form $y = mx^2$ $(m \neq 0)$, X-axis, Y-axis and infinitely many other ways. So in two variable case by the statement '$f(\mathbf{x})$ tends to l as $\mathbf{x}$ approaches a' we will mean that the values $f(x)$ becomes '*arbitrarily close*' to l as $\mathbf{x}$ approaches a along any direction (which are *infinite* in this case). Hence in case we find that the limiting value of $f(\mathbf{x})$ depends on the path through which x approaches $\mathbf{a}$ we will certainly conclude that the limit does not exist. We now give the following definition:

Definition Suppose $f: D \to \mathbb{R}$ where $D \subseteq \mathbb{R}^2$ and let $\mathbf{a} = (a_1, a_2) \in \mathbb{R}^2$ be a limit point of D. For a real number l we say that limit of $f(\mathbf{x})$ as x tends to a is l and write $\lim\limits_{x \to a} f(x) = l$ (or $\lim\limits_{(x, y) \to (a_1, a_2)} f(x, y)$) if given for any given $\varepsilon > 0$ there exists a open neighbourhood U_a of $\mathbf{a}$ such that,

$$|f(\mathbf{x}) - l| < \varepsilon$$

for all
$$\mathbf{x} \neq \mathbf{a} \text{ in } U_a.$$

Remarks

1. As in of one variable case $f(\mathbf{a})$ may or may not be defined and even if $f(\mathbf{a})$ is defined it may or may satisfy the relation $|f(\mathbf{x}) - l| < \varepsilon$, however we will not be concerned about the point $\mathbf{a}$ while discussing limit of $f(\mathbf{x})$ at $\mathbf{x} = \mathbf{a}$.

Note that by definition every point of an open set is an interior point hence open neighbourhood

2. U_a in the above definition can be replaced by a circular neighbourhood (or by a rectangular neighbourhood as they are equivalent).

3. We also observe that $\lim_{x \to a} f(\mathbf{x})$ is equivalent to $\lim_{u \to 0} f(\mathbf{u})$ where $\mathbf{u} = \mathbf{x} - \mathbf{a}$.

Example 9.1   Show that $\displaystyle\lim_{(x,\,y) \to (0,\,0)} \frac{x^2}{\sqrt{x^2 + y^2}} = 0$

Sol: See that $f(0, 0)$ does not exist in this case. Let $f(x, y) = \dfrac{x^2}{\sqrt{x^2 + y^2}}$

Now observe that,

$$x^2 \le x^2 + y^2$$

$$\Rightarrow \qquad \frac{x^2}{\sqrt{x^2 + y^2}} \le \sqrt{x^2 + y^2}$$

$$f(x, y) \le \sqrt{x^2 + y^2}$$

For $\varepsilon > 0$ we choose $\delta = \varepsilon$ then we get from above inequality that,

$$|f(x, y) - 0| < \varepsilon \quad \text{whenever } \sqrt{x^2 + y^2} < \delta \text{ i.e. } x^2 + y^2 < \delta^2$$

This proves $\displaystyle\lim_{(x,\,y) \to (0,\,0)} \frac{x^2}{\sqrt{x^2 + y^2}} = 0.$

Example 9.2   Show that $\displaystyle\lim_{(x,\,y) \to (0,\,0)} xy = 0$

Sol: First note that this case $f(0, 0)$ is defined (equals 0). Let $\varepsilon > 0$ be any real number. Now,

$$|xy| = |x|\,|y| \le \sqrt{x^2 + y^2}\,\sqrt{x^2 + y^2} \quad \left(\text{as } |x| \le \sqrt{x^2 + y^2} \ \& \ |y| \le \sqrt{x^2 + y^2}\right)$$

$$= x^2 + y^2$$

If we choose $\delta = \varepsilon$ then,

$$|xy| < \varepsilon \text{ whenever } x^2 + y^2 < \delta$$

Thus for $\varepsilon > 0$ we have got a circular neighbourhood of $\mathbf{0} = (0, 0)$ namely $B(\mathbf{0}, \delta)$ such that for every $\mathbf{x} \ne (0, 0)$ in $B(\mathbf{a}, \delta)$ we have $|f(x, y)| < \varepsilon$ hence we get $\displaystyle\lim_{(x,\,y) \to (0,\,0)} xy = 0.$

Observe that in this case $f(0, 0)$ satisfies the relation $|f(x, y)| < \varepsilon$.

Example 9.3 Show that $\displaystyle\lim_{(x,\,y)\to(0,\,0)}\frac{x^2-y^2}{x^2+y^2}$ does not exist.

Proof: Let us approach $(0, 0)$ by line of the form $y = mx$. See that as x tends to zero on this line y also tends to zero hence,

$$\lim_{(x,\,y)\to(0,\,0)}\frac{x^2-y^2}{x^2+y^2}=\lim_{x\to 0}\frac{x^2-m^2x^2}{x^2+m^2x^2}$$

$$=\lim_{x\to 0}\frac{1-m^2}{1+m^2}=\frac{1-m^2}{1+m^2}$$

We see that the required limit is not unique as it depends on m hence we conclude that

$$\lim_{(x,\,y)\to(0,\,0)}\frac{x^2-y^2}{x^2+y^2}\text{ does not exist.}$$

Suppose $f\colon D \to \mathbb{R}^2$ be a function and let $P(a_1, a_2)$ be a limit point of D. Let us keep x fixed and consider $f(x, y)$ as a function of y then if $f(x, y)$ tends to a finite limit as $y \to a_2$ then the limit is a function of x let us denote it by $\psi(x)$ that is,

$$\psi(x) = \lim_{y\to a_2} f(x, y)$$

Now if, $\displaystyle\lim_{x\to x_0}\psi(x) = l_1$ exists we call l_1 the ***repeated limit of f as*** $y\to a_2$ ***and*** $x\to a_1$ & write,

$$\lim_{x\to a_1}\left(\lim_{y\to a_2} f(x, y)\right) = l_1$$

Similarly we can define the ***repeated limit of f as*** $x\to a_1$ ***and*** $y\to a_2$ & write,

$$\lim_{x\to a_2}\lim_{y\to a_1} f(x, y) = l_2$$

We first show that the two repeated limit may not be equal by following example,

Example 9.4 Let f be a function defined by

$$f(x, y) = \frac{x^2-y^2}{x^2+y^2},\ (x, y) \neq (0, 0)$$

Show that $\displaystyle\lim_{x\to 0}\left(\lim_{y\to 0} f(x, y)\right)$ & $\left(\displaystyle\lim_{y\to 0}\lim_{x\to 0} f(x, y)\right)$ are not equal.

Sol: We first calculate $\displaystyle\lim_{y\to 0} f(x, y) = \lim_{y\to 0}\frac{x^2-y^2}{x^2+y^2}$

$$= \lim_{y\to 0}\frac{x^2}{x^2} = 1 \quad \text{(treating } x \text{ as a constant)}$$

Hence
$$\lim_{x \to 0} \left(\lim_{y \to 0} f(x, y) \right) = \lim_{x \to 0} 1 = 1$$

Similarly
$$\lim_{y \to 0} \left(\lim_{x \to 0} f(x, y) \right) = \lim_{y \to 0} \left(\lim_{x \to 0} \frac{x^2 - y^2}{x^2 + y^2} \right)$$

$$\lim_{y \to 0} \frac{-y^2}{y^2} = -1$$

Hence we conclude that the two repeated limits are not equal.

Next we show even if the repeated limits of a function at a certain point are equal the limit of the function at that point may not exist.

Example 9.5   For the function f defined by,

$$f(x, y) = xy \, \frac{x^2 - y^2}{x^4 + y^4}, \ (x, y) \neq (0, 0)$$

show that both the repeated limits at $(0, 0)$ exists and are equal but $\lim_{(x, y) \to (0, 0)} f(x, y)$ does not exist.

Sol: It is easy to see that $\lim_{x \to 0} \left(\lim_{y \to 0} f(x, y) \right) = 0 = \lim_{y \to 0} \left(\lim_{x \to 0} f(x, y) \right)$

Now to find $\lim_{(x, y) \to (0, 0)} f(x, y)$ we approach $(0, 0)$ by line $y = mx$ then we have,

$$\lim_{(x, y) \to (0, 0)} f(x, y) = \lim_{x \to 0} mx^2 \, \frac{x^2 - m^2 x^2}{x^4 + m^4 x^4}$$

$$= \lim_{x \to 0} m \, \frac{1 - m^2}{1 + m^4} = \frac{m(1 - m^2)}{1 + m^4}$$

As we see that $\lim_{(x, y) \to (0, 0)} f(x, y)$ depends on m we conclude that the limit does not exist.

Following theorem tells us that there exists some relation between limit of a function at a point and repeated limits at that point.

Theorem 9.3.1

Suppose $f\colon D \subseteq (\mathbb{R}^2) \to \mathbb{R}$ and (a_1, a_2) be a limit point of D. If we assume

$\lim_{(x, y) \to (a_1, a_2)} f(x, y)$ & $\lim_{x \to a_1} \left(\lim_{y \to a_2} f(x, y) \right)$ both exist then they must be equal. Hence if limit of f exists at a

point then repeated limits at that point cannot exist without being equal.

Proof: Let us write $\lim_{x \to a_1} \left(\lim_{y \to a_2} f(x, y) \right) = \lim_{x \to a_1} \psi(x) = l_1$ where $\psi(x) = \lim_{y \to a_2} f(x, y)$. Now by definition of

limit of a function of one variable, for $\varepsilon > 0$ we will obtain some $\delta_1 > 0$ such that,

$$|f(x, y) - \psi(x)| < \frac{\varepsilon}{3} \quad \text{whenever } 0 < |y - a_2| < \delta_1 \tag{9.3.1.1}$$

Also we will obtain some $\delta_1 > 0$ such that,

$$|f(x, y) - l_1| < \frac{\varepsilon}{3} \quad \text{whenever } 0 < |x - a_1| < \delta_2 \tag{9.3.1.2}$$

If we assume $\lim\limits_{(x, y)\to(a_1, a_2)} f(x, y) = l$ then from definition we will obtain some $\delta_3 > 0$ such that,

$$|f(x, y) - l| < \frac{\varepsilon}{3} \text{ whenever } 0 < |x - a_1| < \delta_3,\ 0 < |y - a_2| < \delta_3 \tag{9.3.1.3}$$

Now if $\delta = \min(\delta_1, \delta_2, \delta_3)$ and choose x, y such that $0 < |x - a_1| < \delta$, and $0 < |y - a_2| < \delta$ then they satisfy all above three inequalities. Hence we have,

$$|l - l_1| = |l - f(x, y) + f(x, y) - \psi(x) + \psi(x) - l_1|$$

$$\leq |l - f(x, y)| + |f(x, y) - \psi(x)| + |\psi(x) - l_1|$$

$$< \frac{\varepsilon}{3} + \frac{\varepsilon}{3} + \frac{\varepsilon}{3} = \varepsilon \quad \text{(using 9.3.1.1, 9.3.1.2 \& 9.3.3)}$$

This implies $\quad l = l_1$

Similarly if $\lim\limits_{y\to a_2} \left(\lim\limits_{x\to a_1} f(x, y) \right) = l_2$ we can show that $l = l_2$. That is $l_1 = l_2$. Hence we conclude that if the limit of f exists at a point then repeated limits at that point cannot exist without being equal.

As an immediate consequence we obtain the following

Corollary 9.3.2

Suppose $f : D \subseteq (\mathbb{R}^2) \to \mathbb{R}$ and $P(a_1, a_2)$ be a limit point of D. If the two repeated limits at P are unequal then the limit of the function f at (a_1, a_2) does not exist.

Example 9.6 If $f(x, y) = x^2 \sin \dfrac{1}{y^2} + y^2 \sin \dfrac{1}{x^2}$, $(x, y) \neq (0, 0)$

then show that $\lim\limits_{(x, y)\to(0, 0)} f(x, y) = 0$ but either of the two repeated limit does not exist.

Sol: We have,

$$\left| x^2 \sin \frac{1}{y^2} + y^2 \sin \frac{1}{x^2} \right| \leq \left| x^2 \sin \frac{1}{y^2} \right| + \left| y^2 \sin \frac{1}{x^2} \right|$$

$$\leq x^2 + y^2 \quad \left(\text{As } \left| \sin \frac{1}{x^2} \right| \leq 1 \ \& \ \sin \frac{1}{y^2} \leq 1 \right)$$

Hence for $\varepsilon > 0$ if we choose $\delta = \sqrt{\varepsilon}$ we have,

$$|f(x, y)| < \varepsilon \quad \text{whenever } x^2 + y^2 < \delta^2$$

So that we conclude $\lim\limits_{(x, y)\to(0, 0)} f(x, y) = 0$.

Now,

$$\lim_{x \to 0} \left(x^2 \sin \frac{1}{y^2} + y^2 \sin \frac{1}{x^2} \right)$$

$$= \lim_{x \to 0} \left(y^2 \sin \frac{1}{x^2} \right) \text{ does not exist} \left(\text{as } \lim_{x \to 0} \sin \frac{1}{x^2} \text{ does not exist} \right).$$

Hence $\lim_{y \to 0} (\lim_{x \to 0} f(x, y))$ does not exist.

Similarly as $\lim_{x \to 0} \sin \frac{1}{y^2}$ does not exist $\lim_{x \to 0} \left(\lim_{y \to 0} f(x, y) \right)$ also does not exist.

Thus we see that both the repeated limit does not exist in this case.

Following theorem is what is known as Algebra of limits and can be proved using similar arguments given in one variable case

Theorem 9.3.3

Let f, g be two functions defined on $D \subseteq \mathbb{R}^2$ and let $a = (a_1, a_2) \in \mathbb{R}^2$ be a limit point of D. If $\lim_{(x, y) \to (a_1, a_2)} f(x, y) = l$ & $\lim_{(x, y) \to (a_1, a_2)} g(x, y) = m$ then,

(i) $\quad \lim_{(x, y) \to (a_1, a_2)} (f(x, y) \pm g(x, y)) = l \pm m$

(ii) $\quad \lim_{(x, y) \to (a_1, a_2)} (f(x, y) g(x, y)) = lm$

(iii) $\quad \lim_{(x, y) \to (a_1, a_2)} \left(\frac{1}{g(x, y)} \right) = \frac{1}{m}$, *provided $m \neq 0$ and $g(x, y) \neq 0$ for all $(x, y) \in D$.*

Example 9.7   Evaluate, $\lim_{(x, y) \to (0, 0)} \dfrac{x^2 y + 1}{1 + 2xy}$

Sol:

$$\lim_{(x, y) \to (0, 0)} \frac{x^2 y + 1}{1 + 2xy} = \frac{\lim_{(x, y) \to (0, 0)} (x^2 y + 1)}{\lim_{(x, y) \to (0, 0)} (1 + 2xy)}$$

$$= \frac{\lim_{(x, y) \to (0, 0)} (x^2 y) + 1}{\lim_{(x, y) \to (0, 0)} (2xy) + 1}$$

$$= \frac{0 + 1}{0 + 1} = 1$$

9.4 CONTINUITY OF FUNCTIONS

Definition Suppose $f: D \subseteq (\mathbb{R}^2) \to \mathbb{R}$ let $\mathbf{a} = (a_1, a_2)$ a point of D. We say f is continuous at $\mathbf{a}$ if given any if given any $\varepsilon > 0$ there exists an open neighbourhood U_a of $\mathbf{a}$ such that,

$$|f(x, y) - f(a_1, a_2)| < \varepsilon \text{ for all } x \in D \cap U_a$$

We say that f is continuous in D if it is continuous at every point of D.

Remarks

1. U_a can be taken as a circular neighbourhood $(x - a_1)^2 + (y - a_2)^2 < \delta^2$ or a rectangular neighbourhood $|x - a_1| < \delta, |y - a_2| < \delta$.

2. From definition it is clear that if f is continuous at $\mathbf{a} = (a_1, a_2)$ then $f(a, b)$ is defined moreover $\lim\limits_{(x,\, y) \to (a,\, b)} f(x, y)$ exists and is equal to $f(a, b)$. Hence it follows that if $f(a, b)$ does not exist then f cannot be continuous at $\mathbf{a}$.

3. It is not very difficult to see that if $f(x, y)$ is continuous at $\mathbf{a} = (a_1, a_2)$ then it is continuous in each variable that is, the function $\alpha(x) = f(x, a_2)$ is continuous at $x = a_1$ and the function $\beta(y) = f(a_1, y)$ is continuous at $y = a_2$.

Example 9.8 Every constant function defined in a subset of $\mathbb{R}^2$ is continuous

Sol: Let $f(x, y) = c$, $(c \in \mathbb{R})$, $(x, y) \in D$. Then for any point $\mathbf{a} = (a_1, a_2) \in D$ we have,

$$|f(x, y) - f(a_1, a_2)| = |c - c| = 0 < \varepsilon$$

for any ε & $\forall (x, y) \in D$. Hence we conclude that f is continuous at every point of D.

Example 9.9 Show that the function $f: \mathbb{R}^2 \to \mathbb{R}$ defined by,

$$f(x, y) = |x| + |y|$$

is continuous at $(0, 0)$.

Sol: Observe that $f(0, 0) = 0$. Then for $\varepsilon > 0$ choose $\delta = \dfrac{\varepsilon}{2}$. Then for x & y such that,

$$|x| < \delta \text{ \& } |y| < \delta \text{ we have,}$$

$$|f(x, y) - 0| = ||x| - |y|| \le |x| + |y| < \varepsilon$$

Thus we have got a rectangular neighbourhood U_0: $|x| < \delta$ & $|y| < \delta$ of $\mathbf{0} = (0, 0)$ such that for $(x, y) \in U_0$, $|f(x, y) - f(0, 0)| < \varepsilon$ that is in other words f is continuous at $\mathbf{0} = (0, 0)$.

Example 9.10 Define a function $f: \mathbb{R} \to \mathbb{R}^2$ as follows,

$$f(x, y) = \begin{cases} x^2 \sin\left(\dfrac{1}{y}\right) + y^2 \cos\left(\dfrac{1}{x}\right) & (x, y) \ne (0, 0) \\ 0 & (x, y) = (0, 0) \end{cases}$$

Show that f is continuous at $(0, 0)$.

Sol: We see that,

$$|f(x, y) - f(0, 0)| \leq \left| x \sin \frac{1}{y} + y \sin \frac{1}{x} \right|$$

$$\leq x + y$$

Given any $\varepsilon > 0$ if we choose $\delta = \dfrac{\varepsilon}{2}$ we have,

$|f(x, y) - f(0, 0)| < \varepsilon$ whenever x & y lies in the rectangular neighbourhood $|x| < \delta$, $|y| < \delta$.

Theorem 9.4.1

Suppose f, g: $D(\subseteq \mathbb{R}^2) \to \mathbb{R}$ be two functions continuous at $\mathbf{a} \in D$ and let r be any real number then,

> *(i) $f \pm g$ is continuous at $\mathbf{a}$*

> *(ii) rf is continuous at $\mathbf{a}$*

> *(iii) $\dfrac{1}{f}$ is continuous at $\mathbf{a}$ provided $f(\mathbf{x}) \neq 0$ for $\mathbf{x} \in D$.*

> *(iv) $|f|$ is continuous at $\mathbf{a}$*

Proof: Proofs are easy and are kept as exercises.

9.5   DIRECTIONAL DERIVATIVES & PARTIAL DERIVATIVES

Suppose $y = f(x)$ represents a function of one variable then we have seen that the increment of the dependent variable y at a point a given by $\Delta y = f(a + h) - f(x)$ is approximately equal to the derivative of the function $f'(a)$ for small increment h of the independent variable x in a direction either to the right or to the left of a. Now consider a function $f(x, y)$ of two variables defined in a domain $D \subseteq \mathbb{R}^2$. Let $\mathbf{a} = (a_1, a_2)$ be an interior point of D. As we know a point in $\mathbb{R}^2$ can move in infinite directions we fix a direction given by an arbitrary point $\mathbf{b} = (b_1, b_2)$ of $\mathbb{R}^2$. When a point moves from $\mathbf{a}$ to a point $\mathbf{a} + h\mathbf{b}$ for small values of $h \in \mathbb{R}$, we will be interested in the quantity $f(\mathbf{a} + h\mathbf{b}) - f(\mathbf{a})$ called the increment of f in the direction of $\mathbf{b}$. It is not difficult to see that the point $\mathbf{a} + h\mathbf{b}$ lies on the line through $\mathbf{a}$ and parallel to $\mathbf{b}$ and also for small values h the point lies inside D. This study gives rise to the notion of what are called **directional derivatives** defined below.

Definition   Suppose f be a real valued function with domain $D \subseteq \mathbb{R}^2$ and let $\mathbf{a} = (a_1, a_2)$ be an interior point of D. For any arbitrary point $\mathbf{b} = (b_1, b_2)$ of $\mathbb{R}^2$ the ***directional derivative*** of f at $\mathbf{a}$ in the direction $\mathbf{b}$ denoted by $D_a(f; \mathbf{b})$ is defined by the equation,

$$D_a(f; \mathbf{b}) = \lim_{h \to 0} \frac{f(\mathbf{a} + h\mathbf{b}) - f(\mathbf{a})}{h}$$

provided the limit on the right exists.

In particular, if we choose $\mathbf{b} = (1, 0)$ the above limit is called the ***first order partial derivative of f with respect to*** x denoted by $f_x(\mathbf{a})$ (or $f_x(a_1, a_2)$ also sometimes by $\left(\dfrac{\partial f}{\partial x} \right)_{(a_1, a_2)}$ that is,

$$f_x(a_1, a_2) = \lim_{h \to 0} \frac{f(a_1 + h, a_2) - f(a_1, a_2)}{h}$$

if the limit on the right exists.

Similarly if we choose $\mathbf{b} = (0, 1)$, the above limit is called the ***first order partial derivative of f with respect to*** y denoted by $f_y(\mathbf{a})$ (or $f_y(a_1, a_2)$ also sometimes by $\left(\dfrac{\partial f}{\partial y}\right)_{(a_1, a_2)}$ is defined by the equation,

$$f_y(a_1, a_2) = \lim_{k \to 0} \frac{f(a_1, a_2 + k) - f(a_1, a_2)}{k}$$

if the limit on the right exists.

From the definition it is clear that the partial derivatives f_x & f_y represents the derivatives of the function $f(x, y)$ treating it as a function of x (keeping y fixed) and y (keeping x fixed) respectively. If partial derivatives exists for all values of (x, y) in the domain of definition of f then f_x & f_y are functions of (x, y) with same domain of definition.

Suppose for a function f with domain $D \subseteq \mathbb{R}^2$ the directional derivative $D_a(f; \mathbf{b})$ exists for an interior point $\mathbf{a} = (a_1, a_2)$ of D and for some point $b = (b_1, b_2)$ of $\mathbb{R}^2$ then we have,

$$\lim_{h \to 0} [f(\mathbf{a} + h\mathbf{b}) - f(\mathbf{a})] = \lim_{h \to 0} \frac{f(\mathbf{a} + h\mathbf{b}) - f(\mathbf{a})}{h} \lim_{h \to 0} h$$

$$= D_a(f; \mathbf{b}) \times 0 = 0$$

This means that $f(\mathbf{x}) \to f(\mathbf{a})$ along a straight line through $\mathbf{a} = (a_1, a_2)$ parallel to $\mathbf{b}$. If we assume $D_a(f; \mathbf{b})$ exists for the any $\mathbf{b}$ then $f(\mathbf{x}) \to f(\mathbf{a})$ along any line through $\mathbf{a}$ thus one may naturally enquire whether existence of $D_a(f; \mathbf{b})$ for any $\mathbf{b}$ implies the continuity of f at $\mathbf{a}$. This is not necessarily true as shown by the following example,

Example 9.11   Define $\qquad\qquad f: \mathbb{R}^2 \to \mathbb{R}$ as,

$$f(x, y) = \frac{x^2 y^2}{x^3 + y^6} \quad \text{if } (x, y) \neq (0, 0) / 0 \text{ if } (x, y) = (0, 0)$$

Show that the directional derivative of f at $(0, 0)$ exists for any direction $\mathbf{b}$ but f is not continuous at $(0, 0)$.

Sol: First we prove that f is not continuous at $(0, 0)$. For that we approach $(0, 0)$ through parabola of the form $\qquad\qquad\qquad y^2 = mx$ then,

$$\lim_{(x, y) \to (0, 0)} f(x) = \lim_{x \to 0} \frac{x^2(mx)}{x^3 + (m^3 x^3)}$$

$$= \frac{m}{1 + m^3}$$

As $\lim\limits_{(x,\,y)\to(0,\,0)} f(x)$ depends on m we conclude the limit does not exist and hence f is not continuous at $(0, 0)$.

Now let $\mathbf{b} = (b_1, b_2)$ be any direction in $\mathbb{R}^2$ then we have,

$$D_0(f;\, b) = \lim_{h \to 0} \frac{f(\mathbf{0} + h\mathbf{b}) - f(\mathbf{0})}{h}$$

$$= \lim_{h \to 0} \frac{f(hb_1,\, hb_2)}{h}$$

$$= \lim_{h \to 0} \frac{h^2 b_1^2 h^2 b_2^2}{h(h^3 b_1^3 + h^6 b_2^6)} = \lim_{h \to 0} \frac{b_1^2 b_2^2}{b_1^3 + h^3 b_2^6} = \begin{cases} \dfrac{b_2^2}{b_1} & \text{if } b_1 \neq 0 \\[2mm] 0 & \text{if } b_1 = 0 \end{cases}$$

This shows that directional derivative of f at $(0, 0)$ exists for any direction $\mathbf{b}$ in particular, by choosing $\mathbf{b} = (1, 0)$ & $(0, 1)$ we obtain,

$$f_x(0, 0) = f_y(0, 0) = 0$$

Note that the above example also shows that existence of derivatives (partial) at any point for functions in two variable does not imply the continuity of the function at that point contrary to what we have seen in one variable case. But we shall see that it is possible to ensure continuity by assuming some more conditions on partial derivatives together with their existence. Before that we prove a theorem which will be useful for this purpose.

Theorem 9.5.1

Suppose f be a function with domain S an open subset of $\mathbb{R}^2$ and let (a_1, a_2) be a point of S. Further let f_x exists for all points of S and $f_y(a_1, a_2)$ exists then if $(a_1 + h, a_2 + k) \in S$ we have,

$$f(a_1 + h, a_2 + k) - f(a_1, a_2) = h f_x(a_1 + \theta h, a_2 + k) + k f_y(a_1, a_2) + \eta(k),$$

for some θ satisfying $0 < \theta < 1$ and $\eta(k)$ a function of k satisfying $\eta(k) \to 0$ as $k \to 0$.

Proof: Let $\varphi : [a_1, a_1 + h] \to \mathbb{R}$ be defined as,

$$\varphi(x) = f(x, a_2 + k)$$

As f_x exists for all points of S we conclude φ is differentiable and hence continuous for $x \in [a_1, a_1 + h]$. By using Mean Value Theorem for one variable (Theorem 2.5.11) we will get some θ satisfying $0 < \theta < 1$, such that,

$$\varphi(a_1 + h) - \varphi(a_1) = h\varphi'(a_1 + h)$$

or,

$$f(a_1 + h, a_2 + k) - f(a_1, a_2 + k) = h f_x(a_1 + \theta h, a_2 + k) \tag{9.5.1.1}$$

Again as $f_y(a_1, a_2) = \lim\limits_{k \to 0} \dfrac{f(a_1, a_2 + k) - f(a_1, a_2)}{k}$ exists we have,

$$f(a_1, a_2 + k) - f(a_1, a_2) = k f_y(a_1, a_2) + \eta(k) \tag{9.5.1.2}$$

where $\eta(k)$ is a function of k and tends to zero as $k \to 0$.

Thus,

$$f(a_1 + h, a_2 + k) - f(a_1, a_2) = [f(a_1 + h, a_2 + k) - f(a_1, a_2 + k)] + [f(a_1, a_2 + k) - f(a_1, a_2)]$$
$$= hf_x(a_1 + \theta h, a_2 + k) + kf_y(a_1, a_2) + \eta(k)$$

$$\text{(using (9.5.1.1) \& (9.5.1.2))}$$

Using above theorem we get a set of sufficient conditions for a function to be continuous at a point which we note as a theorem below,

Theorem 9.5.2

Suppose f be a function with domain S an open subset of $\mathbb{R}^2$ and let $\mathbf{a} = (a_1, a_2)$ be a point of S. If any one of the partial derivatives of f say f_x exists and bounded for all points of S and the other say f_y exists at (a_1, a_2) then f is continuous at a.

Proof: We assume $|f_x(x, y)| < M \ \forall (x, y) \in S$ for some $M > 0$ and $f_y(a_1, a_2)$ exists. We choose h, k small enough so that $(a_1 + h, a_2 + k) \in S$ then by Theorem 9.4.1 we get,

$$f(a_1 + h, a_2 + k) - f(a_1, a_2) = hf_x(a_1 + \theta h, a_2 + k) + kf_y(a_1, a_2) + \eta(k) \qquad (9.5.2.2)$$

where
$$0 < \theta < 1 \text{ and } \eta(k) \to 0 \text{ as } k \to 0$$

That is,

$$|f(a_1 + h, a_2 + k) - f(a_1, a_2)| \le |h||f_x(a_1 + \theta h, a_2 + k)| + |k||f_y(a_1, a_2)| + |\eta(k)|$$
$$\le |h| M + |k||f_y(a_1, a_2)| + |\eta(k)| \qquad (9.5.2.3)$$

Clearly right side of (9.4.2.3) tends to zero as $(h, k) \to (0, 0)$ thus,

$$\lim_{(h, k) \to (0, 0)} f(a_1 + h, a_2 + k) = f(a_1, a_2)$$

In other words f is continuous at a.

Theorem 9.5.3 (Chain Rule)

Let U be an open subset of $\mathbb{R}^2$ and $x = f_1(u, v)$ & $y = f_2(u, v)$ be two real valued functions with domain U. Suppose $V \subseteq \mathbb{R}^2$ be open such that for $(u, v) \in U (f_1(u, v), f_2(u, v)) \in V$ and let f be a function with domain V so that $f(x, y)$ is defined for every $(u, v) \in U$. If we assume that the first order partial derivatives of $f_1(u, v), f_2(u, v)$ and $f(x, y)$ are continuous in their respective domains then,

$$\frac{\partial f}{\partial u} = \frac{\partial f}{\partial x}\frac{\partial x}{\partial u} + \frac{\partial f}{\partial y}\frac{\partial y}{\partial u}$$

$$\frac{\partial f}{\partial v} = \frac{\partial f}{\partial x}\frac{\partial x}{\partial v} + \frac{\partial f}{\partial y}\frac{\partial y}{\partial v}$$

Proof: We choose an arbitrary point $(u_0, v_0) \in U$ and put $x_0 = f_1(u_0, v_0)$ and $y_0 = f_2(u_0, v_0)$.

Let us prove the first formula. For that consider,

$$f(f_1(u_0 + \Delta u, v_0), f_2(u_0 + \Delta u, v_0)) - f(f_1(u_0, v_0), f_2(u_0, v_0))$$
$$= f(x_0 + \Delta x, y_0 + \Delta y) - f(x_0, y_0) \qquad (9.5.3.1)$$

Where

$$\Delta x = f_1(u_0 + \Delta u, v_0) - f_1(u_0, v_0)$$

$$\Delta y = f_2(u_0 + \Delta u, v_0) - f_2(u_0, v_0)\} \qquad (9.5.3.2)$$

Again,

$$f(x_0 + \Delta x, y_0 + \Delta y) - f(x_0, y_0)$$

$$= [f(x_0 + \Delta x, y_0 + \Delta y) - f(x_0, y_0 + \Delta y)] + [f(x_0, y_0 + \Delta y) - f(x_0, y_0)] \qquad (9.5.3.3)$$

See that the functions φ & ζ defined by

$$\varphi(x) = f(x, y_0 + \Delta y) \ \& \ \zeta(y) = f(x_0, y)$$

satisfies all conditions of Mean Value Theorem in the intervals $[x_0, x_0 + \Delta x]$ and $[y_0, y_0 + \Delta y]$ (as the first order partial derivatives of f are continuous) hence there must exists $0 < \theta_1, \theta_2 < 1$ such that,

$$f(x_0 + \Delta x, y_0 + \Delta y) - f(x_0, y_0 + \Delta y) = f_x(x_0 + \theta_1 \Delta x, y_0 + \Delta y)\, \Delta x$$

and
$$f(x_0, y_0 + \Delta y) - f(x_0, y_0) = f_y(x_0, y_0 + \theta_2 \Delta y)\Delta y$$

So that from (9.4.3.1),(9.4.3.2) & (9.4.3.3) we get

$$\Delta f = f(f_1(u_0 + \Delta u, v_0), f_2(u_0 + \Delta u, v_0)) - f(f_1(u_0, v_0), f_2(u_0, v_0))$$

$$= f(x_0 + \Delta x, y_0 + \Delta y) - f(x_0, y_0)$$

$$= f_x(x_0 + \theta_1 \Delta x, y_0 + \Delta y)\, \Delta x + f_y(x_0, y_0 + \theta_2 \Delta y)\Delta y$$

$$= f_x(x_0 + \theta_1 \Delta x, y_0 + \Delta y)[f_1(u_0 + \Delta u, v_0) - f_1(u_0, v_0)] + f_y(x_0, y_0 + \theta_2 \Delta y)$$

$$[f_2(u_0 + \Delta u, v_0) - f_2(u_0, v_0)]$$

That is,

$$\frac{\Delta f}{\Delta u} = f_x(x_0 + \theta_1 \Delta x, y_0 + \Delta y)\, \frac{[f_1(u_0 + \Delta u, v_0) - f_1(u_0, v_0)]}{\Delta u}$$

$$+ f_y(x_0, y_0 + \theta_2 \Delta y)\, \frac{[f_2(u_0 + \Delta u, v_0) - f_2(u_0, v_0)]}{\Delta u} \qquad (9.5.3.3)$$

Now again using the fact that the first order partial derivatives of $f_1(u, v)$, $f_2(u, v)$ and $f(x, y)$ are continuous it is clear that as $\Delta u \to 0$ Δx & Δy also tends to zero and hence

$$f_x(x_0 + \theta_1 \Delta x, y_0 + \Delta y) \to f_x(x_0, y_0) \ \& \ f_y(x_0, y_0 + \theta_2 \Delta y) \to f_y(x_0, y_0)$$

and from $\qquad\qquad\qquad\qquad\qquad\qquad\qquad\qquad\qquad\qquad\qquad (9.5.3.3)$

taking
$$\Delta u \to 0$$

we get,

$$\frac{\partial f}{\partial u} = \frac{\partial f}{\partial x}\frac{\partial x}{\partial u} + \frac{\partial f}{\partial y}\frac{\partial y}{\partial u}$$

Similarly we can prove the second formula.

Other forms of chain rules can similarly be deduced for example:

If $f(x, y)$ is defined in some domain $D \subseteq \mathbb{R}^2$ and suppose $x = f_1(t)$ & $y = f_2(t)$ for $t \in T$ such that $f(f_1(t), f_2(t))$ is defined $\forall t \in T$. Then observe that ultimately f is function of single variable t. If all appropriate first order derivatives are continuous then we have,

$$\frac{df}{dt} = \frac{\partial f}{\partial x}\frac{dx}{dt} + \frac{\partial f}{\partial y}\frac{dy}{dt}$$

9.6 HOMOGENEOUS FUNCTIONS AND EULER'S THEOREM

In this section we define an important class of functions called homogeneous functions in two variables. We will also give a characterization of the class of homogenous functions which has continuous partial derivatives. We begin with the following definition:

Definition Let $D \subseteq \mathbb{R}^2$ such that for any $\mathbf{x} \in D$, $\lambda\mathbf{x} \in D$ for $\lambda > 0$. A function $f: D \to \mathbb{R}$ is said to be a homogeneous function of degree n if,

$$f(\lambda\mathbf{x}) = \lambda^n f(\mathbf{x}) \;\; \forall x \in D \text{ and for } \lambda > 0$$

Note that here n can be any real number not necessarily a positive integer.

Example 9.12 Let $f(x, y) = \cos\left(\frac{x}{y}\right) + \sin^{-1}\left(\frac{y}{x}\right)$

Then see that $f(\lambda x, \lambda y) = f(x, y) = \lambda^0 f(x, y)$ hence here f is homogeneous function of degree 0.

Example 9.13 $f(x, y) = \sqrt{x^2 + y^2}\, \cot^{-1}\left(\frac{y}{x}\right)$ is homogeneous of degree 1 as,

$$f(\lambda x, \lambda y) = \sqrt{\lambda^2 x^2 + \lambda^2 y^2}\, \cot^{-1}\left(\frac{\lambda y}{\lambda x}\right) = \lambda\sqrt{x^2 + y^2}\, \cot^{-1}\left(\frac{y}{x}\right) = \lambda f(x, y) \text{ for } \lambda > 0$$

Example 9.14 $f(x, y) = \dfrac{\sqrt{x}}{x^2 + y^2}$ is homogeneous of degree $-\dfrac{3}{2}$ as,

$$f(\lambda x, \lambda y) = \frac{\sqrt{\lambda}\sqrt{x}}{\lambda^2(x^2 + y^2)} = \lambda^{\frac{1}{2}-2} f(x, y) = \lambda^{\frac{-3}{2}} f(x, y) \text{ for } \lambda > 0$$

Theorem 9.6.1

Let $D \subseteq \mathbb{R}^2$ such that for any $(x, y) \in D$, $(\lambda x, \lambda y) \in D$ for $\lambda > 0$ and suppose f is a function with domain D. If f has continuous first order partial derivatives then it is homogeneous of degree n if and only if it satisfies,

$$xf_x + yf_y = nf \;\; \forall (x, y) \in D \tag{9.6.1.1}$$

Proof: Suppose f is homogeneous of degree n then by definition,

$$f(\lambda x, \lambda y) = \lambda_n f(x, y), \; \lambda > 0 \tag{9.6.1.2}$$

Differentiating (9.5.1.1) with respect to λ we get

$$x\,\frac{\partial f(u,\,v)}{\partial u} + y\,\frac{\partial f(u,\,v)}{\partial v} = \lambda^{n-1} f(x,\,y) \qquad (9.6.1.3)$$

putting $\lambda x = u$ & $\lambda y = v$ in the left hand side of (9.6.1.2) and using chain rule.

Note that (9.6.1.3) is true for all $\lambda > 0$. If we now put $\lambda = 1$ we obtain $x = u$ & $y = v$ hence   (9.6.1.3) becomes

$$x = \frac{\partial f(x,\,y)}{\partial x} + y\,\frac{\partial f(x,\,y)}{\partial y} = f(x,\,y) \cap (x,\,y) \in D$$

This proves the necessary condition.

(Note that the necessary condition is sometimes known as ***Euler's Theorem***).

Suppose f satisfies (9.6.1.1). For any $(x',\,y') \in D$ let,

$$\psi(\lambda) = f(\lambda x',\,\lambda y') \qquad (9.6.1.4)$$

Differentiating (9.6.1.4) with respect to λ using chain rule we get,

$$\frac{x'\,\partial f(u,\,v)}{\partial u} + \frac{y'\,\partial f(u,\,v)}{\partial v} = \psi'(\lambda) \qquad (9.6.1.5)$$

where $\lambda x' = u$ & $\lambda y' = v$. Multiplying (9.5.1.5) by λ we get,

$$u\frac{\partial f(u,\,v)}{\partial u} + v\,\frac{\partial f(u,\,v)}{\partial v} = \lambda\psi(\lambda)$$

Hence

$$\lambda\psi(\lambda) = nf(u,\,v) = n\psi(\lambda)$$

or,
$$\frac{\psi(\lambda)}{\psi(\lambda)} = \frac{n}{\lambda}$$

$\Rightarrow$
$$\frac{\psi(\lambda)}{\lambda^{n}} = A(\text{constant}) \qquad (9.6.1.6)$$

Now
$$A = \psi(1) = f(x',\,y') \text{ putting } \lambda = 1 \text{ in} \qquad (9.6.1.6)$$

Hence,

$$f(\lambda x',\,\lambda y') = \psi(\lambda) = \lambda^{n} f(x',\,y') \text{ (from (9.6.1.6))}$$

Since $(x',\,y')$ is arbitrary we have,

$$f(\lambda x,\,\lambda y) = \lambda^{n} f(x,\,y) \cap (x,\,y) \in D \; \lambda > 0$$

This proves the sufficient part.

9.7 DIFFERENTIABILITY

We now introduce the notion of differentiability for functions of two variables. We recall that for functions of one variable we have observed that a function is differentiable at a point of the increment function at that can be written as sum of a linear function and an error term near that point. In the case of function of two variables we retain the same concept as in case of one variable. Suppose a function f is defined in a certain open neighbourhood D of a point $\mathbf{a} = (a_1, a_2)$. If $(a_1 + h, a_2 + k) \in D$ then $f(a_1 + h, a_2 + k) - f(a_1, a_2)$ is called the increment of the function at $\mathbf{a}$. We say that f is differentiable at $\mathbf{a}$ if the increment at that point can be written as sum of a linear function and an error term. We give the precise definition below.

Definition   A function f is defined in a certain open neighbourhood D of a point $\mathbf{a} = (a_1, a_2)$ is said to be differentiable at that point if for every point $(a_1 + h, a_2 + k) \in D$ we have,

$$f(a_1 + h, a_2 + k) - f(a_1, a_2) = Ah + Bk + h\phi(h, k) + k\psi(h, k) \tag{A}$$

where A & B are constants which does not depend on h & k and $\phi(h, k)$ & $\psi(h, k)$ are functions of h & k satisfying,

$$\lim_{(h,\, k)\to(0,\, 0)} \phi(h, k) = \lim_{(h,\, k)\to(0,\, 0)} \psi(h, k) = 0.$$

From above definition it is clear that if the function is differentiable at $\mathbf{a} = (a_1, a_2)$ we can write,

$$f(a_1 + h, a_2 + k) - f(a_1, a_2) = \lambda(h, k) + E(h, k)$$

where $\lambda(h, k) = Ah + Bk$ (such functions are called 'linear functions' in $\mathbb{R}^2$) and $E(h, k) = h\phi(h, k) + k\psi(h, k)$ which represents the error in approximating the increment of the function by the linear function $\lambda(h, k)$.

An immediate consequence of the above definition we obtain the following important theorem which can be treated as a necessary condition for differentiability.

Theorem 9.7.1

Suppose $f: D \to \mathbb{R}$ where D is an open subset of $\mathbb{R}^2$ and let $\mathbf{a} = (a_1, a_2) \in D$. If f is differentiable at $\mathbf{a}$ then, both $f_x(\mathbf{a})$ & $f_y(\mathbf{a})$ exists and the function is continuous at that point. Moreover, if $\mathbf{h} = (h, k)$ is such that $\mathbf{a} + \mathbf{h} \in D$ we have,

$$f(\mathbf{a} + \mathbf{h}) - f(\mathbf{a}) = f_x(\mathbf{a})h + f_y(\mathbf{a})k + \|h\|E(h, k) \tag{9.7.1.1}$$

for some function $E: \mathbb{R}^2 \in \mathbb{R}$ satisfying $E(h, k) \to 0$ as $(h, k) \to (0, 0)$, where $\|h\| = \sqrt{h^2 + k^2}$ is the Euclidean norm of $\mathbf{h} = (h, k)$.

Proof: If f is differentiable at $\mathbf{a} = (a_1, a_2)$ and $\mathbf{h} = (h, k)$ be such that $\mathbf{a} + \mathbf{h} \in D$ from definition we can write,

$$f(\mathbf{a} + \mathbf{h}) - f(\mathbf{a}) = f(a_1 + h, a_2 + k) - f(a_1, a_2) = Ah + Bk + h\phi(h, k) + k\psi(h, k) \tag{9.7.1.2}$$

where A & B are constants which does not depend on h & k and $\phi(h, k)$ & $\psi(h, k)$ are functions of h & k satisfying,

$$\lim_{(h,\, k)\to(0,\, 0)} \phi(h, k) = \lim_{(h,\, k)\to(0,\, 0)} \psi(h, k) = 0$$

From (9.7.1.2) we get,

$$\lim_{(h,\, k)\to(0,\, 0)} [f(a_1 + h, a_2 + k) - f(a_1, a_2)] = \lim /(h, k)\to(0, 0) [Ah + Bk + h\phi(h, k) + k\psi(h, k)] = 0$$

This shows that $f(a + h) \to f(a)$ as $(h, k) \to (0, 0)$. That is f continuous at a.

Again, if we put $k = 0$ in (9.7.1.2) and taking limit of both sides as $h \to 0$ we obtain, $A = f_x(a)$

Similarly by taking $h = 0$ in (9.7.1.2) and taking limit of both sides as $k \to 0$ we obtain $B = f_y(a)$

So that we can rewrite equation (9.6.1.2) as,

$$f(a + h) - f(a) = f_x(a)h + f_y(a)k + h\phi(h, k) + k\psi(h, k)$$

$$= f_x(a)h + f_y(a)k + \|h\| \, \frac{h\phi(h, k) + k\psi(h, k)}{\sqrt{h^2 + k^2}}$$

$$= f_x(a)h + f_y(a)k + \|h\| E(h, k)$$

where
$$E(h, k) = \frac{h\phi(h, k) + k\psi(h, k)}{\sqrt{h^2 + k^2}}. \text{ Now we observe that,}$$

$$|E(h, k)| \le \left| \frac{h}{\sqrt{h^2 + k^2}} \right| |\phi(h, k)| + \left| \frac{k}{\sqrt{h^2 + k^2}} \right| |\psi(h, k)|$$

$$\le |\phi(h, k)| + |\psi(h, k)| \to 0 \text{ as } (h, k) \to (0, 0)$$

$$\left(\because \lim_{(h, k)\to(0, 0)} \phi((h, k) = \lim_{(h, k)\to(0, 0)} \psi(h, k) = 0 \right)$$

This concludes the theorem.

Remark

1. From above theorem it is clear that if f is differentiable at $a = (a_1, a_2)$

   and $\qquad h = (h, k)$ be such that $a + h \in D$ then,

   $$\frac{f(a_1 + h, a_2 + k) - f(a_1, a_2) - f_x(a)h - f_y(a)k}{\sqrt{h^2 + k^2}} \to 0 \text{ as } (h, k) \to (0, 0).$$

2. See that in view of above theorem if f is differentiable at any point (x, y) we can write,

   $$\Delta f = f(x + \Delta x, y + \Delta y) - f(x, y) = f_x \Delta x + f_y \Delta y + \phi \Delta x + \psi \Delta y$$

   where Δf denotes the increment of the dependent variable $y = f(x, y)$ whereas Δx & Δv denote the increment of independent variables x & y and ϕ & ψ both tend to zero as $(\Delta x, \Delta y) \to (0, 0)$. Above equation in short can be written as,

   $$\Delta f = df + \phi dx + \psi dy$$

   where $\qquad df = f_x dx + f_y dy$ is known as the ***total differential of the function*** f.

   Next example shows that existence of partial derivatives is only a necessary condition for differentiability but no way it is sufficient.

Example 9.15 Let $f: \mathbb{R}^2 \to \mathbb{R}$ be defined as,

$$f(x, y) = \begin{cases} \dfrac{xy}{\sqrt{x^2 + y^2}} & \text{if } (x, y) \neq (0, 0) \\[2mm] 0 & \text{if } (x, y) = (0, 0) \end{cases}$$

Show that $f_x(0, 0)$ & $f_y(0, 0)$ both exist but the function is not differentiable at $(0, 0)$.

Sol: From definition,

$$f_x(0, 0) = \lim_{h \to 0} \frac{f(0 + h, 0) - f(0, 0)}{h}$$

$$= \lim_{h \to 0} \frac{f(h, 0) - f(0, 0)}{h} = \lim_{h \to 0} \frac{0}{h} = 0$$

Similarly, $\qquad\qquad f_y(0, 0) = 0$

Suppose if possible f is differentiable at $(0, 0)$. Then by the remark following Theorem 9.6.1 we must have,

$$\lim_{(h, k) \to (0, 0)} \frac{f(0 + h, 0 + k) - f(0, 0) - f_x(0, 0)h - f_y(0, 0)k}{\sqrt{h^2 + k^2}} = \lim_{(h,k) \to (0,0)} \frac{hk}{h^2 + k^2} = 0$$

But,

$$\lim_{(h, k) \to (0, 0)} \frac{hk}{h^2 + k^2} = \lim_{k \to 0} \frac{mh^2}{h^2(1 + m^2)} \quad \text{(putting } k = mh\text{)}$$

$$= \frac{m}{1 + m^2}$$

This shows that above limit depends on m as $(h, k) \to (0, 0)$ through different lines $k = mh$ that is the limit does not exist. As this is a contradiction we conclude f is not differentiable at $(0, 0)$.

Next example shows continuity of a function at a point does not imply differentiability

Example 9.16 Consider $f: \mathbb{R}^2 \to \mathbb{R}$ be defined as,

$$f(x) = |x| + |y|$$

Even as the function is continuous it is not differentiable at $(0, 0)$.

Sol: We have already shown (Example 9) that the function is continuous at $(0, 0)$.

Now,

$$f_x(0, 0) = \lim_{h \to 0} \frac{f(0 + h, 0) - f(0, 0)}{h} = \lim_{h \to 0} \frac{|h|}{h} \text{ does not exist.}$$

Similarly, $f_y(0, 0)$ also does not exist hence by Theorem 9.5.1 the function cannot be differentiable at $(0, 0)$.

Finally we show that assuming existence of partial derivatives as well as continuity of the function at a point does not guarantee differentiability.

Example 9.17   Consider $f: \mathbb{R}^2 \to \mathbb{R}$ be defined as,

$$f(x, y) = \begin{cases} y & \text{if } |x| \geq |y| \\ -y & \text{if } |x| < |y| \end{cases}$$

Show that

(i) f is continuous at $(0, 0)$

(ii) $f_x(0, 0), f_y(0, 0)$ exists

(iii) f is not differentiable at $(0, 0)$

Sol: We see that $\qquad\qquad f(0, 0) = 0$ and,

$$|f(x, y) - f(0, 0)| = |y|$$

Hence for any $\varepsilon > 0$ if we choose $\delta = \varepsilon$ we have from above equation

$$|f(x, y)| = |y| < \varepsilon \quad \text{whenever } x \ \& \ y \text{ satisfies } |x| < \delta \ \& \ |y| < \delta.$$

That is f is continuous at $(0, 0)$

Again,

$$f_x(0, 0) = \lim_{h \to 0} \frac{f(0 + h, 0) - f(0, 0)}{h} = \lim_{h \to 0} \frac{0}{h} = 0$$

$$f_y(0, 0) = \lim_{h \to 0} \frac{f(0, 0 + k) - f(0, 0)}{k} = \lim_{h \to 0} \frac{-k}{k} = -1$$

Suppose if possible f is differentiable at $(0, 0)$ then there exists functions $\phi(h, k)$ & $\psi(h, k)$ satisfying

$$\lim_{(h, k) \to (0, 0)} \phi(h, k) = \lim_{(h, k) \to (0, 0)} \psi(h, k) = 0 \text{ such that,}$$

$$f(0 + h, 0 + k) - f(0, 0) = f_x(0, 0)h + f_y(0, 0)h + h\phi(h, k) + k\psi(h, k)$$

In particular if $h = k$ we have,

$$f(h, h) - f(0, 0) = f_x(0, 0)h + f_y(0, 0)h + h_f((h, k) + h\psi(h, k)$$

or, $\qquad\qquad h - 0 = 0h + (-1)h + h\phi(h, k) + h\psi(h, k)$

$\Rightarrow \qquad\qquad 1 = -1 + \phi(h, k) + \psi(h, k) \quad \text{(canceling } h \text{ through out)}$

Now as $(h, k) \to (0, 0)$ we get,

$1 = -1$ which is a contradiction. Hence we conclude that f is not differentiable at $(0, 0)$.

Next theorem gives a sufficient condition in terms of partial derivatives under which a function becomes differentiable at a point.

Theorem 9.7.2

Suppose $f: D \to R$ where D is an open subset of $\mathbb{R}^2$ and let $a = (a_1, a_2) \in D$. If

    (i) any one of the partial derivatives (say f_x) exists and continuous at a

    (ii) other partial derivative (say f_y) exists at a

       then ϕ is differentiable at a.

Proof: As f_x is continuous at a there exists an open neighbourhood of the point $D_a \subset D$ such that f_x for all $(x, y) \in D_a$. We choose $(h, k) \in R^2$ such that points $(a_1 + h, a_2 + k)$, $(a_1 + h, a_2)$ & $(a_1, a_2 + k)$ all lies in D_a. Consider the function,

$$\varphi(x) = f(x, a_2 + k) \; x \in [a_1, a_1 + h]$$

We see that φ satisfies all conditions of mean value theorem hence there exists $\theta \in (0, 1)$ such that,

$$\varphi(a_1 + h) - \varphi(a_1) = h\varphi'(a_1 + \theta h)$$

$$\Rightarrow \qquad f(a_1 + h, a_2 + k) - f(a_1, a_2 + k) = hf_x(a_1 + \theta h, a_2 + k) \text{ for some } \theta \in (0, 1) \qquad (9.7.2.1)$$

Also as f_y exists we have,

$$f(a_1, a_2 + k) - f(a_1, a_2) = kf_y(a_1, a_2) + k\psi(k) \qquad (9.7.2.2)$$

where, $\qquad\qquad \psi(k) \to 0$ as $k \to 0$.

Now,

$$f(u_1 + h, u_2 + k) - f(a_1, a_2) = [f(a_1 + h, a_2 + k) - f(a_1, a_2 + k)] + [f(a_1, a_2 + k) - f(a_1, a_2)]$$

$$= hf_x(a_1 + \theta h, a_2 + k) = kf_y(a_1, a_2) + k\psi(k) \qquad (\text{Using}(9.6.2.1)\,\&\,(9.6.2.2))$$

$$= hf_x(a_1, a_2) + kf_y(a_1, a_2) + h[f_x(a_1 + \theta h, a_2 + k) - f_x(a_1, a_2)] + k\psi(k)$$

$$= hf_x(a_1, a_2) + kf_y(a_1, a_2) + h\phi(h, k) + k\psi(k) \qquad (9.7.2.3)$$

where $\phi(h, k) = f_x(a_1 + \theta h, a_2 + k) - f_x(a_1, a_2) \to 0$ as $(h, k) \to (0, 0)$ (as f_x is continuous at a) and we have already observed $\psi(k) \to 0$ as $k \to 0$. Thus from (9.7.2.3) we conclude f is differentiable at a.

Example 9.18 If $f: \mathbb{R}^2 \to \mathbb{R}$ be defined as,

$$f(x, y) = \begin{cases} \dfrac{xy(x^2 - y^2)}{x^2 + y^2} & \text{if } (x, y) \neq (0, 0) \\ 0 & \text{if } (x, y) = (0, 0) \end{cases}$$

then f is differentiable at $(0, 0)$.

    Sol It is easy to see that $\qquad\qquad f_x(0, 0) = 0 = f_y(0, 0)$.

When $(x, y) \neq (0, 0)$ we have,

$$f_x(x, y) = \frac{y(x^4 + 4x^2 y^2 - y^4)}{(x^2 + y^2)^2}$$

Hence,

$$f_x(r \cos \theta, r \sin \theta) = \frac{r \sin \theta \, (r^4 \cos^4 \theta + 4r^4 \cos^2 \theta \sin^2 \theta - r^4 \sin^4 \theta)}{r^4}$$

So that $\qquad\qquad |f_x(x, y)| \leq 6|r|$ as $|\sin \theta|, |\cos \theta| \leq 1$

That is $\qquad\qquad |f_x(x, y)| \leq 6\sqrt{x^2 + y^2}$

So that for any $\varepsilon > 0$ if we choose $0 < \delta < \dfrac{\varepsilon}{6}$ then,

$|f_x(x, y)| < \varepsilon$ whenever $\sqrt{x^2 + y^2} < \delta$ or in other words $f_x(x, y)$ is continuous at $(0, 0)$. So by above theorem the function is differentiable at $(0, 0)$.

Next example shows above conditions are sufficient but not necessary

Example 9.19   If the function $f \colon \mathbb{R}^2 \to \mathbb{R}$ is defined as,

$$f(x, y) = \begin{cases} (x^2 + y^2)\cos\left(\dfrac{\sqrt{2}}{\sqrt{x^2 + y^2}}\right) & \text{if } (x, y) \neq (0, 0) \\[4ex] 0 & \text{if } (x, y) = (0, 0) \end{cases}$$

then although it is differentiable at $(0, 0)$ but neither of its partial derivatives are continuous at that point.

Sol: We see that,

$$f_x(0, 0) = \lim_{h \to 0} \frac{f(h, 0) - f(0, 0)}{h} = \lim_{h \to 0} \frac{h^2 \cos\left(\dfrac{\sqrt{2}}{|h|}\right)}{h}$$

$$= \lim_{h \to 0} h \cos \frac{\sqrt{2}}{|h|} = 0$$

Similarly,

$$f_y(0, 0) = 0$$

Again if $(x, y) \neq (0, 0)$

$$f_x(x, y) = 2x \cos\left(\frac{\sqrt{2}}{\sqrt{x^2 + y^2}}\right) + (x^2 + y^2) \sin\left(\frac{\sqrt{2}}{\sqrt{x^2 + y^2}}\right) \times \frac{\sqrt{2}}{2} (x^2 + y^2)^{\frac{3}{2}} \times 2x$$

$$= 2x \cos \frac{\sqrt{2}}{\sqrt{x^2 + y^2}} + \frac{\sqrt{2}x}{\sqrt{x^2 + y^2}} \sin \frac{\sqrt{2}}{\sqrt{x^2 + y^2}} \qquad\qquad \text{(A)}$$

We show that $\lim\limits_{(x,\,y)\to(0,\,0)} f_x(x, y)$ does not exist. Consider

$$\lim_{(x,\,y)\to(0,\,0)} \frac{\sqrt{2}x}{\sqrt{x^2 + y^2}} \sin \frac{\sqrt{2}}{\sqrt{x^2 + y^2}}$$

$$= \lim_{x\to 0} \left[\left(\frac{x}{|x|}\right) \sin \left(\frac{1}{|x|}\right) \right] \quad \text{(putting } y = x)$$

But $\lim\limits_{x\to 0^+} \left(\dfrac{x}{|x|}\right) \sin \left(\dfrac{1}{|x|}\right) = \lim\limits_{x\to 0+} \sin \left(\dfrac{1}{x}\right)$ does not exist.

So that $\lim\limits_{(x,\,y)\to(0,\,0)} \dfrac{\sqrt{2}x}{\sqrt{x^2 + y^2}} \sin \dfrac{\sqrt{2}}{\sqrt{x^2 + y^2}}$ does not exist and hence from (A) we conclude

$$\lim_{(x,\,y)\to(0,\,0)} \left[\frac{\sqrt{2}x}{\sqrt{x^2 + y^2}} \sin \left(\frac{\sqrt{2}}{\sqrt{x^2 + y^2}}\right) \right] \text{ does not exist. Thus } f_x(x, y) \text{ is not continuous at } (0, 0).$$

Similarly we can show $f_y(x, y)$ is not continuous at $(0, 0)$. Next we show that f is differentiable at $(0, 0)$. Now,

$$\left| \frac{f(0 + h, 0 + k) - hf_x(0, 0) - kf_y(0, 0)}{\sqrt{h^2 + k^2}} \right| = \left| \frac{f(h, k)}{\sqrt{h^2 + k^2}} \right| = \left| \sqrt{h^2 + k^2} \cos \left(\frac{\sqrt{2}}{\sqrt{h^2 + k^2}}\right) \right| \le \sqrt{h^2 + k^2}$$

Thus,

$$\lim_{(h,\,k)\to(0,\,0)} \left[\frac{f(0 + h, 0 + k) - hf_x(0, 0) - kf_y(0, 0)}{\sqrt{h^2 + k^2}} \right] = 0$$

Hence we conclude f is differentiable at $(0, 0)$.

9.8   HIGHER ORDER PARTIAL DERIVATIVES AND TAYLOR'S THEOREM

We have already defined first order partial derivatives $f_x(x, y)$ & $f_y(x, y)$ and noted that they are themselves functions of x, y. Hence we can talk about the first order partial derivatives of f_x, f_y if they exists. The

partial derivative of f_x with respect to x is denoted by, f_{xx} or $\dfrac{\partial^2 f}{\partial x^2}$ that is $\dfrac{\partial}{\partial x} \left(\dfrac{\partial f}{\partial x}\right) = \dfrac{\partial^2 f}{\partial x^2}$ or $\dfrac{\partial^2 f}{\partial x^2}$ and f_{yx}

or $\dfrac{\partial^2 f}{\partial y \partial x}$ denotes the partial derivative of f_x with respect to y that is $\dfrac{\partial}{\partial y} \left(\dfrac{\partial f}{\partial x}\right) = \dfrac{\partial^2 f}{\partial y \partial x}$. Similarly the partial

derivatives of f_y with respect to x and y is denoted by $f_{xy} \left(\dfrac{\partial^2 f}{\partial x \partial y}\right)$ and $f_{yy} \left(\dfrac{\partial^2 f}{\partial y^2}\right)$. In fact, for positive integer $i > 1$, $\dfrac{\partial^i f}{\partial x^i}$ is called the ith order partial derivate of f with respect to x and is defined as $\dfrac{\partial}{\partial x}\left(\dfrac{\partial^{i-1} f}{\partial x^{i-1}}\right)$.

Similarly $\dfrac{\partial^i f}{\partial y^i}$ is called the ith order partial derivate of f with respect to y and is defined as $\dfrac{\partial}{\partial y}\left(\dfrac{\partial^{i-1} f}{\partial y^{i-1}}\right)$. For

$i > 1 \ \& \ j > 1$ we can define $\dfrac{\partial^{i+j} f}{\partial x^i \partial y^j}$ as $\dfrac{\partial^i}{\partial x^i}\left(\dfrac{\partial^j f}{\partial y^j}\right)$. In the same way one can define $\dfrac{\partial^{i+j} f}{\partial y^j \partial x^i}$ $i, j > 1$ as $\dfrac{\partial^i}{\partial y^j}\left(\dfrac{\partial^j f}{\partial x^i}\right)$

. Note that $\dfrac{\partial^{i+j} f}{\partial x^i \partial y^j}$ and $\dfrac{\partial^{i+j} f}{\partial y^j \partial x^i}$ are called $(i+j)^{\text{th}}$ order mixed derivative of f and in general are not equal as

shown by the following example.

Example 9.20   Define $f \colon \mathbb{R}^2 \to \mathbb{R}$ as follows,

$$f(x, y) = \begin{cases} \dfrac{xy(x^2 - y^2)}{x^2 + y^2} & \text{if } (x, y) \neq (0, 0) \\[2ex] 0 & \text{if } (x, y) = (0, 0) \end{cases}$$

Show that $\qquad f_{xy}(0, 0) \neq f_{yx}(0, 0)$

Sol: It follows from definition that,

$$f_{xy}(0, 0) = \lim_{h \to 0} \frac{f_y(0 + h, 0) - f_y(0, 0)}{h} = \lim_{h \to 0} \frac{f_y(h, 0) - f_y(0, 0)}{h} \tag{A}$$

and

$$f_{yx}(0, 0) = \lim_{k \to 0} \frac{f_x(0, 0 + k) - f_x(0, 0)}{k} = \lim_{k \to 0} \frac{f_x(0, k) - f_x(0, 0)}{k} \tag{B}$$

Again,

$$f_y(0, 0) = \lim_{k \to 0} \frac{f(0, k) - f(0, 0)}{k} = \lim_{k \to 0} \frac{0}{k} = 0 \tag{C}$$

$$f_x(0, 0) = \lim_{h \to 0} \frac{f(h, 0) - f(0, 0)}{h} = \lim_{h \to 0} \frac{0}{h} = 0 \tag{D}$$

$$f_y(h, 0) = \lim_{k \to 0} \frac{f(h, k) - f(h, 0)}{k} = \lim_{k \to 0} \frac{\dfrac{hk(h^2 - k^2)}{h^2 + k^2}}{k} = h \tag{E}$$

$$f_x(0, k) = \lim_{h \to 0} \frac{f(h, k) - f(0, k)}{h} = \lim_{h \to 0} \frac{\dfrac{hk(h^2 - k^2)}{h^2 + k^2}}{h} = -k \tag{F}$$

Using (A), (C) & (E) we get,

$$f_{xy}(0, 0) = \lim_{h \to 0} \frac{f_y(h, 0) - f_y(0, 0)}{h} = \lim_{h \to 0} \frac{h}{h} = 1 \qquad \text{(G)}$$

Similarly Using (B),(D)& (F) we get,

$$f_{xy}(0, 0) = \lim_{k \to 0} \frac{f_x(0, k) - f_x(0, 0)}{k} = \lim_{k \to 0} \frac{-k}{k} = -1 \qquad \text{(H)}$$

From (G) & (H) we get

$$f_{xy}(0, 0) \neq f_{yx}(0, 0)$$

Next two theorems gives sufficient conditions under which the two mixed derivatives of order two are equal.

Theorem 9.8.1

Let D be an open subset of $\mathbb{R}^2$ and f be a real valued function with domain D such that f_x, f_y & f_{yx} exists $\forall (x, y) \in D$. If further we assume f_{yx} is continuous at $(a, b) \in D$ then, $f_{xy}(a, b)$ exists and moreover we have,

$$f_{xy}(a, b) = f_{yx}(a, b)$$

Proof: As f_{yx} is continuous at (a, b) then there exists an open neighbourhood $D_1 \subset D$ of the point such that f_{yx} exits for all $(x, y) \in D_1$. Hence we can choose h & k such that the rectangle R with vertices (a, b), $(a + h, b)$, $(a + h, b + k)$ & $(a, b + k)$ lies within D_1. Now by definition we have,

$$f_{xy}(a, b) = \lim_{h \to 0} \frac{f_y(a + h, b) - f_y(a, b)}{h} \qquad (9.8.1.1)$$

$$f_y(a, b) = \lim_{k \to 0} \frac{f(a, b + k) - f(a, b)}{k} \qquad (9.8.1.2)$$

$$f_y(a + h, b) = \lim_{k \to 0} \frac{f(a + h, b + k) - f(a + h, b)}{k} \qquad (9.8.1.3)$$

Now using (9.8.1.2) & (9.8.1.3) we get,

$$\frac{f_y(a + h, b) - f_y(a, b)}{n} = \frac{1}{h} \lim_{k \to 0} \left[\frac{f(a + h, b + k) - f(a + h, b)}{k} - \frac{f(a, b + k) - f(a, b)}{k} \right]$$

$$= \frac{1}{h} \lim_{k \to 0} \frac{\varphi(h, k)}{k} \qquad (9.8.1.4)$$

where, $$\varphi(h, k) = f(a + h, b + k) - f(a + h, b) - f(a, b + k) + f(a, b)$$

We define,

$$\psi(x) = f(x, b + k) - f(x, b), \; x \in [a, a + h]$$

So that,

$$\varphi(h, k) = \psi(a + h) - \psi(a) \tag{9.8.1.5}$$

Now,

$$\psi'(x) = f_x(x, b + k) - f_x(x, b)$$

As f_x exists for all $(x, y) \in D_1$ it follows that ψ satisfies all conditions of Mean Value Theorem hence there exists $\theta_1 \psi \ (0, 1)$ which is a function of h as well as such that,

$$\psi(a + h) - \psi(a) = \eta \psi'(a + \theta_1 h)$$

$$\Rightarrow \qquad \varphi(h, k) = h[f_x(a + \theta_1 h, b + k) - f_x(a + \theta_1 h, b)] \tag{9.8.1.6}$$

Again as f_{yx} exists in D_1 we see that the function

$$\zeta(y) = f_x(a + \theta_1 h, y), \ y \in [b, b + k]$$

satisfies conditions of Mean Value Theorem hence there exists $\theta_2 \in (0, 1)$ such that,

$$\zeta(b + k) - \zeta(b) = k\zeta'(b + \theta_2 k) = kf_{yx}(a + \theta_1 h, b + \theta_2 k)$$

$$\Rightarrow \qquad f_x(a + \theta_1 h, b + k) - f_x(a + \theta_1 h, b) = kf_{yx}(a + \theta_1 h, b + \theta_2 k) \tag{9.8.1.7}$$

Using (9.8.1.6) & (9.8.1.7) we get,

$$\varphi(h, k) = hkf_{yx}(a + \theta_1 h, b + \theta_2 k) \tag{9.8.1.8}$$

Again (9.8.1.4) & (9.8.1.8) gives,

$$\frac{f_y(a + h, b) - f_y(a, b)}{h} = \lim_{k \to 0} f_{yx}(a + \theta_1 h, b + \theta_2 k)$$

$$\Rightarrow \qquad \lim_{h \to 0} \left[\frac{f_y(a + h, b) - f_y(a, b)}{h} \right] = \lim_{h \to 0} \left[\lim_{k \to 0} f_{yx}(a + \theta_1 h, b + \theta_2 k) \right] \tag{9.8.1.9}$$

See that as $0 < \theta_1, \theta_2 < 1$ $|\theta_1 h| < |h|$ & $|\theta_2 k| < |k|$ hence as $h \to 0$, $a + \theta_1 h \to a$ and as $k \to 0$, $b + \theta_2 k \to b$. We now use above fact and the condition that f_{yx} is continuous at (a, b) so that (9.8.1.9) yields,

$$\Rightarrow \qquad \lim_{h \to 0} \frac{f_y(a + h, b) - f_y(a, b)}{h} = f_{yx}(a, b) \tag{9.8.1.10}$$

As left hand side of (9.8.1.10) is by definition $f_{xy}(a, b)$ we conclude that $f_{xy}(a, b)$ exists and is equal to $= f_{yx}(a, b)$.

Theorem 9.8.2

Let D be an open subset of $\mathbb{R}^2$ and f be a real valued function with domain D such that f_x & f_y are both differentiable at $(a, b) \in D$. Then $f_{xy}(a, b) = f_{yx}(a, b)$.

Proof: From the given condition it is clear that all second order partial derivatives namely f_{xx}, f_{yy}, f_{xy} & f_{yx} all exists in a certain open neighbourhood of (a, b) and hence without loss of generality we may assume they all exist $\forall (x, y) \in D$. We choose h such that points $(a + h, b)$, $(a, b + h)$ and $(a + h, b + h)$ are in D. We define the quantity φ as,

$$\varphi(h, h) = f(a + h, b + h) - f(a + h, b) - f(a, b + h) + f(a, b)$$

and the functions ψ & ξ as follows:

$$\psi(x) = f(x, b + h) - f(x, b), \ x \in [a, a + h]$$

$$\xi(y) = f(a + h, y) - f(a, y), \ y \in [b, b + h]$$

Hence we get,

$$\varphi(h, h) = \psi(a + h) - \psi(a) = \xi(b + h) - \xi(b) \tag{9.8.2.1}$$

We see that ψ satisfies all conditions of Mean Value Theorem (Theorem 2.5.11) hence there exists $\theta_1 \in (0, 1)$ such that,

$$f(a + h, b + h) - f(a, b) = h[f_x(a + \theta_1 h, b + h) - f_x(a + \theta_1 h, b)] \tag{9.8.2.2}$$

Now as f_x is differentiable at (a, b) we have,

$$f_x(a + \theta_1 h, b + h) = f_x(a, b) + (\theta_1 h) f_{xx}(a, b) + h f_{yx}(a, b) + (\theta_1 h)\lambda_1 + h\lambda_2 \tag{9.8.2.3}$$

and

$$f_x(a + \theta_1 h, b) = f_x(a, b) + (\theta_1 h) f_{xx}(a, b) + (\theta_1 h)\lambda_3 \tag{9.8.2.4}$$

where λ_1, λ_2 & λ_3 all tend to zero as $h \to 0$. Now using (9.8.2.1) (9.8.2.2),(9.8.2.3) & (9.8.2.4) we get,

$$\frac{\varphi(h, h)}{h^2} = f_{yx}(a, b) + (\theta_1)\lambda_1 + \lambda_2 + (\theta_1)\lambda_3$$

Hence,

$$\lim_{h \to 0} \frac{\varphi(h, h)}{h^2} = f_{yx}(a, b) \tag{9.8.2.5}$$

Similarly using the function $\xi(y)$ instead of $\psi(x)$ in the above argument we will arrive at

$$\lim_{h \to 0} \frac{\varphi(h, h)}{h^2} = f_{xy}(a, b) \tag{9.8.2.6}$$

From (9.8.2.4) & (9.8.2.5) we conclude $f_{yx}(a, b) = f_{xy}(a, b)$.

Our next theorem is what is called Taylor's Theorem of two variables but before stating the theorem we will introduce a notation. For h & k real numbers and n a positive integer we write,

$$\left(h \frac{\partial}{\partial x} + k \frac{\partial}{\partial y} \right)^n \equiv h^n \frac{\partial^n}{\partial x^n} + {}^nC_1 h^{n-1} k \frac{\partial^n}{\partial x^{n-1}\partial y} + {}^nC_2 h^{n-2} k^2 \frac{\partial^n}{\partial x^{n-2}\partial y^2} + \dots + {}^nC_n k^n \frac{\partial^n}{\partial y^n}$$

So that if f is a function with appropriate domain in $\mathbb{R}^2$ having continuous partial derivatives of all orders up to n then,

Theorem 9.8.3 (Taylor's Theorem)

Let D be an open subset of $\mathbb{R}^2$ and f be a real valued function with domain D possessing continuous partial derivatives of all orders up to n where n is a positive integer. Suppose (a, b) & $(a + h, b + k) \in D$ then there exists $\theta \in (0, 1)$ such that,

$$f(a + h, b + k) = f(a, b) + \left(h\frac{\partial}{\partial x} + k\frac{\partial}{\partial y}\right)f(x, y) + \frac{1}{2!}\left(h\frac{\partial}{\partial x} + k\frac{\partial}{\partial y}\right)^2 f(x, y) + \frac{1}{3!}\left(h\frac{\partial}{\partial x} + k\frac{\partial}{\partial y}\right)^3 f(x, y) + \ldots$$

$$+ \ldots + \frac{1}{(n-1)!}\left(h\frac{\partial}{\partial x} + k\frac{\partial}{\partial y}\right)^{n-1} f(x, y) + \frac{1}{n!}\left(h\frac{\partial}{\partial x} + k\frac{\partial}{\partial y}\right)^n f(a + \theta h, b + \theta k). \tag{9.8.3.1}$$

Proof: Let $x = a + th$ & $y = b + tk$ and write

$$\phi(t) = f(x, y) = f(a + th, b + tk)$$

Then using chain rule we get,

$$\frac{d\phi}{dt} = \frac{\partial f}{\partial x}\frac{dx}{dt} + \frac{\partial f}{\partial y}\frac{dy}{dt}$$

$$= h\frac{\partial f}{\partial x} + k\frac{\partial f}{\partial y} = \left(h\frac{\partial}{\partial x} + k\frac{\partial}{\partial y}\right)f$$

Again,

$$\frac{d^2\phi}{dt^2} = \frac{d}{dt}\left(\frac{d\phi}{dt}\right) = \frac{d}{dt}\left(h\frac{\partial f}{\partial x} + k\frac{\partial f}{\partial y}\right)$$

$$= h\frac{d}{dt}\left(\frac{\partial f}{\partial x}\right) + k\frac{d}{dt}\left(\frac{\partial f}{\partial y}\right)$$

$$= h\left(h\frac{\partial}{\partial x}\left(\frac{\partial f}{\partial x}\right) + k\frac{\partial}{\partial y}\left(\frac{\partial f}{\partial x}\right)\right) + k\left(h\frac{\partial}{\partial x}\left(\frac{\partial f}{\partial y}\right) + k\frac{\partial}{\partial y}\left(\frac{\partial f}{\partial y}\right)\right)$$

$$= h^2\frac{\partial^2 f}{\partial x^2} + 2hk\frac{\partial^2 f}{\partial x\partial y} + k^2\frac{\partial^2 f}{\partial y^2} \quad \text{(As the mixed derivatives of order two are equal)}$$

$$= \left(h\frac{\partial}{\partial x} + k\frac{\partial}{\partial y}\right)^2 f$$

Using mathematical induction we can conclude,

$$\frac{d^k\phi}{dt^k} = \left(h\frac{\partial}{\partial x} + k\frac{\partial}{\partial y}\right)^k f \quad 1 \le k \le n \tag{9.7.3.2}$$

Now using Taylor's Theorem to the function $\phi(t)$ in the interval $[0, 1]$, we get,

$$\phi(t) = \phi(0) + t\phi'(0) + \frac{t^2}{2!}\phi''(0) + \ldots + \frac{t^{n-1}}{(n-1)!}\phi^{(n-1)}(0) + \frac{t^n}{n!}\phi(n)(\theta t) \quad 0 \le t \le 1 \tag{9.8.3.3}$$

for some $\theta \in (0, 1)$. In particular if we put $t = 1$ (9.7.3.2) in we get,

$$\phi(1) = \phi(0) + \phi'(0) + \frac{1}{2!}\,\phi''(0) + \ldots + \frac{1}{(n-1)!}\,\phi^{(n-1)}(0) + \frac{1}{n!}\,\phi(n)(\theta) \tag{9.8.3.4}$$

Writing ϕ in terms of f in (9.7.3.4) and using (9.7.3.2) we get,

$$f(a + h, b + k) = f(a, b) + \left(h\frac{\partial}{\partial x} + k\frac{\partial}{\partial y}\right)f(x, y) + \frac{1}{2!}\left(h\frac{\partial}{\partial x} + k\frac{\partial}{\partial y}\right)^2 f(x, y) + \frac{1}{3!}\left(h\frac{\partial}{\partial x} + k\frac{\partial}{\partial y}\right)^3 f(x, y) + \ldots$$

$$+ \ldots + \frac{1}{(n-1)!}\left(h\frac{\partial}{\partial x} + k\frac{\partial}{\partial y}\right)^{n-1} f(x, y) + \frac{1}{n!}\left(h\frac{\partial}{\partial x} + k\frac{\partial}{\partial y}\right)^n f(a + \theta h, b + \theta k).$$

Remarks

1. Putting $x = a + h$ & $y = b + h$ in (9.8.3.1) we get Taylor's Theorem in the following form ,

$$f(x, y) = f(a, b) + \left((x - a)\frac{\partial}{\partial x} + (y - b)\frac{\partial}{\partial y}\right)f(x, y) + \frac{1}{2!}\left((x - a)\frac{\partial}{\partial x} + (y - b)\frac{\partial}{\partial y}\right)^2 f(x, y) +$$

$$\ldots + \frac{1}{n!}\left((x - a)\frac{\partial}{\partial x} + (x - b)\frac{\partial}{\partial y}\right)^n f(a + \theta h, b + \theta k)$$

2. If we put $n = 1$ in (9.8.3.1) theorem we obtain,

   $f(a + h, b + k) - f(a, b) = hf_x(a + \theta h, b + \theta k) + kf_y(a + \theta h, b + \theta k)$ which is known as **Mean Value Theorem** for function two variables.

3. If we assume $n = 2$ then (9.8.3.1) yields,

   $f(a + h, b + k) - f(a, b) = hf_x(a +, b) + kf_y(a, b) + R_n$ where $\frac{1}{2!}\left(h\frac{\partial}{\partial x} + k\frac{\partial}{\partial y}\right)^2 f(a + \theta h, b + \theta k)$. The

   quantity $hf_x(a +, b) + kf_y(a, b)$ is called the ***first order or linear approximation*** of the increment

   $f(a + h, b + k) - f(a, b)$.

4. Similarly,

   $$hf_x(a, b) + kf_y(a, b) + h^2 f_{xx}(a, b) + 2hk f_{xy}(a, b) + k^2 f_{yy}(a, b)$$

   quadratic or second order approximation of the increment

   $$f(a + h, b + k) - f(a, b).$$

Example 9.20 Find the quadratic approximation of the function

$$f(x, y) = e^x \sin xy$$

at the point $(1, 0)$.

Sol: We see that,

$$f_x(x, y) = e^x \sin xy + ye^x \cos xy,$$

$$f_{xx}(x, y) = e^x \sin xy + ye^x \cos xy + y[-e^x \sin xy + e^x \cos xy]$$

$$f_y(x, y) = xe^x \cos xy, \quad f_{yy}(x, y) = xe^x(-x \sin xy)$$

$$f_{xy}(x, y) = xe^x \cos xy + e^x \cos xy + y(e^x \cos xy - e^x \sin xy)$$

So that,

$$f_x(1, 0) = 0, f_{xx}(1, 0) = 0, f_y(1, 0) = e, f_{yy}(1, 0) = 0, f_{xy}(1, 0) = 2e$$

Hence,

$$f(1 + h, k) = ke + 4ehk = ke(1 + 4h)$$

Example 9.21 Expand $f(x, y) = \cos xy$ in powers of $(x - \pi)$ and $\left(y - \dfrac{1}{2}\right)$ up to second order terms.

Sol:

$$f_x(x, y) = -y \sin xy, f_{xx}(x, y) = -y^2 \cos xy$$

$$f_y(x, y) = -x \sin xy, f_{yy}(x, y) = -x^2 \cos xy$$

$$f_{xy}(x, y) = -\sin xy - xy \cos xy$$

$$f_x\left(\pi, \frac{1}{2}\right) = \frac{1}{2}, f_{xx}\left(\pi, \frac{1}{2}\right) = 0, f_y\left(\pi, \frac{1}{2}\right) = -\pi, f_{yy}\left(\pi, \frac{1}{2}\right) = 0$$

By Taylor's Theorem keeping only second order terms,

$$\cos xy = \cos\left(\frac{\pi}{2}\right) + (x - \pi)\left(-\frac{1}{2}\right) + \left(y - \frac{1}{2}\right)(0) + \frac{1}{2}\left[(x - \pi)^2 \times 0 + 2(x - \pi)\left(y - \frac{1}{2}\right)(-1) + \left(y - \frac{1}{2}\right)^2 \times 0\right]$$

$$\cos xy = -\frac{(x - \pi)}{2} - (x - \pi)\left(y - \frac{1}{2}\right) = -xy + \pi y$$

9.9 SOME APPLICATIONS OF PARTIAL DERIVATIVES: IMPLICIT FUNCTION THEOREM, MAXIMA MINIMA

In this section we will give two important and useful theorems which involves the concept of partial derivatives. We will begin with what is called *Implicit Function Theorem* for functions of two variables. Suppose $f(x, y)$ a function of two variables and consider the equation

$$f(x, y) = 0 \tag{A}$$

Apart from the query about the solution set of above equation one may also enquire whether it is possible to express y as function of x that is if we can write,

$$y = \phi(x) \tag{B}$$

from the above equation (A). Let us illustrate by some examples.

(i) If we take $f(x, y) = y - ax$ we know that in this case the solution of $f(x, y) = 0$ geometrically gives a straight line passing through the origin given by $y = ax$. Hence in this case it is possible to solve y in terms of x.

(ii) If $f(x, y) = x^2 + 3y^2 + 2$ then we see that the solution set of $f(x, y) = 0$ is empty.

(iii) If $f(x, y) = x^2 + y^2 - 2$ then we know the solution set of $f(x, y) = 0$ is the circle with centre origin and radius $\sqrt{2}$ but see that for each given x we get two values y namely

$$y = \sqrt{\frac{2-x^2}{3}} \text{ and } y = -\sqrt{\frac{2-x^2}{3}}.$$ Hence in this case we cannot express y as function of x.

From above examples it is clear that even if the solution set of $f(x, y) = 0$ is non empty it may not be always possible to express y as function of x from (A). Implicit function theorem gives a set of sufficient conditions under which we can write $y = \phi(x)$ for some function ϕ from the given equation $f(x, y) = 0$.

Theorem 9.9.1 (Implicit Function Theorem)

Suppose f be a function of two variables and let (a, b) be a point of $\mathbb{R}^2$. If

   (a) $f(a, b) = 0$

   (b) f_x, f_y *are continuous in* $N(\delta)$ *where* $N(\delta) = \{(x, y): |x - a| \leq \delta, |y - b| \leq \delta\}$

   (c) $f_y(a, b) \neq 0$

Then there is a unique function $\phi(x)$ and $\lambda > 0$ such that

   (i) $b = \phi(a)$

   (ii) $f(x, \phi(x)) = 0$, *for* $|x - a| < \lambda$

   (iii) $\phi(x)$ *is continuously differentiable function with*

$$\phi'(x) = \frac{f_x(x, y)}{f_y(x, y)} \text{ whenever } |x - a| < \lambda.$$

Proof: We first note that as f_x, f_y are continuous in $N(\delta)$, f is differentiable and hence continuous there. Again see that the function $-f$ also satisfies the conditions (a) to (c) and proving conclusions (i) to (ii) for f and $-f$ are equivalent. Hence we may assume without loss of generality that $f_y(a, b) > 0$ otherwise we may replace the function f by $-f$ and proceed. Now as $f_y(x, y)$ is continuous at (a, b) corresponding to $\varepsilon = \dfrac{f_y(a, b)}{2} > 0$ we will get some δ_1 such that,

$$|f_y(x, y) - f_y(a, b)| < \frac{f_y(a, b)}{2}$$

whenever $(x, y) \in N(\delta_1) = \{(x, y): |x - a| \leq \delta_1, |y - b| \leq \delta_1\}$.

That is,

$$f_y(x, y) > \frac{f_y(a, b)}{2} > 0, \forall (x, y) \in N(\delta_1) \tag{9.9.1.1}$$

Now if we choose $\delta' = \min[\delta, \delta_1]$ then condition (b) and (9.9.1.1) both holds for $\forall (x, y) \in N(\delta')$.

Hence $f(a, y)$ is an increasing function in $|y - b| \leq \delta'$ by Theorem 2.5.5. Hence from definition $f(a, b - \delta') < f(a, b) = 0$ and $f(a, b + \delta') > f(a, b) = 0$. We now use the continuity of the function f and conclude that there exists $\lambda > 0$ such that $f(x, b - \delta') < 0$ & $f(x, b + \delta') > 0 \to x \in |x - a| < \lambda$. Now by Theorem 2.3.4 for every x' satisfying $|x' - a| < \lambda$ there exists some $y' \in (b - \delta', b + \delta')$ such that $f(x', y') = 0$. We now claim that this y is unique. If $f(x', y'') = 0$ for $y'' \in (b - \delta, b + \delta)$ & $y' \neq y''$ then we use Rolle's Theorem to the function $f(x', y)$ in the interval $[b - \delta'b + \delta']$ to conclude $f_y(x', \xi) = 0$ for some $\xi \in (b - \delta', b + \delta')$ contradicting (9.9.1.1). So that y' corresponding to x' is unique as claimed. We denote this y' by $\phi(x')$. Hence we see that there exists a function ϕ such that, $f(x, \phi(x)) = 0$ for $|x - a| < \lambda$ and also $b = f(a)$ this proves (i) and (ii). We now show (iii).

We choose $x, x + h \in (a - \lambda, a + \lambda)$ then we have $y = \phi(x)$, $y + k = \phi(x + h)$ and $f(x + h, y + k) = f(x, y) = 0$. Now as f is differentiable we have by, $0 = f(x + h, x + k) - f(x, y) = hf_x(x + \theta h, y + \theta k) + kf_y(x + \theta h, y + \theta k)$

$$(9.9.1.2)$$

Above equation shows that $k \to 0$ as $h \to 0$. This is because $hf_x(x + \theta h, y + \theta k) \to 0$ as $h \to 0$ (since $f_x(x + \theta h, y + \theta k)$ is finite) and so $kf_y(x + \theta h, y + \theta k)$ must also tend to zero. But we know $f_y(x + \theta h, y + \theta k) > 0$ by (9.9.1.1) hence k must tend to zero. Again from (9.9.1.2) we get,

$$\phi'(x) = \lim_{h \to 0} \frac{\phi(x + h) - \phi(x)}{h} = \lim_{h \to 0} \frac{k}{h} = -\frac{f_x(x, y)}{f_y(x, y)}$$

As ϕ' is quotient of two continuous functions and $f_y \neq 0$ for $x \in (a - \lambda, a + \lambda)$ we conclude that ϕ' is continuously differentiable in $|x - a| < \lambda$. Thus the proof of the theorem is complete.

Example 9.23   Show that the equation $x^2 e^{xy} - xy + y = 0$ defines y as a function of x in the neighborhood of $(0, 0)$. Find the derivative of this function at $x = 0$

Sol: Let $\qquad\qquad f(x, y) = x^2 e^{xy} - xy + y$. Then we see that $f(0, 0) = 0$ also

$f_x(x, y) = 2xe^{xy} + x^2 ye^{xy}$ and $f_y(x, y) = x^3 e^{xy} - x + 1$ which are both continuous. Also $t f_x(0, 0) = 0$ and $f_y(0, 0) = 1 \neq 0$. Hence by above theorem there is a function ϕ such that $y = \phi(x)$ in a suitable neighborhood of $(0, 0)$. Also we have,

$$\phi'(0) = -\frac{f_x(0, 0)}{f_y(0, 0)} = 0$$

Next we turn our attention to the discussion of maximum and minimum of a functions of two variables. In fact we will derive a sufficient condition of a function two variables having maximum or minimum at some point of it's domain. Before we proceed we will need the following definitions:

Definition   Let f is a real valued function defined in a domain $D \subset \mathbb{R}^2$. Suppose $\mathbf{a} \in D$ then f is said to have,

(i) a *local minimum* at $\mathbf{a}$ if there exists an open neighbourhood of $\mathbf{a}$ say U_a such that $f(\mathbf{x}) > f(\mathbf{a})$ $\forall \mathbf{x} \in U_a$.

(ii) a *local maximum* at $\mathbf{a}$ if there exists an open neighbourhood of a say U_a such that $f(\mathbf{x}) < f(\mathbf{a})$ $\forall x \in U_a$.

The function f is said to have a local extremum at $a \in D$ if it has either a local minimum or local maximum at that point.

Suppose f is a continuous real valued function defined in a domain $D \subset \mathbb{R}^2$ with the partial derivatives existing in a certain neighbourhood of $\mathbf{a} = (a, b) \in D$. Then if f has a minimum at $\mathbf{a} = (a, b)$ it follows from definition that the function $\phi(x) = f(x, b)$ has a minimum at $x = a$ and hence $\phi'(a) = f_x(a, b) = 0$ by Theorem 2.5.9. Similarly $f_y(a, b) = 0$. By same argument one can prove that if f has a maximum at $\mathbf{a} = (a, b)$ then $f_x(a, b) = 0$ & $f_y(a, b) = 0$.

The points where the first order partial derivatives of a function vanishes are called ***critical values*** of the functions.

Theorem 9.9.2

Suppose f has continuous second order partial derivatives in some open neighbourhood at U of the point $\mathbf{a} = (a, b) \in \mathbb{R}^2$ *with* $f_x(a, b) = f_y(a, b) = 0$. *If further,*

(a) $(f_{xy}(a, b))^2 - f_{xx}(a, b)f_{yy}(a, b) < 0$

(b) $f_{xx}(a, b) < 0$

then, f has a local maximum at $\mathbf{a} = (a, b)$.

Proof: First notice that the given conditions of continuous second order derivatives implies that there exists some open ball with centre $\mathbf{a}$ say $B(\mathbf{a}, \delta) \subset U$ such that condition (a) & (b) holds for all $(x, y) \in B(\mathbf{a}, \delta)$. Also we will have $f_{xy} = f_{yx}$ for all points of $B(\mathbf{a}, \delta)$. We choose $(x, y) \in B(\mathbf{a}, \delta)$ and let $h = x - a$, $k = y - b$ (so that $h^2 + k^2 < \delta^2$). By using Taylor's Theorem we obtain,

$f(x, y) - f(a, b)$

$$= hf_x(a, b) + kf_y(a, b) + \frac{1}{2}[h^2 f_{xx}(a + \theta h, b + \theta k) + 2hkf_{xy}(a + \theta h, b + \theta k) + k^2 f_{yy}(a + \theta h, b + \theta k)]$$

for some $\theta \in (0, 1)$

$$= \frac{1}{2}[h^2 f_{xx}(a + \theta h, b + \theta k) + 2hkf_{xy}(a + \theta h, b + \theta k) + k^2 f_{yy}(a + \theta h, b + \theta k)]$$

$$\text{(using } f_x(a, b) = f_y(a, b) = 0)$$

$$= \frac{1}{2A}[(Ah + Bk)^2 + (AC - B^2)k^2](9.9.2.1)$$

where $A = f_{xx}(a + \theta h, b + \theta k)$, $B = f_{xy}(a + \theta h, b + \theta k)$ & $C = f_{yy}(a + \theta h, b + \theta k)$

As $(a + \theta h, b + \theta k) \in B(\mathbf{a}, \delta)$ we have $AC - B^2 > 0$ and $A < 0$ hence right hand side of (9.9.2.1) is negative. Thus we obtain $f(x, y) < f(a, b) \ \forall (x, y) \in B(\mathbf{a}, \delta)$. That is f has a local maximum at $\mathbf{a} = (a, b)$.

By similar arguments we can prove the following theorem:

Theorem 9.9.3

Suppose f has continuous second order partial derivatives in some open neighbourhood at U of the point $\mathbf{a} = (a, b) \in \mathbb{R}^2$ *with* $f_x(a, b) = f_y(a, b) = 0$. *If further,*

(a) $(f_{xy}(a, b))^2 - f_{xx}(a, b)f_{yy}(a, h) < 0$

(b) $f_{xx}(a, b) > 0$

then, f has a local minimum at $\mathbf{a} = (a, b)$.

Example 9.24   Find the extreme values of the function defined by,

$$f(x, y) = x^3 - 3x^2 + y^2$$

Sol: We have, $f_x(x, y) = 3x^2 - 6x, f_y(x, y) = 2y, f_{xx}(x, y) = 6x - 6, f_{yy}(x, y) = 2$

$f_{xy}(x, y) = 0$. Now for critical points we must have $f_x(x, y) = 0$ and $f_y(x, y) = 0$

Now $f_x(x, y) = 0 \Rightarrow x = 0, 2$ and $f_y(x, y) = 0 \Rightarrow y = 0$. Hence the critical points are $(0, 0)$ and $(2, 0)$. Now as,

$$(f_{xy}(0, 0))^2 - f_{xx}(0, 0)\, f_{yy}(0, 0) = 12 > 0$$

we conclude $(0, 0)$ is not an extreme value.

Again as,

$$(f_{xy}(2, 0))^2 - f_{xx}(2, 0)\, f_{yy}(2, 0) = -12 < 0$$

and $f_{xx}(2, 0) = 12 > 0$ we conclude f attains minimum at $(2, 0)$.

Example 9.25 Find minimum and maximum of the function,

$$f(x, y) = y^2 + 4xy + 3x^2 + x^3$$

Sol: $\quad f_x(x, y) = 4y + 6x + 3x^2, f_{xx}(x, y) = 6 + 6x, f_y(x, y) = 2y + 4x, f_{yy}(x, y) = 2$

$f_{yx}(x, y) = 4$. For critical values,

$f_x(x, y) = 4y + 6x + 3x^2 = 0 \ \& \ f_y(x, y) = 2y + 4x = 0$

Solving above two equations we get the critical points as, $(0, 0)$, $(0, -3)$, $\left(\dfrac{3}{2}, 0\right) \ \& \ \left(\dfrac{3}{2}, -3\right)$

As

$$(f_{xy}(0, 0))^2 - f_{xx}(0, 0)\, f_{yy}(0, 0) = f_{xy}(0, -3))^2 - f_{xx}(0, -3) f_{yy}(0, -3) = 16 - 12 = 4 > 0$$

$(0, 0) \ \& \ (0, -3)$ is not an extreme value.

Similarly we can show

$$\left(\frac{3}{2}, 0\right) \ \& \ \left(\frac{3}{2}, -3\right) \text{ are points for minimum.}$$

PROBLEMS

1. Using definitions of limit show the following:

   (i) $\displaystyle\lim_{(x, y)\to(1, 1)} (x^2 + y^2) = 2$

   (ii) $\displaystyle\lim_{(x, y)\to(1, 3)} (xy) = 3$

   (iii) $\displaystyle\lim_{(x, y)\to(0, 0)} xy\left(\frac{x^2 - y^2}{x^2 + y^2}\right) = 0$

   (iv) $\displaystyle\lim_{(x, y)\to(0, 0)} e^{-\left(\frac{1}{x^2 + y^2}\right)} = 0$

   (v) $\displaystyle\lim_{(x, y)\to(0, 0)} \left(y \sin\frac{1}{x} + x \cos\frac{1}{y}\right) = 0$

   (vi) $\displaystyle\lim_{(x, y)\to(0, 0)} \frac{x^2 y^2}{x^2 + y^2} = 0$

2. Show that the following limits does not exist:

   (i) $\displaystyle\lim_{(x, y)\to(0, 0)} \left(\frac{x^2 y}{x^4 + y^2}\right)$

   (ii) $\displaystyle\lim_{(x, y)\to(0, 0)} \left(\frac{xy}{x^2 + y^2}\right)$

(iii) $\lim\limits_{(x,\,y)\to(0,\,0)} \left(\dfrac{x^2 - y^2}{x^2 + y^2}\right)$

(iv) $\lim\limits_{(x,\,y)\to(0,\,0)} \left(\dfrac{x - y}{x + y}\right)$

3. Find the repeated limits and double limits of the following functions at origin in case they exist and compare their values.

(i) $f(x, y) = \begin{cases} \dfrac{xy}{x^2 + y^2} & \text{if } (x, y) \neq (0, 0) \\ 0 & \text{if } (x, y) \neq (0, 0) \end{cases}$

(ii) $f(x, y) = \begin{cases} x \cos \dfrac{1}{y} + y & \text{if } y \neq 0 \\ 0 & \text{if } y = 0 \end{cases}$

(iii) $f(x, y) = \begin{cases} \dfrac{x^2 - y^2}{x^2 + y^2} & \text{if } (x, y) \neq (0, 0) \\ 0 & \text{if } (x, y) \neq (0, 0) \end{cases}$

(iv) $f(x, y) - \begin{cases} \sin\left(\dfrac{x}{y}\right) & \text{if } (x, y) \neq (0, 0) \\ 0 & \text{if } (x, y) \neq (0, 0) \end{cases}$

(v) $f(x, y) = \begin{cases} xy\left(\dfrac{x^2 - y^2}{x^2 + y^2}\right) & \text{if } (x, y) \neq (0, 0) \\ 0 & \text{if } (x, y) \neq (0, 0) \end{cases}$

(vi) $f(x, y) = \begin{cases} x \cos \dfrac{1}{x} + y \sin \dfrac{1}{y} & \text{if } (x, y) \neq (0, 0) \\ 0 & \text{if } (x, y) = (0, 0) \end{cases}$

4. If $f(x, y) = \begin{cases} 1 & \text{if } xy \neq 0 \\ 0 & \text{if } xy = 0 \end{cases}$

then show that although the repeated limits of the above function exist at the origin but the double does not exist at that point.

5. Show that the following functions are continuous at the origin:

   (i) $f(x, y) = \begin{cases} \dfrac{\sin xy}{\sqrt{x^2 + y^2}} & \text{if } (x, y) \neq (0, 0) \\ 0 & \text{if } (x, y) \neq (0, 0) \end{cases}$

   (ii) $f(x, y) = \begin{cases} \dfrac{x^2 y^2}{x^2 + y^2} & \text{if } (x, y) \neq (0, 0) \\ 0 & \text{if } (x, y) \neq (0, 0) \end{cases}$

6. Let $f(x, y)$ be the function as defined in problem 4 above then show that the functions $\phi(x)$ and $\psi(y)$ defined by, $\phi(x) = f(x, 0)$ & $\psi(y) = f(0, y)$ are continuous at $x = 0$ and $y = 0$ respectively but $f(x, y)$ is not continuous at $(0, 0)$.

7. Let $f: \mathbb{R}^2 \to \mathbb{R}$ be defined by,

   $$f(x, y) = 2x + 3y$$

   Find $D_a(f; \mathbf{b})$ where $\mathbf{a} = (1, 1)$ & $\mathbf{b} = (2, 3)$

8. Let $f: \mathbb{R} \to \mathbb{R}$ be defined by,

   $$f(x, y) = (x^2 + y^2)^2$$

   Find all $\mathbf{b} = (x, y)$ such that $D_a(f; \mathbf{b}) = 6$, where $\mathbf{a} = (2, 6)$

9. If $f(x, y)$ is homogenous degree n with continuous second order partial derivatives show that

   $$x^2 f_{xx} + 2xy f_{xy} + y^2 f_{yy} = n(n - 1)f$$

11. $f(x, y)$ is a homogenous equation of degree $n > 2$ then show that the second order partial derivatives are homogenous of degree $n - 2$.

12. If $f(x, y)$ be defined as follows,

   $$f(x, y) = \begin{cases} \dfrac{x^3 - y^3}{x^2 + y^2} & \text{if } (x, y) \neq (0, 0) \\ 0 & \text{if } (x, y) \neq (0, 0) \end{cases}$$

   then show that f is continuous at $(0, 0)$ and possesses first order partial derivatives but is not differentiable there.

12. Show that,

   $$f(x, y) = |x - 1| + |y - 1|$$

   is continuous but not differentiable at $(1, 1)$.

13. Show that the following functions are differentiable at $(0, 0)$

    (i) $f(x, y) = |x^2 - y^2|$

    (ii) $f(x, y) = \begin{cases} x^2 y^2 \log(x^2 + y^2) & \text{if } (x, y) \neq (0, 0) \\ 0 & \text{if } (x, y) \neq (0, 0) \end{cases}$

14. Show that $f(x, y) = \sqrt{|xy|}$ is not differentiable at $(0, 0)$

15. Let

$$f(x, y) = \begin{cases} x^2 \sin \dfrac{1}{x} + y^2 \sin \dfrac{1}{y} & \text{if } (x, y) \neq (0, 0) \\[2mm] x^2 \sin \dfrac{1}{x} & \text{if } y = 0, x \neq 0 \\[2mm] y^2 \sin \dfrac{1}{y} & \text{if } x = 0, y \neq 0 \\[2mm] 0 & \text{if } (x, y) = (0, 0) \end{cases}$$

    Show that

    (a) $f_x(0, 0) \,\&\, f_y(0, 0)$ exists

    (b) $f_x(x, y) \,\&\, f_y(x, y)$ are not continuous at $(0, 0)$

    (c) $f(x, y)$ is not differentiable at $(0, 0)$

16. If

$$f(x, y) = \begin{cases} (x^2 + y^2) \tan^{-1} \dfrac{y}{x} & \text{if } x \neq 0 \\[2mm] \dfrac{\pi y^2}{2} & \text{if } x = 0 \end{cases}$$

    Show that $f_{xy}(0, 0) \neq f_{yx}(0, 0)$

17. Verify whether the second order mixed derivatives at $(0, 0)$ are equal or not for the following functions:

    (i) $f(x, y) = e^{xy}(\sin x + x \cos y)$

    (ii) $f(x, y) = \begin{cases} x^2 \sin \left(\dfrac{y}{x}\right) & \text{if } x \neq 0 \\[2mm] 0 & \text{if } x = 0 \end{cases}$

$$\text{(iii)} \quad f(x, y) = \begin{cases} xy^2 & \text{if } y > 0 \\ -xy^2 & \text{if } y \le 0 \end{cases}$$

(vi) $f(x, y) = |x^2 - y^2|$

18. Find the linear approximation of the function

$$f(x, y) = \sin xy$$

at the point $\left(1, \dfrac{\pi}{4}\right)$.

19. Find the quadratic approximation of the function

$$f(x, y) = x^2 \cos y + y^2$$

at the point $\left(\dfrac{\pi}{2}, 0\right)$.

20. Expand the function $f(x, y) = x^2 y \cos x$ in powers of $\left(x - \dfrac{\pi}{4}\right)$ and $(y - 1)$.

21. Verify whether from the following equations y can be solved explicitly in terms of x in the neighbourhood of the indicated points:

(i) $x^3 + y^3 - 3xy + y = 0$ at $(0, 0)$

(ii) $y^2 - x^2 y - 2x^5 = 0$ at $(1, -1)$

(iii) $x + y + \sin xy = 0$ at $(0, 0)$

(iv) $xe^{xy} + y = 0$ at $(0, 0)$

22. Find the critical values of the following functions and find whether they are the extreme values of the function or not

(i) $f(x, y) = x^3 - y^3 - 3x - 12y + 20$

(ii) $f(x, y) = 2x^4 - x^2 y + y^2$

(iii) $f(x, y) = e^x \cos y$

(iv) $f(x, y) = \sqrt{x^2 + y^2}$

Chapter 10

Double and Triple Integrals

10.1 INTRODUCTION

We have seen in Chapter 3 that the symbol $\int_a^b f(x)\, dx$ called the integral of the function, whenever exists, geometrically represents the area of under the curve $y = f(x)$ between $x = a$ and $x = b$ provided, $f(x) \geq 0$. In this chapter we will be interested in generalizing the above notion of integrals to bounded functions defined on closed rectangles of $\mathbb{R}^2$ and rectangular parallelepiped of $\mathbb{R}^3$ which are natural analogues of closed intervals of $\mathbb{R}$. This leads to the definitions of double and triple integrals respectively. Finally, we will extend the definition of double and triple integrals to any bounded domains of $\mathbb{R}^2$ and $\mathbb{R}^3$ respectively.

10.2 DOUBLE (TRIPLE) INTEGRAL OVER A CLOSED RECTANGLE (BOX)

We have to formulate certain definitions and notions as in Chapter 3 before coming to the actual definition of double integral. We will see that definition of triple integral over a closed box is analogous to the definition of a double integral over a closed rectangle.

Suppose $R = [a, b] \times [c, d]$ be a closed rectangle and let f be a bounded function defined on R. A **partition** P of R is of the form, $P = P_1 \times P_2$ where, P_1 and P_2 are partitions of the intervals $[a, b]$ and $[c, d]$ respectively that is,

$P_1 = \{x_0, x_1, x_2, ..., x_r, ... x_n\}$ and $P_2 = \{y_0, y_1, y_2, ..., y_k, ... y_m\}$ are finite collections of points on X-axis and on Y-axis respectively, satisfying,

$$a = x_0 < x_1 < x_2 < ... < x_{r-1} < x_r < ... < x_n = b$$

$$c = y_0 < y_1 < y_2 < ... < y_{k-1} < y_k < ... < y_m = d$$

P_1 & P_2 will be called partitions in X and Y dimensions respectively.

We see that the partition P above divides the rectangle R into mn sub-rectangles (Fig. 1) of the form,

$$R_{rk} = [x_{r-1}, x_r] \times [y_{k-1}, x_k] \quad r = 1, 2, ... n \ \& \ k = 1, 2 ...m.$$

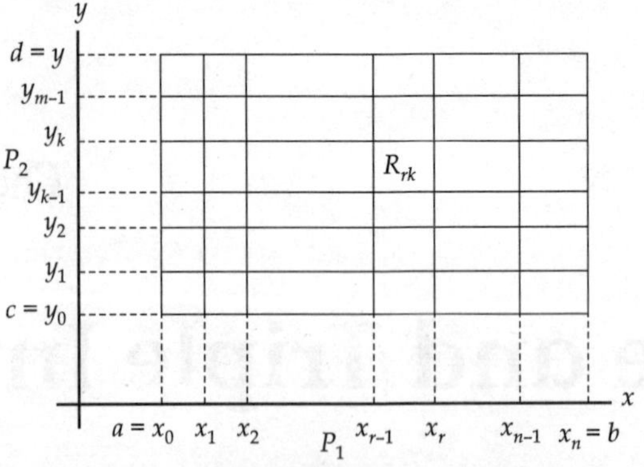

Fig. 10.1

A partition P' will be called a **refinement** of the partition P if $P \subseteq P'$. Let $\Delta_{rk} = (x_r - x_{r-1})(y_k - y_{k-1})$ and $\delta_{rk} = \sqrt{(x_r - x_{r-1})^2 + (y_k - y_{k-1})^2}$ be the area and the length of the diagonal of R_{rk} respectively for $r = 1, 2, \ldots n$ & $k = 1, 2, \ldots m$. The **norm** of the partition P denoted by $\|P\|$ is defined as $\|P\| = \max_{r,k} \delta_{rk}$. It is clear that if P' is a refinement of P then $\|P\| \leq \|P'\|$.

Definition   Let f be a bounded function defined on $R = [a, b] \times [c, d]$ and $P = P_1 \times P_2$ be any partition of R sub dividing it into mn sub-rectangles of the form $R_{rk} = [x_{r-1}, x_r] \times [y_{k-1}, x_k]$ as described above. For any point $(\xi_{rk}, \eta_{rk}) \in R_{rk}$ the sum of the form,

$$S(P, f) = \sum_{r=1}^{n} \sum_{k=1}^{m} f(\xi_{rk}, \eta_{rk}) \Delta_{rk}$$

is called the Riemann Approximating Sum. We say that above sums tends to a finite limit A as $\|P\| \to 0$ if, given any $\varepsilon > 0$ there always exists a real number $\delta(\varepsilon) > 0$ such that for any partition P with $\|P\| < \delta$ we have,

$$|S(P, f) - A| < \varepsilon$$

and in that case we write $\lim_{\|P\| \to 0} S(P, f) = A$

We say that the double integral of f over R exists if Riemann Approximating Sums tends to a finite limit and in that case we write

$$\iint_R f \, dx \, dy \text{ or } \int_a^b \int_c^b f(x, y) \, dx \, dy = \lim_{\|P\| \to 0} S(P, f) = A$$

Suppose $f \geq 0$ then, for a partition P of $R = [a, b] \times [c, d]$ we note that $f(\xi_{rk}, \eta_{rk}) \Delta_{rk}$ denotes the volume of the cylinder with base R_{rk} and $S(P, f) = \sum_{r=1}^{n} \sum_{k=1}^{m} f(\xi_{rk}, \eta_{rk}) \Delta_{rk}$ is the approximate volume of the region under the surface f. Hence it is natural to define the volume V of the region under the surface f as $\lim_{\|P\| \to 0} S(P, f)$. In other words, $\int_a^b \int_c^d f(x, y) \, dx \, dy$ denotes the volume of the region under the surface f if it exists.

For the remaining section we will assume that f is a bounded function on R. Now for a partition P the **Upper Sum** and **Lower Sum** of f denoted by $U(P, f)$ and $L(P, f)$ are defined as,

$$L(P, f) = \sum_{r=1}^{n} \sum_{k=1}^{m} m_{rk} \, \Delta_{rk}$$

$$U(P, f) = \sum_{r=1}^{n} \sum_{k=1}^{m} M_{rk} \, \Delta_{rk}$$

where, M_{rk} and m_{rk} are supremum and infimum of f on the rectangle R_{rk} for $r = 1, 2, ..., n$ & $k = 1, 2, ... m$. We note that M_{rk} and m_{rk} are finite as f is bounded on R and hence on its subset Δ_{rk}.

It easily follows as in Chapter 3 (Lemma 3.3.1) that for a bounded function f defined on a closed rectangle $[a, b] \times [c, d]$ the upper and lower sums corresponding to any partition is bounded in fact we have the following lemma,

Lemma 10.2.1

If M & m are supremum and infimum of f then,

$$mS \le L\,(P, f) \le U(P, f) \le MS$$

where, P is a partition and S = (b – a) (d – c) is area of the rectangle R.

Proof: One can proceed exactly same way as Lemma 3.33.1 to prove above lemma and is left an exercise.

We define,

$$I = \sup_{p \in p} L(P, f)$$

$$J = \inf_{p \in p} U(P, f)$$

where, P is the set of all partitions of the rectangle R.

Then again it easily follows as in Chapter 3 that $I \le J$. We have the following theorem called Darboux Theorem whose proof similar to the proof of Theorem 3.5.1 and is left as an exercise.

Theorem 10.2.1   Given any $\varepsilon > 0$ there exist a partition P such that,

$$U(P, f) < J + \varepsilon \quad \text{and}$$

$$L(P, f) > I - \varepsilon$$

We have the following important theorem which essentially provides us necessary and sufficient conditions for f to be integrable.

Theorem 10.2.2

The following are equivalent,

    (i) f is integrable

    (ii) *Given any $\varepsilon > 0$ there exists a partition P such that,*

        $U(P, f) - L(P, f) < \varepsilon$

(iii) Given any $\varepsilon > 0$ there exists $\delta > 0$ such that for any partition P with $\|P\| < \delta$ we have,

$$U(P, f) - L(P, f) < \varepsilon$$

(iv) $I = J$

Proof of above theorem can be deduced modifying suitably the arguments given in Theorem 3.6.1 Theorem 3.7.1 and Theorem 3.7.2.

Remark

In view of above theorem we get another equivalent definition of integrability, we can say that a bounded function f is integrable (double) on R if $I = J$.

Next we will now define triple integral of a bounded function f defined on a subset of R^3 of the form $R = [a, b] \times [c, d] \times [e, f]$ known as rectangular parallelepiped. For this we consider partitions P_1, P_2 and P_3 of $[a, b]$, $[c, d]$ and $[e, f]$ as follows,

$$P_1 : a = x_0 < x_1 < x_2 < \dots < x_{r-1} < x_r < \dots < x_n = b$$

$$P_2 : c = y_0 < y_1 < y_2 < \dots < y_{k-1} < y_k < \dots < y_m = d$$

$$P_3 : c = z_0 < z_1 < z_2 < \dots < z_{s-1} < z_s < \dots < z_l = d$$

as in the case of the double integral, $P = P_1 \times P_2 \times P_3$ defines a partition of R which divides it into *mnl* smaller box type regions of the form $R_{rks} = [x_{r-1}, x_r] \times [y_{k-1}, y_k] \times [z_{s-1}, z_s]$ $r = 1, 2, \dots n, k = 1, 2, \dots m$ and $s = 1, 2, \dots l$. Volume of R_{rks} will be denoted by Δ_{rks} whose value is given by, $(x_r - x_{r-1}) (y_k - y_{k-1})$ $(z_s - z_{s-1})$ and let $\delta_{rks} = \sqrt{(x_r - x_{r-1})^2 + (y_k - y_{k-1})^2 + (z_s - z_{s-1})^2}$. Then the norm of P denoted by $\|P\|$ is defined as $\|P\| = \max_{r,k,s} \delta_{rks}$. The Riemann Approximating sum of the function f for the partition P will be defined in this case as,

$$S(P, f) = \sum_{r=1}^{n} \sum_{k=1}^{m} \sum_{s=1}^{l} f(\xi_{rks}, \eta_{rks}, \zeta_{rks}) \, \Delta_{rks}$$

where $(\xi_{rks}, \eta_{rks}, \zeta_{rks})$ is any point of R_{rks} for $r = 1, 2, \dots n, k = 1, 2, \dots m$ and $s = 1, 2, \dots l$. We say that the triple integral of f exists and is equal to A if for every $\varepsilon > 0$ we can find a real number $\delta(\varepsilon) > 0$ such that for any partition P with $\|P\| < \delta$ we have,

$$|S(P, f) - A| < \varepsilon$$

In symbol we write,

$$\int_{a}^{b} \int_{c}^{d} \int_{e}^{f} f(x, y, z) \, dx \, dy \, dz = \iiint_R I f(x, y, z) \, dx \, dy \, dz$$

$$= \lim_{\|P\| \to 0} S(P, f) = A$$

For partition P we define the **Upper Sum** and **Lower Sum** of f denoted by $U(P, f)$ and $L(P, f)$ are defined as,

$$L(P, f) = \sum_{r=1}^{n} \sum_{k=1}^{m} \sum_{s=1}^{l} m_{rks} \, \Delta_{rks}$$

$$U(P,f) = \sum_{r=1}^{n} \sum_{k=1}^{m} \sum_{s=1}^{l} M_{rks}\, \Delta_{rks}$$

where, M_{rks} and m_{rks} denote supremum and infimum of f on R_{rks} for $r = 1, 2, ..., n$, $k = 1, 2, ..., m$ and $s = 1, 2, ..., l$. As in the two variable case we can show that for any partition P, $U(P,f)$ is always more than $L(P,f)$ and more over they are always bounded hence as before we define,

$$I = \sup_{p\in p} L(P,f) \quad \text{and}$$

$$J = \inf_{p\in p} U(P,f)$$

It is not hard to see that an analogue of Theorem 10.2.2 is true in this case so that we can formulate an equivalent definition of integrability as follows :

A bounded function f on R exists if $I = J$ and in that case,

$$\iiint_{R} f(x, y, z)\, dx\, dy\, dz = I = J.$$

We remark that that as the definitions of double and triple integrals are similar any theorem on double integral will have an similar corresponding theorem for triple integral hence in the subsequent sections we will discuss mostly results of double integrals readers are expected to see the particular analogue in the triple integral case.

10.3 CLASSES OF INTEGRABLE FUNCTIONS

In this section we will determine some classes of functions which are integrable. As in one variable case the first theorem shows that every continuous function is integrable.

Theorem 10.3.1

If f is a continuous on $R = [a, b] \times [c, d]$ then it is integrable.

Proof: Let $\varepsilon > 0$ be any arbitrary positive number. As R is compact f is uniformly continuous and hence corresponding to this ε there exists $\delta(\varepsilon) > 0$ such that whenever $(x - x')^2 + (y - y')^2 < \delta^2$ for (x, y) & (x', y') $\in R$ we have,

$$|f(x, y) - f(x', y')| < \frac{\varepsilon}{(b - a)\,(d - c)} \tag{10.3.1.1}$$

We now choose a partition P of R such that $\|P\| < \delta(\varepsilon)$. As f is continuous on R and hence on each R_{rk} which is compact, there exist points (x_{rk}, y_{rk}) & (x'_{rk}, y'_{rk}) in R_{rk} such that,

$$M_{rk} = f(x_{rk}, y_{rk}) \quad \text{and} \quad m_{rk} = f(x'_{rk}, y'_{rk}) \quad 1 \leq r \leq n \ \& \ 1 \leq k \leq m$$

Now,

$$U(P,f) - L(P,f) = \sum_{r=1}^{n} \sum_{k=1}^{m} (M_{rk} - m_{rk})\, \Delta_{rk}$$

$$= \sum_{r=1}^{n} \sum_{k=1}^{m} (f(x_{rk}, y_{rk}) - f(x'_{rk}, y'_{rk}))\, \Delta_{rk} \tag{10.3.1.2}$$

Again as $\|P\| < \delta(\varepsilon)$ we have,

$$(x_{rk} - x'_{rk})^2 + (y_{rk} - y'_{rk})^2 < \delta^2 \; 1 \le r \le n \;\&\; 1 \le k \le m$$

Now from (10.3.1.1) we get,

$$f(x_{rk}, y_{rk}) - f(x'_{rk}, y'_{rk}) < \frac{\varepsilon}{(b-a)(c-d)} \; 1 \le r \le n \;\&\; 1 \le k \le m \qquad (10.3.1.3)$$

(10.3.1.2) and (10.3.1.3) together implies,

$$U(P,f) - L(P,f) < \frac{\varepsilon}{(b-a)(c-d)} \sum_{r=1}^{n} \sum_{k=1}^{m} \Delta_{rk}$$

$$= \frac{\varepsilon}{(b-a)(c-d)} (b-a)(c-d) = \varepsilon$$

By Theorem 10.2.2 f is integrable.

Example 10.1   Compute the upper and lower integral of the function defined as,

$$f(x, y) = x + y \; (x, y) \in [a, b] \times [c, b] \; a, b, c, d > 0$$

Hence find the value of the integral $\displaystyle\int_{a}^{b} \int_{c}^{d} (x + y) \, dx \, dy$

Sol: Consider any partition $P = P_1 \times P_2$ of R $= [a, b] \times [c, d]$ as,

$$P_1 : a = x_0 < x_1 < x_2 < \ldots < x_{r-1} < x_r < \ldots < x_n = b$$

$$P_2 : c = y_0 < y_1 < y_2 < \ldots < y_{k-1} < y_k < \ldots < y_m = d$$

dividing R into mn sub-rectangles of the form $R_{rk} = [x_{r-1}, x_r] \times [y_{k-1}, x_k] \; r = 1, 2, \ldots n \;\&\; k = 1, 2, \ldots m$. For this partition P if M_{rk} and denote the supremum and infimum of f on the rectangle R_{rk} then, $M_{rk} = x_r + y_k$ and $m_{rk} = x_{r-1} + y_{k-1}$ for $r = 1, 2, \ldots, n \;\&\; k = 1, 2, \ldots, m$. Hence,

$$L(P,f) = \sum_{r=1}^{n} \sum_{k=1}^{m} m_{rk} \Delta_{rk} = \sum_{r=1}^{n} \sum_{k=1}^{m} (x_r + y_k) \Delta_{rk}$$

$$U(P,f) = \sum_{r=1}^{n} \sum_{k=1}^{m} M_{rk} \Delta_{rk} = \sum_{r=1}^{n} \sum_{k=1}^{m} (x_{r-1} + y_{k-1}) \Delta_{rk}$$

Now,

$$x_{r-1} + y_{k-1} \le \frac{1}{2}(x_r + x_{r-1}) + \frac{1}{2}(y_k + y_{k-1}) \le x_r + y_k$$

Multiplying above inequality by Δ_{rk} and summing over r, k we get,

$$L(P,f) \le \sum_{r=1}^{n} \sum_{k=1}^{m} \frac{1}{2}(x_r + x_{r-1}) \Delta_{rk} + \sum_{r=1}^{n} \sum_{k=1}^{m} \frac{1}{2}(y_k + y_{k-1}) \Delta_{rk} \le U(P,f)$$

Now as $\Delta_{rk} = (x_r - x_{r-1})(y_k - y_{k-1})$ we have,

$$\sum_{r=1}^{n} \sum_{k=1}^{m} \frac{1}{2}(x_r + x_{r-1})\,\Delta_{rk} + \sum_{r=1}^{n} \sum_{k=1}^{m} \frac{1}{2}(y_k + y_{k-1})\,\Delta_{rk}$$

$$= \frac{1}{2}\sum_{r=1}^{n}(x_r^2 - x_{r-1}^2)\sum_{k=1}^{m}(y_k - y_{k-1}) + \frac{1}{2}\sum_{k=1}^{m}(y_k^2 - y_{k-1}^2)\sum_{r=1}^{n}(x_r - x_{r-1})$$

$$= \frac{1}{2}(b^2 - a^2)(d - c) + \frac{1}{2}(d^2 - c^2)(b - a)$$

$$= \frac{1}{2}(a + b + c + d)(b - a)(d - c)$$

so that we obtain,

$$L(P, f) \le \frac{1}{2}(a + b + c + d)(b - a)(d - c) \le U(P, f)$$

As the above inequality is true for all partition P we get,

$$I \le \frac{1}{2}(a + b + c + d)(b - a)(a - c) \le J$$

Now observe that $f(x, y) = x + y$ is a continuous function in R hence integrable so that $I = J$. Hence from above inequality we conclude,

$$I = J + \int_{a}^{b}\int_{c}^{d}(x + y)\,dx\,dy = \frac{1}{2}(a + b + c + d)(b - a)(d - c)$$

As in the one dimension case next theorem will shows that a function will be integrable if and only if it is "close" to a continuous function. That In fact there is an analogue of Theorem 3.13.4 in case of double integrals. For that we need to define the concept of negligible set in $\mathbb{R}^2$.

A set of the form $I = (a, b) \times (c, d)$ is called an ***open rectangle*** of $\mathbb{R}^2$ with ***length*** $l(I) = (b - a)(d - c)$.

Definition   A subset N of $\mathbb{R}^2$ is said to be a ***negligible set*** if, for any $\varepsilon > 0$ there exists a sequence of open intervals $\{I_n\}$ such that, $N \subset \bigcup_{n=1}^{\infty} I_n$ and $\sum_{n=1}^{\infty} l(I_n) < \varepsilon$.

Finite sets are natural trivial examples of negligible sets. Again it is not hard to see that countable union of negligible sets is negligible hence any countable subset $\mathbb{R}^2$ is also negligible.

Interestingly, ***graph*** of any curve that is, sets of the from,

$$\{(x, y) : x = \phi(t), y = \psi(t), a \le t \le b\}, \{(x, y): y = \psi(x)\} \text{ or, } \{(x, y) : x = \phi(y)\}$$

ϕ & ψ being continuous functions is negligible.

Next theorem tells us in what sense an integrable function is 'close' to a continuous function.

Theorem 10.3.2

A bounded function defined on the rectangle $R = [a, b] \times [c, d]$ is integrable if and only if it's set of points of discontinuities is negligible.

Proof: The proof can obtained by suitable modification of the arguments Theorem 3.13.4 we can get and kept as an exercise.

As an immediate application of the above theorem we get the following corollaries whose proofs are omitted.

Corollary 10.3.3

A bounded function defined on the rectangle $R = [a, b] \times [c, d]$ with at most countable number of discontinuities is integrable. In particular if f has finite number of discontinuities then it is integrable.

Corollary 10.3.4

If for a bounded function f defined on the rectangle $R = [a, b] \times [c, d]$ the set of its discontinuities lies on finite number of curves then the function is integrable.

10.4 PROPERTIES OF DOUBLE INTEGRAL

Double integral has similar properties as that of one variable Riemann Integrals and is given as the following theorem whose proof is omitted and kept as an exercise.

Theorem 10.4.1

Suppose f, g are bounded integrable functions on the closed rectangle R and let k be any real number then,

(i) $f \pm g, fg, kf \ \& \ |f|$ are integrable

(ii) $\displaystyle\iint\limits_{R}(f \pm g)\,dx\,dy = \iint\limits_{R} f\,dx\,dy \pm \iint\limits_{R} g\,dx\,dy \ \& \ \iint\limits_{R} kf\,dx\,dy = k\iint\limits_{R} f\,dx\,dy$

(iii) $\displaystyle\iint\limits_{R} f\,dx\,dy = \iint\limits_{R_1} f\,dx\,dy + \iint\limits_{R_2} f\,dx\,dy$ where $R_1 \ \& \ R_2$ are disjoint closed rectangles such that,

$R = R_1 \cup R_2$

10.5 EVALUATION OF DOUBLE INTEGRALS: CAVALIERI'S PRINCIPLE

We will now try to see how a double integral can be evaluated using a simple technique known as 'Cavalieri's Principle'. We have already seen that if $f \geq 0$ then $\displaystyle\int_{a}^{b}\int_{c}^{d} f(x)\,dx$ denotes the volume of solid under the surface $z = f(x, y)$. Cavalier's Principle gives us a technique by which we can calculate the volume of this solid. By this principle we first divide the solid by cutting it by thin planes perpendicular to X-axis and calculate the area of this cross section of the solid using the concept of Riemann Integral. The volume of this section is obtained by multiplying the area with the thickness. Finally we add volumes of all these infinitesimally thin areas to obtain the total volume of the solid and hence the value of the double integral. Same process can be done by dividing the solid with planes parallel to Y-axis. We now illustrate this principle, in the following Fig. consider

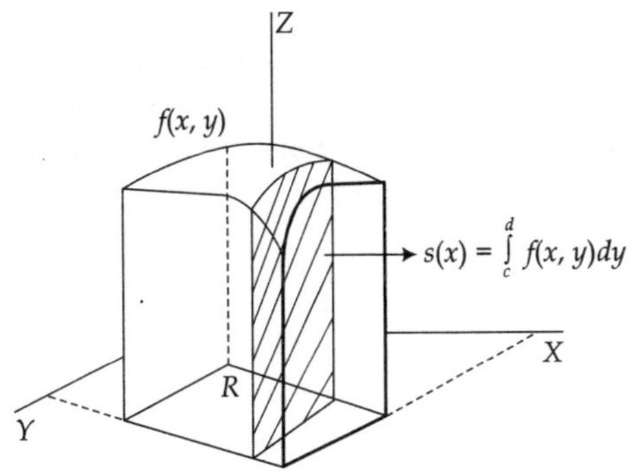

Fig. 10.2

the cross section of the solid under the surface $z = f(x, y)$ formed by cutting it by a plane perpendicular to X-axis of thickness say dx. The area of this cross section is given by $S(x) = \int_c^d f(x, y)\, dy$ (x is fixed while we integrate f with respect to y) hence the volume of this cross section is $S(x)\, dx$. Now the total volume is the sum of the volumes of these cross sections of infinitesimal thickness which is given by

$$\int_a^b S(x)\, dx = \int_a^b \left(\int_c^d f(x, y)\, dy \right) dx$$

Hence if the double integral $\int_a^b \int_c^d f(x)\, dx$ exists it is equal to $\int_a^b \left(\int_c^d f(x, y)\, dy \right) dx$. Similarly by cutting the solid with planes perpendicular to Y-axis one can show that the double integral is equal to $\int_c^d \left(\int_a^b f(x, y)\, dx \right) dy$. The integrals $\int_a^b \left(\int_c^d f(x, y)\, dy \right) dx$ and $\int_c^d \left(\int_a^b f(x, y)\, dx \right) dy$ are called iterated integrals which are also some times denoted by $\int_a^b dx \int_c^d f(x, y)\, dy$ and $\int_c^d dy \int_a^b f(x, y)\, dx$ respectively.

Note that what we have observed is that, if both the double integral and iterated integral exist then by Cavalier's Principle they must be equal. But existence of the double integral does not imply the existence of the iterated integrals and vice-versa. We illustrate this by following examples:

Example 10.2 Let $R = [0, 1] \times [0, 1]$ be the unit square and f be the function defined on R as follows:

$$f(x, y) = \begin{cases} \dfrac{1}{n^2} & \text{if } x = \dfrac{m}{n}\ m, n \text{ are positive integers in lowest terms \& } y \text{ is rational} \\ 0 & \text{if either } x \text{ or } y \text{ is irrational} \end{cases}$$

We first show that $\int_0^1 \int_0^1 f \, dx \, dy$ exist. For an arbitrary positive number $\varepsilon > 0$ there exist finitely many rational numbers say $x_1, x_2, \ldots, x_p$ on X-axis of the form $x_i = \dfrac{m_i}{n_i}$ m_i, n_i are positive integers in lowest terms $i = 1, 2, \ldots p$ such that for any rational y

$$f(x_i, y) > \frac{\varepsilon}{2} \quad i = 1, 2, \ldots p$$

We enclose these 'P' points on X-axis by p many non overlapping closed intervals such that the total length of these intervals is less than ε. Let P_1 be the partition of $[0, 1]$ on X-axis induced by these intervals and let P_2 be any partition of $[0, 1]$ on Y-axis then consider the partition $P = P_1 \times P_2$ of R. It is clear that $m_{rk} = 0$ for $r = 1, 2, \ldots n$ & $k = 1, 2, \ldots m$. Hence

$$U(P, f) - L(P, f) = \sum_{r=1}^{n} \sum_{k=1}^{m} M_{rk} \, \Delta_{rk}$$

$$= \sum_{J} \sum M_{rk} \, \Delta_{rk} + \sum_{J^c} \sum M_{rk} \, \Delta_{rk}$$

where $J = \{r, k\}: R_{rk}$ contains x_j for some $j\}$.

Clearly $\sum_{J^c} \sum M_{rk} \Delta_{rk} < \dfrac{\varepsilon}{2} \sum_{J^c} \sum \Delta_{rk} < \dfrac{\varepsilon}{2}$. We see that $M_{rk} < 1$ for $r = 1, 2, \ldots, n$ & $k = 1, 2, \ldots, m$. Hence $\sum_{J} \sum M_{rk} \Delta_{rk} < \sum_{J} \sum \Delta_{rk} < \dfrac{\varepsilon}{2}$ as length of each Δ_{rk} $(r, k) \in J$ is less than $\dfrac{\varepsilon}{2}$ by construction. This shows that

$$U(P, f) - L(P, f) < \frac{\varepsilon}{2} + \frac{\varepsilon}{2} = \varepsilon$$

That is $\int_0^1 \int_0^1 f \, dx \, dy$ exists. Now as $L(P, f) = 0$ for all partition of R we have $\int_0^1 \int_0^1 f \, dx \, dy = 0$.

Suppose y is a fixed rational number then,

$$f(x, y) = \begin{cases} \dfrac{1}{n^2} & \text{if } x = \dfrac{m}{n} \\[2mm] 0 & \text{if } x \text{ is irrational} \end{cases}$$

Hence $\int_0^1 f(x, y) \, dx$ exists and equals 0 in this case. Also if y is a fixed irrational number then $f(x, y) = 0 \; \forall x$, hence again in this case $\int_0^1 f(x, y) \, dx = 0$. So combining above two cases we conclude that $\int_0^1 \int_0^1 \left(\int_0^1 f(x, y) \, dx \right) dy = 0$. But it is easy to see that if $x = \dfrac{m}{n}$ is a fixed rational number then,

$$f(x, y) = \begin{cases} \dfrac{1}{n^2} & \text{if } y \text{ is rational} \\[2mm] 0 & \text{if } y \text{ is irrational} \end{cases}$$

Hence $\int_0^1 f(x, y)\, dx$ does not exist for x rational number.

Thus existence of double integral does not imply the existence of iterated integrals.

Example 10.4 Let $R = [0, 1] \times [0, 1]$ be the unit square and f be the function defined on R as follows:

$$f(x, y) = \begin{cases} \dfrac{1}{2} & \text{if } y \text{ is rational} \\[2mm] x & \text{if } y \text{ is irrational} \end{cases}$$

First let us calculate the iterated integral $\int_0^1 \left(\int_0^1 f(x, y)\, dx \right) dy$. Now we see that

$$\int_0^1 \left(\int_0^1 f(x, y)\, dx \right) dy = \begin{cases} \int_0^1 \left(\int_0^1 x\, dx \right) dy = \dfrac{1}{2} & \text{if } y \text{ is irrational} \\[4mm] \int_0^1 \left(\int_0^1 \dfrac{1}{2}\, dx \right) dy = \dfrac{1}{2} & \text{if } y \text{ is rational} \end{cases}$$

hence in any case we find $\int_0^1 \int_0^1 f(x, y)\, dx\, dy$ exists and is equal to $\dfrac{1}{2}$. We now show in this case the double integral does not exists. For that consider partitions P_1 and P_2 of $[0, 1]$ along X-axis and Y- axis respectively as follows:

$$P_1 : 0 = x_0 < \frac{1}{2n} = x_1 < \frac{2}{2n} = x_2 < \dots \frac{n-1}{2n} = x_{n-1} < \frac{1}{2} = x_n < \dots \frac{2n-1}{2n} = x_{2n-1} < 1 = x_{2n}$$

$$P_2 : 0 = y_1 < \dots y_{n-1} < y_n < \dots y_{n+1} < y_{2n} = 1$$

where $y_i - y_{i-1} = \dfrac{1}{2n}$, $i = 1, 2, \dots 2n$.

Hence $P = P_1 \times P_2$ forms a partition of R. By our usual notation area of each of the $(2n)^2$ sub-rectangles of the partition is given by $\Delta_{rs} = \dfrac{1}{(2n)^2}$. Also note that as each of these sub-rectangles contains infinite number of points with rational and irrational Y-coordinates we have,

$$m_{rk} = \begin{cases} x_r = \dfrac{r}{2n} & 0 \le r \le n \\[3mm] \dfrac{1}{2} & n+1 \le r \le 2n \end{cases} \qquad k = 1, 2, \dots, 2n$$

Again it is easy to see that,

$$M_{rk} = \begin{cases} \dfrac{1}{2} & 0 \le r \le n \\[3mm] x_r = \dfrac{r}{2n} & n+1 \le r \le 2n \end{cases} \qquad k = 1, 2, \dots, 2n$$

Now,

$$L(P, f) = \sum_{r=1}^{n} \sum_{k=1}^{2n} \left(\frac{r}{2n}\right)\Delta_{rk} + \sum_{r=n=1}^{2n} \sum_{k=1}^{2n} \left(\frac{1}{2}\right)\Delta_{rk}$$

$$= \frac{1}{(2n)^3} \sum_{r=1}^{n} \sum_{k=1}^{2n} r + \frac{1}{2} \cdot \frac{1}{2}$$

$$= \frac{1}{(2n)^2} \sum_{r=1}^{n} r + \frac{1}{4}$$

$$= \frac{1}{(2n)^2} \frac{n(n+1)}{2} + \frac{1}{4}$$

Similar calculation will yeild,

$$U(P, f) = \frac{1}{4} + \frac{1}{(2n)^2} \frac{n(3n+1)}{2}$$

Next we observe that as $n \to \infty$ $\|P\| \to 0$ hence using Darboux's Theorem we get,

$$J = \lim_{\|P\| \to 0} U(P, f) = \lim_{n \to \infty} U(P, f) = \frac{1}{4} + \frac{3}{8} \quad \text{and}$$

$$I = \lim_{\|P\| \to 0} L(P, f) = \lim_{n \to \infty} L(P, f) = \frac{1}{4} + \frac{1}{8}$$

Clearly $I \neq J$ hence the double integral does not exist.

Hence we conclude that existence of the iterated integrals does not imply the existence of double integral.

Next theorem gives the sufficient conditions under which the iterated integrals exists and are equal to the double integral.

Theorem 10.5.1

Let f be a bounded function defined on a closed rectangle $R = [a, b] \times [c, d]$ such that

(i) $\displaystyle\iint_R f \, dx \, dy$ exists and

(ii) $\displaystyle\int_c^d f(x, y) \, dy$ exists for every $x \in [a, b]$

Then, the iterated integral $\displaystyle\int_a^b \left(\int_c^d f(x, y) \, dy\right) dx$ exists and moreover,

$$\iint_R f \, dx \, dy = \int_a^b \int_c^d f(x, y) \, dy \, dx$$

Proof: As usual let $P_1 : a = x_0 < x_1 < x_2 < \ldots < x_{r-1} < x_r < \ldots < x_n = b$ be a partition of $[a, b]$ and $P_2 : c = y_0 < y_1 < y_2 < \ldots < y_{k-1} < y_k < \ldots < y_m = d$ be a partition of $[c, b]$ so that $P_1 \times P_2$ is a partition of $R = [a, b] \times [c, d]$. In the sub rectangle R_{rk} if M_{rk} and m_{rk} be the supremum and infimum of f then,

$$M_{rk} \leq f(x, y) \leq m_{rk} \ \forall (x, y) \in R_{rk}$$

Now for a fixed $x \in [x_{r-1}, x_r]$ we have,

$$m_{rk}(y_k - y_{k-1}) \leq \int_{y_{k-1}}^{y_k} f(x, y) \, d_y \leq M_{rk}(y_k - y_{k-1}) \tag{10.5.1.1}$$

Using Theorem 3.10.1 we get,

$$m_{rk}(y_k - y_{k-1})(x_r - x_{r-1}) \leq \int_{x_{r-1}}^{x_r} \left(\int_{y_{k-1}}^{y_k} f(x, y) \, dy \right) dx \leq M_{rk}(y_k - y_{k-1})(x_r - x_{r-1})$$

That is,

$$m_{rk} \Delta_{rk} \leq \int_{x_{r-1}}^{x_r} \left(\int_{y_{k-1}}^{y_k} f(x, y) \, dy \right) dx \leq M_{rk} \Delta_{rk} \tag{10.5.1.2}$$

Summing above equation first with respect to k then with respect to r we obtain,

$$\sum_{r=1}^{n} \sum_{k=1}^{m} m_{rk} \Delta_{rk} \leq \sum_{r=1}^{n} \int_{x_{r-1}}^{\overline{x_r}} \left(\sum_{k=1}^{m} \int_{y_{k-1}}^{y_k} f(x, y) \, dy \right) dx \leq \sum_{r=1}^{n} \sum_{k=1}^{m} M_{rk} \Delta_{rk}$$

or,
$$L(P, f) \leq \int_a^{\overline{b}} \left(\int_c^d f(x, y) \, dy \right) dx \leq U(P, f)$$

Hence,

$$I \leq \int_a^{\overline{b}} \left(\int_c^d f(x, y) \, dy \right) dx \leq J \tag{10.5.1.3}$$

By similar argument we can show that,

$$I \leq \int_{\underline{a}}^{b} \left(\int_c^d f(x, y) \, dy \right) dx \leq J \tag{10.5.1.4}$$

As the double integral $\iint_R f \, dx \, dy$ exists $I = J$ so that from (10.5.1.3) & (10.5.1.4) we conclude that

That is $\iint_R f \, dx \, dy = \int_a^b \left(\int_c^d f(x, y) \, dy \right) dx$.

Remarks

1. It is clear from the above proof that if the double integral exists and any of the iterated integral exists then they must be equal. Hence we conclude that if the double integral exists then the two iterated integrals cannot exist without being equal.

2. Note that even if the two iterated integrals are equal the double integral may not exist as is shown by the following example.

Example 10.4 Let $R = [0, 1] \times [0, 1]$ and Q' be the set of points of R with rational coordinates in lowest terms in the form $\left(\dfrac{p_1}{q}, \dfrac{p_2}{q}\right)$ where p_1, p_2 & q are positive integers. In other words Q' consists of those points of R with rational coordinates having same denominators when they are expressed in lowest terms. We observe that Q' is dense in R as, given any real numbers $a, b, c, d \in [0, 1]$ one can choose a positive integer q so large that there will exist positive integers p_1 & p_2 such that $aq \le p_1 \le bq$ and $cq \le p_2 \le dq$, that is $a \le \dfrac{p_1}{q} \le b$ and $c \le \dfrac{p_2}{q} \le d$. Now define a function f on R as follows,

$$f(x, y) = \begin{cases} 1 & \text{if } (x, y) \in R \cap Q'^c \\ 0 & \text{if } (x, y) \in Q' \end{cases}$$

Let $y_0 \in [0, 1]$, then there will be several cases,

Case1: y_0 is irrational, in this case,

$$f(x, y_0) = 1 \text{ for all } x \in [0, 1]$$

Hence, $\displaystyle\int_0^1 f(x, y_0)\, dx = 1$

Case2: y_0 is rational number of the form $\dfrac{p_0}{q_0}$. In this case we note that there exists only finitely many rational numbers in $[0, 1]$ say $x_1, x_2, \ldots x_n$ such that $(x_i, y_0) \in Q'$ $i = 1, 2, \ldots, n$. So in this case, $f(x, y_0) = 1$ for all but finitely points of $[0, 1]$

This implies

$$\int_0^1 f(x, y_0)\, dx = 1$$

So combining either case we conclude that $\displaystyle\int_0^1 f(x, y)\, dx = 1$ for any $y \in [0, 1]$. Hence,

$\displaystyle\int_0^1 \left(\int_0^1 f(x, y)\, dx\right) dy = 1$. Similar arguments will show that

Next we show that the double integral for this function does not exist. As Q' is dense it is clear that for any partition P, $m_{rk} = 0$ $r = 1 \ldots n$, $k = 1 \ldots m$ hence $L(P, f) = 0$. But as $M_{rk} = 1$ $r = 1 \ldots n$, $k = 1 \ldots m$ we have $U(P, f) = 1$. This clearly shows that $I \ne J$ and hence the double integral of the function does not exist.

As an easy application of the above theorem we get the following corollary,

Corollary 10.5.2

Let $\phi(x)$ and $\psi(y)$ be bounded functions on $[a, b]$ and $[c, d]$ respectively so that $f(x, y) = \phi(x)\,\psi(y)$ defines a bounded function the rectangle $R = [a, b] \times [c, d]$ then if the double integral $\displaystyle\iint_R f\, dx\, dy$ and the iterated integrals exists we have,

$$\iint_R f\, dx\, dy = \int_a^b \phi(x)\, dx \times \int_c^d \psi(x)\, dx$$

Cavalier's Principle can we extended to evaluate the triple integrals as follows:

Let f be a bounded function defined on $R = [a, b] \times [c, d] \times [e, f]$ then if the triple integral of f over R and the iterated integrals exists then,

$$\iiint\limits_{R} f \, dx \, dy \, dz = \int_{e}^{f} dz \int_{c}^{d} dy \int_{a}^{b} f(x, y, z) \, dx = \int_{e}^{f} dz \int_{a}^{b} dx \int_{c}^{d} f(x, y, z) \, dy$$

$$= \int_{a}^{b} dx \int_{e}^{f} dz \int_{c}^{d} f(x, y, z) \, dy = \int_{a}^{b} dx \int_{c}^{d} dy \int_{e}^{f} f(x, y, z) \, dz$$

$$= \int_{c}^{d} dy \int_{a}^{b} dx \int_{e}^{f} f(x, y, z) \, dz = \int_{c}^{d} dy \int_{e}^{f} dz \int_{a}^{b} f(x, y, z) \, dx$$

10.6 DOUBLE INTEGRALS OF FUNCTIONS DEFINED ON ANY ARBITRARY BOUNDED DOMAIN

We now extend the definition of double integral of a function which is defined on any arbitrary bounded domain other than rectangles. Let f be a bounded function defined on a bounded domain E. Now as E is bounded we can find a closed rectangle R such that $E \subset R$ we now define a function f' on R as follows,

$$f'(x, y) = \begin{cases} f(x, y) & \text{if } (x, y) \in E \\ 0 & \text{if } (x, y) \in R \bigcap E^{c} \end{cases}$$

Clearly, f' defined as above is a bounded function on the rectangle R. We say that f is integrable over E if the function f' defined above is integrable over R and in that case we define,

$$\iint\limits_{E} f(x, y) \, dx \, dy = \iint\limits_{R} f'(x, y) \, dx \, dy$$

Note that if we put $f(x, y) = 1$ on E then $\iint\limits_{E} f(x, y) \, dx \, dy = \iint\limits_{E} dx \, dy$ if exists will give the area of the region E.

Similarly, one can extend the definition of triple integral of a function over any bounded domain other than the rectangular parallelepiped and we note $\iiint\limits_{E} dx \, dy \, dz$ if exists will give the volume of the bounded region E.

10.7 ITERATED INTEGRALS OVER ARBITRARY BOUNDED DOMAINS CHANGE IN THE ORDER OF INTEGRATION

Suppose f be a bounded function defined on a bounded domain E and suppose the iterated integrals exist. We will evaluate the integral by calculating the iterated integrals. We will do this for simple regions. A region E will be called a ***Type I*** region if it is of the form,

$$E = \{(x, y) : x \in [a, b], \psi(x) \le y \le \phi(x)\}$$

where $\psi(x)$, $\phi(x)$ are two continuous functions satisfying $\psi(x) \le \phi(x)$, $x \in [a, b]$. Note that for such regions, for any $x' \in [a, b]$ the line $x = x$ parallel to X-axis intersects E at a line segment joining the curves $y = \psi(x)$ and $y = \phi(x)$ (see Fig. 10.3). Also see that such a region is always bounded as $\psi(x)$, $\phi(x)$ are continuous.

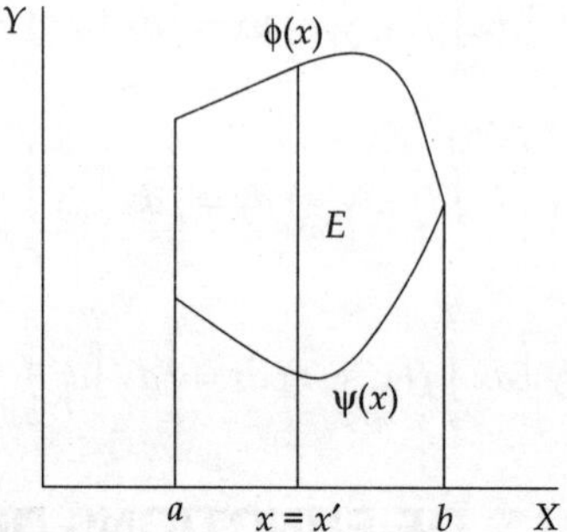

Fig. 10.3   (Type I region)

Similarly we define a region to be of **Type II** if it has the form, $E = \{x, y): y \in [c, d], \zeta(y) \le x \le \eta(y)\}$ where $\zeta(y)$, $\eta(y)$ are two continuous functions satisfying $\zeta(y) \le \eta(y)$, $x \in [a, b]$. Again it is easy to see that for such regions, for any $y' \in [c, d]$ the line $x = x'$ parallel to Y-axis intersects E at a line segment joining the curves $x = \zeta(y)$ and $x = \eta(y)$ (Fig. 10.4)

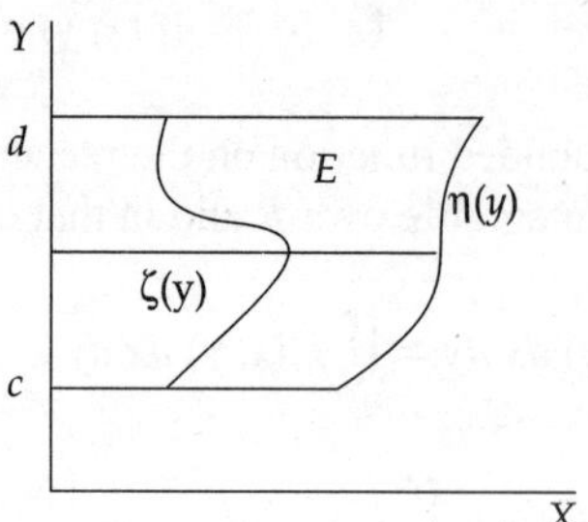

Fig. 10.4   (Type II region)

Finally a region is called **Type III** if it can be expressed both as Type I and Type II regions (Fig. 10.5)

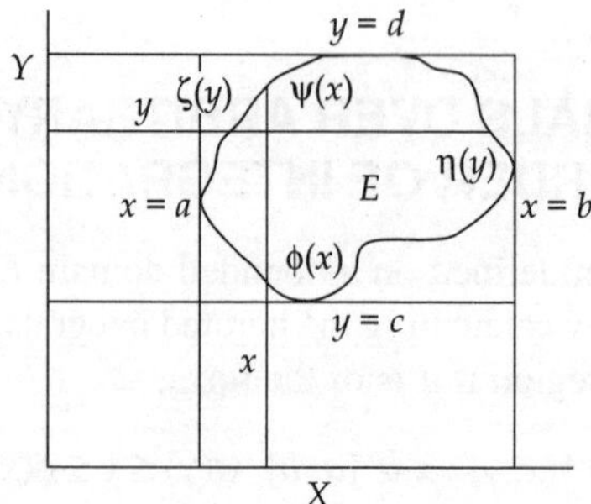

Fig. 10.5   (Type III region)

We now try to find a method how to integrate a function defined on a Type I region. For that let f be a bounded function on a Type I region E

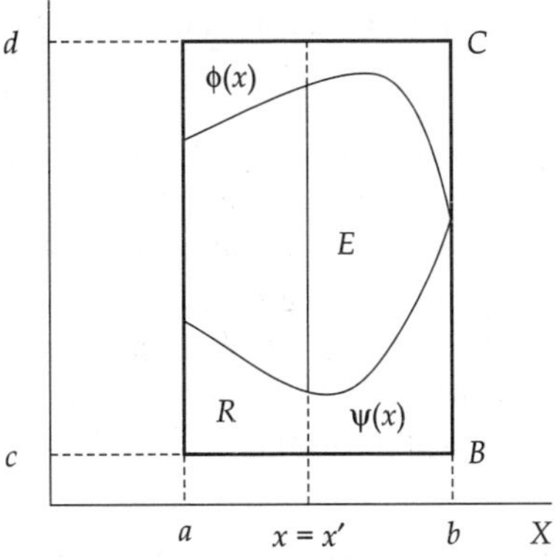

Fig. 10.6

and $R = [a, b] \times [c, d]$ be a rectangle containing E. We define f' on R as follows,

$$f'(x, y) = \begin{cases} f(x, y) & \text{if } (x, y) \in E \\ 0 & \text{if } (x, y) \in R \cap E^c \end{cases}$$

We assume that f' is integrable and all its iterated integrals exists then from the definition given in previous section we have,

$$\iint_E f(x, y)\, dx\, dy = \iint_R f'(x, y)\, dx\, dy$$

$$= \int_a^b \left(\int_c^d f'(x, y)\, dy \right) dx$$

Now observe that any line parallel to X-axis cuts the region at line segment joining the curves $y = \psi(x)$ and $y = \phi(x)$ (Fig.) and also we know that $f' = 0$ outside E hence we have,

$$\int_c^d f'(x, y)\, dy = \int_{\psi(x)}^{\phi(x)} f(x, y)\, dy$$

so that we get,

$$\iint_E f(x, y)\, dx\, dy = \int_a^b \left(\int_{\psi(x)}^{\phi(x)} f(x, y)\, dy \right) dx$$

Similarly, for a function f defined in Type II region E we will obtain,

$$\iint_E f(x, y)\, dx\, dy = \int_c^d \left(\int_{\zeta(y)}^{\eta(y)} f(x, y)\, dx \right) dy$$

Finally, for a function defined on a Type III region we obtain,

$$\iint\limits_{E} f(x,\,y)\,dx\,dy = \int\limits_{a}^{b} dx \int\limits_{\phi(x)}^{\psi(x)} f(x,\,y)\,dy = \int\limits_{c}^{d} dy \int\limits_{\zeta(y)}^{\eta(y)} f(x,\,y)\,dx$$

Most of the regions that we that we encounter are mostly Type III or regions which can be sub-divided into Type III regions. In the second case we can evaluate the integrals on each Type III sub regions by method shown above and final result is obtained by adding those integrals.

We follow the same procedure in calculating the triple integrals of functions that is we integrate the function with respect to any one of the variable say x in appropriate range treating other two y, z as constants and then integrate the result (which is a function of with y, z) with respect to any of y or z say y treating z as constant and finally we integrate with respect to the remaining variable (z in this case).

Example 10.5   Evaluate $\iint\limits_{E} (x+y)\,dx\,dy$ where E: $\{(x,\,y) : 0 \le x \le a,\, 0 \le y \le b\}$

Sol: The region of integration is the rectangle $ABCD$ as the figure (Fig. 10.7).

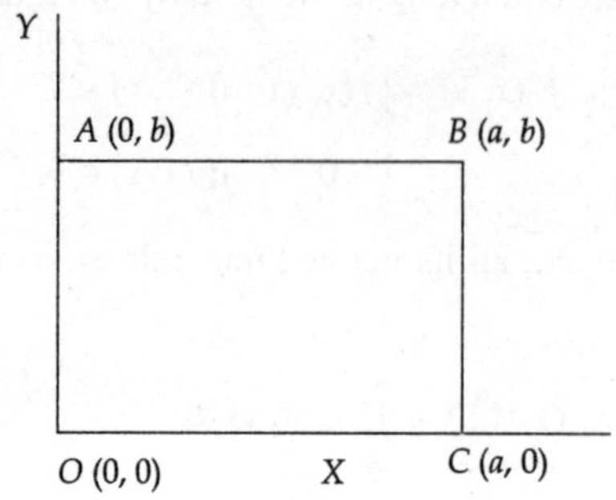

Fig. 10.7

Now,

$$\iint\limits_{E} (x+y)\,dx\,dy = \int\limits_{0}^{a} \left(\int\limits_{0}^{b} (x+y)\,dy \right) dx$$

$$= \int\limits_{0}^{a} \left[xy + \frac{y^2}{2} \right]_{y=0}^{y=b} dx$$

$$= \int\limits_{0}^{a} \left(bx + \frac{b^2}{2} \right) dx = \left[\frac{bx^2}{2} + \frac{b^2 x}{2} \right]_{0}^{a} = \frac{1}{2}\, ab(a+b)$$

Example 10.6   Evaluate $\iint\limits_{E} (xy)\, dx\, dy$ where E the positive quadrant of the circle $x^2 + y^2 = 1$.

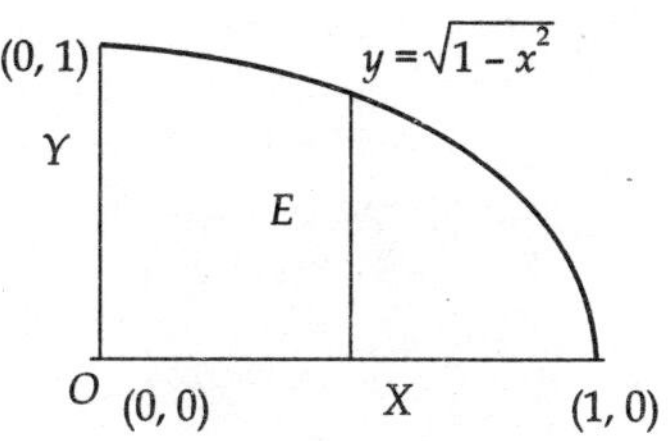

Fig. 10.8

Sol: From the above figure it is clear that,

$$\iint\limits_{E} (xy)\, dx\, dy = \int\limits_{0}^{1} dx \left(\int\limits_{0}^{\sqrt{1-x^2}} xy\, dy \right)$$

$$= \int\limits_{0}^{1} \frac{1}{2} x\, (1 - x^2) = \frac{1}{2}\left[\frac{1}{2} x^2 + \frac{1}{4} x^4 \right]_{0}^{1} = \frac{3}{8}$$

Example 10.7   Find the area of the shaded region E indicated in the figure below.

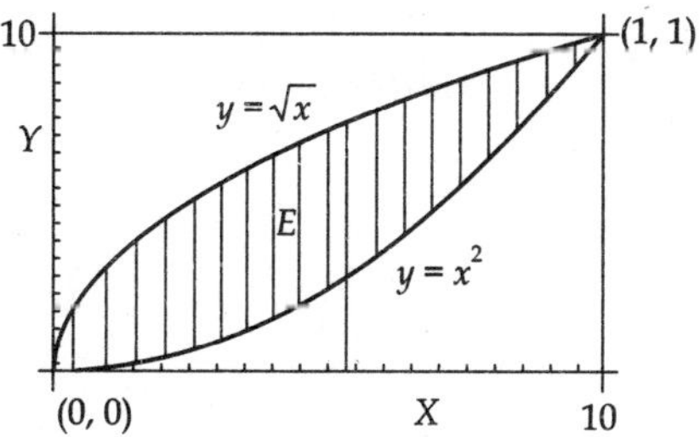

Fig. 10.9

Sol: The area A of E is given by,

$$A = \iint\limits_{v} dx\, dy = \int\limits_{0}^{1} dx \int\limits_{x^2}^{\sqrt{x}} dy$$

$$= \int\limits_{0}^{1} [y]_{x^2}^{\sqrt{x}}\, dx = \int\limits_{0}^{1} (\sqrt{x} - x^2)\, dx$$

$$= \left[\frac{2}{3} x^{3/2} - \frac{1}{3} x^3 \right]_{0}^{1} = \frac{1}{3} \text{ sq. units}$$

Example 10.7   Evaluate $\iiint\limits_{E} (x^2 + y^2 + z^2)\, dx\, dy\, dz$ where E is the unit cube that is, $E : \{(x, y, z) : 0 \le x \le 1, 0 \le y \le 1, 0 \le z \le 1\}$

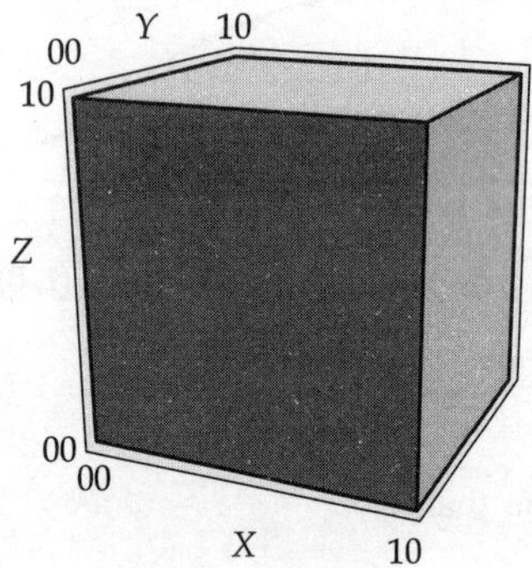

Fig. 10.10

Sol: As in the case of double integral we have,

$$\iiint\limits_{E} (x^2 + y^2 + z^2)\, dx\, dy\, dz = \int\limits_0^1 dx \int\limits_0^1 dy \int\limits_0^1 (x^2 + y^2 + z^2)\, dz$$

$$= \int\limits_0^1 dx \int\limits_0^1 dy \left[\left(x^2 z + y^2 z + \frac{z^3}{3} \right) \right]_{z=0}^{z=1}$$

$$= \int\limits_0^1 dx \int\limits_0^1 \left(x^2 + y^2 + \frac{1}{3} \right) dy = \int\limits_0^1 \left(x^2 + \frac{1}{3} + \frac{1}{3} \right) dx$$

$$= \left[\frac{x^3}{3} + \frac{x}{3} + \frac{x}{3} \right]_0^1 = 1$$

We now illustrate that some times it is easier to evaluate one of the above integrals than the other hence one uses the method of change in order of integration.

Example 10.8   Evaluate $\int\limits_0^1 dy \int\limits_0^{\sqrt{y}} \cos x^3\, dx$

Sol: First note that this is iterated integral for the double integral $\iint\limits_{E} \cos x^2\, dx\, dy$ where E is as shown as shaded part in the figure.

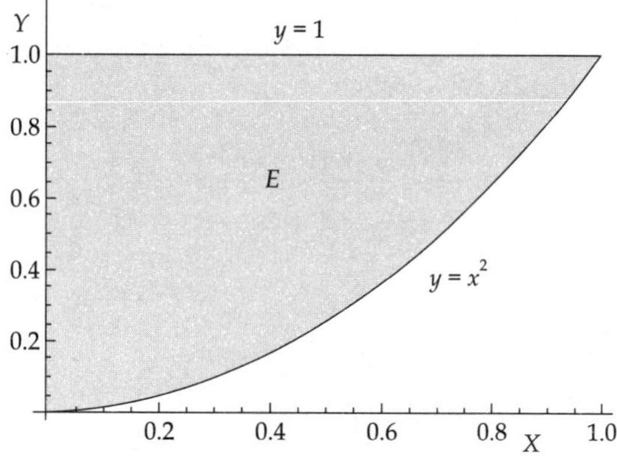

Fig. 10.11

To find E we proceed as follows, from the above figure it is clear that for a fixed y, x value in the region varies between $x = 0$ and $x = \sqrt{y}$ also as the limits of y are given to be 0 and 1 we conclude that E is the region bounded by the curves $y = x^2$, $x = 0$, $y = 1$. As cosine function is a continuous function the double integral together with the iterated integrals exists and are all equal. Now we see that $\int \cos\left(x - \dfrac{1}{3}x^3\right) dx$ cannot be calculated easily in terms of simple functions hence instead of directly finding the given iterated integral we find the other iterated integral that is we change the order of integration and get,

$$\int_0^1 dy \int_0^{\sqrt{y}} \cos\left(x - \frac{1}{3}x^3\right) dx = \int_0^1 dx \int_{x^2}^1 \cos\left(x - \frac{1}{3}x^3\right) dy$$

$$= \int_0^1 (1 - x^2) \cos\left(x - \frac{1}{3}x^3\right) dx$$

$$= \int_0^{\frac{2}{3}} \cos u\, du \text{ putting } u = x - \frac{1}{3}x^3$$

$$= \sin\left(\frac{2}{3}\right).$$

We give some more examples to exhibit how to change the order of integration for different regions of integrations

Example 10.9   By changing the order of integration evaluate,

$$\int_0^1 dy \int_{1-\sqrt{1-y}}^{1+\sqrt{1-y}} \frac{2y}{(2-x)^2}\, dx$$

Sol: As for fixed y the limit of x is given to be $x = 1 - \sqrt{1-y}$ and $x = 1 + \sqrt{1-y}$ also as range of y is given as $y = 0$ and $y = 1$ the region of integration is the area lying between the parabola $y = x\,(2-x)$ and X-axis (shaded region in Fig. 10.12).

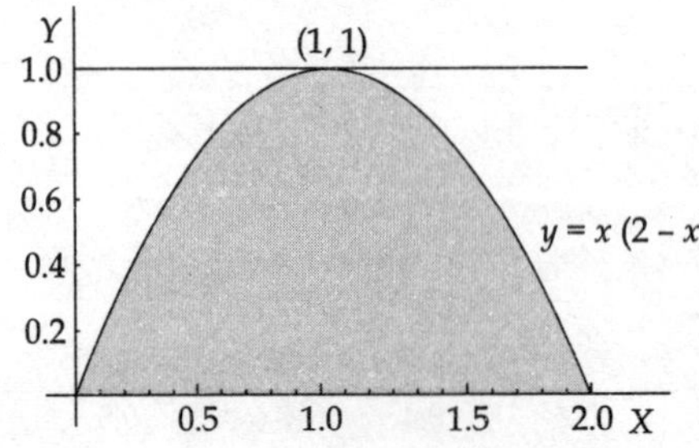

Fig. 10.12

Hence, by changing the order of integration we get

$$\int\limits_{0}^{1} dy \int\limits_{1-\sqrt{1-y}}^{1+\sqrt{1-y}} \frac{2y}{(2-x)^2}\, dx = \int\limits_{0}^{2} dx \int\limits_{0}^{x(2-x)} \frac{2y}{(2-x)^2}\, dy$$

$$= \int\limits_{0}^{2} \frac{1}{(2-x)^2}\, [y^2]_{0}^{x(2-x)}\, dx = \int\limits_{0}^{2} x^2\, dx = \frac{8}{3}$$

Example 10.10 Change the order of integration and evaluate

$$\int\limits_{0}^{7} dy \int\limits_{1}^{\sqrt{8-y}} (2xy)\, dx.$$

Sol: We see that for a fixed y, $1 \le x \le \sqrt{8-y}$ hence the region of integration E is the area bounded by curves $x^2 = 8 - y$, $x = 1$ and $x = 0$ (as shown in the Fig. 10.12)

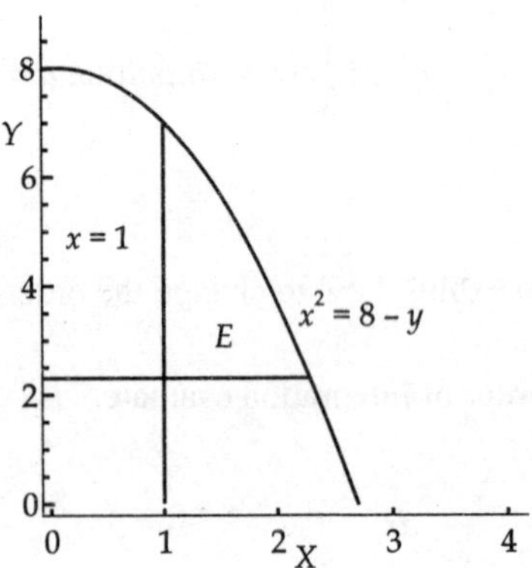

Fig. 10.13

Now by interchanging the limits we get,

$$\int_0^7 dy \int_1^{\sqrt{8-y}} (2xy)\, dx = \int_1^{2\sqrt{2}} \left(\int_0^{8-x^2} 2xy\, dy \right) dx$$

$$= \int_1^{2\sqrt{2}} x\, [y^2]_0^{8-x^2}\, dx$$

$$= \int_1^{2\sqrt{2}} x\, (8-x^2)^2\, dx = \frac{6079}{60} = \frac{256\sqrt{2}}{5}$$

Example 10.11   Change the order of integration $\int_0^1 dx \int_0^x f(x, y)\, dy$

Sol: The region integration E is as shown in the Fig. is the triangular region bounded by $x = y$, $x = 1$, $x = 0$. Hence by changing the order of integration the above integral becomes,

$$\int_0^1 dy \int_y^1 f(x, y)\, dy.$$

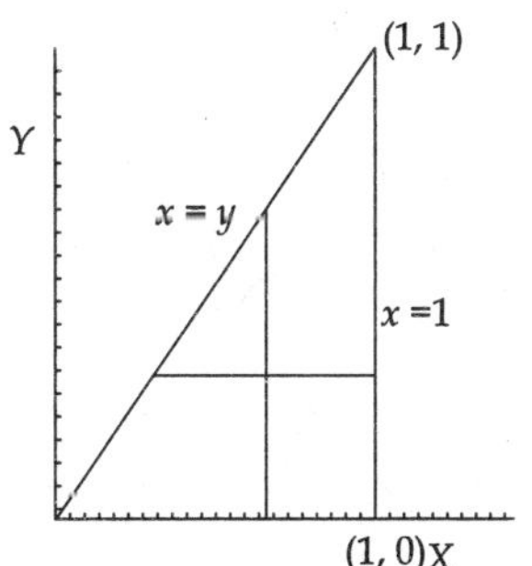

Fig. 10.14

Example 10.12   We express $\iint_E f(x, y)\, dx\, dy$ in terms of two iterated integrals where E is the region AOB shown in the figure below:

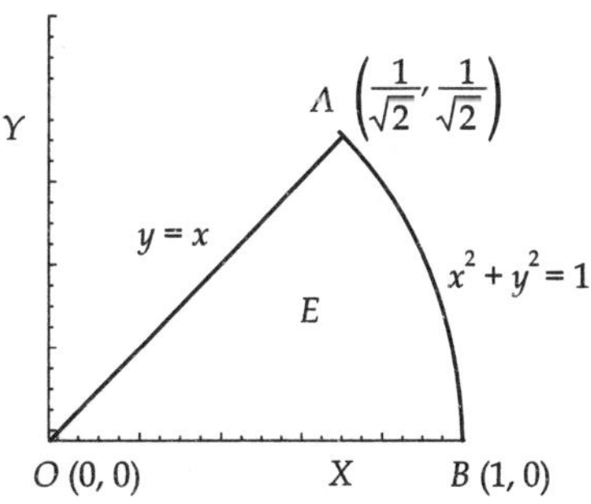

Fig. 10.15

We see that if any line parallel to the X-axis at a height y cuts the region at (y, y) and $\left(\sqrt{1 - y^2}, y\right)$ as shown in Fig. 10.15.

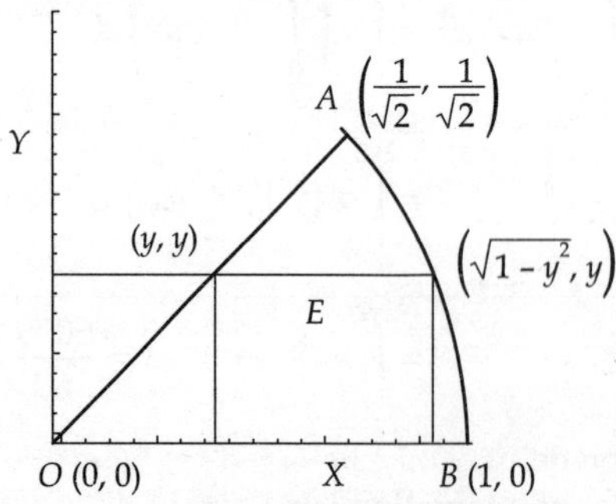

Fig. 10.16

Hence integrating the function first with respect then with respect to y yields

$$\iint_{v} f(x, y)\, dx\, dy = \int_{0}^{\frac{1}{\sqrt{2}}} dy \int_{y}^{\sqrt{1 - y^2}} f(x, y)\, dx$$

To find the other iterated integral we divide the region E into two parts E_1 and E_2 as shown in the figure

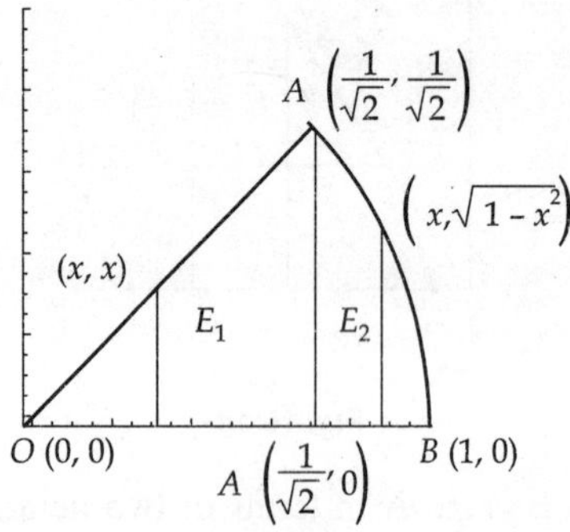

Fig. 10.17

Now it can be seen easily from Fig. 10.17 E_1 for any x the line passing through $(x, 0)$ cuts the region at (x, x) and in E_2 such a line cuts the region at $\left(x, \sqrt{1 - x^2}\right)$ hence we have,

$$\iint_{E} f(x, y)\, dx\, dy = \iint_{E_1} f(x, y)\, dx\, dy + \iint_{E_2} f(x, y)\, dx\, dy$$

$$= \int_{0}^{\frac{1}{\sqrt{2}}} dx \int_{0}^{x} f(x, y)\, dy + \int_{\frac{1}{\sqrt{2}}}^{1} dx \int_{0}^{\sqrt{1 - x^2}} f(x, y)\, dy$$

10.8   CHANGE OF VARIABLE FORMULA IN DOUBLE INTEGRALS

Recall that if $f \in R\,[a, b]$ and g be a strictly monotone differentiable function on $[\alpha, \beta]$ such that $g(\alpha) = a$, $g(\beta) = b$ with $f \circ g$ and $g' \in R[\alpha, \beta]$ then,

$$\int_a^b f(x)\, dx = \int_\alpha^\beta f\{g(t)\}\, g'(t)\, dt$$

and above equation is called the method of substitution or change of variable formula in one dimensional integral. We expect similar kind of formula in double integrals. Before going into actual formula let us give a geometrical motivation of how change of variables affects a double integral.

Let $f(x, y)$ be a bounded function defined on a bounded set E and suppose that $\iint_E f\, dx\, dy$ exists.

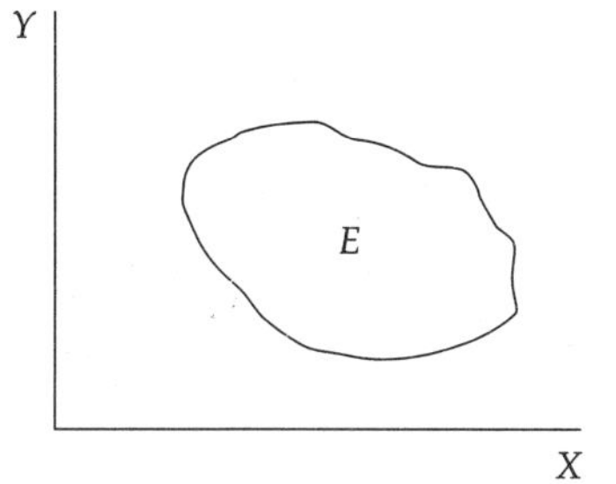

Fig. 10.18

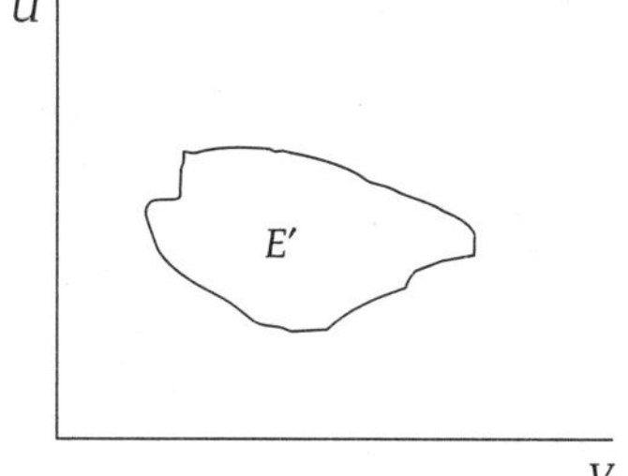

Fig. 10.19

Let x, y be transformed by the rule,

$$T(x, y) = (\phi(x, y), \psi(x, y))$$

such that T represents a one-one function and ϕ, ψ are functions of x, y possessing continuous first order partial derivatives. If we write $u = \phi(x, y)$ and $v = \psi(x, y)$ then T transform the region E of X–Y plane to a region E' of $U - V$ plane in one-one fashion. Again as T is a one-one function T^{-1} exists or in other words we can solve x, y in terms of u, v say, $x = \xi(u, v)$ and $y = \zeta(u, v)$ as ϕ, ψ has continuous first order partial ξ, ζ will also have the same property above. Now let R be a small rectangle of dimension Δu, Δ on E' and suppose (u_0, v_0) be the bottom left corner point of R. Now under T^{-1}, R will be transformed into a curvilinear R' (say) in the X–Y plane with bottom left point as $(x_0, y_0) = (\phi'(u_0, v_0), \psi'(u_0, v_0))$ (Fig. 10.20 and Fig. 10.21).

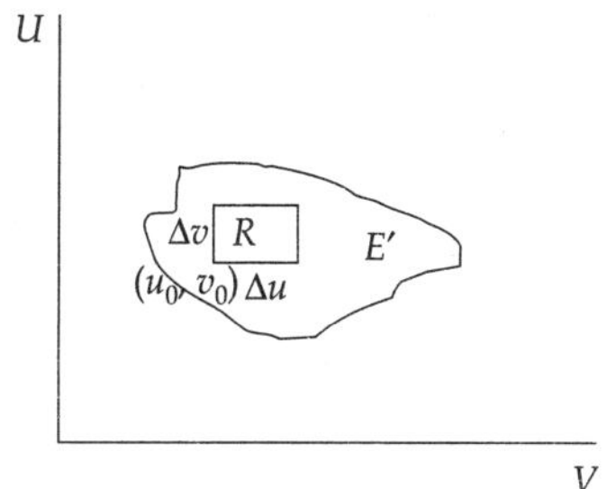

Fig. 10.20

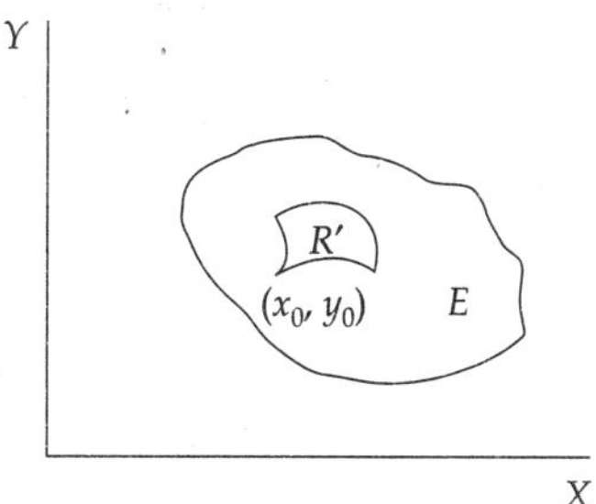

Fig. 10.21

Let $\vec{r}(u, v)$ denote the position vector of image of the point (u, v) then,

$$\vec{r}(u, v) = \xi(u, v)\,\vec{i} + \zeta(u, v)\,\vec{j}$$

Now the lower side of R is represented by the curve $v = v_0$ and its image in X–Y plane is given by $\vec{r}(u, v_0)$. The tangent vector to this image curve at (u_0, v_0) is given by,

$$\frac{\partial \vec{r}}{\partial u} = \left(\frac{\partial \xi}{\partial u}\right)_{(u_0,\, v_0)}\vec{i} + \left(\frac{\partial \zeta}{\partial u}\right)_{(u_0,\, v_0)}\vec{j} = \frac{\partial x}{\partial u}\vec{i} + \frac{\partial y}{\partial u}\vec{j}$$

Similarly the tangent vector to the image curve of the left side of R (whose equation is $u = u_0$) at (u_0, v_0) is given by,

$$\frac{\partial \vec{r}}{\partial v} = \left(\frac{\partial \xi}{\partial v}\right)_{(u_0,\, v_0)}\vec{i} + \left(\frac{\partial \zeta}{\partial v}\right)_{(u_0,\, v_0)}\vec{j} = \frac{\partial x}{\partial v}\vec{i} + \frac{\partial y}{\partial v}\vec{j}$$

Now as Δu, Δv are small the approximate area of the curvilinear parallelogram R' is given by the cross product $\vec{A} \times \vec{B}$ where $\vec{A} = \vec{r}(u_0 + \Delta u, v_0) - \vec{r}(u_0, v_0)$ and $\vec{B} = \vec{r}(u_0 + v_0, \Delta v_0) - \vec{r}(u_0, v_0)$. Now as,

$$\frac{\partial \vec{r}}{\partial u} = \lim_{\Delta u \to 0} \frac{\vec{r}(u_0 + \Delta u, v_0) - \vec{r}(u_0, v_0)}{\Delta u}$$

we write $\vec{A} = \vec{r}(u_0 + \Delta u, v_0) - \vec{r}(u_0, v_0) \approx \dfrac{\partial \vec{r}}{\partial u}\,\Delta u$. By similar argument we write,

$$\vec{B} = \vec{r}(u_0, v_0 + \Delta v) - \vec{r}(u_0, v_0) \approx \frac{\partial \vec{r}}{\partial u}\,\Delta v$$

Hence approximate area $\Delta R'$ of R' is given by,

$$\left|\frac{\partial \vec{r}}{\partial u}\Delta u \times \frac{\partial \vec{r}}{\partial u}\Delta v\right| = \left|\frac{\partial \vec{r}}{\partial u} \times \frac{\partial \vec{r}}{\partial u}\right|\Delta u\,\Delta v$$

We have,

$$\left|\frac{\partial \vec{r}}{\partial u} \times \frac{\partial \vec{r}}{\partial u}\right| = \begin{vmatrix} i & j & k \\ \dfrac{\partial x}{\partial u} & \dfrac{\partial y}{\partial u} & 0 \\ \dfrac{\partial x}{\partial v} & \dfrac{\partial y}{\partial v} & 0 \end{vmatrix} = \begin{vmatrix} \dfrac{\partial x}{\partial u} & \dfrac{\partial x}{\partial u} \\ \dfrac{\partial y}{\partial v} & \dfrac{\partial y}{\partial v} \end{vmatrix}\vec{k}$$

So that $\Delta R' \approx \begin{vmatrix} \dfrac{\partial x}{\partial u} & \dfrac{\partial y}{\partial u} \\ \dfrac{\partial x}{\partial v} & \dfrac{\partial y}{\partial v} \end{vmatrix}\Delta u\Delta v = \left|\dfrac{\partial x}{\partial u}\dfrac{\partial y}{\partial v} - \dfrac{\partial y}{\partial u}\dfrac{\partial x}{\partial v}\right|\Delta u\Delta v = |J|\,\Delta u\Delta v$

Where the determinant J is called the Jacobian of the transformation and is also some times denoted by $\dfrac{\partial(x, y)}{\partial(u, v)}$. Next we divide the region E' of $U - V$ plane into mn sub-rectangles denoted by R_{rk} and let the image of R_{rk} under the transformation T^{-1} be denoted as R'_{rk}. Then using approximation for each R'_{rk} we get,

$$\iint\limits_{E} f(x, y)\, dx\, dy \approx \sum_{r=1}^{m} \sum_{k=1}^{n} f(x_r, y_k)\, \Delta R_{rk} \approx \sum_{r=1}^{m} \sum_{k=1}^{n} f(\xi(u_r, v_k),\, \zeta(u_r, v_k))\, |J|\, \Delta u\, \Delta v$$

$$\approx \iint\limits_{E} f(\xi(u, v),\, \zeta(u, v))\, |J|\, du\, dv.$$

Above discussion suggests the following theorem,

Theorem 10.8.1

Let $f(x, y)$ be a bounded function on a bounded domain E of $\mathbb{R}^2$ and suppose that the transformation given by, $x = \xi(u, v), y = \zeta(u, v), (u, v) \in E'$ be one-one mapping of the region E' of $u - v$ plane and the region of $x - y$ plane where functions ξ and ζ have continuous first order partial derivatives with the Jacobian of the transformation $J = \dfrac{\partial(x, y)}{\partial(u, v)} \neq 0$ at any point of E' then if $\iint\limits_{E} f\, dx\, dy$ exists we have,

$$\iint\limits_{E} f(x, y)\, dx\, dy = \iint\limits_{E'} f(\xi(u, v))\, |J|\, du\, dv$$

Note that the above theorem can be extended to the case of triple integrals as,

Let $f(x, y, z)$ be a bounded function on a bounded domain E of $\mathbb{R}^3$ and suppose that the transformation given by, $x = \xi(u, v, w), y = \zeta(u, v, w), z = \psi(u, v, w), (u, v, w) \in E'$ be one-one mapping of the region E' and E where functions ξ, ζ, ψ have continuous first order partial derivatives with the Jacobian of the transformation $J = \dfrac{\partial(x, y, z)}{\partial(u, v, w)} \neq 0$ at any point of E' then if $\iint\limits_{E} f\, dx\, dy$ exists we have,

$$\iiint\limits_{E} f(x, y, z)\, dx\, dy\, dz = \iiint\limits_{E} f(\xi(u, v, w),\, \zeta(u, v, w),\, \psi(u, v, w))\, |J|\, du\, dv\, dw$$

where in this case,

$$J = \frac{\partial(x, y, z)}{\partial(u, v, w)} = \begin{vmatrix} \dfrac{\partial x}{\partial u} & \dfrac{\partial y}{\partial u} & \dfrac{\partial z}{\partial u} \\[2mm] \dfrac{\partial x}{\partial v} & \dfrac{\partial y}{\partial v} & \dfrac{\partial z}{\partial v} \\[2mm] \dfrac{\partial x}{\partial w} & \dfrac{\partial y}{\partial w} & \dfrac{\partial z}{\partial w} \end{vmatrix}$$

As an application of above theorem we prove Theorem 6.6.1

Proof of Theorem 6.6.1

We will show,

$$B(m, n) = \frac{\Gamma(m)\,\Gamma(n)}{\Gamma(m+n)} \quad \text{for } m > 0 \ \& \ n > 0$$

In the gamma integral $\Gamma(n) = \int_0^\infty e^{-x} x^{n-1}\, dx \ \ n > 0$ we substitute $x = r^2$ and get

$$\int_0^\infty e^{-r^2} r^{2n-1}\, dr = \frac{1}{2}\,\Gamma(n) \tag{6.6.1.1}$$

Also we know from Theorem 6.3.1(iii)

$$\int_0^{\frac{\pi}{2}} \sin^{2m-1}\theta \ \cos^{2n-1}\theta \ d\theta = \frac{1}{2}\,B(m, n) \tag{6.6.1.2}$$

Let us now consider three regions in the plane E_1, E_2 & E_3 where E_1 is the positive quadrant of the e circle $x^2 + y^2 = a^2$, E_2 is the square of side a touching E_1 at the points $(0, a)$ & $(a, 0)$ and E_3 is the positive quadrant of the e circle $x^2 + y^2 = 2a^2$ (see Fig. 10.22 below)

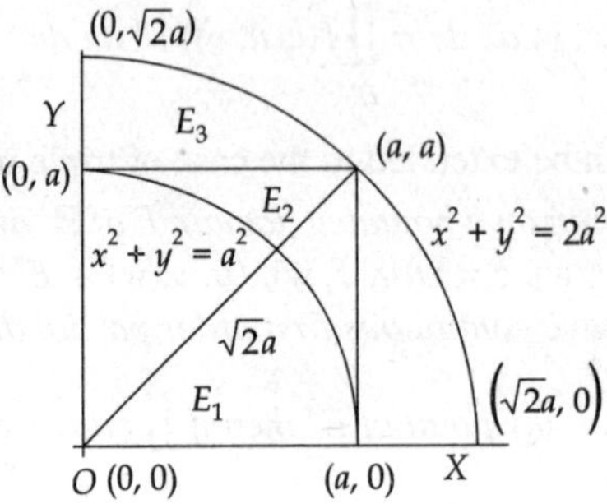

Fig. 10.22

Let us now consider the function $f(x, y) = x^{2m-1}\, y^{2n-1}\, e^{-(x^2 + y^2)}$ which is non-negative for $x \geq 0$, $y \geq 0$. Hence we have,

$$\iint_{E_1} f(x, y)\, dx\, dy \leq \iint_{E_2} f(x,y)\, dx\, dy \leq \iint_{E_3} f(x, y)\, dx\, dy \tag{6.6.1.3}$$

Now,

$$\iint_{E} f(x, y)\, dx\, dy = \int_0^a x^{2m-1} e^{-x^2}\, dx \int_0^a y^{2n-1} e^{-y^2}\, dy \to \frac{1}{2}\,\Gamma(m)\,\frac{1}{2}\,\Gamma(n) \text{ as } a \to \infty \text{ by } (6.6.1.1)$$

Again consider,

$$\iint_{E_1} f(x, y)\, dx\, dy = \iint_{E_2} x^{2m-1}\, y^{2n-1}\, e^{-(x^2 + y^2)}\, dx\, dy$$

If we write, $x = r\cos\theta$, $y = r\sin\theta$, then the Jacobian of the transformation $J = \dfrac{\partial(x, y)}{\partial(r, \theta)} = r$ and the above integral becomes,

$$\iint_{E_1} x^{2m-1}\, y^{2n-1}\, e^{-(x^2 + y^2)}\, dx\, dy = \int_{r=0}^{a} \int_{\theta=0}^{\theta=\frac{\pi}{2}} r^{2m+2n-2} \cos^{2m-1}\theta \, \sin^{2m-1}\theta\, r\, dr\, d\theta$$

$$= \int_{r=0}^{r=a} r^{2m+2n-1}\, e^{-r^2}\, dr \int_{\theta=0}^{\theta=\frac{\pi}{2}} \cos^{2m-1}\theta \, \sin^{2m-1}\theta\, d\theta$$

$$= \frac{1}{2}\, B(m, n) \int_{r=0}^{r=a} r^{2m+2n-1}\, e^{-r^2}\, dr$$

$$\rightarrow \quad \frac{1}{2}\, B(m, n) \int_{0}^{\infty} r^{2m+2n-1}\, e^{-r^2}\, dr = \frac{1}{2}\, B(m, n)\, \frac{1}{2}\, \Gamma(m + n) \text{ as } a \to \infty \quad \text{(using (6.6.1.1))}$$

Similarly, by using same transformation we can show

$$\iint_{E_3} f(x, y)\, dx\, dy = \frac{1}{2}\, B(m, n) \int_{r=0}^{r=\sqrt{2a}} r^{2m+2n-1}\, e^{-r^2}\, dr$$

$$\rightarrow \quad = \frac{1}{2}\, B(m, n)\, \frac{1}{2}\, \Gamma(m + n) \text{ as } a \to \infty$$

Hence, taking in (6.6.1.3) we get

$$\frac{1}{2}\, B(m, n)\, \frac{1}{2}\, \Gamma(m + n) \leq \frac{1}{2}\, \Gamma(m)\, \frac{1}{2}\, \Gamma(n) \leq \frac{1}{2}\, B(m, n)\, \frac{1}{2}\, \Gamma(m + n) \qquad (6.6.1.4)$$

As the two extreme quantities of the above inequality are same we conclude all the quantities of the above inequality are same and hence we have,

$$B(m, n) = \frac{\Gamma(m)\, \Gamma(n)}{\Gamma(m + n)}$$

Similar technique is used in the next example

Example 10.13 Evaluate the Euler's integral $\displaystyle\int_{-\infty}^{\infty} e^{-x^2}\, dx$ without using Beta-Gamma relation.

Sol: As the integrant is an even function of x it follows that,

$$\int_{-\infty}^{\infty} e^{-x^2}\, dx = 2 \int_{0}^{\infty} e^{-x^2}\, dx$$

Hence it is enough to calculate $I = \int\limits_0^\infty e^{-x^2}\,dx$. For that we consider the function $f(x, y) = e^{-(x^2 + y^2)}$ which

is non-negative for $x \geq 0$, $y \geq 0$. Hence we have,

$$\iint\limits_{E_1} f(x, y)\,dx\,dy \leq \iint\limits_{E_2} f(x, y)\,dx\,dy \leq \iint\limits_{E_3} f(x, y)\,dx\,dy$$

where E_1, E_2, & E_3 are as defined in proof of Theorem 6.6.1 above (Fig. 10.22).

Now,

$$\iint\limits_{E_2} f(x, y)\,dx\,dy = \int\limits_0^a \int\limits_0^a e^{-(x^2, y^2)}\,dx\,dy$$

$$= \int\limits_0^a e^{-x^2}\,dx \int\limits_0^a e^{-y^2}\,dy = \left(\int\limits_0^a e^{-x^2}\,dx \right)^2$$

In $\iint\limits_{E_1} f(x, y)\,dx\,dy$ if we put $x = r \cos\theta$, $y = r \sin\theta$, then the Jacobian of the transformation $J = \dfrac{\partial(x, y)}{\partial(r, \theta)}$

$= r$ and the above integral becomes,

$$\iint\limits_{E_1} f(x, y)\,dx\,dy = \int\limits_0^{\frac{\pi}{2}} d\theta \int\limits_0^a r e^{-r^2}\,dr = \frac{\pi}{4}(1 - e^{-a^2})$$

Similarly,

$$\iint\limits_{E_3} f(x, y)\,dx\,dy = \frac{\pi}{4}(1 - e^{-2a^2})$$

Hence by what we have deduced we get,

$$\frac{\pi}{4}(1 - e^{-a^2}) \leq \left(\int\limits_0^a e^{-x^2}\,dx \right)^2 \leq \frac{\pi}{4}(1 - e^{-2a^2})$$

Now as $a \to \infty$ both $\frac{\pi}{4}(1 - e^{-a^2})$ & $\frac{\pi}{4}(1 - e^{-2a^2})$ tends to $\frac{\pi}{4}$ also $\left(\int\limits_0^a e^{-x^2}\,dx \right)^2 \to I^2$ hence from above

inequality we get,

$$\frac{\pi}{4} \leq I^2 \leq \frac{\pi}{4}$$

which implies $I^2 = \frac{\pi}{4}$ that is $I = \frac{\pi}{2}$ (as $I \geq 0$)

Hence the value of Euler's integral $\int\limits_{-\infty}^\infty e^{-x^2}\,dx$ is $\sqrt{\pi}$.

10.9 SOME MISCELLANEOUS PROBLEMS

Example 10.14 Evaluate $\iint\limits_{E}\left(1 - \dfrac{x^2}{a^2} - \dfrac{y^2}{b^2}\right)dx\,dy$ where,

$$E = \left\{(x, y) : x \geq 0,\ y \geq 0;\ \frac{x^2}{a^2} + \frac{y^2}{b^2} \leq 1\right\}$$

Sol: It is clear that E is part of the elliptical disc lying in the positive quadrant of the plane as shown in

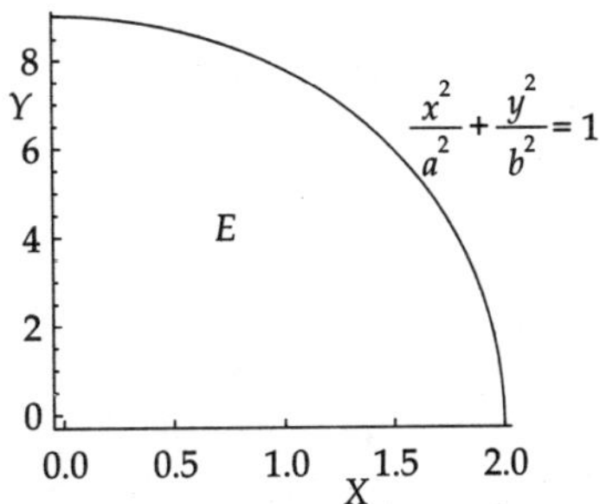

Fig. 10.23

the figure. To evaluate the integral we first substitute in the above integral $x = au$, $y = bv$ so that the Jacobian for this transformation is given by $J = \dfrac{\partial(x, y)}{\partial(u, v)} = ab$. Now note that by above substitution the given region E transforms to the region E' given by

$$E' = \{(u, v\} : u \geq 0,\ v \geq 0;\ u^2 + v^2 \leq 1\}$$

that is E' is the part of the unit disc lying in the positive quadrant of the u-v plane. Hence we have

$$\iint\limits_{E}\left(1 - \frac{x^2}{a^2} - \frac{y^2}{b^2}\right)dx\,dy = \iint\limits_{E'}(1 - u^2 - v^2)\,ab\,du\,dv$$

In the right hand side integral of the above equation we use usual polar transformation as $u = r\cos\theta$, $v = r\sin\theta$ so that Jacobian of the transformation is given by $J = \dfrac{\partial(u, v)}{\partial(r, \partial)} = r$.

Also the region E' in $r - \theta$ plane is given by $E'' = \left\{(r, \theta) : 0 \leq r \leq 1,\ 0 \leq \theta \leq \dfrac{\pi}{2}\right\}$. Hence

$$\iint\limits_{E}(1 - u^2 - v^2)\,ab\,du\,dv = ab\int\limits_{r=0}^{1}\ \int\limits_{\theta=1}^{\frac{\pi}{2}}(1 - r^2\cos^2\theta - r^2\sin^2\theta)\,rd\,rd\,\theta$$

$$= ab \int_0^1 r(1 - r^2)\, dr \int_0^{\frac{\pi}{2}} d\theta = \frac{\pi ab}{8}$$

So that $\displaystyle\iint_{E_1}\left(1 - \frac{x^2}{a^2} - \frac{y^2}{b^2}\right) dx\, dy = \frac{\pi ab}{8}$

Example 10.15 Evaluate $\displaystyle\iint_E \frac{dx\, dy}{(1 + x^2 + y^2)^2}$ where E is the triangle with vertices $(0, 0)$, $(2, 0)$ & $(1, \sqrt{3})$.

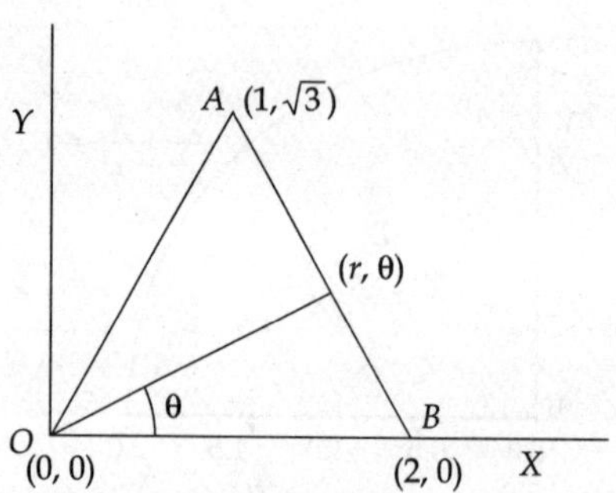

Fig. 10.24

Sol: We notice that triangle ABC is an equilateral triangle hence each angle of the triangle is equal to $\frac{\pi}{3}$. As the term $x^2 + y^2$ is present in the above integral we consider the following transformation,

$$x = r \cos \theta \quad y = r \sin \theta$$

Now from the above figure (Fig. 10.24) it is clear that for a fixed value of θ minimum value r is 0 where as the maximum value of r will be lying on the line joining points A & B whose equation is given by,

$$y = -\sqrt{3}\,(x - 2)$$

Putting $x = r \cos \theta$ and $y = r \sin \theta$ in the above line we obtain,

$$r = \frac{2\sqrt{3}}{\sqrt{3}\cos\theta + \sin\theta} = \frac{\sqrt{3}}{\sin\left(\theta + \frac{\pi}{3}\right)}$$

Hence the above integral I is given by,

$$I = \int_0^{\frac{\pi}{3}} d\theta \int_0^{\frac{\sqrt{3}}{\sin\left(\theta + \frac{\pi}{3}\right)}} \frac{r\, dr}{(1 + r^2)^2} = \frac{1}{2} \int_0^{\frac{\pi}{3}} \frac{3\, d\theta}{3 + \sin^2\left(\theta + \frac{\pi}{3}\right)} = \frac{\sqrt{3}}{2} \tan^{-1}\left(\frac{1}{2}\right)$$

Example 10.16 Find the volume enclosed by the ellipsoid $\dfrac{x^2}{a^2} + \dfrac{y^2}{b^2} + \dfrac{z^2}{c^2} = 1$.

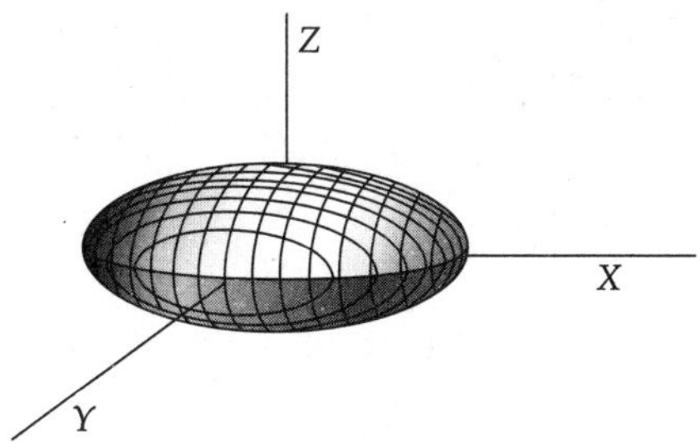

Fig. 10.25

Sol: The volume V is given by,

$$V = \iiint\limits_{E} dx\, dy\, dz$$

where E is the region enclosed by the given ellipsoid.

As usual to evaluate the above integral we first substitute $x = au$, $y = bv$, $z = cw$ so that the region E chances to the region E' enclosed by the unit sphere $u^2 + v^2 + w^2 = 1$. So that,

$$V - \iiint\limits_{E} dx\, dy\, dz - \iiint\limits_{E'} abc\, du\, dv\, dw$$

Now in the above integral put,

$$u = r \sin\theta \cos\phi$$

$$v = r \sin\theta \sin\phi$$

$$w = r \cos\theta$$

then Jacobian of the transformation $r^2 \sin\theta$ hence,

$$V = abc \iiint\limits_{E'} du\, dv\, dw = abc \int\limits_{r=0}^{1} \int\limits_{\theta=0}^{\pi} \int\limits_{\phi=0}^{2\pi} r^2 \sin\theta\, dr\, d\theta\, d\phi$$

$$= abc \left[\frac{r^3}{3}\right]_0^1 [-\cos\theta]_0^\pi [\phi]_0^{2\pi}$$

$$= \frac{2\pi}{3} abc \left\{ - [\cos\pi - \cos 0] \right\} = \frac{4\pi}{3} abc$$

Example 10.17   Evaluate $\iint\limits_{E} \sqrt{16 - x^2 - y^2}\, dx\, dy$ where E is the upper half of the circle $x^2 + y^2 - 4x = 0$.

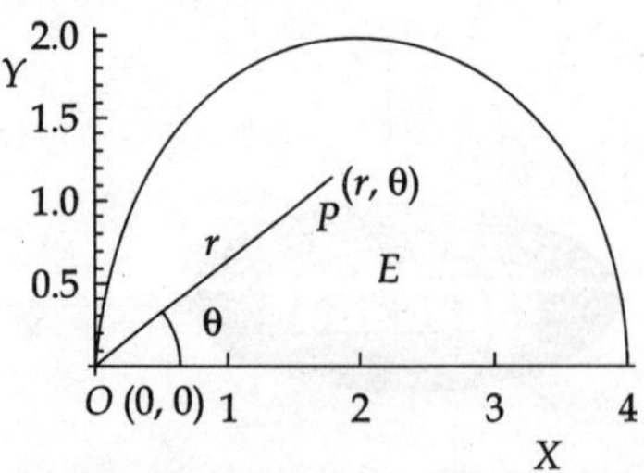

Fig. 10.26

Sol: We put $x = r\cos\theta$, $y = r\sin\theta$. Note that it is clear from the given equation, for fixed value of ϕ range of r is given as $0 \le r \le 4\cos\theta$ also the total range of θ in E is given by $0 \le \theta \le \dfrac{\pi}{2}$.

Hence,

$$\iint\limits_{E} \sqrt{16 - x^2 - y^2}\, dx\, dy = \int\limits_{\theta=0}^{\frac{\pi}{2}} \int\limits_{r=0}^{4\cos\theta} r\sqrt{16 - r^2}\, dr\, d\theta$$

$$= \int\limits_{0}^{\frac{\pi}{2}} \frac{1}{2} \cdot \frac{2}{3} \left[-u^{\frac{3}{2}} \right]_{16}^{16\sin^2\theta} d\theta \quad \text{putting } u = 16 - r^2$$

$$= \frac{1}{3} \int\limits_{0}^{\frac{\pi}{2}} (4^3 - 4^3 \sin^3\theta)\, d\theta$$

Now using Corollary 6.6.2 we get $\displaystyle\int\limits_{0}^{\frac{\pi}{2}} \sin^3\theta\, d\theta = \frac{1}{2} \frac{\Gamma(2)\Gamma\left(\frac{1}{2}\right)}{\Gamma\left(\frac{5}{2}\right)} = \frac{2}{3}$

So that $\iint\limits_{E} \sqrt{16 - x^2 - y^2}\, dx\, dy = \dfrac{4^3}{3} \left(\dfrac{\pi}{2} - \dfrac{2}{3} \right).$

Example 10.18   Using the transformation,

$$x + y + z = u$$
$$x + y = uv$$
$$x = uvw$$

evaluate the triple integral

$$\iint\limits_{E} x^a y^b z^c\, (1 - x - y - z)^d\, dx\, dy\, dz \quad a, b, c, d > -1$$

where E is the region bounded by the planes $x = 0$, $y = 0$, $z = 0$ and $x + y + z = 1$.

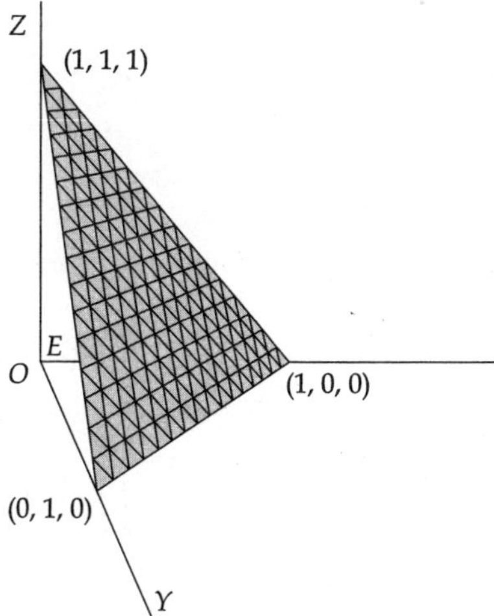

Fig. 10.27

Sol: If we employ the above transformation then we see that,

$$x = uvw$$

$$y = uv\,(1 - w)$$

$$z = u\,(1 - v)$$

and the Jacobian of the transformation $J = \dfrac{\partial(x,\,y,\,z)}{\partial(u,\,v,\,w)} = -\,u^2 v$. Also the region E in $(x,\,y,\,z)$ plane changes

to E' in $(u,\,v,\,w)$ plane which is unit cube. Hence,

$$\iiint\limits_{E} x^a y^b z^c \,(1 - x - y - z)^d \; dx\,dy\,dz$$

$$= \iiint\limits_{E'} (uvw)^a \,[uv(1-w)]^b \,[u(1-v)]^c \,(1-u)^d \,u^2 v \; du\,dv\,dw$$

$$= \int_0^1 u^{a+b+c+2} \,(1-u)^d \,du \int_0^1 v^{a+b+1} \,(1-v)^c \,dv \int_0^1 w^a \,(1-w)^b \,dw$$

$$= B(a+b+c+3,\, d+1)\cdot B(a+b+2,\, c+1)\cdot B(a+1,\, b+1)$$

$$= \frac{\Gamma(a+b+c+3)\,\Gamma(d+1)}{\Gamma(a+b+c+d+4)}\;\frac{\Gamma(a+b+2)\,\Gamma(c+1)}{\Gamma(a+b+c+3)}\;\frac{\Gamma(a+1)\,\Gamma(b+1)}{\Gamma(a+b+2)}$$

$$= \frac{\Gamma(a+1)\,\Gamma(b+1)\,\Gamma(c+1)\,\Gamma(d+1)}{\Gamma(a+b+c+d+4)}$$

PROBLEMS

1. Prove Lemma 10.2.1

2. Prove Theorem 10.2.2

3. Prove Theorem 10.2.3

4. Evaluate the following double integrals (assuming the integrals exist) using definition only:

(i) $\displaystyle\int_0^1\int_0^1 x\,dx\,dy$

(ii) $\displaystyle\int_0^1\int_0^1 xy\,dx\,dy$

(iii) $\displaystyle\int_0^1\int_0^2 (x-y)\,dx\,dy$

(iv) $\displaystyle\int_0^1\int_0^1 (x^2+y^2)\,dx\,dy$

5. Evaluate:

(i) $\displaystyle\int_0^1\int_0^1 xe^{x+y}\,dx\,dy$

(ii) $\displaystyle\int_{-1}^1\int_{-1}^1 \frac{x^2}{1+y^2}\,dx\,dy$

(iii) $\displaystyle\int_0^1\int_0^1 \left(\frac{x-y}{x+y}\right)dx\,dy$

(iv) $\displaystyle\int_{\frac{1}{2}}^1\int_{\frac{1}{2}}^1 \frac{xe^{xy}}{y}\,dx\,dy$

6. Evaluate:

(i) $\displaystyle\int_0^1 dx\int_0^x (xy+1)\,dy$

(ii) $\displaystyle\int_{-1}^1 dx\int_0^{\sqrt{1-x^2}} 2y\,dy$

(iii) $\displaystyle\int_{-1}^1 dx\int_0^{\sqrt{1-x^2}} 2y\,dy$

(iv) $\displaystyle\int_0^1 dx\int_0^{x^2} 2y\,dy$

7. Show that following double integrals does not exist

(i) $\displaystyle\int_0^1\int_0^1 \frac{y-x}{(x+y)^3}\,dx\,dy$

(ii) $\displaystyle\int_0^1\int_0^1 \frac{y^2-x^2}{(x^2+y^2)^2}\,dx\,dy$

(iii) $\displaystyle\int_0^1\int_0^1 f(x,y)\,dx\,dy$ where $f(x,y)=\begin{cases}\dfrac{1}{y^2} & \text{if } 0<x<y<1 \\[2mm] -\dfrac{1}{x^2} & \text{if } 0<y<x<1\end{cases}$

(iv) $\displaystyle\int_0^1\int_0^1 f(x,y)\,dx\,dy$ where,

$$f(x,y)=\begin{cases} 1 & \text{if } x\ \&\ y \text{ are both rational and satisfies } 0\le x,\,y\ge 1 \\ -1 & \text{if atleast one of } x \text{ or } y \text{ is irrational and satisfies } 0\le x,\,y\le 1\end{cases}$$

(v) $\int_0^1 \int_0^2 f(x, y) \, dx \, dy$ where,

$$f(x, y) = \begin{cases} 2 & \text{if } y \in [0, 2] \text{ is rational} \\ x & \text{if } y \in [0, 2] \text{ is irrational} \end{cases}$$

8. Change the order of integration in the following double integrals indicating the domain of integration:

(i) $\int_0^1 dx \int_{x^2}^{x} f(x, y) \, dy$

(ii) $\int_0^a dx \int_{\sqrt{a^2 - x^2}}^{x + 2a} f(x, y) \, dy$

(iii) $\int_0^{2a} dx \int_{\frac{x^2}{4a}}^{3a - x} f(x, y) \, dy$

9. Show that:

(i) $\iint_R \sqrt{xy(1 - x - y)} \, dx \, dy = \dfrac{2\pi}{105}$

where R is the region in the plane bounded by the co-ordinate axes and the line $x + y = 1$.

(ii) $\int_0^1 dx \int_0^x \sqrt{x^2 + y^2} \, dy = \dfrac{1}{6} [\sqrt{2} + \log(1 + \sqrt{2})]$

(iii) $\int_0^{2a} dx \int_{\sqrt{2ax - x^2}}^{\sqrt{4ax - x^2}} \left(1 + \dfrac{y^2}{x^2}\right) dy = \left(\pi + \dfrac{8}{3}\right)$

(iv) $\int_0^1 dy \int_0^{1 - y^2} \{(x - 1)^2 + y^2\} \, dx = \dfrac{44}{105}$

(v) $\iint_R \sqrt{4a^2 - x^2 - y^2} \, dx \, dy = \dfrac{4a^3 (3\pi - 4)}{9} (x, y) \, dy$ where $R : \{x, y) : y \ge 0 \ \& \ x^2 + y^2 \le 2ax\}$

10. Evaluate the following:

(i) $\iint_R \sqrt{x(2a - x) + y(2b - y)} \, dx \, dy$ where R is the region in the plane bounded by the circle

$x^2 + y^2 - 2ax - 2by = 0.$

(ii) $\iint_R e^{\left(\frac{x - y}{x + y}\right)} \, dx \, dy$ where R is the region in the plane bounded by the co-ordinate axes and

the line $x + y = 1$. [Hint: Use the transformation $u = x - y, v = x + y$].

(iii) $\displaystyle\iint\limits_{R} \sqrt{1-x^2-y^2}\; dx\, dy$ where $R : \{x, y\} : x \geq 0, y \geq 0\; \&\; x^2 + y^2 \leq 1\}$

(iv) $\displaystyle\iint\limits_{R} \frac{x^2}{y^2}\; dx\, dy$ where R is the region in the plane bounded by $y = x$, $y = 2$ and $xy = 1$.

(v) $\displaystyle\iint\limits_{E} (x^2 + y^2)\; dx\, dy$ where $R : \{(x, y) : x^2 + y^2 \leq 2y\}$.

(vi) $\displaystyle\iint\limits_{R} \sqrt{(x^2 + y^2)}\; dx\, dy$ where $R : \{(x, y) : x^2 + y^2 \leq 9\; \&\; x^2 + y^2 \geq 4\}$.

(vii) $\displaystyle\iint\limits_{R} \left(\frac{1 - x^2/a^2 - y^2/b^2}{1 + x^2/a^2 + y^2/b^2} \right) dx\, dy$ where $R : \{(x, y) : \dfrac{x^2}{a^2} + \dfrac{y^2}{b^2} \leq 1, x \geq 0\; \&\; y \geq 0\}$

(viii) $\displaystyle\iint\limits_{R} \sqrt{(xy - y^2)}\; dx\, dy$ where R is the triangular region in the plane bounded by the vertices

$\quad\quad (0, 0),\ (1, 1)\ \&\ (3, 1)$.

(ix) $\displaystyle\iint\limits_{R} \sqrt{(x^2 + y^2)}\; dx\, dy$ where $R : \{(x, y) : x^2/y^2 + y^2 \leq 1\}$.

(x) $\displaystyle\iint\limits_{R} (3\sqrt{(x^2 + y^2)} + 1)\; dx\, dy$ where $R : \left\{ (x, y) : x^2 + y^2 \leq \dfrac{1}{4} \right\}$.

11. Evaluate following triple integrals:

(i) $\displaystyle\iiint\limits_{V} (ax^2 + by^2 + cz^2)\; dx\, dy\, dz$ where $V : \{x, y, z) : x^2 + y^2 + z^2 \leq r^2\}$

(ii) $\displaystyle\iiint\limits_{V} \frac{dx\, dy\, dz}{\sqrt{x^2 + y^2 + (z-2)^2}}$ where $V : \{(x, y, z) : x^2 + y^2 + z^2 \leq 1\}$

(iii) $\displaystyle\iiint\limits_{V} \frac{xyz\; dx\, dy\, dz}{\sqrt{(x^2 + y^2 + z^2)}}$ where $V : \{(x, y, z) : x^2 + y^2 + z^2 \leq 1, x \geq 0, y \geq 0, z \geq 0\}$

(iv) $\displaystyle\iiint\limits_{V} \left(1 - \frac{x^2}{a^2} - \frac{y^2}{b^2} - \frac{z^2}{c^2} \right)^{-\frac{1}{4}} (xyz)^{-\frac{1}{2}}\; dx\, dy\, dz$

$\quad\quad$ where $V : \left\{ x, y, z) : \left(1 - \dfrac{x^2}{a^2} - \dfrac{y^2}{b^2} - \dfrac{z^2}{c^2} \right) \leq 1, x \geq 0, y \geq 0, z \geq 0 \right\}$

(v) $\iiint\limits_{V} (x^2 + y^2)\, dx\, dy\, dz$ where $V: \{x, y, z\} : a^2 \le x^2 + y^2 + z^2 \le b^2\}$

12. **Show that**

(i) $\iiint\limits_{R} x^{l-1}\, y^{m-1}\, z^{n-1}\, (1 - ax - by - cz)^{p-1}\, dx\, dy\, dz = \dfrac{\Gamma(l)\,\Gamma(m)\,\Gamma(n)\,\Gamma(p)}{\Gamma(l + m + n + p)},$

$a, b, c, l, m, n, p > 0$ where,

$V : \{(x, y, z) : ax + by + cz \le 1,\, x \ge 0,\, y \ge 0,\, z \ge 0\}$

(ii) $\iiint\limits_{V} \log(x + y + z)\, dx\, dy\, dz = \dfrac{1}{8}$ where,

$V : \{(x, y, z) : x + y + z \le 1,\, x \ge 0,\, y \ge 0,\, z \ge 0\}$

(iii) $\iiint\limits_{V} (ax + by + cz)^2\, dx\, dy\, dz = \dfrac{4\pi\,(a^2 + b^2 + c^2)}{15}$ where,

$V : \{(x, y, z) : x^2 + y^2 + z^2 \le 1\}$

(iv) $\iiint\limits_{V} z\, dx\, dy\, dz = \dfrac{\pi}{8}$ where V is the region in space bounded by the surfaces $x^2 + y^2 + z^2 \le$

$1, x^2 + y^2 + z^2 \le z^2$ and $z \ge 0$.

(v) $\iiint\limits_{V} e^{x+y+z}\, dx\, dy\, dz = \dfrac{1}{2}\,(e - 2)$ where $V : \{(x, y, z) : x + y + z \le 1,\, x \ge 0,\, y \ge 0,\, z \ge 0\}$

Chapter 11

Line and Surface Integrals

11.1 INTRODUCTION

The integration theory so far developed in the previous chapters dealt with the concept of integral of a bounded functions on a closed interval (in case of single variable) or the double or triple of functions on a bounded domain of $\mathbb{R}^2$ and $\mathbb{R}^3$ (in case of two and three variables respectively). In this chapter we will study the concept of integrals of functions defined on a curve (either on plane $\mathbb{R}^2$ or space $\mathbb{R}^3$) or functions defined on surfaces (on $\mathbb{R}^{3'}$) leading to the definitions of line and surface integrals. Our ultimate goal in this chapter is to formulate some sort of generalizations of Fundamental Theorem of Integral calculus known as Green's, Stokes' and Gauss' Theorem.

Line and surface integrals has lots of applications in various branches of mathematics and particularly plays a significant role in Physics. We here assume the reader is familiar with the 'analytic' approach of defining **vectors** in terms of its components and also we will assume that he is acquainted with the idea of 'Dot product' and 'Cross Product' of vectors.

11.2 SCALAR AND VECTOR FIELD

Suppose $f: D \to \mathbb{R}^m$ is a function where $D \subset \mathbb{R}^n$. In general if $m > 1$ we say f is vector valued and will be denoted by $\vec{f}$. If $n = m = 1$ then we say that f is a real valued function of real variable, if $n = 1$ & $m > 1$ we say f is **vector function of real variable**. If $n > 1$ and $m = 1$ f is called a **scalar field**, finaly if both n & $m > 1$ we call f a **vector field**.

Example 11.1 Let $D \subset \mathbb{R}^3$ and suppose,

$$T : D \to \mathbb{R}$$

be a function which associates to every point of D the temperature of that point that is,

$$T(x,\, y,\, z) = \text{tempurature of } P(x,\, y,\, z)$$

where $P(x, y, z)$ is a point of D. Then T is a scalar field.

Example 11.2 Let $\vec{v}: [0,\, T] \to \mathbb{R}^2$ be a function which denotes the velocity of a moving particle in plane at any instant $t \in [0,\, T]$ then v is a vector function of a real variable.

Example 11.3   We know from elementary physics that the force of attraction between two charges q_1 & q_2 separated by a distance r is given by $k \dfrac{q_1 q_2}{r^2}$ (k is some constant) which is called **Coulomb's Law**. Let us think that the charge q_1 is placed at the origin 0 of the coordinate system and suppose that q_2 is placed at a point $P(x, y, z)$ at a distance r from origin that is $\vec{r} = x\vec{i} + y\vec{j} + z\vec{k}$ then by **Coulomb's Law** the force $\vec{F}(x, y, z)$ experienced by q_2 due to q_1 is given by,

$$\vec{F}(x, y, z) = k \frac{q_1 q_2}{r^3} \vec{r}$$

Here $\vec{F}$ is an example of vector field.

11.3   DIVERGENCE AND CURL OF A VECTOR FIELD

Let ϕ be a scalar field defined on some open subset U of $\mathbb{R}^3$ whose partial derivatives exists at every point then ***Gradient of*** ϕ (or grad ϕ in short) is defined as,

$$\nabla \phi = \frac{\partial \phi}{\partial x} \vec{i} + \frac{\partial \phi}{\partial y} \vec{j} + \frac{\partial \phi}{\partial z} \vec{k}$$

The operator $\nabla \equiv \left(\dfrac{\partial}{\partial x} \vec{i} + \dfrac{\partial}{\partial y} \vec{j} + \dfrac{\partial \phi}{\partial z} \vec{k} \right)$ is known as ***Del Operator.***

If $\vec{F} = (f_1, f_2, f_3)$ is a vector field defined on some open subset U of $\mathbb{R}^3$ such that the partial derivatives f_1, f_2 & f_3 of exists at every point of U then we define ***Divergence*** of $\vec{F}$ (div $\vec{F}$) as,

$$\nabla . \vec{F} = \frac{\partial f_1}{\partial x} + \frac{\partial f_2}{\partial y} + \frac{\partial f_3}{\partial z}$$

and ***Curl*** of $\vec{F}$ (curl $\vec{F}$) as,

$$\nabla \times \vec{F} = \left(\frac{\partial f_3}{\partial y} - \frac{\partial f_2}{\partial z} \right) \vec{i} + \left(\frac{\partial f_1}{\partial z} - \frac{\partial f_3}{\partial x} \right) \vec{j} + \left(\frac{\partial f_2}{\partial x} - \frac{\partial f_1}{\partial y} \right) \vec{k}$$

[Note that Curl of $\vec{F}$ can be calculated by evaluating the determinant

$$\begin{bmatrix} \vec{i} & \vec{j} & \vec{k} \\ \dfrac{\partial}{\partial x} & \dfrac{\partial}{\partial y} & \dfrac{\partial}{\partial z} \\ f_1 & f_2 & f_3 \end{bmatrix}$$

Example 11.4   Let $\phi(x, y, z) = xy + e^x \cos z + y^2$ be a scalar field defined on $\mathbb{R}^3$ then, grad $\phi (\nabla \phi)$

$$= \frac{\partial}{\partial x} (xy + e^x \cos z + y^2) \vec{i} + \frac{\partial}{\partial y} (xy + e^x \cos z + y^2) \vec{j} + \frac{\partial}{\partial z} (xy + e^x \cos z + y^2) \vec{k}$$

$$= (y + e^x \cos z) \vec{i} + (x + 2y) \vec{j} + (-e^x \cos z) \vec{k}$$

Example 11.5   If $\vec{F} = (xy^2 + yz^2, yx^2 + xz^2, yx^2 + zy^2)$ then,

$$\text{div } \vec{F}\,(\nabla . \vec{F}) = \frac{\partial}{\partial x}(xy^2 + yz^2) + \frac{\partial}{\partial y}(yx^2 + xz^2) + \frac{\partial}{\partial z}(yx^2 + zy^2)$$

$$= y^2 + x^2 + y^2 = 2y^2 + x^2$$

$$\text{curl } \vec{F}\,(\nabla \times \vec{F}) = \begin{vmatrix} \vec{i} & \vec{j} & \vec{k} \\[4pt] \dfrac{\partial}{\partial x} & \dfrac{\partial}{\partial y} & \dfrac{\partial}{\partial z} \\[8pt] xy^2 + yz^2 & yx^2 + xz^2 & yx^2 + zy^2 \end{vmatrix}$$

$$= \left(\frac{\partial}{\partial y}(yx^2 + zy^2) - \frac{\partial}{\partial z}(yx^2 + xz^2) \right)\vec{i} + \left(\frac{\partial}{\partial z}(xy^2 + yz^2) - \frac{\partial}{\partial x}(yx^2 + zy^2) \right)\vec{j}$$

$$+ \left(\frac{\partial}{\partial x}(yx^2 + xz^2) - \frac{\partial}{\partial y}(xy^2 + yz^2) \right)\vec{k}$$

$$= (x^2 - 2zy - 2zx)\,\vec{i} + (2yz - 2xy)\,\vec{j} + (2xy + z^2 - 2xy - z^2)\,\vec{k}$$
$$= (x^2 - 2zy - 2zx)\,\vec{i} + (2yz - 2xy)\,\vec{j}$$

Example 11. 6   If ϕ is a scalar field defined in some open set of $\mathbb{R}^3$ having continuous second order partial derivatives then show that $\nabla \times \nabla\phi = 0$

Sol: $$\nabla \times \nabla\phi = \begin{vmatrix} \vec{i} & \vec{j} & \vec{k} \\[4pt] \dfrac{\partial}{\partial x} & \dfrac{\partial}{\partial y} & \dfrac{\partial}{\partial z} \\[8pt] \dfrac{\partial\phi}{\partial x} & \dfrac{\partial\phi}{\partial y} & \dfrac{\partial\phi}{\partial z} \end{vmatrix}$$

$$= \left(\frac{\partial^2\phi}{\partial y\,\partial z} - \frac{\partial^2\phi}{\partial z\,\partial y} \right)\vec{i} + \left(\frac{\partial^2\phi}{\partial z\,\partial x} - \frac{\partial^2\phi}{\partial x\,\partial z} \right)\vec{j} + \left(\frac{\partial^2\phi}{\partial x\,\partial y} - \frac{\partial^2\phi}{\partial y\,\partial x} \right)\vec{k} = 0$$

(as mixed partial derivatives are equal by assumption)

11.4   CURVES IN PLANE AND SPACE

We are all familiar with the equation $x^2 + y^2 = a^2$ which is a planar curve representing a circle. In general an equation of the form $f(x, y) = 0$ represents a curve in $\mathbb{R}^2$. Also note that the above equation of the circle can be written as,

$$x = a \cos\theta, \; y = a \sin\theta, \; \theta \in [0, 2\pi]$$

which is called the parametric representation of the circle. The parametric representation tells us that the circle formed by *deforming* of an interval. It is easy to visualize the situation, take a piece of string representing an interval join its two ends and we end up getting a circle! In fact, any curve can be thought of as a deformation of an interval. We now give the formal definition of a curve.

Definition Let $[a, b]$ be a closed interval of $\mathbb{R}$ then a continuous function $\vec{c}\colon [a, b] \to \mathbb{R}^2$ or $\mathbb{R}^3$ is called a plane curve or in space curve respectively.

Note that by definition a curve c is vector valued function of a real variable hence we will denote it by $\vec{c}$. The set $C = \{\vec{c}\,(t) : t \in [a, b]\}$ is called the **graph** of the curve. Some time by abuse of language we will denote a curve by C (its graph) or by $\vec{c}$.

A curve is said to be a **simple** curve or a **Jordan** curve if c is one-one function. The curve is said to be a closed curve if $\vec{c}\,(a) = \vec{c}\,(b)$.

If $\vec{c}$ is a curve in $\mathbb{R}^2$ then we can write,

$$\vec{c}\,(t) = (x(t), y(t))\ t \in [a, b]$$

where $x(t)$ & $y(t)$ are called components of the curve. Such a representation of a curve is called a **parametric** representation of the curve. If c' exists and is continuous in the open interval (a, b) then the curve is called a **smooth** curve. Note that a curve can have one than one representation for example $c'(t) = \left(t, \sqrt{1 - t^2}\right)$, $t \in [-1, 1]$ represents the unit semi-circle in $\mathbb{R}^2$ which is also represented by the function $\vec{g}\,(\theta) = (\cos\theta, \sin\theta)$, $\theta \in [0, \pi]$.

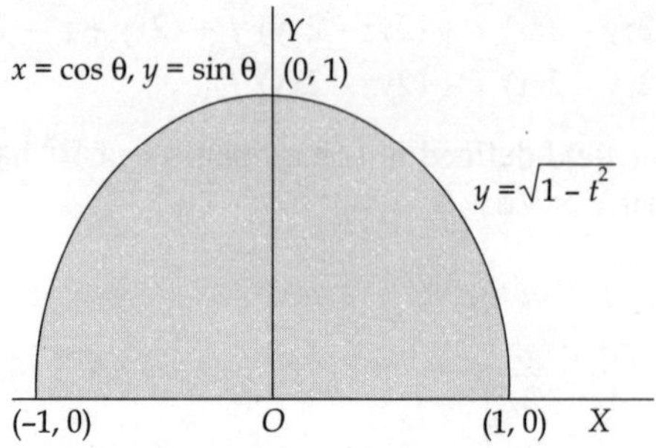

Fig. 11.1

Also note that $\vec{g}\,(\theta) = \vec{c}\,(\cos\theta)$. Let $\vec{c}_1(t)$, $t \in [a, b]$ and $\vec{c}_2(\theta)$, $\theta \in [c, d]$ be two curves and suppose that there exists a differentiable onto function $h\colon [c, d] \to [a, b]$ such that $h'(\theta) \neq 0\ \theta \in [c, b]$ and $\vec{c}_2(\theta) = \vec{c}_1(h(\theta))$ then we say that the two curves $\vec{c}_1$ & $\vec{c}_2$ are **equivalent**.

Example 11.7 The function $\vec{c}\,(\theta) = (a\cos\theta, b\sin\theta)$, $\theta \in [0, 2\pi]$ represents a closed smooth Jordan curve which is the ellipse.

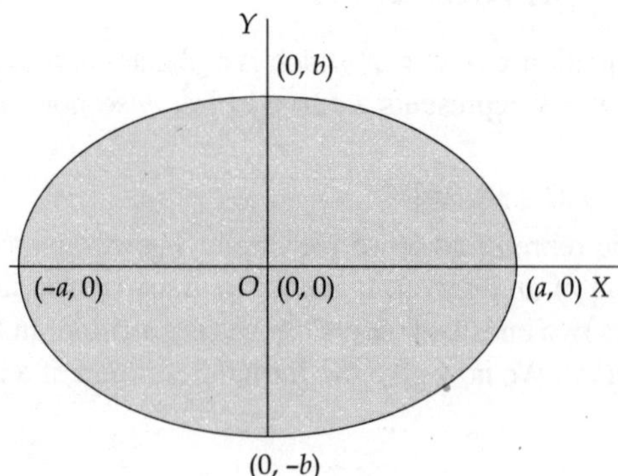

Fig. 11.2

The curve Jordan curve is said to be *piecewise smooth* if the interval $[a, b]$ can be partitioned into finite number of subintervals in each of which the curve is smooth. The figure (Fig. 11.3) below gives the graph of a piecewise smooth curve.

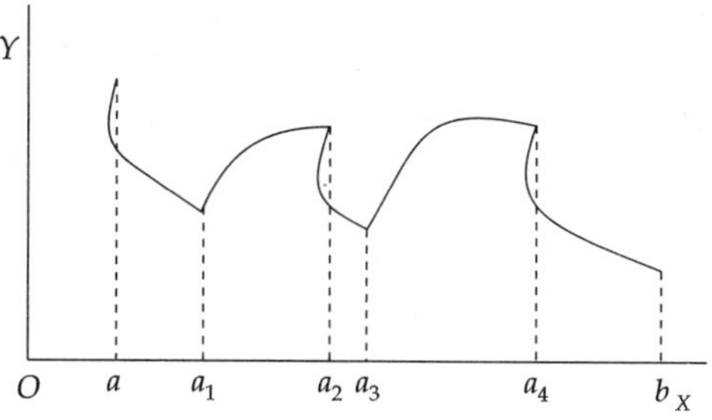

Fig. 11.3

If we closely look at the above figure we see that the interval $[a, b]$ is divided into smaller sub-intervals inside which the curve is smooth. Note that at both the upper end points and lower end points of these sub-intervals the curve has 'corners' that is at each of these points the tangent to the curve takes an abrupt turn. But at all other points of the sub-interval except the end points the tangent takes continuous turn. Now suppose,

$$\vec{c}(t) = (x(t), y(t), t \in [a, b]$$

denotes a smooth curve in $\mathbb{R}^2$ then, the direction of the tangent at any point of the curve is given by either, $\tan\theta = \dfrac{y'(t)}{x'(t)}$ or $\cot 0 = \dfrac{x(t)}{y(t)}$ hence if $x'(t), y'(t)$ are not simultaneously equal to zero we will get the direction of the tangent. Hence we can say that *piecewise smooth* Jordan curves are those curves for which the interval $[a, b]$ can be partitioned into finite number of subintervals in each of which $(x'(t))^2 + (y'(t))^2 > 0$. In fact, a piecewise smooth Jordan curve is obtained by joining finite number of smooth Jordan curves producing 'corners' at joining points.

Note that *Smooth* and *piecewise smooth* Jordan space curves can be defined on similar lines as defined in case of plane curves.

Example 11.8 The function $\vec{c}(t) = (3t, 4t^2, t^3)$, $t \in [0, 1]$ is a smooth space curve. See Fig. (11.4) below

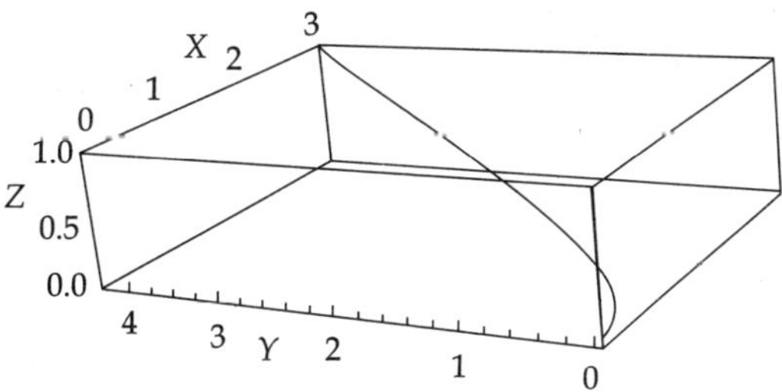

Fig. 11.4

Example 11. 9   The function $\vec{c}(t) = (2 + 3\cos t, 2 + 3\sin t, 3)\, t \in [-\pi, \pi]$ defines a smooth closed space curve. See Fig. (11.5) below.

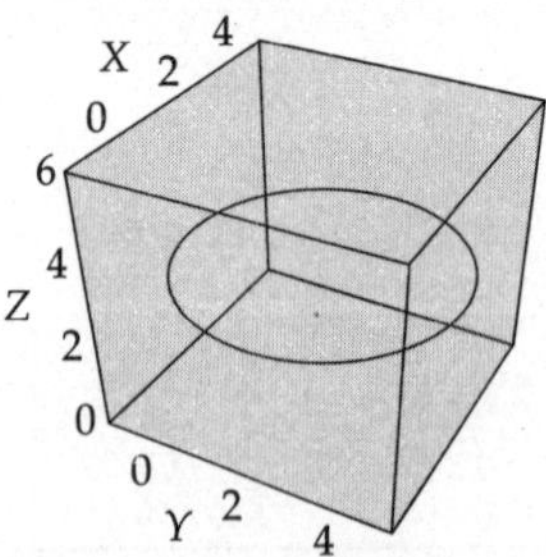

Fig. 11.5

Suppose, $\vec{c}(t) = (x(t), y(t))$, $t \in [a, b]$ be a closed Jordan plane curve then the curve can be traversed in two directions namely anticlockwise or clockwise direction conventionally we choose anticlockwise direction to be the positive direction. At any 'instant' t the coordinate of a point P on the curve is given by $P(t) = (x(t), y(t))$. We say that above curve is ***positively oriented*** if a point P on the curve moves in an anticlockwise direction starting from the initial position $P(a)$ and reaching the same position as the value of t changes from $t = a$ to $t = b$. Otherwise we say that the curve is ***negatively oriented***.

As an example consider the closed circle parameterized by the equations,

$$\left. \begin{array}{l} x = \cos\theta \\ y = \sin\theta \end{array} \right\} \quad \theta \in [0, 2\pi]$$

If we look at Fig. 6 we see that as θ (which is the angle that the radius vector of any point P makes with the positive direction of the X-axis) increases from 0 to 2π a point P starts from the point A travels in an anti-clock wise direction and reaches the same point A again. Hence in this case we say that the curve is positively oriented.

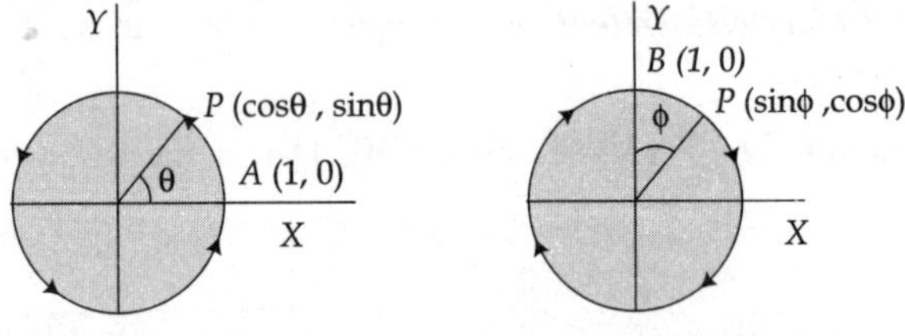

Positively oriented circle   Negatively oriented circle

Fig. 11.6   **Fig. 11.7**

Again if we consider the same curve with parametric equations as

$$\left. \begin{array}{l} x = \cos\phi \\ y = \sin\phi \end{array} \right\} \quad \phi \in [0, 2\pi]$$

we see that (Fig. 11.7) as ϕ (which is the angle that the radius vector of any point P makes with the positive direction of the Y-axis) increases from 0 to 2π a point P starts from the point B travels in a clock wise direction and reaches the same point B again. Hence in this case we say that the curve is negatively oriented. Also we note a very interesting point here, if we denote by $\vec{c}_1(\theta)$ the positively oriented circle and if $\vec{c}_2(\phi)$ denotes the negatively oriented circle then we see that,

$$\vec{c}_2(\theta) = \vec{c}_1(\pi/2 - \theta)$$

or in other words, $\vec{c}_1$ & $\vec{c}_2$ are equivalent by definition where h is given by,

$$h(\theta) = \frac{\pi}{2} - \theta$$

so that $h'(\theta) = -1 < 0 \ \forall \theta \in [0, 2\pi]$.

From the above observation we conclude that even if the two curves are equivalent they may not trace the curve in the same direction in fact, the two equivalent curves will have same orientation if $h' > 0$ in which case h is called **orientation preserving** and in case $h' < 0$ the two curves will be oriented in opposite direction and call h **orientation reversing**.

11.5 RECTIFIABLE CURVES (ARC LENGTH)

In this section we want to define the notion of length of a curve. For that let us first recall that for any point $P(x, y)$ the distance Euclidean distance from origin is dented by $\|P\|$ and is defined as $\|P\| = \sqrt{x^2 + y^2}$ Here $\|.\|$ is called the Euclidean norm. Now if $P_1(x_1, y_1)$ and $P_2(x_2, y_2)$ be two points of $\mathbb{R}^2$ then the Euclidean distance between P_1 and P_2 denoted by $\|P_1 - P_2\|$ is defined by the equation $\|P_1 - P_2\| = \sqrt{(x_2 - x_1)^2 + (y_2 - y_1)^2}$. Similarly the Euclidean distance of two points $P_1(x_1, y_1, z_1)^2$ and $P_2(x_2, y_2, z_2)$ of $\mathbb{R}^3$ is defined by the equation $\|P_1 - P_2\| = \sqrt{(x_2 - x_1)^2 + (y_2 - y_1)^2 + (z_2 - z_1)^2}$.

Now suppose $\vec{c}(t) \ t \in [a, b]$ be a (plane or space) curve. Let $P : a = t_0 < t_1 < ... < t_{r-1} < t_r < ... t_n = b$ be a partition of $[a, b]$ with norm δ(say). Let $\Lambda(P) = \sum_{i=1}^{n} \|\vec{x}(t_i) - \vec{x}(t_{i-1})\|$ where $\|.\|$ denotes the Euclidean norm. Note that $\Lambda(P)$ denotes the length of the inscribed polygon with vertices $\vec{x}(t_i)$, $i = 0, 1, 2 ..., n$

We say that the curve is **rectifiable** if $\Lambda_c = \lim_{\delta \to 0} \sum_{i=1}^{n} \|\vec{x}(t_i) - \vec{x}(t_{i-1})\|$ exists and in that case Λ_c is known as the **arc length** of the curve.

Below we mention a result without proof which is useful in finding the arc length of smooth curves in space as well as in plane.

Theorem 11.5.1

Let $\vec{c}(t) = (x(t), y(t), z(t))$, $t \in [a, b]$ be a smooth curve then it is rectifiable and the arc length Λ_c is given by,

$$\Lambda_c = \int_a^b \sqrt{(x'(t))^2 + (y'(t))^2 + (z'(t))^2} \ dt \qquad (11.5.1.1)$$

To give a physical interpretation of above formula, let us think as if a particle is moving along the curve $\vec{c}$ that is at any instant of time $t \in [a, b]$ the position of the particle is given by $\vec{c}(t)$. Then the speed of the particle is denoted by $\|\vec{c}'(t)\|$ at that instant and hence the total distance traversed by the particle from time $t = a$ to $t = b$ or the arc length of the curve is given by,

$$\Lambda_c = \int_a^b \|\vec{c}'(t)\| \, dt$$

where $\|\vec{c}(t)\| = \sqrt{(x'(t))^2 + (y'(t))^2 + (z'(t))^2}$ which is precisely the equation (11.5.1.1).

Also see that we can use equation (11.5.1.1) for finding the arc length of a smooth planar curve by ignoring the term $(z'(t))^2$

We define the **arc length function** $s(\xi)$ of a rectifiable curve $\vec{c}(t)$ $t \in [a, b]$ as,

$$s(\xi) = \int_a^\xi \|\vec{c}'(t)\| \, dt \tag{11.5.1.2}$$

Example 11.10   The curve $\vec{c}$ with parametric equation,

$$x(t) = t, \, y(t) = t^2, \, z(t) = t^3, \, t \in [0, 1]$$

is rectifiable.

We note that each $x(t)$, $y(t)$ & $z(t)$ is a continuously differentiable function hence by Theorem 11.3.1 $\vec{c}$ is rectifiable and the length is given by,

$$\Lambda_c = \int_0^1 \sqrt{1 + 2x + 3x^2} \, dt = \frac{1}{18} \left[-3 + 12\sqrt{6} - 2\sqrt{3} \, \sin h^{-1} \left(\frac{1}{\sqrt{2}} \right) + 2 \sin h^{-1} (2\sqrt{2}) \right]$$

Example 11.11   Find the length of the curve $\vec{c}(t) = (\sin t, \cos t)$ $t \in [0, \pi]$

$$\Lambda_c = \int_0^\pi \sqrt{(\cos t)^2 + (-\sin t)^2 + 1} \, dt$$

$$= \int_0^\pi \sqrt{2} \, dt = \sqrt{2}\pi$$

11.6   SIMPLE AND MULTIPLY CONNECTED REGIONS OF PLANE

An open set of $\mathbb{R}^2$ is said to be **connected** if any two points of the set can be joined by a broken line having finite number of segments all of whose points lies in the set. A **region** is an open connected set with or without its boundary. If it contains whole of its boundary it is called **closed region**.

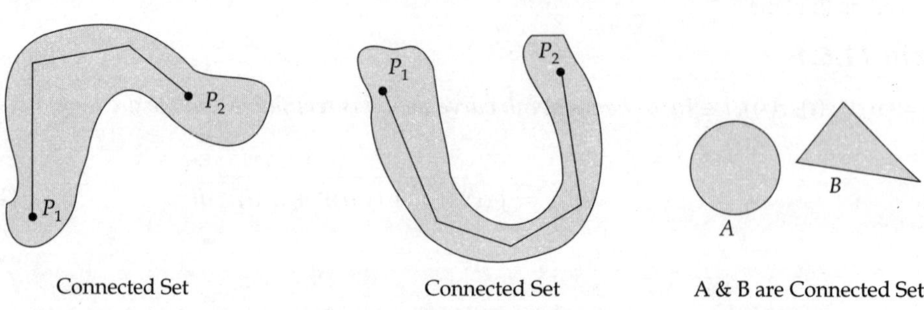

Connected Set	Connected Set	A & B are Connected Set
Fig. 11.8	**Fig. 11.9**	**Fig. 11.10**

A region E in plane is called ***simply connected*** if for every Jordan curve C which lies in E the inner region of C is a subset of E. Basically simply connected regions do not have holes for example the annulus which is the region between two concentric circles is not a simply connected region. A region which is not simply connected is called ***multiply connected*** region.

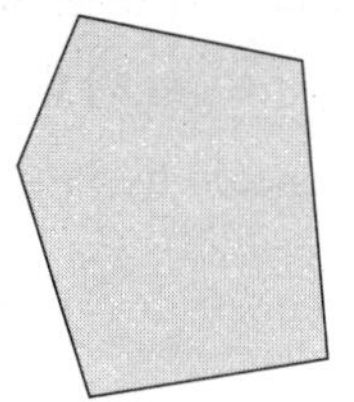

A polygon is simply connected

Fig. 11.11

Region with holes is multiply connected

Fig. 11.12

Consider a region R as shown in the figure (Fig. 11.13) with the boundary curve say C clearly one can

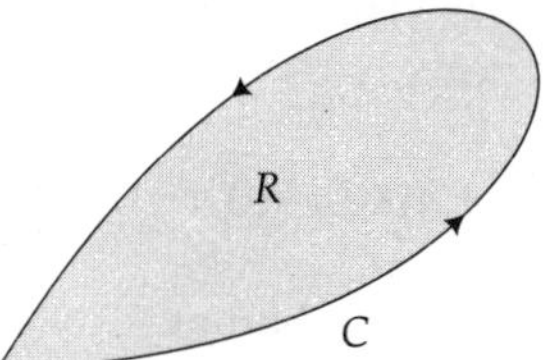

Fig. 11.13

orient the boundary curve C in two directions. Suppose we orient the boundary curve with a direction such that when one goes along the curve in that direction the region is always to left of the person then the curve is said to be ***positively oriented*** (as shown in Fig. 11.13). Now suppose we consider a regions with holes as in Fig. 11.14. Here we see that the region E has two boundary curves C_1 & C_2

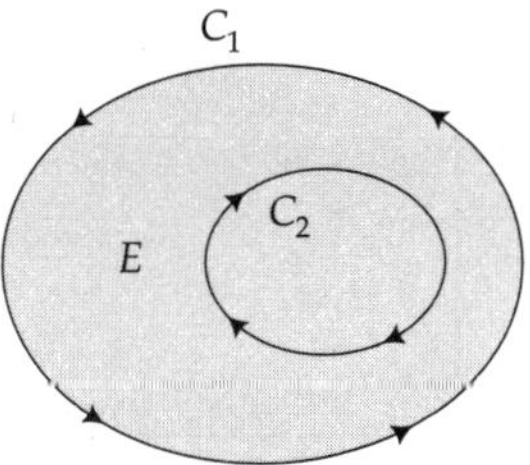

Fig. 11.14

corresponding to the outer and inner boundary respectively. Note that in order that the boundary of this region is positively oriented C_1 & C_2 must be given opposite orientations.

11.7   LINE INTEGRALS

We now introduce the concept of integral of a function which is defined on a curve on plane or space which is called a line integral or sometimes a path integral

Definition   Let $\vec{c}(t)$: $[a, b] \to \mathbb{R}^2$ $(\mathbb{R}^3)$ be a smooth plane (space) curve with graph C and let $\vec{F}(t)$ be a vector field defined and bounded on the curve C (that is for every $t \in [a, b]$ the function $\vec{F}(\vec{c}(t))$ is defined and bounded) then the line integral of $\vec{F}$ along the curve c denoted by $\int_C \vec{F} \cdot d\vec{c}$ is defined by the equation,

$$\int_C \vec{F} \cdot d\vec{c} = \int_a^b \vec{F}(\vec{c}(t)) \cdot \vec{c}(t)\, dt \tag{11.7.1}$$

whenever the integral on the right exists.

If $\vec{c}(t)$ is a smooth space curve with components $(x(t), y(t), z(t))$ $t \in [a, b]$ and suppose $\vec{F} = (f, g, h)$ be a vector field defined and bounded on the curve then the Line integral of $\vec{F}$ along the curve according to the definition is given by,

$$\int_C \vec{F} \cdot d\vec{c} = \int_a^b [\{f(x(t), y(t), z(t)), g(x(t), y(t), z(t)), h(x(t), y(t), z(t))\} \cdot \{x'(t), y'(t), z'(t)\}]\, dt$$

$$= \int_a^b [(f(x(t), y(t), z(t))\, x'(t) + g(x(t), y(t), z(t))\, y'(t) + (h(x(t), y(t), z(t))\, z'(t)]\, dt$$

The above expression is frequently written in short as,

$$\int_C \vec{F} \cdot d\vec{c} = \int_C f dx + g dy + h dz$$

Example 11.12   Find the line integral of $\vec{F} = (xy, y^2)$ along the parabola $x = y^2$ from $A(1, -1)$ to $B(1, 1)$.

Sol: Here the curve is given by, $\vec{c}(t) = (t^2, t)$ $t \in [-1, 1]$

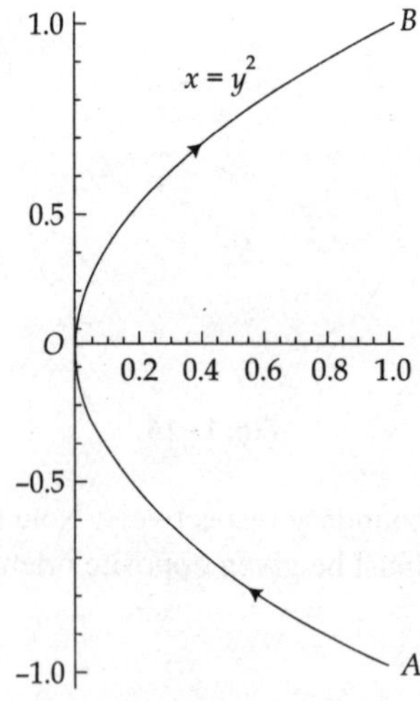

Fig. 11.15

So that $\vec{c}'(t) = (2t, 1)$. Now (11.7.1) becomes,

$$\int_C \vec{F} \cdot d\vec{c} = \int_{-1}^{1} (t^3, t^2) \cdot (2t, 1)\, dt$$

$$= \int_{-1}^{1} (2t^4 + t^2)\, dt = \frac{22}{15}$$

Example 11.13   Find the line integral of $\vec{F} = (x, xy, xz - y)$ over the line segment from $(0, 0, 0)$ to $(1, 2, 4)$.

Sol: We know that the equation of the line segment from $(0, 0, 0)$ to $(1, 2, 4)$ is given by

$$\frac{x - 0}{1} = \frac{y - 0}{2} = \frac{z - 0}{4}$$

hence the curve which is a line segment here is given by $\vec{c}(t) = (t, 2t, 4t)\ t \in [0, 1]$

We get here $\vec{c}'(t) = (1, 2, 4)$ so that,

$$\int_C \vec{F} \cdot d\vec{c} = \int_{-1}^{1} (t, 2t^2, 4t^2 - 2t) \cdot (1, 2, 4)\, dt$$

$$= \int_{-1}^{1} (t + 4t^2 + 16t^2 - 8t)\, dt$$

$$= \int_{-1}^{1} (20t^2 - 7t)\, dt = \frac{19}{6}$$

Example 11.14   Evaluate $\int_C (x^2 - y^2)\, dx + 2x\, dy$

where C is the circle $x^2 + y^2 = a^2$ with positive orientation.

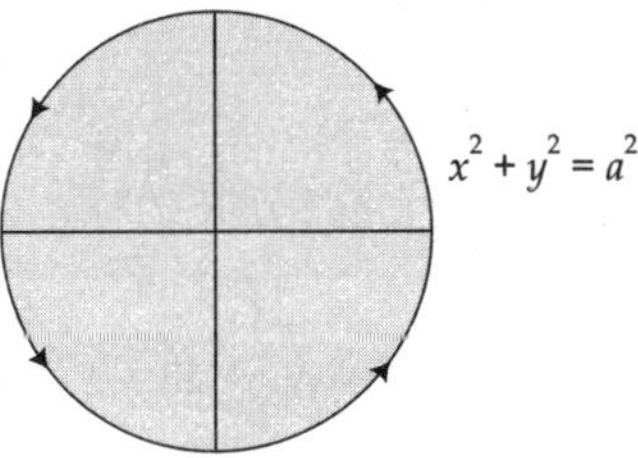

Fig. 11.16

Sol: Here the parameterization of the given circle with positive orientation is given by,

$$\left.\begin{array}{l} x = a \cos \theta \\ y = a \sin \theta \end{array}\right\}\quad 0 \le \theta \le 2\pi$$

Then given line integral reduces to,

$$\int_0^{2\pi} [(\cos^2 \theta - \sin^2 \theta)\,(-a \sin \theta) + (2 \cos \theta)\,(a \cos \theta)]\, d\theta$$

$$= a \int_0^{2\pi} (2 \cos^2 \theta - \cos^2 \theta \sin \theta + \sin^3 \theta)\, d\theta = 2a\pi$$

Example 11.15   Evaluate $\int_C \vec{F} \cdot d\vec{c}$

where $\vec{F} = (\cos z,\, e^x,\, e^{2y})$ and $\vec{c} = (1,\, t,\, e^{2t})\, t \in [0, 4]$

Sol:
$$\int_C \vec{F} \cdot d\vec{c} = \int_0^4 (\cos e^{2t},\, e^1,\, e^{2e^t}) \cdot (0,\, 1,\, 2e^{2t})\, dt$$

$$= \int_0^4 (edt + 2e^{2t}\, e^{2e^t})\, dt$$

$$= 4e + \frac{1}{2}\,[-e^2 + e^{2e^4}\,(-1 + 2e^4)]$$

Definition   Suppose $\vec{c}(t)$ is a smooth curve and $s(\xi)$ denote its arc length function and let f be any scalar field defined and bounded on the curve. We define the line integral of f with respect to the arc length (denoted by $\int_C fds$) as,

$$\int_C fds = \int_a^b f(\vec{c}(t))\, s'(t)\, dt \tag{11.7.2}$$

whenever the equation on the right exists.

Note that by equation (11.5.1.2),

$$s'(t) = \|c'(t)\|$$

hence (11.7.2) can be written as,

$$\int_C fds = \int_a^b f(\vec{c}(t))\, \|\vec{c}(t)\|\, dt \tag{11.7.3}$$

Example 11.16   Evaluate $\int_C yds$ where the curve $\vec{c}$ is given by, $\vec{c}(t) = (a(1 - \sin t),\, a(t - \cos t))t \in [0, 2\pi]$

Sol: Here
$$s'(t) = \|\vec{c}'(t)\| = a \sqrt{(-\cos t)^2 + (1 + \sin t)^2}$$

$$= a \sqrt{2 + 2 \sin t}$$

Hence,

$$\int_C y\, ds = \int_0^{2\pi} a(t - \cos t)\, a \sqrt{2 + 2 \sin t}\, dt$$

$$= \sqrt{2}a \int_0^{2\pi} (t - \cos t) \sqrt{1 + \sin t} \; dt$$

$$= 4\pi \left(3 - \sqrt{2}\right) a^2$$

Example 11.17 Evaluate $\int_C \sqrt{(x^2 + y^2 + z)} \; ds$ where the curve $\vec{c}$ is given by

$$\vec{c}(t) = (t \sin t, \; t \cos t, \; t^2) \; t \in [0, \pi]$$

Sol: As before we calculate

$$s'(t) = \|\vec{c}'(t)\| = \sqrt{(\sin t + t \cos t)^2 + (\cos t - t \sin t)^2 + 4t^2}$$

$$= \sqrt{1 + t^2 + 4t^2} = \sqrt{1 + 5t^2}$$

Now,

$$\int_C \sqrt{(x^2 + y^2 + z)} \; ds = \int_0^{\pi} \sqrt{(t^2 \sin^2 t + t^2 \cos^2 t + t^2)} \; \sqrt{1 + 5t^2} \; dt$$

$$\int_0^{\pi} \sqrt{2}t \, \sqrt{1 + 5t^2} \; dt = \frac{\sqrt{2}}{15} \left[(1 + 5\pi^2)^{3/2} - 1\right]$$

11.8 SOME PROPERTIES OF LINE INTEGRALS

As the line integral is defined in terms of ordinary integral we expect that line integrals should have same type of properties that an ordinary integrals possesses. For example we know that an ordinary integral is a linear transformation that is,

$$\int_a^b (af + bg) \; dx = a \int_a^b f \, dx + b \int_a^b g \, dx$$

the corresponding property of line integral is given by,

$$\int_C (a\vec{F} + b\vec{G}) \cdot d\vec{c} = a \int_C \vec{F} \cdot d\vec{c} + b \int_C \vec{G} \cdot d\vec{c} \qquad (11.8.1)$$

where $\vec{F}$ and $\vec{G}$ are vector fields and $\vec{c}$ is a path.

Again suppose $\vec{c}_1(t) \; t \in [a, c]$ & $\vec{c}_2(t), \; t \in [c, b]$ be two Jordan smooth curves 'joined', at c that is $\vec{c}_1(c) = \vec{c}_2(c)$. We denote the joined curve by $\vec{c}(t)$, that is $\vec{c}(t) = \vec{c}_1(t) \cup \vec{c}_2(t), \; t \in [a, b]$ then,

$$\int_C \vec{F} \cdot d\vec{c} = \int_{C_1} \vec{F} \cdot d\vec{c} + \int_{C_2} \vec{F} \cdot d\vec{c} \qquad (11.8.2)$$

where $\vec{F}$ is a vector field. Note that the above property is similar to the property of the ordinary integral given by Theorem 3.9.8. The above two properties follows as a direct consequence of equation (11.7.1) which expresses line integral as a ordinary integral. In the following theorem we look into a far more important property of the line integral which says that numerical value of a line integral over two equivalent curves of a vector field remains unchanged.

Theorem 11.8.1

Let $\vec{c}_1(t)$, $t \in [a, b]$ and $\vec{c}_2(\theta)$, $\theta \in [c, d]$ be two equivalent smooth curves that is, exists a differentiable onto function $h : [c, d] \to [a, b]$ such that $h'(\theta) \neq 0$ $\theta \in [c, d]$ and $\vec{c}_2(\theta) = \vec{c}_1(h(\theta))$ suppose C denotes the common graph of the curves. Then for a vector function $\vec{F}$ if $\int_C \vec{F} \cdot d\vec{c}_1$ and $\int_C \vec{F} \cdot d\vec{c}_2$ exists we have,

$$\int_C \vec{F} \cdot d\vec{c}_1 = \pm \int_C \vec{F} \cdot d\vec{c}_2$$

where $+$ sign is taken if $h(\theta)$ is orientation preserving and $-$ sign is used if $h(\theta)$ is orientation reversing.

Proof: Note that if $h'(\theta) \neq 0$ then by Theorem () either $h'(\theta) > 0$ or $h'(\theta) < 0$ that is either $h(\theta)$ is monotone increasing or monotone decreasing respectively. Now using chain rule we have,

$$\vec{c}_2(\theta) = \vec{c}_1'(h(\theta))\, h'(\theta) \tag{11.8.1.1}$$

Now from definition we have,

$$\int_C \vec{F} \cdot d\vec{c}_2 = \int_c^d \vec{F}(c_2(\theta)) \cdot (\vec{c}_2'(\theta))\, d\theta$$

$$= \int_c^d [\vec{F}(\vec{c}_1(h(\theta)))] \cdot [\vec{c}_1'(h(\theta))\, h'(\theta)]\, d\theta \quad \text{using (11.8.1.1)}$$

& (11.8.1.2). We substitute in the right of above equation $u = h(\theta)$ then $du = h'(\theta)\, d\theta$ and if we assume that $h(\theta)$ is order preserving then, we have $h(c) = a$ and $h(d) = b$, in that case we get,

$$\int_C \vec{F} \cdot d\vec{c}_2 = \int_C \vec{F} \cdot d\vec{c}_2 = \int_a^b \vec{F}(\vec{c}_1(u))] \cdot [\vec{c}_1(u)]\, du = \int_C \vec{F} \cdot dc_1$$

In case $h(\theta)$ is order reversing we have $h(c) = b$ & $h(d) = a$ and in that case by substituting $u = h(\theta)$ in (11.8.1.2) we get,

$$\int_C \vec{F} \cdot d\vec{c}_2 = \int_a^b \vec{F}(\vec{c}_1(u))] \cdot [\vec{c}_1'(u)]\, du$$

$$= [\vec{c}_1'(u)]\, du = -\int_C \vec{F} \cdot d\vec{c}_1$$

Remark

The above theorem can be easily extended to a piecewise smooth equivalent curves in which case the above theorem can be applied for each of its smooth component and then finally use additive property of line integrals (11.8.2).

Example 11.18   Evaluate $\int_C xy\, dx + (x + y^2)\, dy$ where $\vec{c}$ consists of the straight lines OA and AB joining points $0(0, 0)$, $A(2, 0)$ & $B(2, 4)$ as shown in the figure.

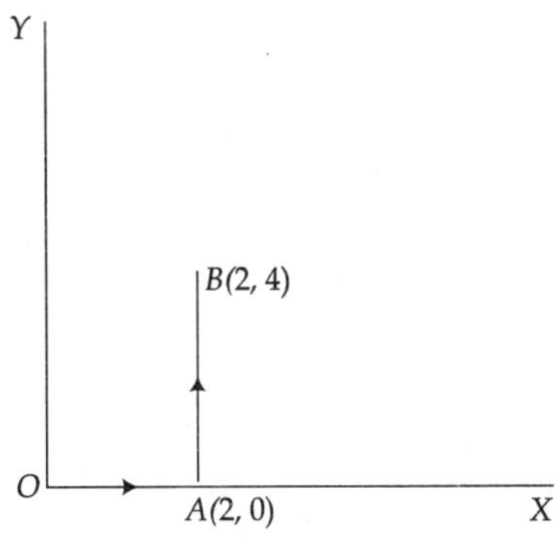

Fig. 11.17

Sol: We can see that $C = \overset{\frown}{OA} \cup \overset{\frown}{AB}$. Using (11.8.2) we obtain

$$\int_C xy\,dx + x\,dy = \int_{\overset{\frown}{OA}} xy\,dx + (x + y^2)\,dy + \int_{\overset{\frown}{AB}} xy\,dx + (x + y^2)\,dy$$

Now see that on $\overset{\frown}{OA}$ $y = 0$ hence,

$$\int_{\overset{\frown}{OA}} xy\,dx + (x + y^2)\,dy = 0$$

Again on $\overset{\frown}{AB}$ $x = 2$ and the limit of the variable y is from $y = 0$ to $y = 4$ hence,

$$\int_{\overset{\frown}{AB}} xy\,dx + (x + y^2)\,dy = 0 + \int_0^4 (2 + y^2)\,dy = \frac{88}{3}$$

Example 11.19 Find the line integral of the vector field

$$F = (x^2y^2, x^2 - y^2)$$

over the closed curve formed by the parts of the line $x = 1$ and the parabola $y^2 = x$ in the clockwise direction.

Clearly the curve C on which the line is calculated comprises of the arc $\overset{\frown}{AOB}$ of the parabola $y^2 = x$ and the straight line $\overline{BA}$ hence using (11.8.2) we get,

$$\int_C \vec{F}\cdot d\vec{c} = \int_C (x^2y^2)\,dx + (x^2 - y^2)\,dy$$

$$= \int_{\overset{\frown}{AOB}} (x^2y^2)\,dx + (x^2 - y^2)\,dy + (x^2y^2)\,dx + \int_{\overline{BA}} (x^2y^2)\,dx + (x^2 - y^2)\,dy$$

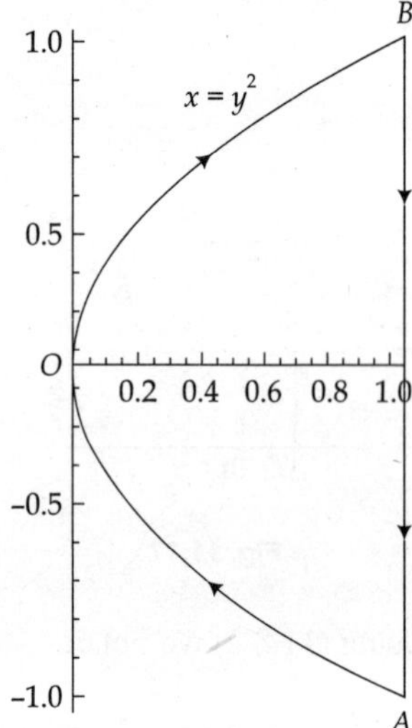

Fig. 11.18

We have the of $\widehat{AOB}$ as $x = t^2$ & $y = t, t \in [-1, 1]$ hence,

$$\int_{\widehat{AOB}} (x^2 y^2)\, dx + (x^2 - y^2)\, dy = \int_{-1}^{1} [(t^4 \cdot t^2)\, 2t\, dt + (t^4 - t^2)\, dt$$

$$= \int_{-1}^{1} (2t^7 + t^4 - t^2)\, dt = -\frac{4}{15} \tag{1}$$

Again on $\overline{BA}$ we have $x = 1$ and limit of y changes from $y = 1$ to $y = -1$ so that we get,

$$\int_{\overline{BA}} (x^2\, y^2)\, dx + (x^2 - y^2)\, dy = 0 + \int_{1}^{-1} (1 - y^2)\, dy = -\frac{4}{3} \tag{2}$$

Adding equations (1) & (2) we have,

$$\int_{C} (x^2 y^2)\, dx + (x^2 - y^2)\, dy = -\frac{4}{15} - \frac{4}{3} = -\frac{24}{15}$$

Example 11.20   Find the line integral $\int_{C} y\, dx - x\, dy$ where C is

(i) arc BA of the circle $x = \cos\theta\ y = \sin\theta,\ \theta \in \left[0, \dfrac{\pi}{2}\right]$

(ii) arc AB of the circle $x = \cos\theta\ y = \sin\theta,\ \theta \in \left[0, \dfrac{\pi}{2}\right]$

(i) Here $\vec{F} = (y, -x)$ and $\vec{c} = (\cos\theta, \sin\theta)$ hence the given line integral can be written as,

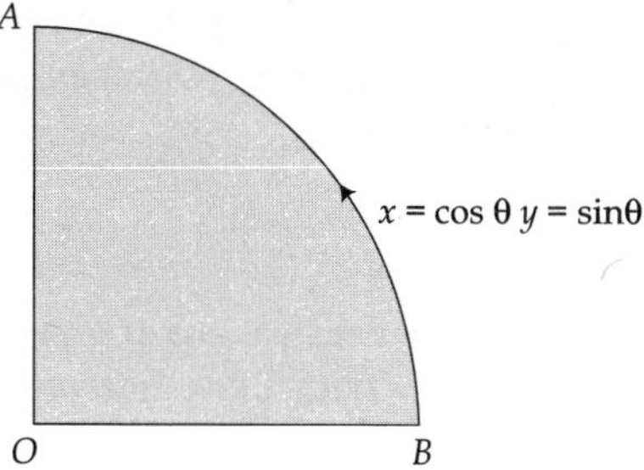

Fig. 11.19

$$\int_C \vec{F} \cdot d\vec{c} = \int_0^{\frac{\pi}{2}} [\sin\theta\,(-\sin\theta) - \cos\theta\,(\cos\theta)]\,d\theta$$

$$= -\frac{\pi}{2}$$

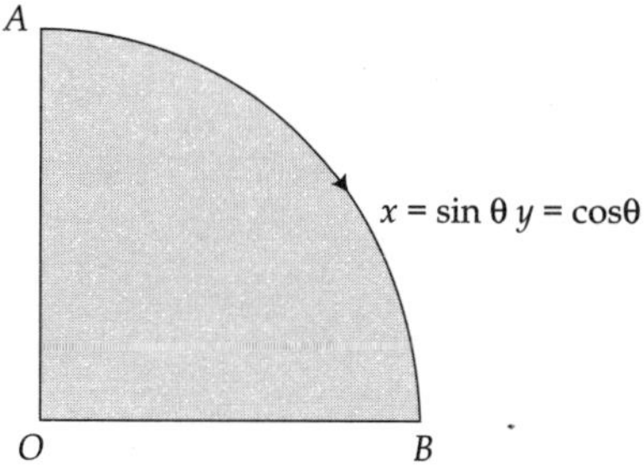

Fig. 11.20

(ii) As in the previous example we have $\vec{F} = (y, -x)$ and $\vec{c} = (\sin\theta, \cos\theta)$

$$\int_C \vec{F} \cdot d\vec{c} = \int_0^{\frac{\pi}{2}} [\cos\theta\,(\cos\theta) - \sin\theta\,(-\sin\theta)]\,d\theta$$

$$= \frac{\pi}{2}$$

Observe that as the direction of the two paths in the above two direction are in opposite direction therefore the value of the line integrals has opposite signs as expected from Theorem 11.5.1 ($\vec{F}$ being same in both the cases).

11.9 GREEN'S THEOREM IN PLANE

We recall that fundamental theorem of integral calculus (Theorem 3.10.10) tells us that if f is a function defined on an interval $[a, b]$ such that f' exists then,

$$\int_a^b F'(x)\,dx = F(b) - F(a).$$

Now the boundary of the interval $[a, b]$ consists of two point set $C = \{a, b\}$. We now orient the interval such that the direction from a to b is taken to be positive. Now the line integral of F over C is the vector sum of the values of F on the two points a & b that is,

$$\int_C F' = F(b) - F(a)$$

Notice that the value of F at a is taken to be negative and at b to be positive as the direction at those points are inward and outward to the interval. With above interpretation the fundamental theorem of integral calculus takes the form,

$$\int_C F = \int_{[a,\,b]} F'$$

The two dimensional analogue of the above theorem is precisely what is called *Green's Theorem In Plane.*

Theorem 11.9.1 (Green's Theorem)

Let R' be a region in $\mathbb{R}^2$ bounded by a smooth closed Jordan curve C. If F_1 and F_2 be two continuously differentiable functions defined on R which the union of R' and C then we have,

$$\int_C F_1\, dx + F_2\, dy = \iint_R \left(\frac{\partial F_2}{\partial x} - \frac{\partial F_1}{\partial y} \right) dx\, dy \tag{11.9.1}$$

Proof: To prove (11.9.1) it is enough to show,

$$\left. \begin{aligned} \iint_R \frac{\partial F_2}{\partial x}\, dx\, dy &= \int_C F_2\, dy \\[2em] \iint_R \frac{\partial F_1}{\partial y}\, dx\, dy &= -\int_C F_1\, dx \end{aligned} \right\} \tag{11.9.2}$$

and

In fact it is easy to see that (11.9.1) & (11.9.2) are equivalent. Note that (11.9.2) implies (11.9.1) and if (11.9.1) is true then by putting $P = 0$ or $Q = 0$ we get (11.9.2). We now prove (11.9.2) for a region R which is Type III (as shown in Fig). As R is Type III we can write,

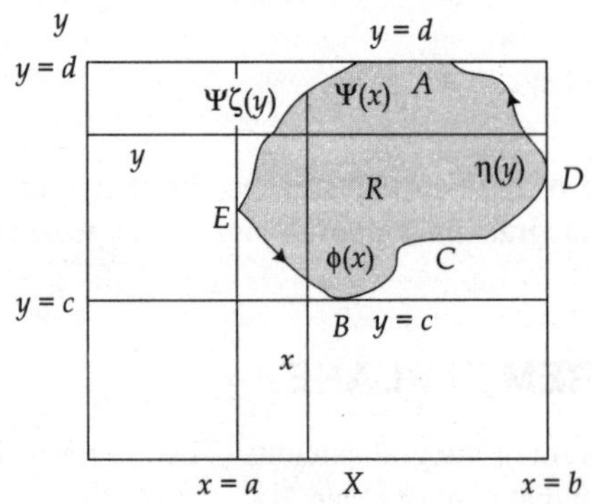

Fig. 11.21

$R = \{(x, y) : y \in [c, d], \zeta(y) \leq x \leq \eta(y)\} = \{(x, y) : x \in [a, b], \phi(x) \leq y \leq \psi(x)\}$. That is the portions *EBD* & *DAE* of the curve is represented by the equations $y = \phi(x)$ & $y = \psi(x)$ respectively. Similarly the parts *BDA* and *AEB* of the curve is represented by the equations $x = \eta(y)$ & respectively. Now we evaluate the double integral $\iint\limits_R \dfrac{\partial F_2}{\partial x}\, dx\, dy$ in terms of repeated integrals as explained in the section 10.7 so that,

$$\iint\limits_R \frac{\partial F_2}{\partial x}\, dxdy = \int\limits_c^d dy \int\limits_{\zeta(y)}^{\eta(y)} \frac{\partial F_2}{\partial x}\, dx$$

$$= \int\limits_c^d (F_2(\eta(y), y) - F_2(\zeta(y), y))\, dy$$

$$= \int\limits_c^d F_2(\eta(y), y)\, dy - \int\limits_c^d F_2(\zeta(y), y))\, dy$$

$$= \int\limits_c^d F_2(\eta(y), y)\, dy + \int\limits_c^d F_2(\zeta(y), y))\, dy \tag{11.9.3}$$

On the other hand we have,

$$\int\limits_C F_2\, dy = \int\limits_{C_1} F_2\, dy + \int\limits_{C_2} F_2\, dy \tag{11.9.4}$$

where C_1 & C_2 denotes the curves $\overparen{BDA}$ & $\overparen{AEB}$. Now the parametric equation of C_1 is given by

$$\vec{c}_1(t) = (\eta(t), t)\ c \leq t \leq d$$

so that,

$$\int\limits_{C_1} F_2\, dy = \int\limits_c^d F_2(\eta(t), t)\, \frac{dy}{dt}\, dt \quad \text{(as } y = t \text{ on the curve)}$$

$$= \int\limits_c^d F_2(\eta(t), t)\, dt$$

$$= \int\limits_c^d F_2(\eta(y), y)\, dy \quad \text{(changing the dummy variable)} \tag{11.9.5}$$

For the portion $\overparen{AEB}$ of the curve we note that we start from the point $(\zeta(d), d)$ and end at the point $(\zeta(c), c)$ hence its parametric equation is given by,

$$\vec{c}_2(t) = (\zeta(c + d - t), c + d - t)\ c \leq t \leq d \text{ hence,}$$

$$\int\limits_{C_2} F_2\, dy = \int\limits_c^d F_2(\zeta(c + d - t), c + d - t)\, \frac{d}{dt}\, (c + d - t)\, dt$$

$$= \int\limits_c^d F_2(\zeta(c + d - t), c + d - t)\, (-1)\, dt$$

$$= \int_d^c F_2(\zeta(u), u)\, du \quad \text{(substituting } u = c + d - t)$$

$$= \int_d^c F_2(\zeta(y), y)\, dy \quad \text{(changing the dummy variable)} \tag{11.9.6}$$

From (11.6.5), (11.6.6) we get,

$$\int_C F_2\, dy = \int_c^d F_2(\eta(y), y)\, dy + \int_d^c F_2(\zeta(y), y)\, dy \tag{11.9.7}$$

Comparing (11.6.3) and (11.6.7) we conclude,

$$\iint_R \frac{\partial F_2}{\partial x}\, dxdy = \int_C F_2\, dy$$

To complete the proof we now show that,

$$\iint_R \frac{\partial F_1}{\partial y}\, dxdy = -\int_C F_1\, dx$$

Again as before we first evaluate the double integral as repeated integrals integrating with respect to y first and then with respect to x as follows,

$$\iint_R \frac{\partial F_1}{\partial y}\, dxdy = \int_a^b dx \int_{\phi(x)}^{\psi(x)} \frac{\partial F_1}{\partial y}\, dy$$

$$= \int_a^b (F_1(x, \psi(x)) - F_1(x, \phi(x)))\, dx$$

$$= -\int_a^b F_1(x, \phi(x))\, dx - \int_b^a F_1(x, \psi(x))\, dx \tag{11.9.8}$$

Now,

$$\int_C F_1\, dx = \int_{C_3} F_1\, dx + \int_{C_4} F_1\, dx$$

where C_3 & C_4 denotes the curves $\widehat{EBD}$ & $\widehat{DAE}$ respectively. As before the parametric representation of $\widehat{EBD}$ & $\widehat{DAE}$ are given by,

$$\vec{c}_3(t) = (t, \phi(t))\ a \leq t \leq b$$
$$\vec{c}_4(t) = (a + b - t, \phi(a + b - t))\ a \leq t \leq b$$

Therefore,

$$\int_{C_3} F_1\, dx = \int_a^b F_1(t, \theta(t)) \frac{dx}{dt}\, dt$$

$$= \int_a^b F_1(t, \phi(t))\, dt$$

$$= \int_a^b F_1(x, \phi(x))\, dx \quad \text{(changing the dummy variable)} \tag{11.9.9}$$

Also,

$$\int\limits_{C_4} F_1\, dx = \int\limits_{a}^{b} F_1\left(a+b-t,\ \psi(a+b-t)\right)\frac{d}{dt}\left(a+b-t\right)dt$$

$$= \int\limits_{a}^{b} F_1(a+b-t,\ \psi(a+b-t))\ (-1)\, dt$$

$$= \int\limits_{b}^{a} F_1(u,\ \psi(u))\, du \quad \text{(putting } 4 = a+b-t)$$

$$= \int\limits_{b}^{a} F_1(t,\ \psi(t))\, dt \quad \text{(changing the dummy variable)} \tag{11.9.10}$$

Using (11.9.8), (11.9.9) & (11.9.10) we get

$$\iint\limits_{R} \frac{\partial F_1}{\partial y}\, dxdy = -\int\limits_{C} F_1\, dx$$

Hence the theorem is proved for Type III regions.

We can now prove the above theorem for regions which can be decomposed into finite number of Type III regions. For these regions we introduce what are called 'crosscuts' to decompose the region into sub regions each of which are of Type III. We then apply Green's Theorem to each of these regions and add them. We note that the line integrals along the crosscuts cancel since each such cut is traversed twice in opposite direction (as shown in Fig. 23)

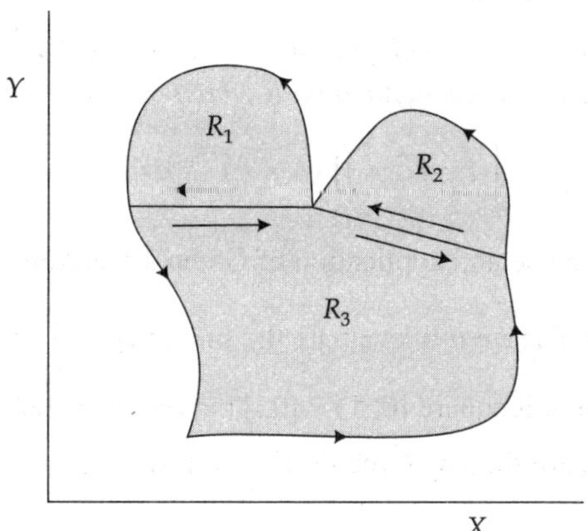

Fig. 11.22

and the double integrals over the sub regions add up to the double integral over the whole region.

Finally we extend Green's theorem to multiply connected region by using the technique as above. We join the external boundary of such regions with the 'holes' by introducing 'crosscuts' and then apply Green's theorem on the closed curve so formed. As before line integrals along the cross cuts will cancel as they are traversed twice in opposite directions (as shown in Fig. 24)

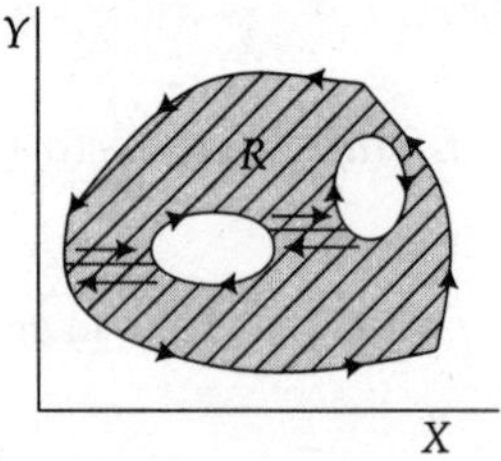

Fig. 11.23

and we will be left with the line integral on the boundary of the region only and the double integral will on the whole region.

We observe that if $\vec{F} = P\vec{i} + Q\vec{j}$ be a vector field and $\vec{c}$ be a smooth curve on $\mathbb{R}^2$ then,

$$\int_C \vec{F}\cdot d\vec{c} = \int_C Pdx + Qdy \text{ also } (\Delta \times \vec{F}) = \begin{vmatrix} \vec{i} & \vec{j} & \vec{k} \\ \dfrac{\partial}{\partial x} & \dfrac{\partial}{\partial y} & \dfrac{\partial}{\partial z} \\ P & Q & 0 \end{vmatrix} = \left(\dfrac{\partial Q}{\partial x} - \dfrac{\partial P}{\partial y}\right)\vec{k} \text{ so that,}$$

$(\Delta \times \vec{F}) \cdot k = \dfrac{\partial Q}{\partial x} - \dfrac{\partial P}{\partial y}$. Hence we get vector form of Green's theorem as follows:

Vector Form of Green's Theorem

Let R' be a region in $\mathbb{R}^2$ bounded by a smooth closed Jordan curve C. If $\vec{F} = P\vec{i} + Q\vec{j}$ where P and Q are two continuously differentiable functions defined on R which the union of R' and C then we have,

$$\int_C \vec{F}\cdot d\vec{c} = \iint_R (\Delta \times \vec{F}) \cdot \vec{k}\, dx\, dy$$

We will now see some of the use and application of Green's Theorem.

Example 11.21   Use Green's Theorem to evaluate the line integral $\int_C (6 - 2xy - y^3)\, dx + (x^2 y - x^3)\, dy$ where C is the boundary of the unit square $[0, 1] \times [0, 1]$ in the plane taken in anti clock wise direction.

Sol: If R denotes the unit square then by Green's Theorem we get,

$$\int_C (6 - 2xy - y^3)\, dx + (x^2 y - x^3)\, dy = \iint_R \frac{\partial}{\partial x}(x^2 y - x^3) - \frac{\partial}{\partial y}(6 - 2xy - y^3)]\, dxdy$$

$$= \int_0^1 \int_0^1 (2xy - 3x^2 + 2x + 3y^2)\, dxdy$$

$$= \int_0^1 2x\, dx \int_0^1 y\, dy - \int_0^1 3x^2\, dx \int_0^1 dy + \int_0^1 2x\, dx \int_0^1 dy + \int_0^1 3x^2\, dy \int_0^1 dx$$

$$= \frac{3}{2}$$

Example 11.22   Verify Green's Theorem for the vector field $\vec{F} = (-y, x)$ defined on the rectangular region with vertices $(0, 0)$, $(3, 0)$, $(3, 2)$, $(0, 2)$.

Sol: The boundary C of the rectangular region is positively oriented as shown below. If we apply

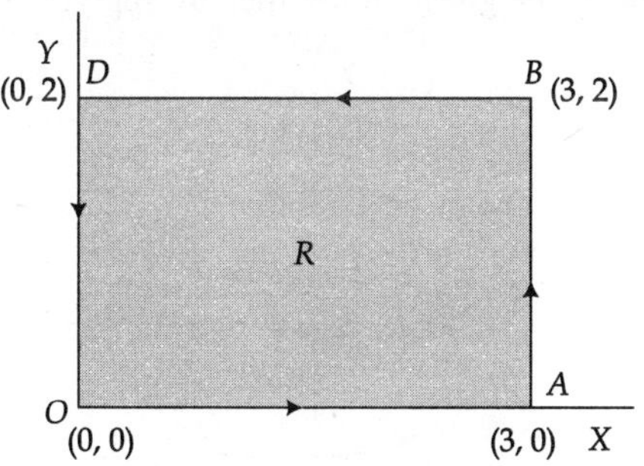

Fig. 11.24

Green's Theorem on the vector field $\vec{F} = (-y, x)$ over the region R we get,

$$\int_C -ydx + xdy = \iint_R \left(\frac{\partial}{\partial x}(x) - \frac{\partial}{\partial y}(-y) \right) dxdy$$

$$= \iint_R 2\, dxdy = 2 \text{ (Area of the rectangle)} = 2 \times 6 = 12.$$

We now calculate the line integral $\int_C -ydx + xdy$. For that observe that, C consists of the directed line segments $\overline{OA}, \overline{AB}, \overline{BD}$ & $\overline{DO}$. Hence,

$$\int_C -ydx + xdy = \int_{\overline{OA}} -ydx + xdy + \int_{\overline{AB}} -ydx + xdy + \int_{\overline{BD}} -ydx + xdy + \int_{\overline{DO}} -ydx + xdy.$$

Now parametric equations of $\overline{OA}, \overline{AB}, \overline{BD}$ & $\overline{DO}$ are given by $y = 0, x = t, 0 \le t \le 3$; $x = 3, y = t, 0 \le t \le 2$, $y = 2, x = 3 - t, 0 \le t \le 3$ & $x = 0, y = 2 - t, 0 \le t \le 2$. So that we have,

$$\int_{\overline{OA}} -ydx + xdy = \int_{\overline{OA}} \left(-y\frac{dx}{dt} + x\frac{dy}{dt} \right) dt = 0 \quad \left(\text{putting } y = 0 \And \frac{dy}{dt} = 0 \right)$$

$$\int_{\overline{AB}} -ydx + xdy = \int_0^2 3\, dt = 6 \quad \left(\text{putting } x = 3, \frac{dx}{dt} = 0 \right)$$

$$\int_{\overline{BD}} -ydx + xdy = \int_0^3 -2\frac{dx}{dt}\, dt = \int_0^3 -2(-1)\, dt = 6 \quad \left(\text{putting } y = 2 \text{ and } \frac{dx}{dt} = -1 \text{ as } x = 3 - t \right)$$

$$\int_{\overline{DO}} -ydx + xdy = \int_{\overline{DO}} \left(-y\frac{dx}{dy} + x\frac{dy}{dt} \right) dt = 0 \quad \left(\text{As } x = 0 \And \frac{dx}{dt} = 0 \right)$$

Adding above integrals we obtain $\int_C -ydx + xdy = 12$ hence Green's Theorem is verified.

Example 11.23 Using Green's Theorem evaluate $\int\limits_C y^2\,dx + 2x\,dy$ where C is the ellipse $\dfrac{x^2}{a^2} + \dfrac{y^2}{b^2} = 1$ oriented in anti clockwise direction.

Sol: If R is the region enclosed by the given ellipse then by applying Green's Theorem we get,

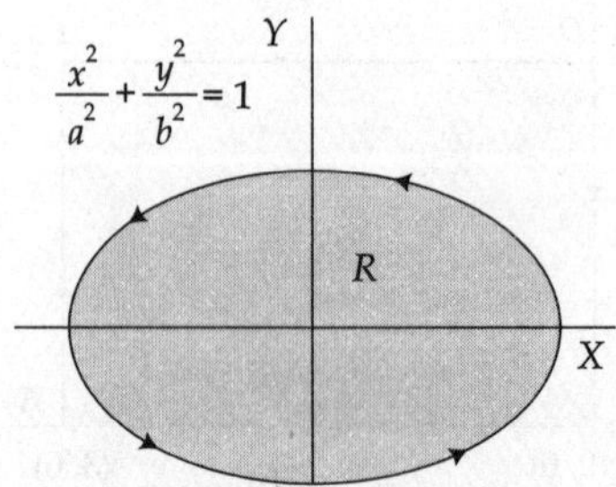

Fig. 11. 25

$$\int\limits_C y^2\,dx + 2x\,dy = \iint\limits_R \left(\frac{\partial}{\partial x}(2x) - \frac{\partial}{\partial y}(y^2) \right) dxdy$$

$$= 2 \iint\limits_R (1 - y)\,dxdy$$

We now evaluate the double integral $\iint\limits_R (1 - y)\,dxdy$. For that we first substitute $x = au$, $y = bv$ then the Jacobian of the transformation is $J = ab$ so that,

$$\iint\limits_R (1 - y)\,dxdy = ab \iint\limits_R (1 - bv)\,dudv \quad \text{where } R' \text{ is the region } u^2 + v^2 \le 1$$

$$= ab \int\limits_{r=0}^{1} \int\limits_{\theta=0}^{2\pi} (1 - br\sin\theta)\,rdrd\theta$$

$$= \pi ab \quad \text{(substituting } u = r\cos\theta \ \& \ v = r\sin\theta)$$

Example 11.24 Let R be the region enclosed by a smooth curve C. If A is the area of R then show that

$$A = \frac{1}{2}\int\limits_C xdy - ydx = \int\limits_C x\,dy$$

Hence find the area of the ellipse $\dfrac{x^2}{a^2} + \dfrac{y^2}{b^2} = 1$.

Sol: By simple application of Green's Theorem we get,

$$\int\limits_C xdy - ydx = \iint\limits_R \left(\frac{\partial}{\partial x}(x) - \frac{\partial}{\partial y}(-y) \right) dxdy$$

$$= 2 \iint\limits_R dxdy = 2A \qquad \left(\text{as we know } \iint\limits_R dxdy \text{ gives the area of the region } R \right)$$

That is,

$$A = \frac{1}{2} \int_C x\,dy - y\,dx$$

Again by Green's Theorem,

$$\int_C x\,dy = \iint_R \frac{\partial}{\partial x}(x)\,dx\,dy = \iint_R dx\,dy = A.$$

So that we get the formula,

$$A = \frac{1}{2} \int_C x\,dy - y\,dx = \int_C x\,dy$$

We now evaluate $\int_C x\,dy - y\,dx$ where C is the ellipse $\dfrac{x^2}{a^2} + \dfrac{y^2}{b^2} = 1$. For that we use the parametric equation of the ellipse as,

$$x = a\cos\theta,\ y = b\sin\theta, \quad 0 \le \theta \le 2\pi$$

Then we have,

$$\int_C x\,dy - y\,dx = \int_0^{2\pi} \left(a\cos\theta\,\frac{dy}{d\theta} - b\sin\theta\,\frac{dx}{d\theta} \right) d\theta$$

$$= ab \int_0^{2\pi} \cos^2\theta + \sin^2\theta\,d\theta = 2\pi ab$$

Hence the area of the ellipse is given by $\dfrac{1}{2}\int_C x\,dy - y\,dx = \pi ab$

Also observe that $\int_C x\,dy = \pi ab$ where

Example 11.25   Verify Green's Theorem for the vector field,

$$F = (y^2 - xy,\ x^2 + y^2)$$

defined over the closed region R bounded by the curves $y^2 = x$ and $y = x^2$.

Sol: We orient the boundary of the region in anti clockwise direction as shown in the figure below. Now applying Green's Theorem we get,

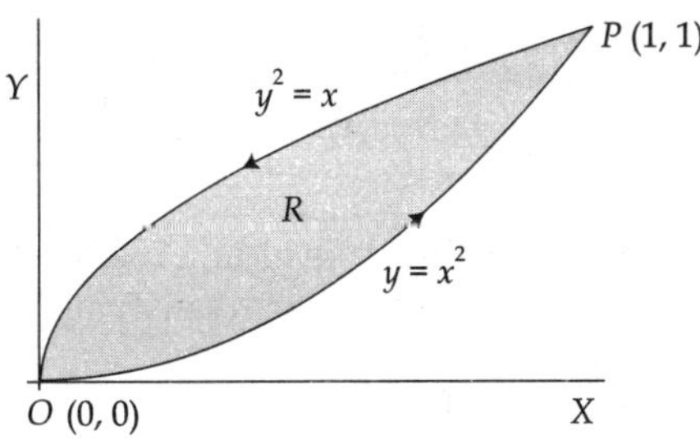

Fig. 11.26

$$\int_C (y - xy)\,dx + (x + y)\,dy = \iint_R \left(\frac{\partial}{\partial x}(x + y) - \frac{\partial}{\partial y}(y - xy) \right) dxdy \qquad \text{(where } C \text{ is the boundary of the region)}$$

$$= \iint_R (1 - 1 + x)\,dxdy$$

$$= \iint_R x\,dxdy$$

We now evaluate both sides and show that they are equal. First we will evaluate the line integral, for that observe that C union of the curves $\overparen{OP}$ and $\overparen{PO}$ whose parametric equations are given by,

$x = t,\ y = t^2\ 0 \le t \le 1$ and $x = (1 - t)^2,\ y = 1 - t\ 0 \le t \le 1$. Hence,

$$\int_{\overparen{OP}} (y - xy)\,dx + (x + y)\,dy = \int_0^1 (t^2 - t^3)\,(1) + (t + t^2)\,2t\,dt$$

$$= \int_0^1 (t^3 + 3t^2)\,dt$$

$$= \frac{5}{4}$$

$$\int_{\overparen{PO}} (y - xy)\,dx + (x + y)\,dy = \int_0^1 \{(1 - t) - (1 - t)^3)\}\,\{(-2)(1 - t)\} + \{(1 - t)^2 + (1 - t)\}\,(-1)\,dt$$

$$= \int_0^1 \{2(1 - t)^4 - 3(1 - t)^2 - (1 - t)\}\,dt$$

$$= -\frac{11}{10}$$

Adding above equations we get,

$$\int_C (y^2 - xy)\,dx + (x^2 + y^2)\,dy = \frac{3}{20}$$

Now,

$$\iint_R x\,dx\,dy = \int_0^1 x\,dx \int_{y=x^2}^{y=\sqrt{x}} dy = \int_0^1 x(\sqrt{x} - x^2)\,dx = \frac{3}{20}$$

Hence Green's Theorem is verified.

Example 11.26   Suppose $\psi(x, y)$ be a function defined in a region R bounded by a smooth curve C having continuous second order partial derivatives and satisfying the partial differential equation $\dfrac{\partial^2 \psi}{\partial x^2} + \dfrac{\partial^2 \psi}{\partial y^2} = 0$. Show that $\int_C \dfrac{\partial \psi}{\partial y}\,dx + \dfrac{\partial \psi}{\partial x}\,dy = 0$.

Sol: By Green's Theorem,

$$\int_C \frac{\partial \psi}{\partial y}\,dx - \frac{\partial \psi}{\partial x}\,dy = \iint_R \left(-\frac{\partial^2 \psi}{\partial x^2} - \frac{\partial^2 \psi}{\partial y^2} \right) dxdy = 0$$

11.10 PARAMETRIC REPRESENTATION OF A SURFACE

We are familiar with the equation $x + y + z - a = 0$ which denotes the simplest surface called 'plane'. If we solve for z from the above equation we get an equivalent representation of the plane as $z = a - x - y$. In general an equation of the form,

$$F(x, y, z) = 0$$

denotes a surface in space and if from above equation it is possible to solve z in terms of the other two variables we can write the equation of a plane as

$$z = f(x, y) \ (x, y) \in R$$

We will be more interested in another representation of a surface called the ***parametric representation*** of surfaces which is more useful in the context of surface integral. Imagine a sheet of paper lying on a table, one lift this sheet in space and give shape of a curved surface! (as shown in the Fig. 11.28) This gives an idea that a surface is formed by deforming some region of $\mathbb{R}^2$.

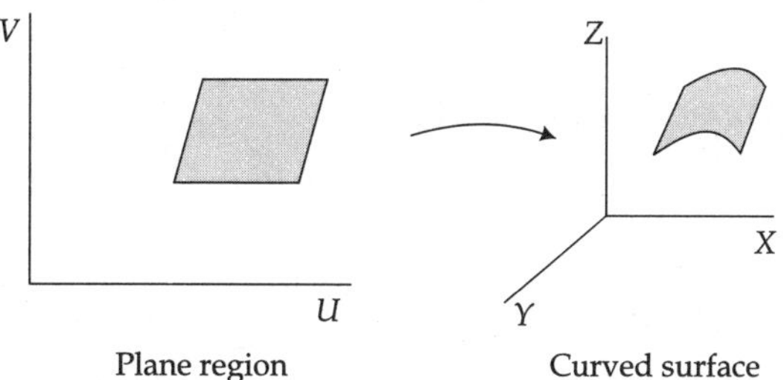

Fig. 11.27

Putting mathematically, suppose R be a region in $\mathbb{R}^2$ then a ***surface*** S is obtained by transforming every point $(u, v) \in R$ is to a point (x, y, z) of $S(\subset \mathbb{R}^3)$ by the rule,

$$x = f_1(u, v), \ y = f_2(u, v), \ z = f_3(u, v) \tag{11.10.1}$$

where f_1, f_2 & f_3 are some functions defined on R. We will assume that first order partial derivatives of f_1, f_2 & f_3 exists. Equation (11.10.1) is known as a ***parametric representation*** of the surface S with parameters u & v. Note that a parametric representation of a surface contains *two* parameters where as we have seen that parametric equation of a curve has *one* parameter.

Let $\vec{r}$ be the radius vector of a point P on a surface S with parametric representation as

$$x = f_1(u, v), \ y = f_2(u, v), \ z = f_3(u, v) \ (u, v) \in R$$

then we can write,

$$\vec{r} = f_1(u, v)\vec{i} + f_2(u, v)\vec{j} + f_3(u, v)\vec{k} \ (u, v) \in R \tag{11.10.2}$$

Above equation is called ***vector equation*** of the surface S.

We now give some examples of parametric representation of surfaces.

Example 11.27 Let us consider the unit $OABC$ square in the $U - V$ plane (Fig. 11.29) and suppose we transform each point (u, v) of the square to a point of $\mathbb{R}^3$ by the rule,

$$x = u, \ y = v, \ z = 1 - u - v$$

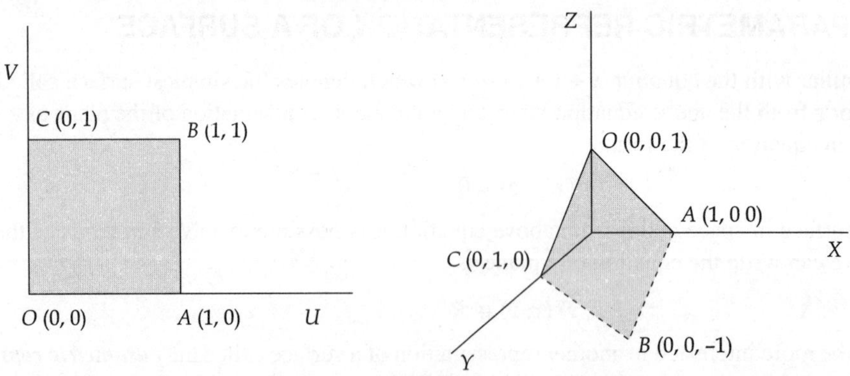

Fig. 11.28

We then obtain the surface $O'\,A'\,B'\,C'$ which is the plane with equation $x + y + z = 1$. Thus the vector equation of the plane $O'\,A'\,B'\,C'$ is given by,

$$\vec{r} = u\vec{i} + v\vec{j} + (1 - u - v)\vec{k}\ (u, v) \in [0, 1] \times [0, 1]$$

We also observe that by the above rule the points $O, A, B\ \&\ C$ of the square transforms to the points $O', A', B'\ \&\ C'$ of the plane.

Example 11.28   Let $R : \left[0, \dfrac{\pi}{2}\right] \times [0, 2\pi]$ be a rectangle in $\theta - \phi$ plane then the transformation,

$$x = \cos\phi\,\sin\theta,\ y = \sin\phi\,\sin\theta,\ z = \cos\theta\ (\phi, \theta) \in R$$

deforms the rectangle into hemisphere of radius 1. Hence the vector equation of the hemisphere of radius 1 is given by

$$\vec{r} = (\cos\phi\,\sin\theta)\vec{i} + (\sin\phi\,\sin\theta)\vec{j} + (\cos\theta)\vec{k},\ (\phi, \theta) \in R$$

Let us try to explore how the deformation takes place. The line OC (whose equation is given by,

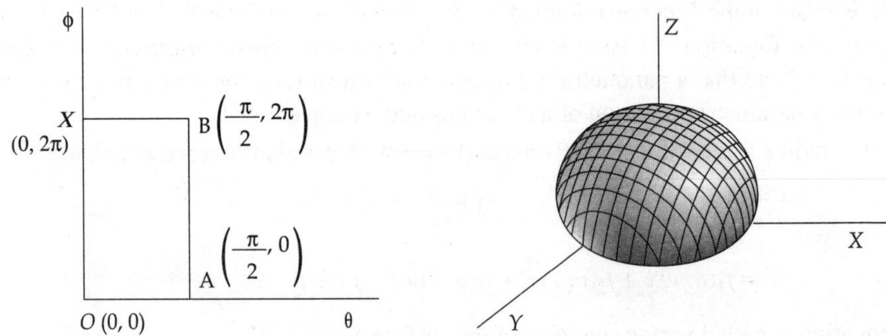

Fig. 11.29

$\theta = 0$) by the given rule collapse to the point $(0, 0, 1)$ and line AB $\left(\text{whose equation is } \theta = \dfrac{\pi}{2}\right)$ is mapped to the circle

$$x = \cos\phi,\ y = \sin\phi,\ z = 0$$

which is the boundary curve of the hemisphere. This in fact shows how we can fold a rectangular piece of paper into a hemisphere.

In the same way the parametric equation of a sphere of radius a is given by,

$$x = a \cos\phi \sin\theta, \ y = a \sin\phi \sin\theta, \ z = a \cos\theta \ (\theta, \phi) \in [0, \pi] \times [0, 2\pi]$$

Example 11.29 Suppose S is a surface with equation $z = f(x, y)$ then the parametric representation of such surface will be given by,

$$x = x, \ y = y, \ z = f(x, y) \ (x, y) \in R$$

that is here we take x & y as parameters and R is called the projection of the surface S on the $X - Y$ plane (Fig. 11.31).

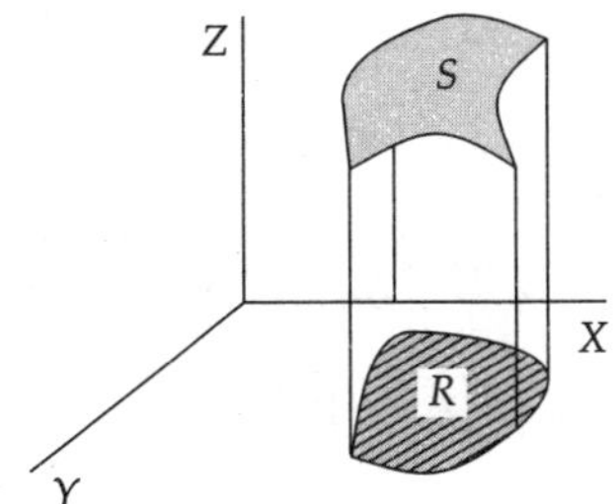

Fig. 11.30

The vector equation of such a surface will be,

$$\vec{r} = x\vec{i} + y\vec{j} + f(x, y)\vec{k}$$

Let us take a concrete example, we know that the equation of the upper hemisphere of unit radius is given by $z = \sqrt{1 - x^2 - y^2}$ thus its parametric representation is given by,

$$x = x, \ y = y, \ z = \sqrt{1 - x^2 - y^2} \ (x, y) \in R$$

where R is the unit disc in this case which is the projection of the hemisphere on the $X - Y$ plane. Note that this example also shows that a surface can have more than one parametric representation.

11.11 NOTION OF SURFACE AREA

Let S be a be surface with vector equation as

$$\vec{r} = f_1(u, v)\vec{i} + f_2(u, v)\vec{j} + f_3(u, v)\vec{k}, \ (u, v) \in R$$

Consider a small rectangle $PABC$ of area $\Delta u \, \Delta v$ in R with $P(u_0, v_0)$ as a corner point as shown in Fig. 11.32. Let the corresponding point on the surface be $p'(x_0, y_0, z_0)$. Now the equation of PA is given by $v = v_0$ which will be mapped to a curve $P'A'$ on the surface and the direction of the tangent to this curve at the point P' is given by $\dfrac{\partial \vec{r}}{\partial u}$ and $\overrightarrow{P'A'}$ is approximately equal to $\dfrac{\partial \vec{r}}{\partial u} \Delta u$. Similarly equation of PC is given by $u = u_0$ which will be mapped to a curve $P'C'$ on the surface and the direction of the tangent to this curve at the point P' is given by $\dfrac{\partial \vec{r}}{\partial v}$ and $\overrightarrow{P'C'}$ is approximately equal to $\dfrac{\partial \vec{r}}{\partial u} \Delta u$. Now the area of the parallelogram spanned by vectors $\overrightarrow{P'A'}$ and

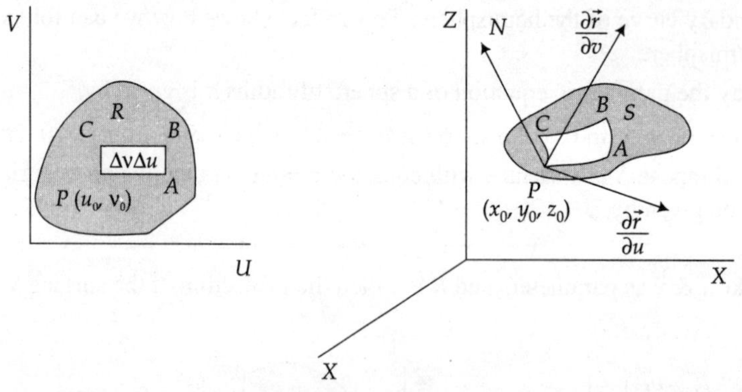

Fig. 11.31

$\overrightarrow{P'C'}$ is given by $\|\overrightarrow{P'A'} \times \overrightarrow{P'C'}\| = \left\| \dfrac{\partial \vec{r}}{\partial u} \times \dfrac{\partial \vec{r}}{\partial v} \right\| \Delta u\, \Delta v$. Thus we see that a small rectangle of area $\Delta u\, \Delta v$

is transformed into a curved parallelogram of area $\left\| \dfrac{\partial \vec{r}}{\partial u} \times \dfrac{\partial \vec{r}}{\partial v} \right\| \Delta u\, \Delta v$. The vector

$\vec{N} = \dfrac{\partial \vec{r}}{\partial u} \times \dfrac{\partial \vec{r}}{\partial v}$ is called the **normal** to the surface. Now if the cross product $\dfrac{\partial \vec{r}}{\partial u} \times \dfrac{\partial \vec{r}}{\partial v}$ vanishes then it

means that the whole rectangle collapses to a single point which produces degenerate surfaces. We define

a point of a surface to be a **singular point** if at that point either $\dfrac{\partial \vec{r}}{\partial u} \times \dfrac{\partial \vec{r}}{\partial v} = 0$ or any one of $\dfrac{\partial \vec{r}}{\partial u}$ & $\dfrac{\partial \vec{r}}{\partial v}$ fails

to be continuous otherwise the point is called a **regular point**. A surface whose every point is a regular
point is called a **smooth surface**.

The **total area** A of a smooth surface S is defined by the equation,

$$A = \iint\limits_{R} \left\| \frac{\partial \vec{r}}{\partial u} \times \frac{\partial \vec{r}}{\partial v} \right\| du\, dv \qquad (11.11.1)$$

Now

$$\frac{\partial \vec{r}}{\partial u} \times \frac{\partial \vec{r}}{\partial v} = \begin{vmatrix} \vec{i} & \vec{j} & \vec{k} \\ \dfrac{\partial f_1}{\partial u} & \dfrac{\partial f_2}{\partial u} & \dfrac{\partial f_3}{\partial u} \\ \dfrac{\partial f_1}{\partial v} & \dfrac{\partial f_2}{\partial v} & \dfrac{\partial f_3}{\partial v} \end{vmatrix}$$

$$= \left(\frac{\partial f_2}{\partial u} \frac{\partial f_3}{\partial v} - \frac{\partial f_3}{\partial u} \frac{\partial f_2}{\partial v} \right) \vec{i} + \left(\frac{\partial f_3}{\partial u} \frac{\partial f_1}{\partial v} - \frac{\partial f_1}{\partial u} \frac{\partial f_3}{\partial v} \right) \vec{j} + \left(\frac{\partial f_1}{\partial u} \frac{\partial f_2}{\partial v} - \frac{\partial f_2}{\partial u} \frac{\partial f_1}{\partial v} \right) \vec{k}$$

$$= \frac{\partial(f_2, f_3)}{\partial(u, v)} \vec{i} + \frac{\partial(f_3, f_1)}{\partial(u, v)} \vec{j} + \frac{\partial(f_1, f_2)}{\partial(u, v)} \vec{k}$$

Thus (11.11.1) can be written as,

$$A = \iint\limits_{R} \sqrt{\left(\frac{\partial(f_2, f_3)}{\partial(u, v)}\right)^2 + \left(\frac{\partial(f_3, f_1)}{\partial(u, v)}\right)^2 + \left(\frac{\partial(f_1, f_2)}{\partial(u, v)}\right)^2} \, du \, dv$$

Example 11.30 Consider the surface with parametric representation

$$x = u, \, y = v, \, z = 1 - u - v, \, (u, v) \in [0, 1] \times [0, 1]$$

as in Example. Here $f_1 = u, f_2 = v \ \& \ f_3 = 1 - u - v$. Then we have in this case,

$$\frac{\partial \vec{r}}{\partial u} \times \frac{\partial \vec{r}}{\partial v} = \begin{vmatrix} \vec{i} & \vec{j} & \vec{k} \\ \frac{\partial f_1}{\partial u} & \frac{\partial f_2}{\partial u} & \frac{\partial f_3}{\partial u} \\ \frac{\partial f_1}{\partial v} & \frac{\partial f_2}{\partial v} & \frac{\partial f_3}{\partial v} \end{vmatrix} = \begin{vmatrix} \vec{i} & \vec{j} & \vec{k} \\ 1 & 0 & -1 \\ 0 & 1 & -1 \end{vmatrix} = \vec{i} + \vec{j} + \vec{k}.$$

Thus total surface area of the surface is given by,

$$A = \iint\limits_{R} \frac{\partial \vec{r}}{\partial u} \times \frac{\partial \vec{r}}{\partial v} \, du \, dv = \int_0^1 \int_0^1 \sqrt{3} \, du \, dv = \sqrt{3}$$

Example 11.31 We now consider the hemispherical surface of radius 1 with parametric representation

$$x = \cos \phi \sin \theta, \, y = \sin \phi \sin \theta, \, z = \cos \theta \qquad (\phi, \theta) \in [0, 2\pi] \times \left[0, \frac{\pi}{2}\right]$$

so that the vector equation is given by,

$$\vec{r} = (\cos \phi \sin \theta)\vec{i} + (\sin \phi \sin \theta)\vec{j} + (\cos \theta)\vec{k}, \quad (\phi, \theta) \in [0, 2\pi] \times \left[0, \frac{\pi}{2}\right]$$

Now,

$$\frac{\partial \vec{r}}{\partial \phi} = (-\sin \phi \sin \theta)\vec{i} + (\cos \phi \sin \theta)\vec{j}$$

$$\frac{\partial \vec{r}}{\partial \theta} = (\cos \phi \cos \theta)\vec{i} + (\sin \phi \cos \theta)\vec{j} - (\sin \theta)\vec{k}$$

Thus,

$$\left\| \frac{\partial \vec{r}}{\partial \phi} \times \frac{\partial \vec{r}}{\partial \theta} \right\| = \sin \theta$$

Hence total area of a hemisphere of radius 1 is given by,

$$\int_0^{2\pi} \left(\int_0^{\frac{\pi}{2}} \sin \theta \, d\theta \right) d\phi = 2\pi$$

Example 11.32   Suppose we have a surface with equation $z = f(x, y)$ then its vector equation is given by,

$$\vec{r} = x\vec{i} + y\vec{j} + f(x, y)\vec{k} \quad (x, y) \in R$$

Recall that R is called the projection of the surface S on the $X - Y$ plane.

Now,

$$\frac{\partial \vec{r}}{\partial x} = i + \frac{\partial f}{\partial x}\vec{j} \quad \& \quad \frac{\partial \vec{r}}{\partial y} = \vec{j} + \frac{\partial f}{\partial y}\vec{k}$$

Hence,

$$\frac{\partial \vec{r}}{\partial x} \times \frac{\partial \vec{r}}{\partial y} = -\frac{\partial f}{\partial x}\vec{i} - \frac{\partial f}{\partial y}\vec{j} + \vec{k}$$

Thus,

$$\left\| \frac{\partial \vec{r}}{\partial x} \times \frac{\partial \vec{r}}{\partial y} \right\| = \sqrt{1 + \left(\frac{\partial f}{\partial x}\right)^2 + \left(\frac{\partial f}{\partial y}\right)^2}$$

So that total are of the surface is given by,

$$\iint_R \sqrt{1 + \left(\frac{\partial f}{\partial x}\right)^2 + \left(\frac{\partial f}{\partial y}\right)^2}\; dx\, dy$$

Also note that if $\vec{n} = \dfrac{\partial \vec{r}}{\partial x} \times \dfrac{\partial \vec{r}}{\partial y}$ denotes the normal vector to the surface and if γ denotes the angle which this normal makes with positive direction of $Z -$ axis then in this case,

$$1 = \vec{n} \cdot \vec{k} = \|\vec{n}\|\, \|\vec{k}\| \cos \gamma$$

Hence,

$$\cos \gamma = \frac{1}{\|\vec{n}\|}$$

So that the total area in this case can also be expressed as

$$\iint_R \frac{1}{\cos \gamma}\, dx\, dy$$

Also we express the above formula as,

$$dS = \frac{1}{\cos \gamma}\, dx\, dy$$

From the above discussion it is clear that if P and Q are the projection of the surface S on $Y - Z$ and $Z - X$ plane and if α and β denotes the inclination of the normal to surface with $\vec{i}$ and $\vec{j}$ vector then we can write,

$$dS = \frac{1}{\cos \alpha}\, dy\, dz \quad \& \quad dS = \frac{1}{\cos \beta}\, dz\, dx$$

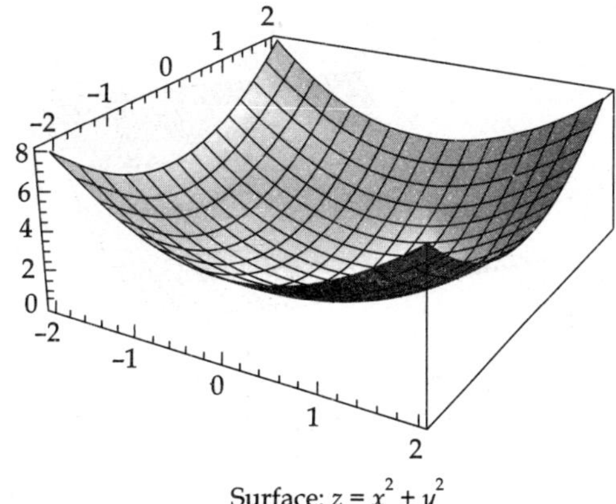

Surface: $z = x^2 + y^2$

Fig. 11.32

As an application of the above formula let us consider the surface given by,

$$z = x^2 + y^2 \ (x, y) \in [-2, 2] \times [-2, 2]$$

Here,

$$f(x, y) = x^2 + y^2$$

So that,

$$\frac{\partial f}{\partial x} = 2x \ \& \ \frac{\partial f}{\partial y} = 2y$$

So the total area is given by,

$$\int_{-2}^{2} \int_{-2}^{2} \sqrt{1 + 4x^2 + 4y^2} \ dx \ dy = \frac{128}{3} \quad \text{(by standard substitutions)}$$

Example 11.33   Suppose the equation of a surface is given by

$$F(x, y, z) = 0 \ (x, y) \in R \subseteq \mathbb{R}^2$$

and further suppose that from the given equation we can solve for z as

$$z = f(x, y)$$

That is,

$$F(x, y, f(x, y)) = 0$$

Taking partial derivatives of the above equation we obtain,

$$\frac{\partial f}{\partial x} = -\frac{\partial F/\partial x}{\partial F/\partial z} \quad \text{and} \quad \frac{\partial f}{\partial y} = -\frac{\partial F/\partial y}{\partial F/\partial z}$$

Using above formulae the total surface area in this case is give by,

$$\iint_R \frac{\sqrt{(\partial F/\partial x)^2 + (\partial F/\partial y)^2 + (\partial F/\partial z)^2}}{|\partial F/\partial z|} \ dx \ dy$$

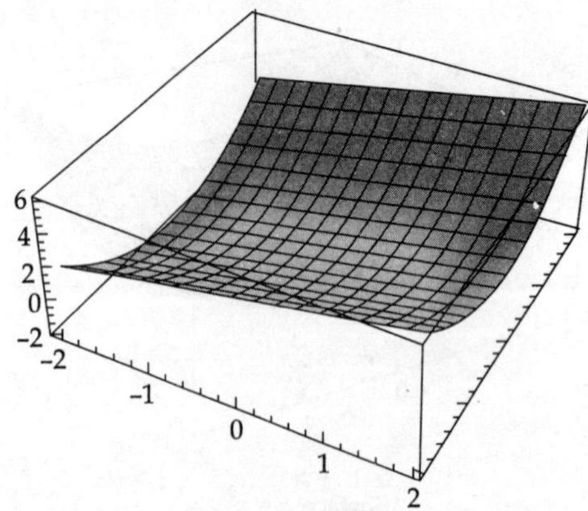

Fig. 11.33

As an application of above formula we consider the surface with equation $x + y^2 - z = 0$ $(x, y) \in$ $[-2, 2] \times [-2, 2]$ (Fig.). Here $F \equiv x + y^2 - z$, so that, $\dfrac{\partial F}{\partial x} = 1, \dfrac{\partial F}{\partial y} = 2y$ & $\dfrac{\partial F}{\partial z} = -1$.

Total area is given by,

$$\int\limits_{-2}^{2} \int\limits_{-2}^{2} \frac{\sqrt{1 + 4y^2 + 1}}{|-1|} \, dx \, dy = \int\limits_{-2}^{2} \int\limits_{-2}^{2} \sqrt{2 + 4y^2} \, dx \, dy$$

$$= 4[6\sqrt{2} + \sinh^{-1}(2\sqrt{2})]$$

11.12   SURFACE INTEGRAL OF FIRST KIND

Suppose f be a reasonably a nice function defined on a smooth surface S with vector equation,

$$\vec{r}(u, v), (u, v) \in R$$

in this section we want to define the 'integral' of the function over this surface called surface integrals. Now see that, as the surface S is a deformation of the region R in $\mathbb{R}^2$ hence we can imagine that as if the function f is defined on the region R so that the concept of surface integral of f over S boils down to the concept of double integral f over R. Thus the ***surface integral of first kind*** of f over S denoted by $\iint\limits_{S} f dS$ is defined by the equation,

$$\iint\limits_{S} f dS = \iint\limits_{R} f(\vec{r}(u, v)) \, \frac{\partial \vec{r}}{\partial u} \times \frac{\partial \vec{r}}{\partial v} \, du \, dv \qquad (11.12.1)$$

provided the integral on the right hand side exists.

Remarks

1. If $f \equiv 1$ then (11.12.1) gives the area formula of the surface.

2. Suppose S is disjoint union of the surfaces say S_i $i = 1, 2, ...n$ then we have,

$$\iint\limits_S f\,dS = \sum_{i=1}^{n} \iint\limits_{S_i} f\,dS$$

For example the surface integral of a function on a sphere can be written as sum of surface integrals on two hemispheres.

Example 11.34 Evaluate $\iint\limits_S xyz\,dS$ where S is the surface given by the equation,

$$x + y + z = 1,\ x \geq 0,\ y \geq 0,\ z \geq 0$$

Sol: Here we have $z = 1 - x - y$ hence the vector representation of the surface is,

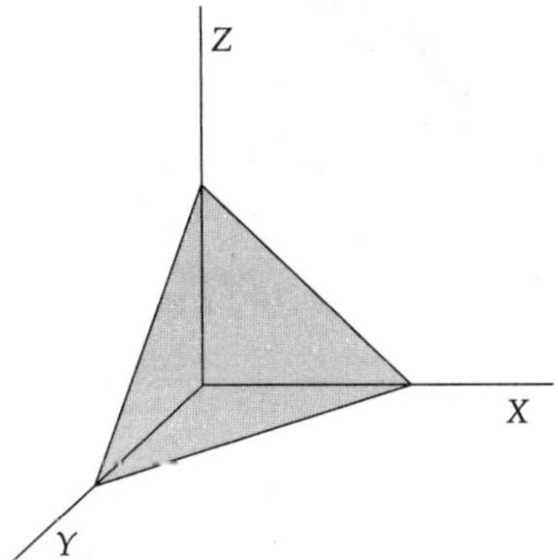

Fig. 11.34

$$\vec{r} = x\vec{i} + y\vec{j} + (1 - x - y)\vec{k}$$

So that $\left\|\dfrac{\partial \vec{r}}{\partial x} \times \dfrac{\partial \vec{r}}{\partial y}\right\| = \sqrt{1 + \left(\dfrac{\partial f}{\partial x}\right)^2 + \left(\dfrac{\partial f}{\partial y}\right)^2}$ (by Example) where $f = 1 - x - y$

$$= \sqrt{1 + (-1)^2 + (-1)^2} = \sqrt{3}$$

Now,

$$\iint\limits_S xyz\,ds = \iint\limits_S xy\sqrt{1 - x - y}\ \sqrt{3}\,dxdy \quad \text{where } R:\ x + y \leq 1,\ x > 0,\ y > 0$$

$$= \sqrt{3} \int\limits_{x=0}^{1} \int\limits_{y=0}^{1-x} xy\,(1 - x - y)\,dxdy.$$

$$= \frac{\sqrt{3}}{120}$$

Example 11.35   Evaluate $\iint\limits_{S} x^2\,dS$ where S denotes the entire surface of the solid cylinder $x^2 + y^2 = 4$ bounded by the planes $z = 0$ and $z = 2x + 3$.

Sol: We refer to the figure below. We see that S comprises of the lateral surface S_1 and upper and lower surfaces S_2 and S_3 respectively. We observe that,

$$S_1 : x^2 + y^2 \le 4, \quad 0 \le z \le 2x + 3$$

$$S_2 : x^2 + y^2 \le 4, \quad z = 2x + 3$$

$$S_3 : x^2 + y^2 \le 4, \quad z = 0$$

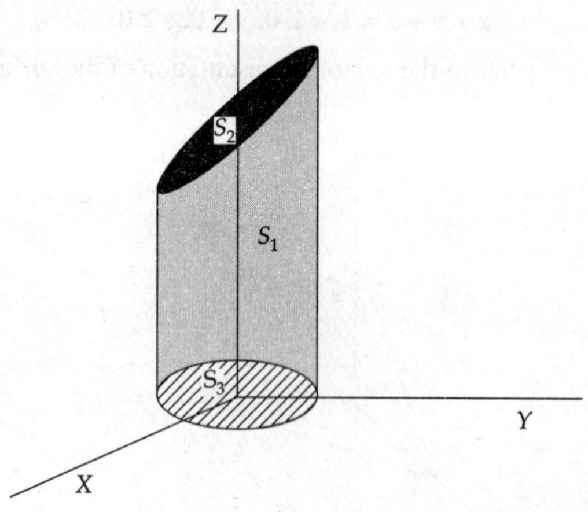

Fig. 11.35

Let D denote the disc $x^2 + y^2 \le 4$. Then S_3 is nothing but the disc D, now the unit normal $\vec{n}$ to D is parallel to $\vec{k}$ so that $\cos \gamma = 1$. Hence,

$$\iint\limits_{S_1} x^2\,dS = \iint\limits_{D} x^2\,dxdy$$

$$= \int_{r=0}^{2} \int_{\theta=0}^{2\pi} r^2 \cos^2 \theta\, rdrd\theta \quad (\text{putting } x = r \cos \theta, \ y = r \sin \theta)$$

$$= \int_{r=0}^{2} \int_{\theta=0}^{2\pi} r^3 \cos^2 \theta\, drd\theta$$

$$= 4\pi$$

Again,

$$\iint\limits_{S_2} x^2\,dS = \iint\limits_{D} x^2 \sqrt{1 + \left(\frac{\partial z}{\partial x}\right)^2 + \left(\frac{\partial z}{\partial y}\right)^2}\, dxdy \quad \text{where } z = 2x + 3$$

$$= \iint\limits_{D} x^2 \sqrt{1 + 4 + 0}\ dxdy$$

$$= \sqrt{5} \iint\limits_{D} x^2\ dxdy = 4\sqrt{5}\,\pi$$

We now parameterize the lateral surface S_1 as follows,

$$x = 2 \cos\theta,\ y = 2 \sin\theta\ z = z - \pi \leq \theta \leq \pi,\ 0 \leq z \leq 4 \cos\theta + 3$$

So that,

$$\frac{\partial \vec{r}}{\partial \theta} = (-2 \cos\theta)\vec{i} + (2 \sin\theta)\vec{j} + 0\vec{k}$$

$$\frac{\partial \vec{r}}{\partial z} = 0\vec{i} + 0\vec{j} + \vec{k}$$

and hence,

$$\left\| \frac{\partial \vec{r}}{\partial \theta} \times \frac{\partial \vec{r}}{\partial z} \right\| = \sqrt{4 \cos^2\theta + 4 \sin^2\theta + 0}$$

Now,

$$\iint\limits_{S_1} x^2\,dS = \int\limits_{\theta=0}^{2\pi} \int\limits_{z=0}^{4\cos\theta+3} \cos^2\theta(2)\ d\theta dz$$

$$= 8 \int\limits_{0-0}^{2\pi} \int\limits_{z=0}^{4\cos\theta+3} \cos^2\theta\ d\theta dz = 24\pi$$

Hence adding the surface integrals on the three surfaces we obtain

$$\iint\limits_{S} x^2\,dS = 4\pi + 4\pi\sqrt{5} + 24\pi = 4\pi(7 + \sqrt{5})$$

Example 11.36   Evaluate $\iint\limits_{S} z^2 \cos\gamma\,dS$

where S is the surface $z = \sqrt{a^2 - x^2 - y^2}$ (upper hemisphere of radius a and γ denotes the inclination of the unit normal to the surface with Z-axis.

Sol: If D denotes the disc $x^2 + y^2 \leq a^2$ then we note that the equation of the given surface can be written as,

$$z = R a^2\ x^2 - y^2\ (x, y) \in D$$

Writing $\cos\gamma\,dS = dxdy$ we get

$$\iint\limits_{S} z^2 \cos\gamma\,dS = \iint\limits_{D} (a^2 - x^2 - y^2)\ dxdy$$

$$= \int\limits_{r=0}^{a} \int\limits_{\theta=0}^{2\pi} (a^2 - r^2)\ rdrd\theta = \frac{\pi a^4}{2}$$

11.13 ORIENTABLE SURFACES

Before proceeding to define surface integral of second kind we need to introduce the concept of orientation of a surface. We note that a surface has in general has two sides, outer and inner surface and in that case the surface is *orientable*. For an orientable surface is if it is impossible to move from one side of the surface to another without crossing the boundary. Not all surfaces are orientable most familiar example of such a surface is a *Möbius strip*. The replica of this surface can be produced by gluing opposite sides of a rectangular piece of paper. It can be easily seen that one can move on Möbius strip from one surface to another without crossing the boundary. Most of the smooth surfaces are orientable. A surface is said to have a positive orientation if the boundary curve of the surface has a *positive orientation* otherwise it has a *negative orientation*.

For example, consider the unit sphere it can be thought of as union of two hemispherical surfaces one upper and lower. Now the boundary curve for both of these surfaces is the unit circle C but

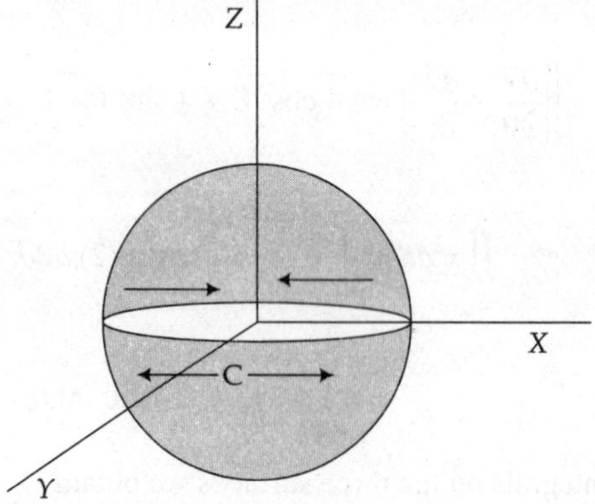

Fig. 11.36

observe that the upper hemisphere is positively oriented if C has an anti-clockwise orientation where as lower hemisphere is positively oriented if C has a clockwise orientation.

Again suppose S is a smooth orientable surface then at any point P the normal to the surface has two direction one inward and other outward. By convention, for a positively oriented surface we assign positive direction to outward drawn normal (Fig. 11.36)

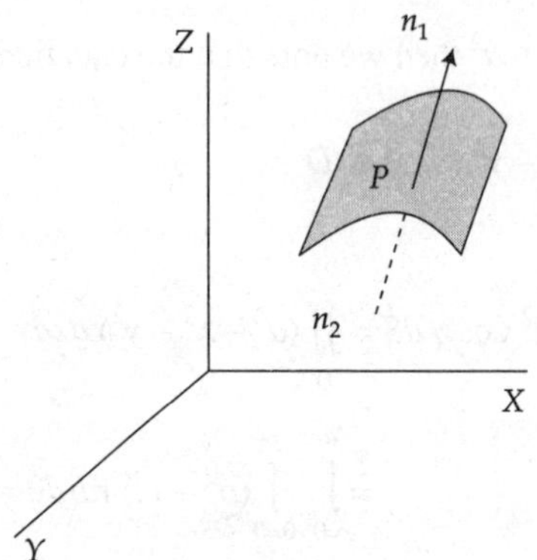

Fig. 11.37

11.14 SURFACE INTEGRAL OF SECOND KIND

We have already discussed the notion of integral of a scalar field defined on a surface. In this section we will define the notion of the integral of a vector field defined on a surface. Let $\vec{F}$ be a vector function defined on a smooth orientable surface $\vec{r}\,(u, v)\,(u, v) \in R$. Further suppose $\vec{n}$ be the unit outward drawn normal at any point P of the surface then $\vec{F} \cdot \vec{n}$ denotes the projection of $\vec{F}$ in the direction of $\vec{n}$ which is a scalar field. Then if $\iint\limits_{S} \vec{F} \cdot \vec{n}\ ds$ exists in the sense defined in section 11.11 is called the **surface integral of second kind**. Thus using (11.11.1) we have,

$$\iint\limits_{S} \vec{F} \cdot \vec{n}\ dS = \iint\limits_{R} \vec{F}\,(\vec{r}(u, v)) \cdot \vec{n}\,(u, v)\,\frac{\partial \vec{r}}{\partial u} \times \frac{\partial \vec{r}}{\partial v} \qquad (11.14.1)$$

We know that $\dfrac{\partial \vec{r}}{\partial u} \times \dfrac{\partial \vec{r}}{\partial v}$ is normal to the surface so that,

$$\vec{n} = \frac{\dfrac{\partial \vec{r}}{\partial u} \times \dfrac{\partial \vec{r}}{\partial v}}{\left\| \dfrac{\partial \vec{r}}{\partial u} \times \dfrac{\partial \vec{r}}{\partial v} \right\|}$$

Hence from (11.14.1) we get,

$$\iint\limits_{S} \vec{F} \cdot \vec{n}\ dS = \iint\limits_{R} \vec{F}\,(\vec{r}(u, v)) \cdot \vec{n}\,(u, v)\,\left\| \frac{\partial \vec{r}}{\partial u} \times \frac{\partial \vec{r}}{\partial v} \right\|\,dudv$$

$$= \iint\limits_{S} \vec{F}\,(\vec{r}(u, v)) \cdot \left(\frac{\partial \vec{r}}{\partial u} \times \frac{\partial \vec{r}}{\partial v} \right)\,dudv \qquad (11.14.2)$$

Again see that,

$$\frac{\partial \vec{r}}{\partial u} \times \frac{\partial \vec{r}}{\partial v} = \frac{\partial(f_2, f_3)}{\partial(u, v)}\,\vec{i} + \frac{\partial(f_3, f_1)}{\partial(u, v)}\,\vec{j} + \frac{\partial(f_1, f_2)}{\partial(u, v)}\,\vec{k}$$

So that (11.14.1) can be written as,

$$\iint\limits_{S} \vec{F} \cdot \vec{n}\ dS = \iint\limits_{R} F_1(\vec{r}(u, v))\,\frac{\partial(f_2, f_3)}{\partial(u, v)}\,dudv + \iint\limits_{R} F_2(\vec{r}(u, v))\,\frac{\partial(f_3, f_1)}{\partial(u, v)}\,dudv$$

$$+ \iint\limits_{R} F_3(\vec{r}(u, v))\,\frac{\partial(f_1, f_2)}{\partial(u, v)}\,dudv \qquad (11.14.3)$$

Again if α, β & γ denotes the inclination of the unit normal to the surface with X, Y & Z axis then we can write, $\vec{n} = \cos\alpha\ \vec{i} + \cos\beta\ \vec{j} + \cos\gamma\ \vec{k}$ so that,

$$\iint_S \vec{F} \cdot \vec{n} \, dS = \iint_S (F_1 \cos \alpha + F_2 \cos \beta + F_3 \cos \gamma) \, dS$$

(where F_1, F_2 & F_3 are components of $\vec{F}$).

$$= \iint_S (F_1 \, dydz + F_2 \, dzdx + F_3 \, dxdy) \qquad (11.14.4)$$

Now if the given surface can be expressed as $z = f(x, y)$ then,

$$\iint_S F_3 (x, y, z) \, dxdy = \iint_{R_1} F_3 (x, y, f(x, y)) \, dxdy$$

(where R_1 is the projection of the surface on $X - Y$ plane.)

Again if $y = g(z, x)$ then,

$$\iint_S F_2 (x, y, z) \, dzdx = \iint_{R_1} F_2 (x, g(x, z), z) \, dzdx$$

(where R_2 is the projection of the surface on $Z - X$ plane).

Finally, if $x = h(y, z)$ then,

$$\iint_S F_1 (x, y, z) \, dydz = \iint_S F_1 (h(y, z), y, z) \, dzdx$$

(where R_3 is the projection of the surface on $Y - Z$ plane)

Example 11.37   Show that $\iint_S (x \, dydz + y \, dzdx + z \, dxdy) = \dfrac{\pi}{2}$ where S is the outer surface of the sphere $x^2 + y^2 + z^2 = 1$ in the first octant.

Sol: Let R_1, R_2 & R_3 be the projection of the surface on the $X - Y$, $Y - Z$ & $Z - X$. Then is it is easy to see that each of them is one fourth of unit disc in those respective planes. Thus we have,

$$\iint_S z \, dxdy = \iint_{R_1} \sqrt{1 - x^2 - y^2} \, dxdy$$

$$= \int_{r=0}^{1} \int_{\theta=0}^{\frac{\pi}{2}} \sqrt{1 - r^2} \, rdrd\theta = \frac{\pi}{6}$$

From symmetry we conclude that,

$$\iint_S x \, dydz = \iint_S y \, dxdz = \frac{\pi}{6}$$

Hence we conclude that,

$$\iint_S (x\,dydz + y\,dzdx + z\,dxdy) = 3 \times \frac{\pi}{6} = \frac{\pi}{2}$$

Example 11.38 Evaluate $\iint_S \vec{F} \cdot \vec{n}\,dS$ where $\vec{F} = x\vec{i} + z^2\vec{j} - y\vec{k}$ and S is the curved surface of the cylinder $x^2 + y^2 = 4$ between $z = 0$ and $z = 1$.

Sol: We parameterize the cylinder by usual cylindrical polar coordinates as

$$\vec{r} = 2\cos\theta\,\vec{i} + 2\sin\theta\,\vec{j} + z\vec{k} \quad (\theta, z) \in [0, 2\pi] \times [0, 1]$$

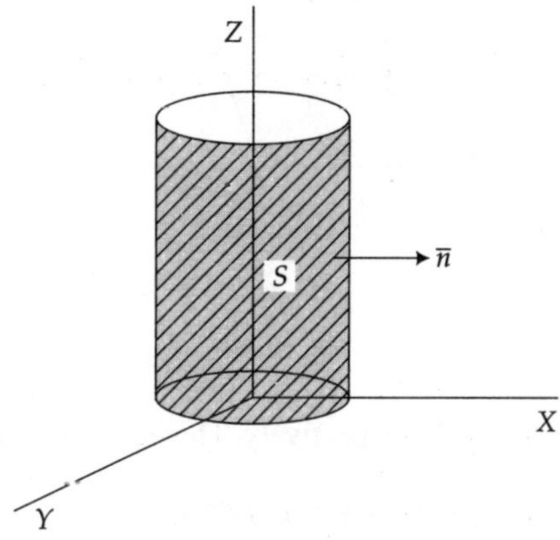

Fig. 11.38

So that,

$$\frac{\partial \vec{r}}{\partial \theta} = (-2\cos\theta)\,\vec{i} + (2\sin\theta)\,\vec{j} + 0\vec{k}$$

$$\frac{\partial \vec{r}}{\partial z} = 0\vec{i} + 0\vec{j} + \vec{k}$$

and hence,

$$\vec{n} = \frac{\partial \vec{r}}{\partial \theta} \times \frac{\partial \vec{r}}{\partial z} = (2\sin\theta)\,\vec{i} + (2\cos\theta)\,\vec{j} + 0\vec{k}$$

Now from (11.14.2) one obtains,

$$= \int_{z=0}^{1} \int_{\theta=0}^{2\pi} (4\sin\theta\cos\theta + 2z^2\cos\theta)\,d\theta dz$$

$$= 0$$

Example 11.39   Evaluate $\iint\limits_{S} (z\,dxdy + x\,dydz + y\,dxdz)$ where S is the entire surface of the cube with

vertices, $(0, 0, 0)$, $(a, 0, 0)$, $(0, a, 0)$, $(0, 0, a)$, $(a, a, 0)$, $(a, 0, a)$, $(0, a, a)$ & (a, a, a)

Sol: Let $OABCDEFG$ (Fig. 11.40) be the cube with faces, $BCDG$, $OEFA$, $DEFG$, $OABC$, $ABGF$,

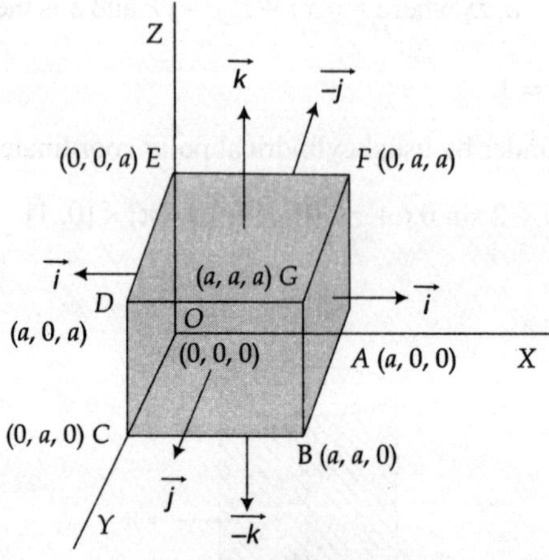

Fig. 11.39

and $OCDE$ denoted by S_1, S_2, S_3, S_4, S_5, S_6 respectively. The equations of these surfaces are given by,

$$S_1: y = a, 0 \le x \le a, 0 \le z \le a$$

$$S_2: y = 0, 0 \le x \le a, 0 \le z \le a$$

$$S_3: z = a, 0 \le x \le a, 0 \le y \le a$$

$$S_4: z = 0, 0 \le x \le a, 0 \le y \le a$$

$$S_5: x = a, 0 \le y \le a, 0 \le z \le a$$

$$S_6: x = 0, 0 \le y \le a, 0 \le z \le a$$

As the outward drawn normal is taken to be positive it is easy to see that the normal to S_1, S_2, S_3, S_4, S_5 & S_6 are given by $\vec{j}$, $-\vec{j}$, $\vec{k}$, $-\vec{k}$, $\vec{i}$ & $-\vec{i}$ respectively. Now,

$$\iint\limits_{S_1} (z\,dxdy + x\,dydz + y\,dxdz) = \int\limits_{0}^{a}\int\limits_{0}^{a} a\,dxdz = a^3$$

$$\iint\limits_{S_2} (z\,dxdy + x\,dydz + y\,dxdz) = -\int\limits_{0}^{a}\int\limits_{0}^{a} 0\,dxdz = 0$$

Similarly,

$$\iint\limits_{S_3} (z\,dxdy + x\,dydz + y\,dxdz) = \int\limits_{0}^{a}\int\limits_{0}^{a} a\,dxdy = a^3$$

$$\iint_{S_4} (z\, dxdy + x\, dydz + y\, dxdz) = -\int_0^a \int_0^a 0\, dxdy = 0$$

$$\iint_{S_5} (z\, dxdy + x\, dydz + y\, dxdz) = \int_0^a \int_0^a a\, dydz = a^3$$

$$\iint_{S_6} (z\, dxdy + x\, dydz + y\, dxdz) = -\int_0^a \int_0^a 0\, dydz = 0$$

Adding above equations we get,

$$\iint_S (z\, dxdy + x\, dydz + y\, dxdz) = 3a^3$$

Example 11.40   Show that $\iint_S \vec{F} \cdot \vec{n}\, dS =$ where $\vec{F} = y^2\vec{i} + z^2\vec{j} + x^2\vec{k}$ and S is the triangular surface on $z = 0$ with $x \geq 0,\, y \geq 0,\, x + y \leq 2$.

Sol:

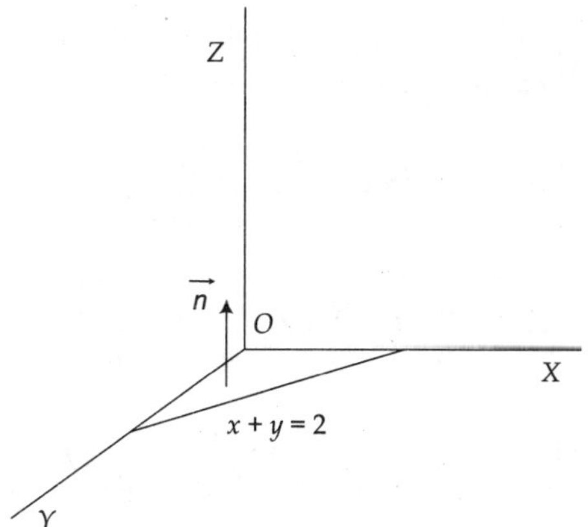

Fig. 11.40

It is clear for this surface the outward drawn unit normal is parallel to $\vec{k}$. Hence,

$$\iint_S \vec{F} \cdot \vec{n}\, dS = \int_{x=0}^{2} \int_{y=0}^{2-x} x^2\, dxdy = \frac{4}{3}$$

11.15 STOKE'S THEOREM

We have seen that Green's Theorem in plane gives a nice relation between double integral and line integral, in this section we will look for a higher dimension analogue of this theorem which is known as Stoke's Theorem. This theorem in fact gives a connection between surface and line integral in space.

Theorem 11.15.1 (Stoke's Theorem)

Let S be a smooth oriented surface with parametric representation as

$$\vec{r}(u, v) = \vec{r} = f_1(u, v)\vec{i} + f_2(u, v)\vec{j} + f_3(u, v)\vec{k} \quad (u, v) \in R$$

where R is a region in U − V plane bounded by a Jordan Curve C_1. We assume that the components of $\vec{r}$ have continuous second order partial derivatives in some open set containing R. If F_1, F_2 & F_3 are continuously differentiable functions defined on S and C is the image of C_1 under $\vec{r}$ then,

$$\int_C F_1 \, dx + F_2 \, dy + F_3 \, dz = \iint_R \left\{ \left(\frac{\partial F_3}{\partial y} - \frac{\partial F_2}{\partial z} \right) dydz + \left(\frac{\partial F_1}{\partial z} - \frac{\partial F_3}{\partial x} \right) dzdx + \left(\frac{\partial F_2}{\partial x} - \frac{\partial F_1}{\partial y} \right) dx\,dy \right\}$$

Proof: From the parameterization of the surface we have,

$$x = f_1(u, v), \; y = f_2(u, v), \; z = f_3(u, v)$$

Also suppose that parametric representation of C_1 is given by

$$u = \varphi(t), \; v = \phi(t)$$

Now,

$$\int_C F_1 \, dx + F_2 \, dy + F_3 \, dz$$

$$= \int_a^b F_1 \left(\frac{\partial x}{\partial u} \frac{du}{dt} + \frac{\partial x}{\partial v} \frac{dv}{dt} \right) + F_2 \left(\frac{\partial y}{\partial u} \frac{du}{dt} + \frac{\partial y}{\partial v} \frac{dv}{dt} \right) + F_3 \left(\frac{\partial z}{\partial u} \frac{du}{dt} + \frac{\partial z}{\partial v} \frac{dv}{dt} \right) dt$$

$$= \int_{C_1} \left(F_1 \frac{\partial x}{\partial u} + F_2 \frac{\partial y}{\partial u} + F_3 \frac{\partial z}{\partial u_3} \right) du + \left(F_1 \frac{\partial x}{\partial v} + F_2 \frac{\partial y}{\partial v} + F_3 \frac{\partial z}{\partial v} \right) dv$$

$$= \iint_R \left[\frac{\partial}{\partial u} \left(F_1 \frac{\partial x}{\partial v} + F_2 \frac{\partial y}{\partial v} + F_3 \frac{\partial z}{\partial v} \right) - \frac{\partial}{\partial v} \left(F_1 \frac{\partial x}{\partial u} + F_2 \frac{\partial y}{\partial u} + F_3 \frac{\partial z}{\partial u} \right) \right] \qquad \text{(by Green's Theorem)}$$

$$= \iint_R \left[\frac{\partial}{\partial u} \left(F_1 \frac{\partial x}{\partial v} - \frac{\partial}{\partial v} \left(F_1 \frac{\partial x}{\partial v} \right) \right) + \frac{\partial}{\partial u} \left(F_2 \frac{\partial y}{\partial v} \right) - \frac{\partial}{\partial v} \left(F_2 \frac{\partial y}{\partial u} \right) + \frac{\partial}{\partial u} \left(F_3 \frac{\partial z}{\partial v} \right) - \frac{\partial}{\partial v} \left(F_3 \frac{\partial z}{\partial v} \right) \right] \quad (11.15.1.1)$$

Now,

$$\frac{\partial}{\partial u} \left(F_1 \frac{\partial x}{\partial v} \right) = F_1 \frac{\partial^2 x}{\partial u \partial v} + \frac{\partial x}{\partial v} \left(\frac{\partial F_3}{\partial x} \frac{\partial x}{\partial u} + \frac{\partial F_3}{\partial y} \frac{\partial y}{\partial u} + \frac{\partial F_3}{\partial z} \frac{\partial z}{\partial u} \right) \qquad (11.15.1.2)$$

and

$$\frac{\partial}{\partial v} \left(F_1 \frac{\partial x}{\partial u} \right) = F_1 \frac{\partial^2 x}{\partial v \partial u} + \frac{\partial x}{\partial u} \frac{\partial F_3}{\partial x} \frac{\partial x}{\partial v} + \frac{\partial F_3}{\partial y} \frac{\partial y}{\partial v} + \frac{\partial F_3}{\partial z} \frac{\partial z}{\partial v} \qquad (11.15.1.3)$$

Subtracting (11.15.1.2) & (11.15.1.3) and using the fact that

$$\frac{\partial^2 x}{\partial u \partial v} = \frac{\partial^2 x}{\partial v \partial u} \quad \text{by assumption we get,}$$

$$\frac{\partial}{\partial u}\left(F_1\frac{\partial x}{\partial v}\right)-\frac{\partial}{\partial v}\left(F_1\frac{\partial x}{\partial u}\right)=\frac{\partial F_1}{\partial z}\frac{\partial(z,x)}{\partial(u,v)}-\frac{\partial F_1}{\partial y}\frac{\partial(x,y)}{\partial(u,v)} \tag{11.15.1.4}$$

Similar calculations will yield,

$$\frac{\partial}{\partial u}\left(F_2\frac{\partial y}{\partial v}-\frac{\partial}{\partial v}\left(F_2\frac{\partial x}{\partial u}\right)\right)=\frac{\partial F_2}{\partial x}\frac{\partial(x,y)}{\partial(u,v)}-\frac{\partial F_1}{\partial z}\frac{\partial(y,z)}{\partial(u,v)} \tag{11.15.1.5}$$

$$\frac{\partial}{\partial u}\left(F_3\frac{\partial y}{\partial v}\right)-\frac{\partial}{\partial v}\left(F_3\frac{\partial x}{\partial u}\right)=\frac{\partial F_3}{\partial y}\frac{\partial(y,z)}{\partial(u,v)}-\frac{\partial F_1}{\partial x}\frac{\partial(z,x)}{\partial(u,v)} \tag{11.15.1.6}$$

Using (11.15.1.1), (11.15.1.4), (11.15.1.5) & (11.15.1.6) we get,

$$\int_C F_1\,dx+F_2\,dy+F_3\,dz=\iint_R\left\{\left(\frac{\partial F_3}{\partial y}-\frac{\partial F_2}{\partial z}\right)\frac{\partial(y,z)}{\partial(u,v)}+\left(\frac{\partial F_1}{\partial z}-\frac{\partial F_3}{\partial x}\right)\frac{\partial(z,x)}{\partial(u,v)}+\left(\frac{\partial F_1}{\partial z}-\frac{\partial F_3}{\partial x}\right)\frac{\partial(x,y)}{\partial(u,v)}\right\}dudv$$

$$=\iint_R\left[\left(\frac{\partial F_3}{\partial y}-\frac{\partial F_2}{\partial z}\right)\frac{\partial(y,z)}{\partial(u,v)}+\left(\frac{\partial F_1}{\partial z}-\frac{\partial F_3}{\partial x}\right)\frac{\partial(z,x)}{\partial(u,v)}+\left(\frac{\partial F_1}{\partial z}-\frac{\partial F_3}{\partial x}\right)\frac{\partial(x,y)}{\partial(u,v)}\right]dudv$$

$$=\iint_S\left[\left(\frac{\partial F_3}{\partial y}-\frac{\partial F_2}{\partial z}\right)dydz+\left(\frac{\partial F_1}{\partial z}-\frac{\partial F_3}{\partial x}\right)dzdx+\left(\frac{\partial F_1}{\partial z}-\frac{\partial F_3}{\partial x}\right)dxdy\right] \tag{11.15.1.7}$$

This concludes the proof.

Suppose $\vec{F}$ be a vector field with components F_1, F_2 & F_3 as defined in the previous theorem then see that Stoke's Theorem can be written as,

$$\int_C \vec{F}\cdot d\vec{c}=\iint_S(\nabla\times\vec{F})\cdot\vec{n}\,dS \tag{11.15.1.8}$$

From above equation it is clear that Green's theorem follows as a special case of Stoke's theorem when the surface S is a region in $X-Y$ plane so that unit normal to the surface is $\vec{k}$ and the vector field $\vec{F}$ is of the form $\vec{F}=F_1\vec{i}+F_2\vec{j}+0\vec{k}$.

Example 11.41 Evaluate the line integral $\iint_S(\nabla\times\vec{F})\cdot\vec{n}\,dS$

where $\vec{F}=y\vec{i}+z\vec{j}+x^2\vec{k}$ and S is the surface with equation $z=16-x^2-y^2$ with $0\le z\le 16$. Use Stoke's Theorem to verify the answer.

Sol: We observe that S is a paraboloid as shown in the figure below. Note that the boundary curve of the surface is given by $x^2+y^2=16$. Now we parametrize the surface as,

$$\vec{r}=x\vec{i}+y\vec{j}+(16-x^2-y^2)\vec{k}\ \ x^2+y^2\le 16$$

Then,

$$\frac{\partial\vec{r}}{\partial x}=\vec{i}-2x\vec{k}\ \ \&\ \ \frac{\partial\vec{r}}{\partial y}=\vec{j}-2y\vec{k}$$

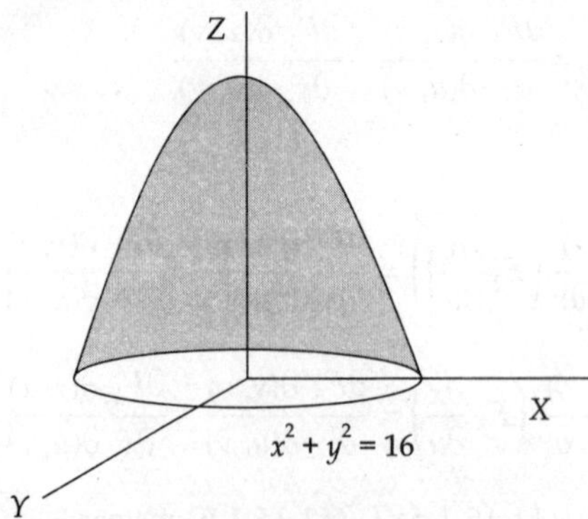

Fig. 11.41

So that,

$$\frac{\partial \vec{r}}{\partial x} \times \frac{\partial \vec{r}}{\partial y} = 2x\vec{i} + 2y\vec{j} + \vec{k}$$

Now $\nabla \times \vec{F} = -\vec{i} - 2x\vec{j} - \vec{k}$ so that $(\nabla \times \vec{F}) \cdot \vec{n} = -2x - 2xy - 1$

Hence,

$$\iint_S (\nabla \times \vec{F}) \cdot \vec{n} \, dS = \iint_R (-2x - 2xy - 1) \, dxdy \quad \text{where } R \text{ is the disc with centre origin and radius 4}$$

$$= \int_{r=0}^{4} \int_{\theta=0}^{2\pi} (-2r \cos \theta - 2r^2 \cos \theta \sin \theta - 1) \, rdrd\theta \quad \text{(putting } x = r \cos \theta, \, y = r \sin \theta\text{)}$$

$$= -16\pi$$

In order to verify above answer using Stoke's Theorem we now evaluate the line integral of the given vector field $\vec{F}$ over the boundary curve C of the surface. If we take the parametric representation of C as $x = 4 \cos \theta, \, y = 4 \sin \theta, \, z = 0$

Hence,

$$\int_C \vec{F} \cdot d\vec{c} = \int_C y \, dx + z \, dy + x^2 \, dz$$

$$= \int_0^{2\pi} (-16 \sin^2 \theta + 0 + 0) \, d\theta$$

$$= -16\pi$$

Thus Stoke's Theorem is verified.

Example 11.42 Use Stoke's Theorem to evaluate the surface integral $\iint_S (\nabla \times \vec{F}) \cdot \vec{n} \, dS$ where $\vec{F} = (x - y)\vec{i} + z\vec{j} + (y + z)\vec{k}$ and S is surface with equation $z^2 = x^2 + y^2$ between $z = 0$ and $z = 4$.

Sol: The given surface is a cone as shown below. We parametrize the surface by the equation

$$x = u \cos \theta, \, y = u \sin \theta, \, z = u, \, (u, \theta) \in [0, 2] \times [0, 2\pi]$$

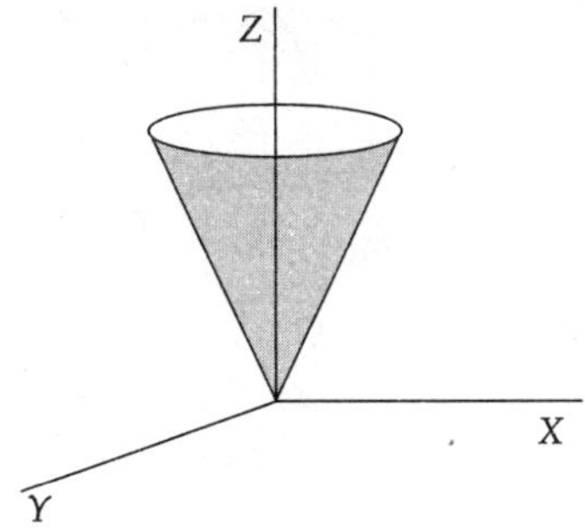

Fig. 11.42

For this the surface,

$$\frac{\partial \vec{r}}{\partial u} \times \frac{\partial \vec{r}}{\partial \theta} = \frac{\partial (f_2, f_3)}{\partial (u, \theta)} \, \vec{i} + \frac{\partial (f_3, f_1)}{\partial (u, \theta)} \, \vec{j} + \frac{\partial (f_1, f_2)}{\partial (u, \theta)} \, \vec{k}$$

$$= (-u \cos \theta) \, \vec{i} + (-u \sin \theta) \, \vec{j} + u \, \vec{k}$$

where $f_1 (u, \theta), = u \cos \theta, f_2 (u, \theta) = u \sin \theta, f_3 (u, \theta) = u$.

Now we see that $\nabla \times \vec{F} = \vec{k}$ so that,

$$(\nabla \times \vec{F}) \cdot \vec{n} - u$$

Thus,

$$\iint_S (\nabla \times \vec{F}) \cdot \vec{n} \, dS = \int_{r=0}^{4} \int_0^{2\pi} r \, dr \, d\theta = 4\pi$$

We verify Stoke's Theorem in this case by evaluating $\int_C \vec{F} \cdot d\vec{c}$ where C is the boundary curve of the above surface with equation $x^2 + y^2 = 4, \, z = 4$. We parametrize the curve as,

$$x = 2 \cos \theta, \, y = 2 \sin \theta, \, z = 4$$

Then,

$$\int_C \vec{F} \cdot d\vec{c} = \int_C (x - y) \, dx + z \, dy + (y + z) \, dz$$

$$= \int_0^{2\pi} \{(2 \cos \theta - 2 \sin \theta) \, (-2 \sin \theta) \, d\theta + 4 \, (2 \cos \theta) \, d\theta\}$$

$$= 4\pi$$

Hence Stoke's Theorem is verified.

Example 11.43 Using Stoke's Theorem evaluate the line integral $\int_C (y \, dx + dy + x \, dz)$ where C is the intersection of the surfaces $x^2 + y^2 + z^2 = a^2$ and $x + y = 0$. The curve starts from the point $P(0, a, 0)$ and goes first below the plane $z = 0$.

Sol: The given curve is nothing but the circle with centre (0, 0, 0) as shown in the figure.

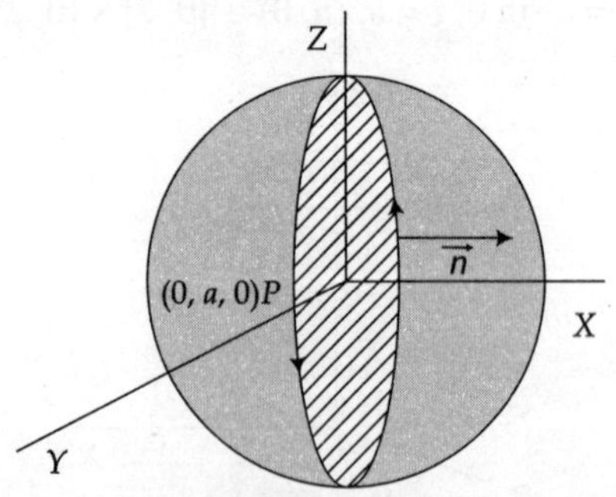

Fig. 11.43

Here the vector field is given by, $\vec{F} = y\vec{i} + \vec{j} + x\vec{k}$ and S is the surface $x + y = 0$ bounded by C. So that the unit normal $\vec{n}$ is perpendicular to $\vec{k}$ and hence $\vec{n} = \dfrac{1}{\sqrt{2}}(\vec{i} + \vec{j})$ also see that,

$$\nabla \times F = 0\vec{i} + \vec{j} - \vec{k}$$

That is,

$$(\nabla \times \vec{F}) \cdot \vec{n} = \frac{1}{\sqrt{2}}.$$

Now by Stoke's Theorem,

$$\int_C (y\,dx + dy + x\,dz) = \iint_S (\nabla \times \vec{F}) \cdot \vec{n}\, dS = \frac{1}{\sqrt{2}} \iint_S dS = \frac{1}{\sqrt{2}} \text{ (area of disc with radius } a)$$

$$= \frac{\pi a^2}{\sqrt{2}}$$

11.16   GAUSS' DIVERGENCE THEOREM

In this section we will see another important and famous integral theorem known as Gauss' Divergence Theorem or simply Gauss' Theorem. This theorem establishes a relationship between surface integral and volume integral of a function.

Theorem 11.16.1 (Gauss' Theorem)

Let V be a region in $\mathbb{R}^3$ enclosed by a smooth, orientable, closed surface S and let $\vec{n}$ denote the unit outward drawn normal to S. If $\vec{F} = F_1\vec{i} + F_2\vec{j} + F_3\vec{k}$ is a vector field defined on an open set containing V and S with continuously differentiable components then,

$$\iint_S (F_1\,dydz + F_2\,dzdx + F_3\,dxdy) = \iiint_V \left(\frac{\partial F_1}{\partial x} + \frac{\partial F_2}{\partial y} + \frac{\partial F_3}{\partial z} \right) dx\,dy\,dz \qquad (11.16.1.1)$$

In vector form (11.16.1.1) can be written as,

$$\iint_S \vec{F} \cdot \vec{n} \, dS = \iiint_V (\vec{\nabla} \cdot \vec{F}) \, dx \, dy \, dz \qquad (11.16.1.2)$$

Proof: To prove Gauss' Theorem in a complete general set up is not easy. We give a proof the

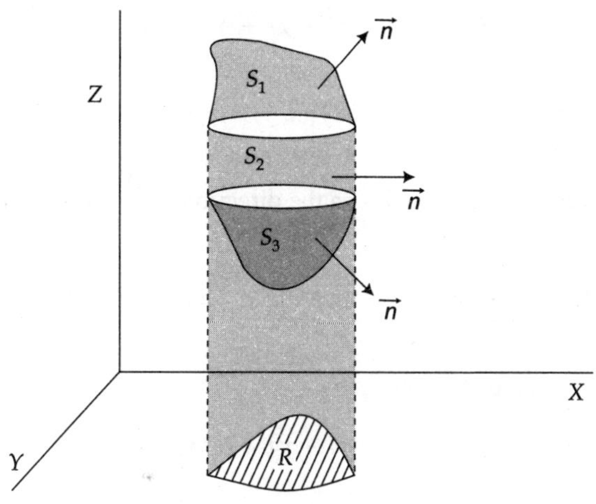

Fig. 11.44

theorem for a surface which comprises of three portions S_1, S_2 & S_3. S_2 is surface of the cylindrical portion with axis parallel to Z – axis and S_1 & S_2 denote the surfaces forming 'caps' of the cylinder with equations $z = \phi_1(x, y)$ & $z = \phi_2(x, y)(x, y) \in R$ respectively where R is the projection of the caps on the X – Y plane. Thus,

$$\phi_2(x, y) \leq z \leq \phi_1(x, y), (x, y) \in R.$$

We also assume that for such a surface every line through R parallel to Z – axis cuts the surface at two points at lower and upper 'caps'. Let α, β & γ be the inclination of the normal $\vec{n}$ with positive direction of X, Y & Z axis respectively. We now evaluate the double integral, $\iiint_V \dfrac{\partial F_3}{\partial z} \, dx \, dy \, dz$ by integrating with respect to z and obtain,

$$\iiint_V \frac{\partial F_3}{\partial z} \, dx \, dy \, dz = \iint_R \left[\int_{\phi_2(x, y)}^{\phi_2(x,y)} \frac{\partial F_3}{\partial z} \, dz \right] dx \, dy$$

$$= \iint_R [(F_3(x, y, \phi_2(x, y)) - F_3(x, y, \phi_1(x, y))) \, dxdy \qquad (11.16.1.3)$$

Now,

$$\iint_S F_3 \, dxdy = \iint_S F_3 \cos \gamma \, dS = \iint_{S_1} F_3 \cos \gamma \, dS + \iint_{S_2} F_3 \cos \gamma \, dS + \iint_{S_3} F_3 \cos \gamma \, dS \qquad (11.16.1.4)$$

We observe that for S_2 the normal to the surface is parallel to $X - Y$ plane hence $\cos \gamma = 0$ so that,

$$\iint_{S_3} F_3 \cos \gamma \, dS = 0 \tag{11.16.1.5}$$

Now the parametric equation of S_1 is given by

$$\vec{r} = x\vec{i} + y\vec{j} + \phi_1(x, y)\vec{k}$$

that is,

$$f_1(x, y) = x, f_2(x, y) = y, f_3(x, y) = \phi_1(x, y)$$

Also note that the direction of the normal is in the direction of $\dfrac{\partial \vec{r}}{\partial x} \times \dfrac{\partial \vec{r}}{\partial y}$ hence,

$$\iint_{S_1} F_3 \cos \gamma \, dS = \iint_{S_1} F_3 \frac{\partial(f_1, f_2)}{\partial(x, y)} \, dxdy = \iint_{R} F_3 \, (x, y, \phi_1(x, y)) \, dxdy \tag{11.16.1.6}$$

Again the parametric equation of S_3 is given by

$$\vec{r} = x\vec{i} + y\vec{j} + \phi_2(x, y)\vec{k}$$

that is in this case,

$$f_1(x, y) = x, f_2(x, y) = y, f_3(x, y) = \phi_2(x, y)$$

But notice that the direction of the normal to this surface is opposite to the direction of $\dfrac{\partial \vec{r}}{\partial x} \times \dfrac{\partial \vec{r}}{\partial y}$

Hence,

$$\iint_{S_3} F_3 \cos \gamma \, dS = -\iint_{S_3} F_3 \frac{\partial(f_1, f_2)}{\partial(x, y)} \, dxdy = -\iint_{R} F_3 \, (x, y, \phi_2(x, y)) \, dxdy \tag{11.16.1.7}$$

Adding (11.16.1.5), (11.16.1.6) & (11.16.1.7) and using (11.16.1.4) we get.

$$\iint_{S} F_3 \cos \gamma \, dS = \iint_{R} F_3(x, y, \phi_1(x, y)) \, dxdy - \iint_{R} F_3(x, y, \phi_2(x, y)) \, dxdy$$

$$= \iint_{R} [F_3(x, y, \phi_1(x, y)) - F_3(x, y, \phi_2(x, y))] \, dxdy \tag{11.16.1.8}$$

Comparing (11.16.1.3) & (11.16.1.8) we get,

$$\iiint_{V} \frac{\partial F_3}{\partial z} \, dx \, dy \, dz = \iint_{S} F_3 \, dxdy \tag{11.16.1.9}$$

By similar argument we can show that,

$$\iiint_{V} \frac{\partial F_1}{\partial z} \, dx \, dy \, dz = \iint_{S} F_1 \, dydz \tag{11.16.1.10}$$

$$\iiint\limits_{V} \frac{\partial F_2}{\partial z}\, dx\, dy\, dz = \iint\limits_{S} F_2\, dzdx \tag{11.16.1.11}$$

Adding (11.16.1.9), (11.16.1.10) & (11.16.1.11) we get,

$$\iiint\limits_{V} \frac{\partial F_1}{\partial x} + \frac{\partial F_2}{\partial y} + \frac{\partial F_3}{\partial z}\, dx\, dy\, dz = \iint\limits_{S} (F_1\, dydz + F_2\, dzdx + F_3\, dxdy)$$

This concludes the proof of divergence theorem.

Example 11.44   Verify Gauss' Theorem for the vector field $\vec{F} = x\vec{i} + y\vec{j} + \vec{k}$ and tetrahedron, $y \geq 0$, $z \geq 0$ & $x + y + z = 1$.

Sol: The given surface S can be decomposed into four surfaces S_1, S_2, S_3 & S_4 with equations

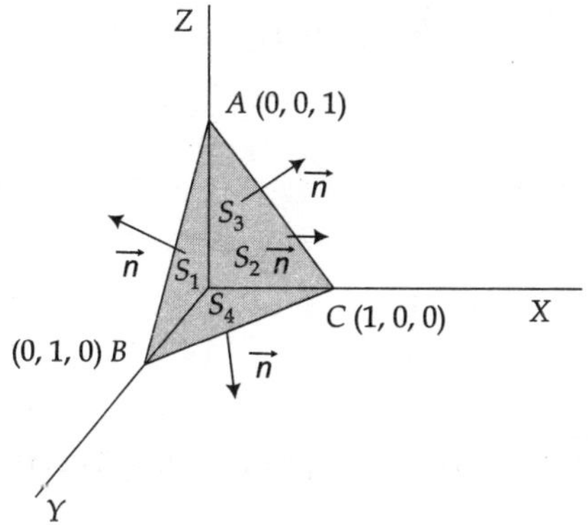

Fig. 11.45

$S_1 : y + z = 1,\ 0 \leq y \leq 1,\ x = 0$ & normal along $-\vec{i}$

$S_2 : x + z = 1,\ 0 \leq z \leq 1,\ y = 0$ & normal along $-\vec{j}$

$S_3 : x + y = 1,\ 0 \leq x \leq 1,\ z = 0$ & normal along $-\vec{k}$

$S_4 : x + y + z = 1,\ 0 \leq x \leq 1,\ 0 \leq y \leq 1 - x$ & normal along $\vec{i} + \vec{j} + \vec{k}$

Now,

$$\iint\limits_{S_1} \vec{F} \cdot \vec{n}\, dS = -\iint\limits_{S_1} x\, dS = 0 \quad (\text{as } x = 0 \text{ on } S_1)$$

$$\iint\limits_{S_2} \vec{F} \cdot \vec{n}\, dS = -\iint\limits_{S_2} y\, dS = 0 \quad (\text{as } y = 0 \text{ on } S_1)$$

$$\iint\limits_{S_3} \vec{F} \cdot \vec{n}\, dS = -\iint\limits_{S_3} dS = -\int\limits_{0}^{1}\int\limits_{0}^{1-y} dxdy = -\frac{1}{2}$$

$$\iint_{S_4} \vec{F} \cdot \vec{n} \, dS = \iint_{S_4} (x + y + 1) \, dS$$

$$= \int_{x=0}^{1} \int_{y=0}^{1-x} (x + y + 1) \, dx\, dy = \frac{5}{6}$$

Hence,

$$\iint_{S} \vec{F} \cdot \vec{n} \, dS = \iint_{S_1} \vec{F} \cdot \vec{n} \, dS + \iint_{S_2} \vec{F} \cdot \vec{n} \, dS + \iint_{S_3} \vec{F} \cdot \vec{n} \, dS + \iint_{S} \vec{F} \cdot \vec{n} \, dS$$

$$\frac{5}{6} - \frac{1}{2} = \frac{1}{3}$$

We see that $(\vec{\nabla} \cdot \vec{F}) = \dfrac{\partial}{\partial x}(x) + \dfrac{\partial}{\partial y}(y) + \dfrac{\partial}{\partial z}(1) = 2$. Hence,

$$\iiint_{V} (\vec{\nabla} \cdot \vec{F}) \, dx\, dy\, dz = 2 \iiint_{V} dx\, dy\, dz$$

$$= 2 \int_{x=0}^{1} \int_{y=0}^{1-x} \int_{z=0}^{1-x-y} dx\, dy\, dz = 2 \times \frac{1}{6} = \frac{1}{3}$$

So that we get $\iint_{S} \vec{F} \cdot \vec{n} \, dS = \iiint_{V} (\vec{\nabla} \cdot \vec{F}) \, dx\, dy\, dz$ that is Gauss' Theorem is verified.

Example 11.45 Evaluate $\iint_{S} \vec{F} \cdot \vec{n} \, dS$

where $\vec{F} = x\vec{i} + y\vec{j} + z\vec{k}$ and S is the surface of the cube centered at the origin with length of side equal to 2.

Sol: By Gauss' Theorem,

$$\iint_{S} \vec{F} \cdot \vec{n} \, dS = \iiint_{V} (\vec{\nabla} \cdot \vec{F}) \, dx\, dy\, dz \text{ where } V \text{ is the volume enclosed by the cube.}$$

$$= \iiint_{V} 3 \, dx\, dy\, dz = 3 \times \text{volume of cube} = 3 \times 2^3 = 24$$

Example 11.46 Show that $\iint_{S} \vec{F} \cdot \vec{n} \, dS =$ where $\vec{F} = (xy^2)\vec{i} + (yx^2)\vec{j} + x\vec{k}$ and S is the surface given by the equation $z = 9 - x^2 - y^2$ and $x^2 + y^2 \leq 9$.

Sol: Let V be volume enclosed by the given closed surface. The figure is similar to Fig. Hence by Gauss' Divergence Theorem we get,

$$\iint_{S} \vec{F} \cdot \vec{n} \, dS = \iiint_{V} (\nabla \cdot \vec{F}) \, dx\, dy\, dz$$

$$= \iiint\limits_{V} (y^2 + x^2)\, dx\, dy\, dz$$

$$= \int\limits_{\theta=0}^{2\pi} \int\limits_{r=0}^{3} \int\limits_{z=0}^{9-r^2} r^2\, r\, dr\, d\theta \quad \text{(Substituting } x = r \cos\theta,\ y = r \sin\theta)$$

$$= \frac{243}{2}\, \pi$$

Example 11.47 Use Gauss' Theorem to show that,

$$\iint\limits_{S} (\vec{\nabla} \times \vec{F}) \cdot \vec{n}\, dS = 0$$

Where S is a closed, smooth, orientable surface and $\vec{F}$ is a vector field defined on an open set containing S and its interior and components having continuous second order derivatives.

Sol: If $\vec{F} = F_1 \vec{i} + F_2 \vec{j} + F_3 \vec{k}$ then,

$$\vec{\nabla} \times \vec{F} = \left(\frac{\partial F_3}{\partial y} - \frac{\partial F_2}{\partial z} \right) \vec{i} + \left(\frac{\partial F_1}{\partial z} - \frac{\partial F_3}{\partial x} \right) \vec{j} + \left(\frac{\partial F_2}{\partial x} - \frac{\partial F_1}{\partial y} \right) \vec{k}$$

Hence,

$$\vec{\nabla} \cdot (\vec{\nabla} \times \vec{F}) = \frac{\partial}{\partial x} \left(\frac{\partial F_3}{\partial y} - \frac{\partial F_2}{\partial z} \right) + \frac{\partial}{\partial y} \left(\frac{\partial F_1}{\partial z} - \frac{\partial F_3}{\partial x} \right) + \frac{\partial}{\partial z} \left(\frac{\partial F_2}{\partial x} - \frac{F_1}{\partial y} \right)$$

$$= \frac{\partial^2 F_3}{\partial x \partial y} - \frac{\partial^2 F_2}{\partial x \partial z} + \frac{\partial^2 F_1}{\partial y \partial z} - \frac{\partial^2 F_3}{\partial y \partial z} + \frac{\partial^2 F_2}{\partial z \partial x} - \frac{\partial^2 F_1}{\partial z \partial y}$$

$$= 0 \quad \text{(as the mixed derivatives are equal by assumption)}$$

By Gauss' Theorem we get,

$$\iint\limits_{S} (\vec{\nabla} \times \vec{F}) \cdot \vec{n}\, dS = \iiint\limits_{V} \vec{\nabla} \cdot (\vec{\nabla} \times \vec{F})\, dx\, dy\, dz$$

$$= 0$$

PROBLEMS

1. Find Divergence and Curl of the following vector fields:

   (a) $\vec{F} = x^2 \vec{i} + y^2 \vec{j} + z^2 \vec{k}$

   (b) $\vec{F} = e^{2x} \vec{i} + xy \vec{j} + zx \vec{k}$

   (c) $\vec{F} = x \sin y\, \vec{i} + y \cos z\, \vec{j} + z \cos x\, \vec{k}$

2. Find Gradient of following scalar fields:

   (a) $\phi(x, y, z) = \dfrac{1}{\sqrt{x^2 + y^2 + z^2}} \quad (x, y, z) \neq (0, 0, 0)$

(b) $\phi(x, y, z) = xye^x$

(c) $\phi(x, y, z) = e^{-\frac{1}{r^2}}$, $r \neq 0$, where $r^2 = x^2 + y^2 + z^2$

3. If $\vec{F}$ & $\vec{G}$ are vector fields and ϕ & ψ are scalar fields prove the following identities

(a) $\vec{\nabla} \cdot (\vec{F} + \vec{G}) = \vec{\nabla} \cdot \vec{F} + \vec{\nabla} \cdot \vec{G}$

(b) $\vec{\nabla} \times (\vec{F} + \vec{G}) = \vec{\nabla} \times \vec{F} + \vec{\nabla} \times \vec{G}$

(c) $\vec{\nabla}(\phi + \psi) = \vec{\nabla}\phi + \vec{\nabla}\psi$

(d) $\vec{\nabla} \times (\vec{\nabla}\phi) = 0$

(e) $\vec{\nabla} \cdot (\vec{\nabla}\phi \times \vec{\nabla}\psi) = 0$

4. Show that the following curves are rectifiable and find their lengths:

(a) $\vec{c}(t) = (a \cos\theta, a \sin\theta, a\theta)$, $a > 0$, $0 \leq \theta \leq 2\pi$

(b) $\vec{c}(t) = (at\,|t|, at^2, a \sin t)$, $a > 0$, $-1 \leq t \leq 1$

(c) $\vec{c}(t) = (\theta - \sin\theta, 1 - \cos\theta, \theta)$, $-\pi \leq \theta \leq \pi$

(d) $\vec{c}(t) = (at^2, 2at, at)$, $0 \leq t \leq 1$

5. Evaluate the line integral $\int_C \vec{F} \cdot d\vec{c}$ for the vector field $\vec{F}$ and curve C as indicated below

(a) $\vec{F} = x^2\,\vec{i} + x^2y^2\,\vec{j}$, consists of the portion of the parabola $y = 2x^2$ between points $(0, 0)$ & $(1, 2)$ and straight line joining points $(1, 2)$ & $(0, 0)$.

(b) $\vec{F} = xy\vec{i} + y\vec{k}$, C is entire the ellipse with equation

$\dfrac{x^2}{a^2} + \dfrac{y^2}{b^2} = 1$ (traversed in anti clockwise direction.

(c) $\vec{F} = y\vec{i} + x\vec{j}$, C consists of the portion of X – axis from $(0, 0)$ to $(a, 0)$, portion of the circle $x^2 + y^2 = a^2$ from $(a, 0)$, $(0, a)$ and part of Y –axis from $(0, a)$ to $(0, 0)$.

(d) $\vec{F} = e^x\,\vec{i} + xe^y\,\vec{j}$, C is the entire square with centre as origin and length of sides is equal to a.

6. If C denotes the curve $x^2 + \dfrac{y^2}{4} = 1$ then show that $\displaystyle\int_C \dfrac{x^2 + y^2}{p}\, ds = \dfrac{41}{8}\pi$ where s denote the length of the curve from a fixed point and p denotes the length of the perpendicular from the origin to the tangent at any point on the curve.

$\left[\text{Hint: use parametric form } x = \cos\theta,\ y = 2\sin\theta \text{ and the relation } \left(\dfrac{ds}{d\theta}\right)^2 = \left(\dfrac{dx}{d\theta}\right)^2 + \left(\dfrac{dy}{d\theta}\right)^2\right]$

7. Verify Green's Theorem for the line integral $\int_C x\,dy - x^2 y\,dx$ where C is the closed curve consisting of the portion of the parabola $y = 4x^2$ between the points $(-1, 4)$ & $(2, 6)$ and the line segment joining them.

8. Find the total area of the following surfaces

   (a) $z = 4x^2 + y^2,\ 0 \le z \le 4$

   (b) $x = (a + b\cos\phi)\cos\psi,\ y = (a + b\cos\phi)\sin\psi,\ z = b\sin\psi,\ 0 \le \phi \le 2\pi,\ 0 \le \psi \le 2\pi$

   (c) $x = \sin\phi,\ y = \cos\phi,\ z = z,\ 0 \le \phi \le 2\pi,\ 0 \le z \le 1$

9. Verify Stoke's Theorem for the line integral $\int_C \vec{F} \cdot d\vec{c}$ in each of the following cases:

   (a) $\vec{F} = z\vec{i} + x\vec{j} + y\vec{k},\ C : x^2 + y^2 + 4,\ z = 0$

   (b) $\vec{F} = x^2\,\vec{i} + xy\vec{j} + z^2\,\vec{k},\ C : z = 1 - x^2 - y^2,\ z \ge 0$

10. If C is a closed curve which forms the boundary of the surface S and ϕ, ψ being scalar fields use Stoke's Theorem to show the following,

    (a) $\int_C (\phi\nabla \cdot \psi + \psi\nabla \cdot \phi) \cdot d\vec{c} = 0$

    (b) $\int_C (xdx + ydy + zdz) = 0$

    (c) $\int_C (\phi\nabla \cdot \psi) \cdot d\vec{c} = \iint_S (\vec{\nabla} \cdot \phi \times \nabla \cdot \psi) \cdot \vec{n}\ dS$

11. Verify Gauss' Theorem for the following vector fields $\vec{F}$ and surface S

    (a) $\vec{F} = x\vec{i} + y\vec{j} + z\vec{k}$, S is the entire surface of a cube with side a and centre origin.

    (b) $\vec{F} = 2x\vec{i} + 3y\vec{j}$, S is the entire surface of a sphere of radius a and centre origin

12. Using Gauss' Theorem evaluate $\iint_S \vec{F} \cdot \vec{n}\ dS$ for following vector fields $\vec{F}$ and surface S given as:

    (a) $\vec{F} = xy\vec{i} + yz\vec{j} + zx\vec{k}$, S is the entire surface of the cylinder $x^2 + y^2 = a^2,\ 0 \le z \le a$

    (b) $\vec{F} = x^2 y\vec{i} + z\vec{j} + z^2 x\vec{k}$, S is the entire surface of the unit sphere with centre origin.

Index

A

Abel's Inequality 3.34

Abel's Test 1.21

Abel's Theorem 5.15

Absolute Convergence 1.19, 5.5, 5.13

B

Bernstein Polynomial 7.34

Beta Function 6.1

Bolzano-Weierstrass Theorem 0.5. 1.9

Bonnet's form 3.35

Bounded 0.4

Bounded below 0.4

C

Cartesian product 0.2

Cauchy Criterion 1.13

Cauchy-Hadamard Formula 8.5

Cauchy Principal Value 5.11

Cauchy Sequence 1.14, 1.13

Cauchy's Root Test 1.18

Cavalieri's Principle 10.8

Chain Rule 2.28, 0.2

Closed rectangle 9.3

Closed set 0.5

Cluster point 1.10

Common refinement 3.4

Compact 0.5

Comparison Test 1.16

Complement 0.2

Conditionally convergent 1.19

Continuity 2.17

Continuously differentiable 7.2

D

Darboux's Theorem 3.8

Del Operator 11.2

Dense 0.5

Density property 0.5

Derivability 2.26

Derived function

Derived set 0.5

Differentiability 1.20

Directional derivative 9.10

Dirichlet's Test 2.12

Dirichlet's Theorem 5.16

Divergence 11.2

Divergent sequence 0.6, 1.5

Domain 6.4

E

Empty set 0.1, 0.2, 0.3, 0.5, 0.6

Euclidean norm 9.2

Euler's integral 10.29

Euler's Theorem 9.16

F

Fibonacci Sequence 1.2

Field 0.2, 0.3

Finite sequence 1.2

First Mean Value Theorem 3.33

Function 0.6

Function of Bounded Variations 4.12

Function or mapping 0.6

Fundamental Theorem 3.28

G

Gamma Function 6.3

Gauss' Divergence Theorem 11.48

Greatest lower bound 11.17

H

Hiene-Borel 0.5

I

Image 0.6

Implicit Function Theorem 9.31

Improper integrals 5.1

Infimum 0.4

Infinite limits 2.5

infinite sequence 1.1, 1.2

Injective 0.6

Integers 0.2, 1.22

Integral Test 5.8

Integration by Parts 3.38

Interior point 0.5, 9.2

Intersection 0.2

Interval of convergence 8.4

Intervals 0.3

Irrational numbers 0.2

J

Jordan 11.4

L

Least upper bound 0.4

Left-hand derivative 2.23

Leibniz Test 1.21

Length Function 11.8

L'Hospital's Rule 2.39

Limit at infinity 2.5

Limit Comparison Test 1.17

Limit inferior 1.11

Limit of Functions 2.1

Limit point 0.5

Limit superior 1.11

Limit Test 5.7

Line integral 11.10

Local extremum 2.31

Local maximum 2.31

Lower bound 0.4

Lower Riemann Integral 3.6

Lower Sum 3.3

M

Maclaurin's Series 2.38

Maximum element 0.4

Mean Value Theorem 2.32

Integral Calculus 3.28

Method of Substitution in an Integral 3.37

Minimum element 0.4

Mn-Test 7.6

Modulus function 0.7

Multiply connected 11.9

N

Natural numbers 0.2, 1.1, 1.3, 1.6, 1.7, 1.9, 1.13

Negatively oriented 11.6

Negligible sets 3.39

Normal 11.30

nth-Term Test 1.16

Null 0.1

Null set 0.1

O

One to one 0.6

onto 0.6

Open 0.5, 9.2

Open cover 0.5

Open neighbourhood 9.2

Open set 0.5

Order complete 0.5

Ordered field 0.2, 0.3

Ordered pair 0.2

Ordered set 0.3

Oscillating Sequences 1.5

Oscillatory 3.15

P

Parametric representation 11.4

Partition 3.3

Path integral 11.10

Piecewise smooth 11.5

Positively oriented 11.6

R

R [a, b] 4.2

Radius of Convergence 8.2

Rational numbers 0.2

Ratio Test 1.18

Real Analytic function 8.8

Rectifiable 11.7

Refinement of the partition 2.23
Regular point 11.30
Riemann Approximating 3.7
Riemann Approximating Sum 3.7
Riemann Integrable 4.1
Riemann-Stieltjes integral 3.7

S

Sandwich Theorem 1.8, 1.22, 1.8
Second Mean Value Theorem 3.35
Set 0.1
Sign preserving property 2.13
Simple 11.4
Simply connected 11.9
Singular point 11.30
Smooth curve 11.4
Stoke's Theorem 11.43, 11.44
Subcover 0.5
Subsequence 1.9
Subset 0.1
Sum 3.15
Superset 0.1
Supremum 0.4
Surface integral of first kind 11.34
Surface integral of second kind 11.39

Surface 11.27
Surjective 0.6

T

Taylor's Series 2.38
Taylor's Theorem 2.36
Total differential of the function 9.18
Triangular inequality 9.2
Type II Integrals 5.3

U

Unbounded below 0.4, 1.2
Uniform Continuity 2.20
Union 0.2
Universal set 0.1
Upper bound 0.4
Upper Riemann Integral 3.7
Upper Sum 3.3

V

Vector field 11.1

W

Weierstrass Approximation Theorem 7.34
Weierstrass' M- Test 7.11
Weierstrass's Form 3.35